P9-AQL-234

Annals of Mathematics Studies

Number 126

An Extension of Casson's Invariant

by

Kevin Walker

PRINCETON UNIVERSITY PRESS

PRINCETON, NEW JERSEY

1992

Princeton University Press books are printed on acid-free paper, and meet the guidelines for permanence and durability of the Committee on Production Guidelines for Book Longevity of the Council on Library Resources

Printed in the United States of America
by Princeton University Press, 41 William Street
Princeton, New Jersey

Library of Congress Cataloging-in-Publication Data

Walker, Kevin, 1963-
An extension of Casson's invariant / Kevin Walker.
p. cm.—(Annals of mathematics studies; no. 126)
Includes bibliographical references.
ISBN 0-691-08766-0 (CL)—ISBN 0-691-02532-0 (PB)
1. Three-manifolds (Topology) 2. Invariants. I. Title.
II. Title: Casson's invariant. III. Series.
QA613.W34 1992
514.3—dc20 91-42226

Contents

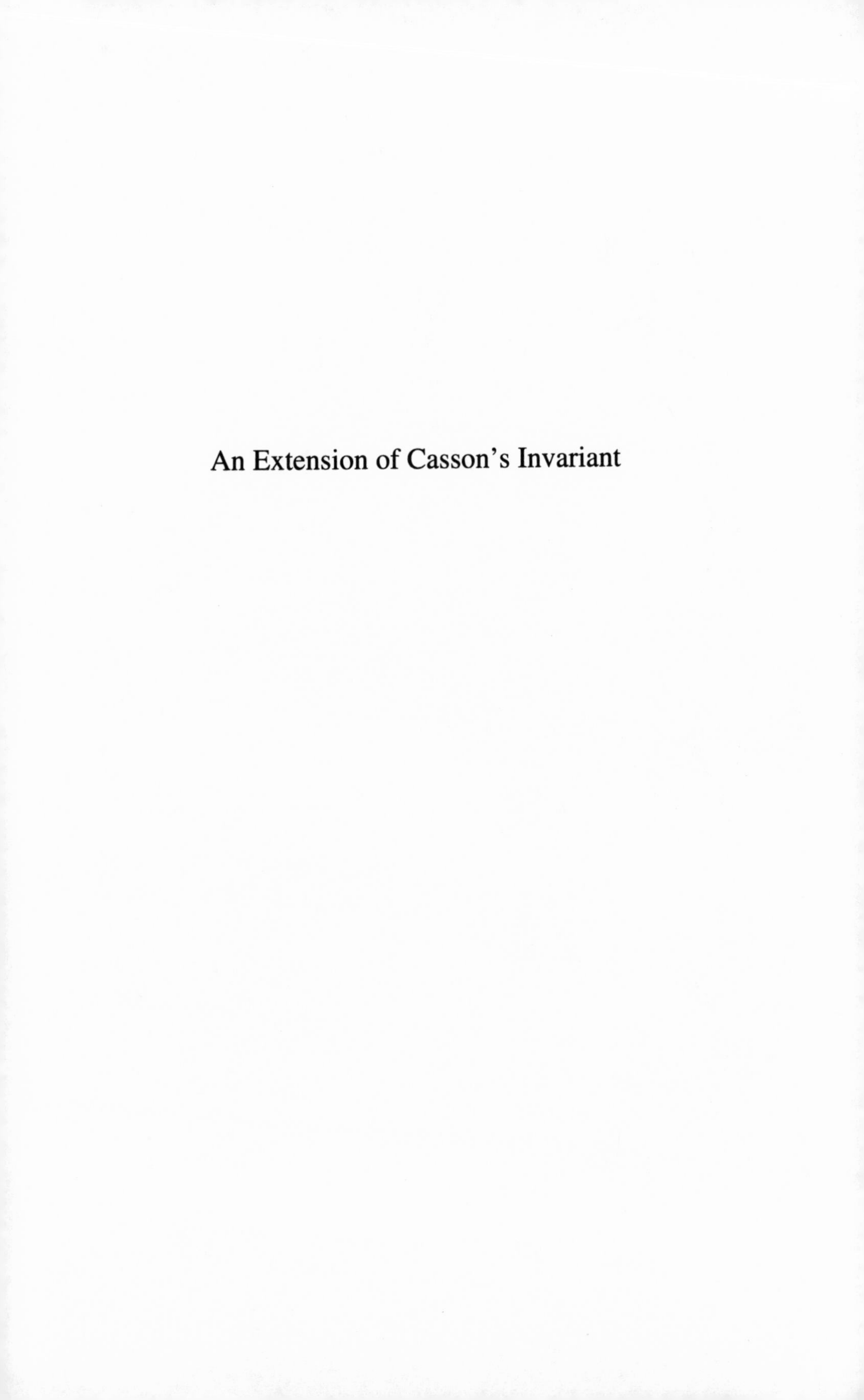

An Extension of Casson's Invariant

§ 0

Introduction

In lectures at MSRI in 1985, Andrew Casson described an integer valued invariant λ of oriented homology 3-spheres (see also [AM]). $\lambda(M)$ can be thought of counting the number of conjugacy classes of representations $\pi_1(M) \to SU(2)$, in the same sense that the Lefschetz number of a map counts the number of fixed points. (Warning: According to Casson's original definition and [AM], $\lambda(M)$ is half this number.)

More precisely, let (W_1, W_2, F) be a Heegaard splitting of M (see (1.A)). For X any space, let $R(X)$ denote the space of conjugacy classes of representations $\pi_1(X) \to SU(2)$. We can make the identification

$$R(M) = R(W_1) \cap R(W_2) \subset R(F).$$

$R(W_1)$ and $R(W_2)$ have complementary dimensions in $R(F)$, and, roughly speaking, $\lambda(M)$ is defined to be the intersection number

$$\lambda(M) \stackrel{\text{def}}{=} \langle R(W_1), R(W_2) \rangle.$$

$R(W_j)$ and $R(F)$ are not manifolds, but singular real algebraic sets, so it might seem that there is some difficulty in defining the above intersection number. But the condition that $H_1(M; \mathbb{Z}) = 0$ guarantees that after excising the trivial representation $R(W_1) \cap R(W_2)$ is a compact subset of the non-singular part of $R(F)$. Thus, after an isotopy supported away from the singularities of $R(F)$, $R(W_1) \cap R(W_2)$ consists of a well-defined (signed) number of non-singular points, and $\langle R(W_1), R(W_2) \rangle$ is defined to be this number. It is not hard to show that this does not depend on the choice of Heegaard splitting.

It is clear from the definition that $\lambda(M) = 0$ if $\pi_1(M) = 1$. In particular,

(0.1) $\lambda(S^3) = 0$.

Casson also showed that λ has the following properties.

(0.2) Let K be a knot in an integral homology sphere and let $K_{1/n}$ denote $1/n$ Dehn surgery on K. Let $\Delta_K(t)$ be the (suitably normalized) Alexander polynomial of K. Then

$$\lambda(K_{1/n}) = \lambda(K_{1/0}) + n\frac{d^2}{dt^2}\Delta_K(1).$$

(0.3) $\lambda(-M) = -\lambda(M)$, where $-M$ denotes M with the opposite orientation.

(0.4) $\lambda(M_1 \# M_2) = \lambda(M_1) + \lambda(M_2)$, where # denotes connected sum.

(0.5) $4\lambda(M) \equiv \mu(M) \pmod{16}$, where $\mu(M)$ denotes the signature of a spin 4-manifold bounded by M.

It is not hard to show that (0.1) and (0.2) uniquely determine λ. Also, (0.3), (0.4) and (0.5) follow easily from (0.1) and (0.2). The proof of (0.2) involves clever exploitation of the isotopy invariance of $\langle R(W_1), R(W_2)\rangle$.

For a more detailed overview of λ for integral homology spheres, see the introduction of [AM].

This paper contains generalizations of the above results to the case where M is a rational homology sphere (i.e. $H_*(M;\mathbf{Q}) \cong H_*(S^3;\mathbf{Q})$, or $|H_1(M;\mathbf{Z})| < \infty$). In this case, $R(W_1) \cap R(W_2)$ contains nontrivial singular points, so finding an isotopy invariant definition of $\langle R(W_1), R(W_2)\rangle$ is more difficult. It can, nevertheless, be done, and one can use the isotopy invariance to prove a generalization (0.2) (see (4.2)), and hence generalizations of (0.3), (0.4) and (0.5) (see (6.5)). For a more detailed introduction, the reader is encouraged to read (1.A) and the beginning of (2.A).

Just as in the integral homology sphere case, the generalized Dehn surgery formula and the fact that $\lambda(S^3) = 0$ uniquely determine λ for all rational homology spheres. In this case, however, it is not very hard to prove that the Dehn surgery formula does not overdetermine λ. In other words, one can use the Dehn surgery formula to define λ and avoid $SU(2)$ representations altogether.

Section 1 contains results on the topology (and symplectic geometry) of certain representation spaces, as well as other miscellaneous background material. Most of the results are summarized in (1.A), and the non-methodical reader may wish to read only this subsection, refering back to the rest as necessary. Section 2 contains the definition of λ (i.e. of $\langle R(W_1), R(W_2)\rangle$). Sections 3 and 4 contain the proof of the Dehn surgery formula. Section 3 contains the parts of the proof involving representation spaces, while Section 4 contains the parts involving the manipulation of surgery diagrams. Some readers may wish to start with Section 4 and refer back to Section 3 as necessary. In Section 5 we use the Dehn surgery formula to give an independent and elementary proof of the existence of λ. This section is independent of sections 1 through 4, and some readers may wish to act accordingly. In Section 6 we use the Dehn surgery formula to prove various things about λ, including its relation to the μ-invariant. Sections A and B are appendices containing background material needed for the statement and proof of the Dehn surgery formula.

This work has previously been announced in [W1]. Certain results found in this paper were obtained independently by S. Boyer and D. Lines [BL] (see (5.1) and (6.5)). A different generalization of Casson's invariant has been described by S. Boyer and A. Nicas [BN].

This research received generous support from the National Science Foundation, the Sloan Foundation and the Mathematical Sciences Research Institute.

It is a pleasure to acknowledge helpful conversations and correspondence with Peter Braam, Dan Freed, William Goldman, Lucien Guillou, Michael Hirsch, Morris Hirsch, Steve Kerckhoff, Gordana Matic, Paul Melvin and Tom Mrowka. Christine Lescop gave an earlier version of this paper a very thorough reading and spotted numerous errors, at least one of which was very non-trivial. Very special thanks are due to Andrew Casson, whose excellent suggestions improved this paper in many ways, and to Rob Kirby, who provided good advice on all sorts of topics. Finally, I would like to thank José Montesinos for suggesting that I work on this problem.

§ 1

Topology of Representation Spaces

A. Summary of Results and Notation.

Let M be an oriented rational homology 3-sphere (QHS) with a genus g Heegaard splitting (W_1, W_2, F). (That is, $M = W_1 \cup W_2$ and $W_1 \cap W_2 = \partial W_1 = \partial W_2 = F$, a surface of genus g.) Let F^* be F minus a disk. We have the following diagram of fundamental groups

$$\begin{array}{ccccccc} & & & & \pi_1(W_1) & & \\ & & & \nearrow & & \searrow & \\ \pi_1(F^*) & \to & \pi_1(F) & & & & \pi_1(M). \\ & & & \searrow & & \nearrow & \\ & & & & \pi_1(W_2) & & \end{array}$$

Note that all maps are surjections. Applying the functor $\mathrm{Hom}(\,\cdot\,, SU(2))$, this diagram becomes

$$\begin{array}{ccccccc} & & & & Q_1^\sharp & & \\ & & & \swarrow & & \nwarrow & \\ R^{*\sharp} & \leftarrow & R^\sharp & & & & Q_1^\sharp \cap Q_2^\sharp \\ & & & \nwarrow & & \swarrow & \\ & & & & Q_2^\sharp & & \end{array}$$

(i.e. $R^\sharp \stackrel{\mathrm{def}}{=} \mathrm{Hom}(\pi_1(F), SU(2))$, $Q_j^\sharp \stackrel{\mathrm{def}}{=} \mathrm{Hom}(\pi_1(W_j), SU(2))$, etc.). Note that all maps are injections and that Van Kampen's theorem implies that $\mathrm{Hom}(\pi_1(M), SU(2)) = Q_1^\sharp \cap Q_2^\sharp$. $SU(2)$ acts naturally on these spaces via the adjoint action. Taking quotients, we get

$$\begin{array}{ccccc} & & & Q_1 & & \\ & & \swarrow & & \nwarrow & \\ R^* & \leftarrow & R & & & Q_1 \cap Q_2 \\ & & \nwarrow & & \swarrow & \\ & & & Q_2 & & \end{array}$$

(i.e. $R = \mathrm{Hom}(\pi_1(F), SU(2))/SU(2)$, etc.).

In the rest of this paper the superscript $\sharp$ will be used, without comment, to denote the "inverse quotient by the adjoint action of $SU(2)$", and vice-versa. That is, if X been defined as $Y/SU(2)$, then $X^\sharp$ will be used to denote Y.

We will see below that R has two singular strata: S, consisting of (equivalence classes of) representations with abelian image, and P, consisting of representations into $\mathbb{Z}_2$, the center of $SU(2)$. Define

$$\begin{aligned} T_j &\stackrel{\text{def}}{=} Q_j \cap S \\ R^- &\stackrel{\text{def}}{=} R \setminus S \\ Q_j^- &\stackrel{\text{def}}{=} Q_j \setminus T_j \\ S^- &\stackrel{\text{def}}{=} S \setminus P \\ T_j^- &\stackrel{\text{def}}{=} T_j \setminus P. \end{aligned}$$

The Zariski tangent space of $R^\sharp$ at a representation ρ can be identified with the cocycle space $Z^1(\pi_1(F); \mathrm{Ad}\,\rho)$, and the tangent space of the orbit through ρ corresponds to the coboundaries $B^1(\pi_1(F); \mathrm{Ad}\,\rho)$. Thus we define the "Zariski tangent space" of R at $[\rho]$ to be $H^1(\pi_1(F); \mathrm{Ad}\,\rho)$.

Let $B : \mathbf{su}(2) \times \mathbf{su}(2) \to \mathbb{R}$ be an Ad-invariant inner product. B induces the cup product

$$\omega_B : H^1(\pi_1(F); \mathrm{Ad}\,\rho) \times H^1(\pi_1(F); \mathrm{Ad}\,\rho) \to H^1(\pi_1(F); \mathbb{R}) \cong \mathbb{R}.$$

It is a result of Goldman that ω_B gives R a symplectic structure. Q_1 and Q_2 turn out to be lagrangian with respect to ω_B.

We will be particularly interested in the normal bundle of S^- in R. Define

$$\begin{aligned} \nu &\stackrel{\text{def}}{=} \text{Zariski normal bundle of } S^- \text{ in } R \\ \eta_j &\stackrel{\text{def}}{=} \text{Zariski normal bundle of } T_j^- \text{ in } Q_j \\ \xi &\stackrel{\text{def}}{=} \text{actual (singular) normal bundle of } S^- \text{ in } R \\ \theta_j &\stackrel{\text{def}}{=} \text{actual (singular) normal bundle of } T_j^- \text{ in } Q_j \end{aligned}$$

It is another result of Goldman that for $p = [\rho] \in S^-$, the fiber ξ_p is diffeomorphic to a quadratic cone in ν_p modulo stab $(\rho) \cong S^1$. For $p \in T_j^-$, $\theta_{j,p}$ is diffeomorphic to $\eta_{j,p}$ modulo stab (ρ).

Let S_0^1 be a fixed, oriented maximal torus of $SU(2)$ and let $\mathbb{Z}_2 \subset S_0^1$ denote the center of $SU(2)$. Let

$$\begin{aligned}
\widetilde{S} &\overset{\text{def}}{=} \operatorname{Hom}(\pi_1(F), S_0^1) \cong (S_0^1)^{2g} \\
\widetilde{S}^- &\overset{\text{def}}{=} \operatorname{Hom}(\pi_1(F), S_0^1) \setminus \operatorname{Hom}(\pi_1(F), \mathbb{Z}_2) \cong (S_0^1)^{2g} \setminus (\mathbb{Z}_2)^{2g} \\
\widetilde{T_j} &\overset{\text{def}}{=} \operatorname{Hom}(\pi_1(W_j), S_0^1) \cong (S_0^1)^g \\
\widetilde{T_j^-} &\overset{\text{def}}{=} \operatorname{Hom}(\pi_1(W_j), S_0^1) \setminus \operatorname{Hom}(\pi_1(W_j), \mathbb{Z}_2) \cong (S_0^1)^g \setminus (\mathbb{Z}_2)^g.
\end{aligned}$$

$\widetilde{S}^-$ $[\widetilde{T_j^-}]$ is a double cover of S^- $[T_j^-]$. Let ν, η_j, ξ and θ_j continue to denote their lifts to $\widetilde{S}^-$ or $\widetilde{T_j^-}$, as the case may be. Over $\widetilde{S}^-$, ν is a symplectic vector bundle, and η_j is an oriented lagrangian subbundle defined over $\widetilde{T_j^-}$. Picking a metric of F induces an hermetian structure on TR compatible with ω_B. This converts ν into a hermetian vector bundle and η_j into a totally real subbundle.

Let $\det^1(\nu)$ denote the unit vectors in the determinant line bundle $\det(\nu)$ of ν. η_j induces a section $\det^1(\eta_j)$ of $\det^1(\nu)$ over $\widetilde{T_j^-}$. $\det(\nu)$ $[\det(\eta_j)]$ extends smoothly and canonically over $\widetilde{S}$ $[\widetilde{T_j}]$.

Now we come to a crucial point. Namely, $c_1(\det(\nu)) = c_1(\nu)$ is represented by a multiple ω of ω_B restricted to S^- and lifted to $\widetilde{S}$. Furthermore, $\widetilde{T_j}$ is lagrangian with respect to ω.

The above results are proved (or referenced) in sections B and C. Section D uses results of Newstead to prove that certain maps of representation spaces act trivially on rational cohomology. Section E is concerned with a class of isotopies of R appropriate to the definition of λ. Section F establishes orientation conventions. Section G establishes conventions for Dehn twists and Dehn surgeries.

B. Fundamental Results of Goldman.

The material in this section is treated in more detail in [G1] and [G2].

Let π be a finitely presented group and G be a Lie group. Let $\rho : \pi \to G$ be a representation and let $\rho_t : \pi \to G$ be a differentiable 1-parameter family of representations such that $\rho_0 = \rho$. Writing

$$\rho_t(x) = \exp(tu(x) + O(t^2))\rho(x)$$

(for $x \in \pi$ and t near 0) and differentiating the homomorphism condition

$$\rho_t(xy) = \rho_t(x)\rho_t(y), \tag{1.1}$$

we find that

$$u(xy) = u(x) + \operatorname{Ad}\rho(x)\,u(y). \tag{1.2}$$

In other words, $u : \pi \to \mathfrak{g}$ is a 1-cocycle of π with coefficients in the π-module $\mathfrak{g}_{\operatorname{Ad}\rho}$. Conversely, solutions of (1.2) lead to maps $\rho_t : \pi \to G$ which satisfy (1.1) to first order in t. Thus the Zariski tangent space of $\operatorname{Hom}(\pi, G)$ at ρ can be identified with $Z^1(\pi; \mathfrak{g}_{\operatorname{Ad}\rho})$.

We now compute the tangent space of the Ad-orbit containing ρ. Let g_t be a path in G with $g_0 = 1$. Let

$$\rho_t(x) = g_t^{-1}\rho(x)g_t.$$

If $g_t = \exp(tu_0 + O(t^2))$, then the cocycle corresponding to ρ_t is

$$u(x) = \operatorname{Ad}\rho(x)u_0 - u_0.$$

In other words, u is the coboundary δu_0. Thus the tangent space of the Ad-orbit through ρ is $B^1(\pi; \mathfrak{g}_{\operatorname{Ad}\rho})$.

The above results suggest that we define the "Zariski tangent space" of $\operatorname{Hom}(\pi, G)$ at $[\rho]$ to be $H^1(\pi; \mathfrak{g}_{\operatorname{Ad}\rho})$. (We will omit the scare quotes from now on.)

We now specialize to the case $\pi = \pi_1(F)$ and G is a Lie group which affords an Ad-invariant, symmetric, non-degenerate bilinear form B on its Lie algebra (e.g. a reductive group with the Cartan-Killing form).

If $X \subset G$, let $Z(X)$ denote the centralizer of X in G. Let $Z(\rho) = Z(\rho(\pi))$.

(1.3) **Proposition (Goldman).** *The dimension of $Z^1(\pi; \mathfrak{g}_{\operatorname{Ad}\rho})$ is*

$$(2g-1)\dim G + \dim Z(\rho).$$

The dimension of $H^1(\pi; \mathfrak{g}_{\operatorname{Ad}\rho})$ is

$$(2g-2)\dim G + 2\dim Z(\rho).$$

□

Note that $\dim Z^1(\pi; \mathbf{g}_{\mathrm{Ad}\rho})$ is minimal, and hence ρ is a nonsingular point of $\mathrm{Hom}(\pi, G)$, if and only if $\dim(Z(\rho)/Z(G)) = 0$. We denote the set of all such points as $\mathrm{Hom}(\pi, G)^-$. If $\rho \in \mathrm{Hom}(\pi, G)$ is a singular point (i.e. if $\dim(Z(\rho)/Z(G)) > 0$) it is still a nonsingular point of the subvariety $\mathrm{Hom}(\pi, Z(Z(\rho)))^-$. Note also that for all $\sigma \in \mathrm{Hom}(\pi, Z(Z(\rho)))^-$, $\mathrm{stab}\,(\sigma(\pi)) = Z(\sigma) = Z(\rho)$. Therefore all points of $\mathrm{Hom}(\pi, Z(Z(\rho)))^-$ have the same orbit type. In fact,

(1.4) **Proposition (Goldman).** *The sets of the form*

$$\mathrm{Hom}(\pi, Z(Z(X)))^- / N_G(Z(Z(X))),$$

where N_G denotes the normalizer in G, give a stratification of $\mathrm{Hom}(\pi, G)/G$. □

Let $[\rho] \in \mathrm{Hom}(\pi, G)/G$ and $[u] \in H^1(\pi; \mathbf{g}_{\mathrm{Ad}\rho})$. We wish to find necessary and sufficient conditions for $[u]$ to be tangent to a path $[\rho_t]$ in $\mathrm{Hom}(\pi, G)/G$. Writing

$$\rho_t(x) = \exp(tu(x) + t^2 v(x) + O(t^3))\rho(x),$$

plugging this into (1.1), and expanding to second order in t, we find

$$v(x) - v(xy) + \mathrm{Ad}\,\rho(x)v(y) = \frac{1}{2}[u(x), \mathrm{Ad}\,\rho(x)u(y)].$$

The left hand side is just the coboundary of the 1-cochain v. The right hand side is the cup product (on the cocycle level) of u with itself, using the Lie bracket $[\,\cdot\,,\,\cdot\,] : \mathbf{g} \times \mathbf{g} \to \mathbf{g}$ as coefficient pairing. Thus a necessary condition for $[u]$ to be tangent to a path in $\mathrm{Hom}(\pi, G)/G$ is

$$[[u], [u]] = 0 \in H^2(\pi; \mathbf{g}_{\mathrm{Ad}\rho}).$$

It turns out that for surface groups this condition is also sufficient:

(1.5) **Theorem (Goldman).** *Let $[\rho] \in \mathrm{Hom}(\pi, G)/G$ and $\alpha \in H^1(\pi; \mathbf{g}_{\mathrm{Ad}\rho})$. Then α is tangent to a path in $\mathrm{Hom}(\pi, G)/G$ if and only if $[\alpha, \alpha] = 0$.* □

Also,

(1.6) **Theorem (Goldman and Millson, [GM]).** *A point*

$$[\rho] \in \mathrm{Hom}(\pi, G)/G$$

has a neighborhood diffeomorphic to

$$\{\alpha \in H^1(\pi; \mathbf{g}_{\mathrm{Ad}\rho}) \mid [\alpha, \alpha] = 0\}/\mathrm{stab}\,(\rho).$$

□

Recall that we have an Ad-invariant, symmetric, nondegenerate bilinear form $B : \mathbf{g} \times \mathbf{g} \to \mathbb{R}$. This induces a cup product

$$\omega_B : H^1(\pi; \mathbf{g}_{\mathrm{Ad}\rho}) \times H^1(\pi; \mathbf{g}_{\mathrm{Ad}\rho}) \to H^2(\pi; \mathbb{R}) \cong \mathbb{R}.$$

Regarding $H^1(\pi; \mathbf{g}_{\mathrm{Ad}\rho})$ as the Zariski tangent space to $\mathrm{Hom}(\pi, G)/G$ at $[\rho]$, ω_B defines a 2-tensor on $\mathrm{Hom}(\pi, G)/G$.

(1.7) **Theorem (Goldman).** *ω_B is a closed, nondegenerate exterior 2-form (i.e. a symplectic structure) on* $\mathrm{Hom}(\pi, G)/G$. □

ω_B is compatible with the stratification of $\mathrm{Hom}(\pi, G)/G$ in the following sense. Let $[\rho] \in \mathrm{Hom}(\pi, G)/G$ and let $\{\rho_j\} \to \rho$. Then there is a natural inclusion

$$i : \lim H^1(\pi; \mathbf{g}_{\mathrm{Ad}\rho_j}) \to H^1(\pi; \mathbf{g}_{\mathrm{Ad}\rho}),$$

and

$$\lim \omega_{[\rho_j]} = i^* \omega_{[\rho]}.$$

(Here ω_x denotes ω_B restricted to $T_x \mathrm{Hom}(\pi, G)/G$.)

We now apply the above results to the case $G = SU(2)$. Let S^1_0 be a fixed maximal torus of $SU(2)$. Up to conjugacy, $SU(2)$ has three subgroups of the form $Z(Z(X))$: $SU(2)$, S^1_0 and $Z(SU(2)) = \mathbb{Z}_2$. Thus, by (1.4), R $(= \mathrm{Hom}(\pi, SU(2))/SU(2))$ has two singular strata:

$$S \stackrel{\mathrm{def}}{=} \mathrm{Hom}(\pi, S^1_0)/\mathbb{Z}_2 \cong (S^1_0)^{2g}/\mathbb{Z}_2$$

(here $\mathbb{Z}_2 = N(S^1_0)/Z(S^1_0)$) and

$$P \stackrel{\mathrm{def}}{=} \mathrm{Hom}(\pi, \mathbb{Z}_2) \cong \mathbb{Z}_2^{2g}.$$

This stratification is compatible with Q_j. That is, Q_j has singular strata $T_j \stackrel{\mathrm{def}}{=} Q_j \cap S \cong (S^1_0)^g/\mathbb{Z}_2$ and $Q_j \cap P \cong \mathbb{Z}_2^g$.

Let $p = [\rho] \in Q_j$. The inclusion $i : F \to W_j$ induces an injection

$$i^* : H^1(\pi_1(W_j); \mathbf{g}_{\mathrm{Ad}\rho}) \to H^1(\pi; \mathbf{g}_{\mathrm{Ad}\rho})$$

which corresponds to the inclusion $T_p Q_j \to T_p R$. Since $H^2(\pi_1(W_j); \mathbb{R}) = 0$, ω_B is zero on $H^1(\pi_1(W_j); \mathbf{g}_{\mathrm{Ad}\rho})$. Furthermore, $\dim H^1(\pi_1(W_j); \mathbf{g}_{\mathrm{Ad}\rho}) = \frac{1}{2} \dim H^1(\pi; \mathbf{g}_{\mathrm{Ad}\rho})$. Hence Q_j is lagrangian with respect to ω_B.

If $p \in P$, then $\mathrm{Ad}\,\rho$ is trivial, and hence

$$H^1(\pi_1(X); \mathbf{g}_{\mathrm{Ad}\rho}) = H^1(\pi_1(X); \mathbb{R}) \otimes \mathbf{g}$$

for $X = F$ or W_j. Since M is a QHS, $H^1(\pi_1(W_1); \mathbb{R})$ and $H^1(\pi_1(W_2); \mathbb{R})$ are transverse in $H^1(\pi_1(F); \mathbb{R})$. Therefore Q_1 and Q_2 are transverse at P.

C. The Normal Bundle of S.

(1.8) The Topology of the Normal Bundle.

Let $p = [\rho] \in S^-$. Let $\mathbf{h}$ be the Lie algebra of S_0^1 and $\mathbf{h}^\perp$ be the orthogonal complement of $\mathbf{h}$ (with respect to B). Then the π-module $\mathbf{g}_{\mathrm{Ad}\rho}$ decomposes as $\mathbf{h}_{\mathrm{Ad}\rho} \oplus \mathbf{h}^\perp_{\mathrm{Ad}\rho}$, and hence

$$H^1(\pi; \mathbf{g}_{\mathrm{Ad}\rho}) = H^1(\pi; \mathbf{h}_{\mathrm{Ad}\rho}) \oplus H^1(\pi; \mathbf{h}^\perp_{\mathrm{Ad}\rho}).$$

The space $H^1(\pi; \mathbf{h}^\perp_{\mathrm{Ad}\rho})$ corresponds to the fiber of the Zariski normal bundle of S^- in R at p. Call this bundle ν. Similarly, if $p \in T_j^-$ also, then $H^1(\pi_1(W_j); \mathbf{h}^\perp_{\mathrm{Ad}\rho})$ is the fiber of the Zariski normal bundle of T_j^- in Q_j at p. Call this bundle η_j.

Let ξ and θ_j denote the actual normal bundles of S^- and T_j^-. By (1.5), we have

$$\begin{aligned} \xi &= \{x \in \nu \mid [x, x] = 0\}/S_0^1 \\ \theta_j &= \eta_j/S_0^1. \end{aligned}$$

(Here S_0^1 should be thought of as stab $(S_0^1) =$ stab $(\rho(\pi))$.) Let $\hat{\xi}$ $[\hat{\theta}_j]$ denote the unit vectors in ξ $[\theta_j]$.

It will be convenient to put a hermetian structure on ν compatible with its symplectic structure. This can be done as follows. Let $\mathbf{h}^\perp_{\mathrm{Ad}\rho}$ denote the flat $\mathbf{h}^\perp$-bundle over F with holonomy $\mathrm{Ad}\,\rho$. (The ambiguity in notation is intentional.) Let $H^1_{dR}(F; \mathbf{h}^\perp_{\mathrm{Ad}\rho})$ be the de Rham cohomology group. Since $\pi_i(F) = 0$ for $i \geq 2$, we have $H^1_{dR}(F; \mathbf{h}^\perp_{\mathrm{Ad}\rho}) = H^1(\pi; \mathbf{h}^\perp_{\mathrm{Ad}\rho})$, and we will identify these two spaces from now on. Choose a metric on F. Note that $H^q(F; \mathbf{h}^\perp_{\mathrm{Ad}\rho})$ can also be identified with the space of harmonic q-forms with coefficients in $\mathbf{h}^\perp_{\mathrm{Ad}\rho}$.

Choose B to be positive definite (e.g. the negative of the Cartan-Killing form). The metric on F, together with B, induces the Hodge star operator

$$* : H^1(\pi; \mathbf{h}^\perp_{\mathrm{Ad}\rho}) \to H^1(\pi; \mathbf{h}^\perp_{\mathrm{Ad}\rho})$$

and the Hodge metric

$$\langle \cdot , \cdot \rangle : H^1(\pi; \mathbf{h}^\perp_{\mathrm{Ad}\rho}) \times H^1(\pi; \mathbf{h}^\perp_{\mathrm{Ad}\rho}) \to \mathbb{R},$$

where

$$\langle \alpha, \beta \rangle \stackrel{\text{def}}{=} \int_F B(\alpha, *\beta).$$

It is easy to see that $*$ and $\langle \cdot, \cdot \rangle$ give a hermetian structure compatible with ω_B. That is,

$$\omega_B(\alpha, \beta) = -\langle \alpha, *\beta \rangle$$

for all $\alpha, \beta \in H^1(\pi; \mathbf{h}^{\perp}_{\mathrm{Ad}\rho})$.

$H^1(\pi; \mathbf{h}^{\perp}_{\mathrm{Ad}\rho})$ has a second complex structure arising from the coefficient module $\mathbf{h}^{\perp}_{\mathrm{Ad}\rho}$. Fix a positively oriented vector $v \in \mathbf{h}$. After possibly rescaling B, there is a map $J : \mathbf{h}^{\perp} \to \mathbf{h}^{\perp}$ such that for all $a, b \in \mathbf{h}^{\perp}$,

$$B([a, b], v) = B(a, Jb). \tag{1.9}$$

The action of $\mathrm{Ad}\,(S^1_0)$ commutes with J, and $J^2 = -1$. Thus $H^1(\pi; \mathbf{h}^{\perp}_{\mathrm{Ad}\rho})$ inherits a complex structure from $\mathbf{h}^{\perp}$, also denoted J. The action of $\mathrm{Ad}\,(S^1_0)$ can be viewed as multiplication by unit complex numbers (with weight 2). However, in order for J to give a complex structure on the *bundle* ν, it must be lifted to $\widetilde{S}^-$, since the action of $N(S^1_0)/Z(S^1_0)$ is *conjugate* linear. For the remainder of this section, regard ν, η_j, etc. as so lifted.

Note that J and $*$ commute. Note also that while η_j is totally real with respect to $*$, it is complex with respect to J. Call such subspaces complex lagrangians.

It follows from (1.9) that

$$[\alpha, \beta] = \langle \alpha, J * \beta \rangle$$

for all $\alpha, \beta \in \nu_p$. Since $(J*)^2 = 1$, we can decompose ν as $A^+ \oplus A^-$, where $J * |_{A^\pm} = \pm 1$. Since η_j is totally real with respect to $*$ and complex with respect to J, we must have $\dim A^{\pm} = \frac{1}{2} \dim \nu = 2g - 2$. Note that $* = -J$ on A^+ and $* = J$ on A^-.

The solutions to $[\alpha, \alpha] = 0$ are the cone on the product of the unit sphere bundles of A^+ and A^-. Thus

$$\hat{\xi}_p \cong X \stackrel{\text{def}}{=} (S^{2g-3} \times S^{2g-3})/S^1,$$

where S^1 acts diagonally via complex multiplication on $S^{2g-3} \subset \mathbf{C}^{g-1}$. Also,

$$\hat{\theta}_{j,p} \cong \mathbf{C}P^{g-2}.$$

We wish to analyze $\mathcal{L}^C_p$, the space of all complex lagrangians of ν_p. Consider a complex lagrangian $L \in \mathcal{L}^C_p$. L must be transverse to A^+ and A^-, and so can be thought of as a graph of a linear map $T : A^+ \to A^-$.

Since L is complex with respect to J, T must be J-linear (and hence $*$-conjugate-linear). Since L is totally real with respect to $*$, we must have, for all $\alpha, \beta \in A^+$,

$$\begin{aligned} 0 &= \langle (\alpha, T\alpha), *(\beta, T\beta) \rangle \\ &= \langle \alpha, *\beta \rangle + \langle T\alpha, *T\beta \rangle \\ &= -\langle \alpha, J\beta \rangle + \langle T\alpha, TJ\beta \rangle. \end{aligned}$$

Therefore T is unitary with respect to J, and $\mathcal{L}_p^C$ can be identified with the set $U(A_p^+, A_p^-)$ of all J-unitary maps from A_p^+ to A_p^-.

(1.10) **Two Homomorphisms.**

The homomorphisms defined in the next two paragraphs will be used in Section 2.

Let $\det(\nu_p) = \bigwedge^{2g-2} \nu_p$, where here ν_p is viewed as a $*$-complex vector space. Let $\det^1(\nu_p) = \det(\nu_p)/\mathbb{R}^+ \cong S^1$. There is a map

$$\det{}^1 : \mathcal{L}_p^C \to \det{}^1(\nu_p) \tag{1.11}$$

defined as follows. Elements of $\det(\nu_p)$ are represented by bases of ν_p (over $\mathbb{C}$), and two such bases are identified if the linear map connecting them has determinant 1. Given $L \in \mathcal{L}_p^C$, choose a basis (over $\mathbb{R}$) of L which is oriented with respect to the orientation coming from the J-complex structure. This will also be a basis of ν_p over $\mathbb{C}$, with respect to $*$. Since L is totally real with respect to $*$, any two such bases differ by an element of $GL^+(2g-2, \mathbb{R}) \subset GL(2g-2, \mathbb{C})$ (well-defined up to conjugacy). Since $\det(GL^+(2g-2, \mathbb{R})) = \mathbb{R}^+$, this basis represents a well-defined element of $\det^1(\nu_p)$. Now define

$$\begin{aligned} \varphi : \quad \pi_1(\mathcal{L}_p^C) &\to \mathbb{Z} \\ [\alpha] &\mapsto \deg(\det{}^1(\alpha)), \end{aligned}$$

where $\det^1(\alpha)$ is viewed as a map from S^1 to S^1, and deg denotes the degree of the map. φ is clearly a homomorphism.

Now we define another homomorphism

$$\psi : \pi_1(\mathcal{L}_p^C) \to \mathbb{Z}.$$

By taking unit vectors and dividing out by the action of $\mathrm{Ad}\,(S_0^1)$, each $L \in \mathcal{L}_p^C$ gives rise to a complex projective space $\mathbb{C}P(L) \subset \hat{\xi}_p$. (e.g. $\mathbb{C}P(\eta_{j,p}) = \hat{\theta}_{j,p}$.) Orient $\mathbb{C}P(L)$ according to its J-complex structure. Orient $\hat{\xi}_p$ so that its orientation followed by an outward pointing radial vector gives the symplectic orientation of ξ_p. Choose $L_0 \in \mathcal{L}_p^C$. Let $[\alpha] \in \pi_1(\mathcal{L}_p^C)$. Define

$$\psi([\alpha]) \stackrel{\text{def}}{=} \langle \mathbb{C}P(\alpha), \mathbb{C}P(L_0)\rangle_{\hat{\xi}_p}.$$

(Here $\langle \cdot , \cdot \rangle_{\hat{\xi}_p}$ denotes the intersection number in $\hat{\xi}_p$.) It is easy to see that $\psi([\alpha])$ depends only on the homotopy class of α and is independent of the choice of L_0.

(1.12) **Lemma.** $\varphi = (-1)^{g-1}\psi$.

Proof: Since $\pi_1(\mathcal{L}_p^C) \cong \pi_1(U(g-1)) \cong \mathbb{Z}$, it suffices to compare φ and ψ on a non-zero element of $\pi_1(\mathcal{L}_p^C)$.

Fix $T \in U(A_p^+, A_p^-) = \mathcal{L}_p^C$. Let U^- be the unitary transformations of A_p^-. Identify $V \in U^-$ with $VT \in U(A_p^+, A_p^-)$. Let $\alpha : S^1 \to U^-$ be the diagonal embedding of S^1. (The orientation is unambiguous, since $J = *$ on A_p^-.) It is easy to verify that

$$\varphi([\alpha]) = g - 1.$$

Now we compute

$$\psi([\alpha]) = \langle \mathbb{C}P(\alpha), \mathbb{C}P(T)\rangle_{\hat{\xi}_p}.$$

Since $\mathbb{C}P(T) \subset \mathbb{C}P(\alpha)$, the above intersection number is equal to the self intersection of $\mathbb{C}P(T)$ inside the normal bundle of $\mathbb{C}P(\alpha)$ restricted to $\mathbb{C}P(T)$. To get the sign right, the orientation of this restricted bundle should be such that it gives the orientation of $\hat{\xi}_p$ when followed by the standard orientation of the "S^1 direction" in $\mathbb{C}P(\alpha)$. This self intersection number is the same as the self intersection of the image of $\mathbb{C}P(T)$ in $\hat{\xi}_p/\alpha(S^1)$. ($\alpha(S^1)$ is a subgroup of U^-, and it acts on $\hat{\xi}_p$ in the obvious way.) It is not hard to see that the pair $(\hat{\xi}_p/\alpha(S^1), \mathbb{C}P(T))$ is diffeomorphic to $(\mathbb{C}P^{g-2} \times \mathbb{C}P^{g-2}, \Delta)$, where Δ is the diagonal. $\hat{\xi}_p/\alpha(S^1)$ can also be identified with $\mathbb{C}P(A_p^+) \times \mathbb{C}P(A_p^-)$. $\mathbb{C}P(A_p^+) \times \mathbb{C}P(A_p^-)$ has three orientations of interest: O_c, the orientation which makes the sign of the self intersection come out correctly; O_J, its orientation as a J-complex manifold; and O_*, its orientation as a $*$-complex manifold. It turns out that $O_c = -O_*$. Also, since $J = -*$ on A_p^+ and $\dim_C(\mathbb{C}P(A_p^+)) = g-2$, $O_* = (-1)^{g-2}O_J$. Since Δ is a J-complex manifold, the self intersection of Δ in $\mathbb{C}P(A_p^+) \times \mathbb{C}P(A_p^-)$ with respect to O_J is equal to $g-1$, the Euler characteristic of $\mathbb{C}P^{g-2}$. Hence

$$\psi([\alpha]) = (-1)^{g-1}(g-1).$$

□

(1.13) **Extending** $\det(\nu)$.

Consider the bundle $\det(\nu)$ over $\widetilde{S}^-$. Our next goal is to show that $\det(\nu)$ extends naturally over all of $\widetilde{S}$ and that the fiber over $\rho \in P$ is naturally identified with $\det(H^1(F;\mathbf{h}^\perp_{\mathrm{Ad}\rho}))$. Furthermore, η_j gives rise to a section $\det^1(\eta_j)$ of $\det^1(\nu)$ over $\widetilde{T_j}^-$ which extends continuously to $\det^1(H^1(W_j;\mathbf{h}^\perp_{\mathrm{Ad}\rho}))$ over $\rho \in \widetilde{T_j} \cap P$. To prove these facts we view $\det(\nu)$ as the determinant line bundle of a family of Fredholm operators (cf. [Q]).

Let $\Omega^q(\mathbf{h}^\perp)$ be the space of q-forms on F with values in $\mathbf{h}^\perp$. The Hodge star operator gives a complex structure on $\Omega^1(\mathbf{h}^\perp)$ and $\Omega^0(\mathbf{h}^\perp) \oplus \Omega^2(\mathbf{h}^\perp)$. Let E be the trivial principal S^1_0 bundle over F. Let $\mathcal{F}$ be the space of flat orthogonal connections on E. $\mathcal{F}$ can also be viewed, via the representation $\mathrm{Ad}\,(S^1_0)$, as the space of flat connections on $F \times \mathbf{h}^\perp$. Let $\mathcal{G}_0$ be the gauge transformations of E which are the identity at some fixed point. $\mathcal{G}_0$ acts freely on $\mathcal{F}$ and $\mathcal{F}/\mathcal{G}_0$ can be identified with $\widetilde{S}$.

For each $A \in \mathcal{F}$, we have the exterior covariant derivative

$$d_A : \Omega^q(\mathbf{h}^\perp) \to \Omega^{q+1}(\mathbf{h}^\perp)$$

and its adjoint

$$d_A^* = - * d_A * : \Omega^{q+1}(\mathbf{h}^\perp) \to \Omega^q(\mathbf{h}^\perp).$$

Let D_A be the operator

$$D_A = d_A + d_A^* : \Omega^1(\mathbf{h}^\perp) \to \Omega^0(\mathbf{h}^\perp) \oplus \Omega^2(\mathbf{h}^\perp).$$

The kernel of D_A consists of the harmonic 1-forms (with respect to the flat structure given by A), which can be identified with $H^1(F;\mathbf{h}^\perp_{\mathrm{Ad}\rho})$, where ρ is the holonomy representation of A. Similarly, $\operatorname{coker} D_A = H^0(F;\mathbf{h}^\perp_{\mathrm{Ad}\rho}) \oplus H^2(F;\mathbf{h}^\perp_{\mathrm{Ad}\rho})$.

Following Quillen ([Q]), we now describe a bundle $\det(D)$ over $\mathcal{F}$, whose fiber at A can be identified with $\det(\ker D_A) \otimes \det(\operatorname{coker} D_A)^*$. (Note that this is non-trivial, since the dimensions of $\ker D_A$ and $\operatorname{coker} D_A$ are not constant.) Let V be any finite dimensional subspace of $\Omega^0(\mathbf{h}^\perp) \oplus \Omega^2(\mathbf{h}^\perp)$. Let U_V be the set of all $A \in \mathcal{F}$ such that D_A is transverse to V. (That is, $\mathrm{im}(D_A) + V = \Omega^0(\mathbf{h}^\perp) \oplus \Omega^2(\mathbf{h}^\perp)$.) For such A there is an exact sequence

$$0 \to \ker D_A \to D_A^{-1}(V) \to V \to \operatorname{coker} D_A \to 0$$

which induces a canonical isomorphism

$$\det(\ker D_A) \otimes \det(\operatorname{coker} D_A)^* \cong \det(D_A^{-1}(V)) \otimes \det(V)^*. \tag{1.14}$$

The set U_V is open, and $D_A^{-1}(V)$ is a smooth vector bundle over U_V. Thus the right hand side of (1.14) is a smooth complex line bundle over U_V. Such bundles over the sets U_V (for various V) fit together to form $\det(D)$ over all of $\mathcal{F}$.

$\mathcal{G}_0$ acts equivariantly on $\det(D)$, yielding a bundle over $\mathcal{F}/\mathcal{G}_0 = \widetilde{S}$, which we also denote by $\det(D)$. If $\rho \in \widetilde{S}^-$, then $H^0(F;\mathbf{h}^\perp_{\mathrm{Ad}\rho}) \cong H^2(F;\mathbf{h}^\perp_{\mathrm{Ad}\rho}) \cong 0$. Hence $\det(D)_\rho = \det(H^1(F;\mathbf{h}^\perp_{\mathrm{Ad}\rho})) = \det(\nu_\rho)$. If $\rho \in P$, then $H^0(F;\mathbf{h}^\perp_{\mathrm{Ad}\rho})$ is a complex lagrangian in $H^0(F;\mathbf{h}^\perp_{\mathrm{Ad}\rho}) \oplus H^2(F;\mathbf{h}^\perp_{\mathrm{Ad}\rho}) = \operatorname{coker} D_\rho$. This gives a canonical element of $\det(\operatorname{coker} D_\rho)$, which allows us to identify $\det(D)_\rho$ with $\det(\ker D_\rho) = \det(H^1(F;\mathbf{h}^\perp_{\mathrm{Ad}\rho}))$. So $\det(D)$ is the desired extension of $\det(\nu)$.

The map $\det^1 : \mathcal{L}^C \to \det^1(\nu)$ of (1.11), applied to η_j, gives a section $\det^1(\eta_j)$ of $\det^1(\nu)$ defined over $\widetilde{T_j}^-$. We wish to show that this section extends smoothly over all of $\widetilde{T_j}$.

Fix a trivial connection $A_0 \in \mathcal{F}$. Let $d = d_{A_0}$. Let H^q be the the harmonic q-forms with coefficients in $\mathbf{h}^\perp$ (with respect to A_0). Let $H^1(F;\mathbf{h})$ be the harmonic 1-forms with coefficients in $\mathbf{h}$. Given $\rho \in \widetilde{S}$ near 1, there is a unique $\alpha_\rho \in H^1(F;\mathbf{h})$ near zero such that for all loops $\gamma \subset F$,

$$\exp(\int_\gamma \alpha_\rho) = \rho(\gamma).$$

Let

$$d_\rho = d + \alpha_\rho.$$

The map $\rho \mapsto d_\rho$ is a section of the fibration $\mathcal{F} \to \widetilde{S}$, for ρ near 1. Let

$$D_\rho = d_\rho + d_\rho^*,$$

and let H^q_ρ be the harmonic q-forms with values in $\mathbf{h}^\perp$ with respect to d_ρ. That is, $H^q_\rho = \ker(D_\rho)$. Note that for $q = 0$ or 2, $H^q_\rho = H^q$ for $\rho = 1$ and $H^q_\rho = 0$ for $\rho \neq 1$.

$D_\rho : \Omega^1(\mathbf{h}^\perp) \to \Omega^0(\mathbf{h}^\perp) \oplus \Omega^2(\mathbf{h}^\perp)$ is transverse to $H^0 \oplus H^2$ for ρ near 1. Hence $\det(\nu)$ can be identified with the determinant of the (finite dimensional) operator

$$D_\rho : D_\rho^{-1}(H^0 \oplus H^2) \to H^0 \oplus H^2$$

for ρ near 1.

Consider the diagram

$$\begin{array}{ccc} D_\rho^{-1}(H^0 \oplus H^2) & \overset{D_\rho}{\rightarrow} & H^0 \oplus H^2 \\ \downarrow \pi & & \downarrow i \\ H^1 & \overset{D'_\rho}{\rightarrow} & \mathbf{h}^\perp \oplus \mathbf{h}^\perp \end{array},$$

where π is orthogonal projection (with respect to the Hodge inner product), i is the isomorphism

$$i(a,b) \overset{\text{def}}{=} \left(\int_F *a, \int_F b\right),$$

and

$$D'_\rho(a) \overset{\text{def}}{=} \left(\int_F \alpha_\rho \wedge *a, \int_F \alpha_\rho \wedge a\right).$$

Using the Hodge decomposition theorem, it is not hard to see that this diagram commutes, and that π is an isomorphism for ρ near 1. Hence, for such ρ, $\det(\nu)$ can be identified with the determinant of D'_ρ.

The lagrangian $H^1(W_1;\mathbf{h}^\perp)\otimes(\mathbf{h}^\perp\oplus 0)^* \subset H^1\otimes(\mathbf{h}^\perp\oplus\mathbf{h}^\perp)^*$ gives rise to a smooth section of $\det^1(D'_\rho) = \det^1(\nu)$ near $\rho = 1$. For $\rho \neq 1$, this section coincides with $\det^1(\eta_1)$. For $\rho = 1$, it coincides with $\det^1(H^1(W_1;\mathbf{h}^\perp))$. Similar things are true if W_1 is replaced by W_2 or if 1 is replaced with another representation in P. Thus we have found the desired extensions of $\det^1(\nu)$ and $\det^1(\eta_j)$.

As a notational contrivance, let us define, for $\rho \in P \subset \widetilde{S}$,

$$\nu_\rho \overset{\text{def}}{=} H^1(F;\mathbf{h}^\perp_{\mathrm{Ad}\rho}) \cong H^1(F;\mathbf{h}^\perp).$$

If $\rho \in P \cap \widetilde{T_j}$, define

$$\eta_{j,\rho} \overset{\text{def}}{=} H^1(W_j;\mathbf{h}^\perp_{\mathrm{Ad}\rho}) \cong H^1(W_j;\mathbf{h}^\perp).$$

So, abusing notation slightly, $\det^1(\nu)$ is a smooth bundle over all of $\widetilde{S}$ and $\det^1(\eta_j)$ is a smooth section defined over all of $\widetilde{T_j}$. This will simplify notation in what follows.

(1.15) **Proposition.** *The first Chern class $c_1(\det(\nu))$ of $\det(\nu)$ is represented by a multiple ω of ω_B. If $T^2 \subset \widetilde{S}$ is a symplectic 2-torus corresponding to a 2-dimensional symplectic summand of $H_1(F;\mathbb{R})$, then $\int_{T^2}\omega = -8$.*

Proof: Let $\mathcal{A}$ be the space of connections on $F \times S_0^1$. The first step is to compute the curvature the universal S_0^1 bundle $\mathcal{U} = \mathcal{A} \times F \times S_0^1$ over $\mathcal{A} \times F$ (cf. [AS]).

Identify $\mathcal{A}$ with $\Omega^1(F;\mathbf{h})$ via $A \mapsto d + A$. Identify $T_A\mathcal{A}$ with $\Omega^1(F;\mathbf{h})$. Let $(A, x, \varphi) \in \mathcal{U}$ and $(a, X, w) \in T_{(A,x,\varphi)}\mathcal{U} = \Omega^1(F;\mathbf{h}) \times T_xF \times \mathbf{h}$. Define the connection 1-form θ by

$$\theta_{(A,x,\varphi)}(a, X, w) = A(X) + w.$$

The connection θ has the property that over the surface $\{A\} \times F$ it restricts to the connection A on $F \times S_0^1$. The curvature of θ is $\Theta = d\theta \in \Omega^2(\mathcal{A} \times F;\mathbf{h})$. Thus, for $(a, X), (b, Y) \in T_{(A,x)}\mathcal{A} \times F$,

$$\Theta((a, X), (b, Y)) = a(Y) - b(X) + dA(X, Y).$$

Let $\mathcal{G}_0$ be the group of smooth maps $g : F \to S_0^1$ such that $g(x_0) = 1$ for some fixed $x_0 \in F$. $\mathcal{G}_0$ acts freely on $\mathcal{U}$. This action preserves the connection θ, and so leads to a connection $\theta^\flat$ on $\mathcal{U}^\flat \overset{\text{def}}{=} \mathcal{U}/\mathcal{G}_0$. Recall that $\widetilde{S}$ can be identified with $\mathcal{F}/\mathcal{G}_0$, where $\mathcal{F} \subset \mathcal{A}$ is the space of flat connection on $F \times S_0^1$. Thus $\theta^\flat$ restricts to a connection on $\widetilde{S} \times F$.

The representation $S_0^1 \to \mathrm{Ad}\,(S_0^1)|_{\mathbf{h}^\perp}$ associates to $\mathcal{U}^\flat$ a $\mathbf{h}^\perp$ bundle E over $\widetilde{S} \times F$ with curvature

$$\Theta' = \begin{bmatrix} 0 & 2\Theta \\ -2\Theta & 0 \end{bmatrix}.$$

Hence $c_1(E \otimes \mathbb{C}) = (2\pi i)^{-1}\mathrm{tr}(\Theta') = 0$ and $c_2(E \otimes \mathbb{C})$ is represented by

$$-\frac{1}{4\pi^2}\det(\Theta') = -\frac{1}{\pi^2}\Theta \wedge \Theta.$$

Thus the Chern character of E is represented by

$$\mathrm{ch}(E) = 2 + \frac{1}{\pi^2}\Theta \wedge \Theta.$$

E can be thought of as a family of flat $\mathbf{h}^\perp$ bundles over F parameterized by $\widetilde{S}$. Corresponding to each flat connection A is the operator D_A, defined above, and $\det(\nu)$ can be identified with the determinant of the index bundle of this family.

A flat connection A on $F \times (\mathbf{h}^\perp \otimes \mathbb{C})$ determines a holomorphic structure A' on $F \times (\mathbf{h}^\perp \otimes \mathbb{C})$, and the index of the family D_A can be identified with the index of the family $\overline{\partial}_{A'}$. (F is given the complex structure compatible with its metric.) So, by Theorem 5.1 of [AS4] and the Riemann-Roch theorem, $c_1(\det(\nu))$ is represented by the 2-dimensional part of

$$\int_F \mathrm{ch}(E)\mathcal{T}(F),$$

where $\int_F$ denotes integration along the fibers (i.e. along F) and $\mathcal{T}(F)$ is a form representing the Todd genus of F. Since $\mathcal{T}(F) = 1 + \alpha$, where $\alpha \in \Omega^2(F)$, this is just

$$\int_F \frac{1}{\pi^2}\Theta \wedge \Theta.$$

Note that, for $a, b \in H^1(F; \mathbf{h}) \cong T_p\widetilde{S}$ and $X, Y \in T_x F$,

$$\begin{aligned} \Theta \wedge \Theta(a, b, X, Y) &= -a(X)b(Y) + a(Y)b(X) + b(X)a(Y) - b(Y)a(X) \\ &= -2a \wedge b(X, Y). \end{aligned}$$

Hence $c_1(\det(\nu))$ is represented by the 2-form ω, where

$$\omega(a, b) \stackrel{\text{def}}{=} -\frac{2}{\pi^2} a \wedge b.$$

(Here we have identified $H^2(F; \mathbf{R})$ with $\mathbf{R}$ via $c \mapsto \int_F c$.) Let $T^2 \subset \widetilde{S}$ be a symplectic 2-torus. Then

$$\int_{T^2} \omega = -\frac{2}{\pi^2}(2\pi)^2 = -8.$$

□

D. Results of Newstead.

In this subsection we use results of Newstead ([N1,N2]) to show that certain diffeomorphisms of representation spaces are homologically trivial.

Let $\gamma \subset F$ be a separating simple closed curve. Define

$$N_\gamma \stackrel{\text{def}}{=} \{[\rho] \in R \mid \rho(\gamma) = -1\}.$$

Let F_1 and F_2 be the components of F cut along γ. Define

$$\begin{aligned} R_j^{*\sharp} &\stackrel{\text{def}}{=} \mathrm{Hom}(\pi_1(F_j), SU(2)) \\ N_j^\sharp &\stackrel{\text{def}}{=} \{\rho \in R_j^{*\sharp} \mid \rho(\gamma) = -1\} \end{aligned}$$

$(j = 1, 2)$. Then we can make the identifications

$$\begin{aligned} R^\sharp &= \{(\rho_1, \rho_2) \in R_1^{*\sharp} \times R_2^{*\sharp} \mid \rho_1(\gamma) = \rho_2(\gamma)\} \\ N_\gamma^\sharp &= N_1^\sharp \times N_2^\sharp. \end{aligned}$$

(In the first equation, some basing of the free loop γ is assumed.)

Let $\tau : F_1 \to F_1$ be a diffeomorphism such that $\tau|_\gamma = \mathrm{id}$ and $\tau_* : H_*(F_1; \mathbf{Z}) \to H_*(F_1; \mathbf{Z})$ is the identity. τ induces diffeomorphisms of F (which is the identity on F_2) and N_γ, which will also be denoted τ. In the proofs of (3.10) and (3.12), we will need the following lemma.

(1.16) **Lemma.** $\tau^* : H^{3g-3}(N_\gamma; \mathbf{Q}) \to H^{3g-3}(N_\gamma; \mathbf{Q})$ *is the identity.*

Proof: Note that since $[\gamma]$ lies in the commutator subgroup of $\pi_1(F)$, $N_\gamma^\sharp = N_1^\sharp \times N_2^\sharp$ consists entirely of irreducible representations. Hence there is a fibration

$$p : N_1^\sharp \times N_2^\sharp \to N_\gamma \tag{1.17}$$

with fiber $SU(2)/Z(SU(2)) \cong SO(3)$. Since $SO(3)$ is a rational homology 3-sphere, there is a Gysin sequence

$$\cdots \to H^{i-4}(N_\gamma) \xrightarrow{\cup \chi} H^i(N_\gamma) \xrightarrow{p^*} H^i(N_1^\sharp \times N_2^\sharp) \to H^{i-3}(N_\gamma) \to \cdots .$$

(All coefficients are in $\mathbf{Q}$. $\chi \in H^4(N_\gamma)$ is the Euler class of the fibration.)

I claim that $p^* : H^i(N_\gamma) \to H^i(N_1^\sharp \times N_2^\sharp)$ is an isomorphism for $i = 2$ or 3. This is obvious for $i = 2$, and injectivity is obvious for $i = 3$. By Proposition 2.6 of [N2], the natural map

$$H^3(N_j) \to H^3(N_j^\sharp)$$

is an isomorphism $(j = 1, 2)$. Furthermore, $H^1(N_j) \cong H^1(N_j^\sharp) \cong 0$ (Theorems 1 and 1′ of [N2]). Therefore, by the Künneth formula, the composite

$$H^3(N_1 \times N_2) \to H^3(N_\gamma) \to H^3(N_1^\sharp \times N_2^\sharp)$$

is an isomorphism. (The first arrow is induced by the fibration $N_\gamma \to N_1 \times N_2$.) Hence $p^* : H^3(N_\gamma) \to H^3(N_1^\sharp \times N_2^\sharp)$ is onto.

By Theorem 1′ of [N2] and the Künneth formula, $H^i(N_1^\sharp \times N_2^\sharp)$ is generated by classes of dimensions 2 and 3 for $i \leq 3g - 3$. I claim that $H^i(N_\gamma)$ is generated by classes of dimensions 2 and 3 and $\chi \in H^4(N_\gamma)$, for $i \leq 3g - 3$. This is clear for $i \leq 3$. Suppose it holds for $i \leq k < 3g - 3$. Let $\alpha \in H^{k+1}(N_\gamma)$. Then there exists $\beta \in H^{k+1}(N_\gamma)$ lying in the subring of

$H^*(N_\gamma)$ generated by classes of dimensions 2 and 3 such that $p^*(\alpha - \beta) = 0$. Hence

$$\alpha = \beta + \eta \cup \chi,$$

where, by inductive assumption, $\eta \in H^{k-3}(N_\gamma)$ lies in the subring of $H^*(N_\gamma)$ generated by classes of dimensions 2 and 3 and χ.

So it suffices to show that τ^* is the identity on $H^2(N_\gamma)$, $H^3(N_\gamma)$, and χ.

Since τ acts on the fibration (1.17) in an orientation preserving fashion, $\tau^*(\chi) = \chi$.

Since $p^* : H^i(N_\gamma) \to H^i(N_1^\sharp \times N_2^\sharp)$ is an isomorphism for $i = 2$ or 3, it suffices to show that τ acts trivially on $H^i(N_1^\sharp \times N_2^\sharp)$. By the Künneth formula and the fact that τ act trivially on $H^i(N_2^\sharp)$, it suffices to show that τ acts trivially on $H^i(N_1^\sharp)$ $(i = 2, 3)$. This is done in [AM] (Lemmas VI.2.1 and VI.2.2). □

Let $x_1, y_1, \ldots, x_k, y_k$ be a symplectic basis of $\pi_1(F_1)$. This gives rise to identifications

$$\begin{aligned} R_1^{*\sharp} &= (SU(2))^{2k} \\ N_1^\sharp &= \{(X_1, Y_1, \ldots, X_k, Y_k) \in R_1^{*\sharp} \mid \prod_1^k [X_i, Y_i] = -1\}. \end{aligned}$$

Define $\sigma : N_1^\sharp \to N_1^\sharp$ by

$$\sigma(X_1, Y_1, \ldots, X_k, Y_k) = (X_1, -Y_1, X_2, Y_2, \ldots, X_k, Y_k).$$

$\sigma \times \mathrm{id} : N_1^\sharp \times N_2^\sharp$ descends to a diffeomorphism of N_γ, which will also be denoted by σ. In the proof of (3.59), we will need the following lemma.

(1.18) **Lemma.** $\sigma^* : H^{3g-3}(N_\gamma; \mathbb{Q}) \to H^{3g-3}(N_\gamma; \mathbb{Q})$ *is the identity.*

Proof: The proof of (1.16) can be adapted to this case almost verbatim. It is left to the reader to check that the proofs of Lemmas VI.2.1 and VI.2.2 of [AM] work with τ replaced by σ. (This amounts to observing that $\sigma^* : H^3(R_1^{*\sharp}) \to H^3(R_1^{*\sharp})$ is the identity and that σ acts on various sphere bundles in an orientation preserving fashion.) □

E. Special Isotopies.

(1.19) *Definition.* An isotopy $\{h_t\}_{0 \le t \le 1}$ of R is called *special* if

be a 3-manifold and let $T \subset \partial N$ be a boundary component of N
iffeomorphic to a torus. Let $a \in H_1(T;\mathbb{Z})$ be a primitive homology
is, one which can be represented by a simple closed curve). Let
$D^2) \to T$ be a diffeomorphism which sends $\{1\} \times \partial D^2$ to a curve
g a.

$$N_a \stackrel{\text{def}}{=} N \cup_f (S^1 \times D^2)$$

e Dehn surgery of N along a. Note that N_a does not depend
of a. If N is oriented, we give N_a the orientation induced from

I is a knot in a 3-manifold M, then Dehn surgery on K means
y on $M \setminus U$, where U is an open tubular neighborhood of K.
$M \setminus U);\mathbb{Z})$ is primitive, then $K_a \stackrel{\text{def}}{=} (M \setminus U)_a$.
neridian of K to be a generator of

$$\ker(H_1(\partial(M \setminus U);\mathbb{Z}) \to H_1(\overline{U};\mathbb{Z})).$$

ude of K to be a generator of

$$\ker(H_1(\partial(M \setminus U);\mathbb{Z}) \to H_1(M \setminus U;\mathbb{Z})).$$

classes are well-defined up to sign. If K is a null-homologous
$0 \in H_1(M;\mathbb{Z})$), then a meridian, m, and a longitude, l, of
of $H_1(\partial(M \setminus U);\mathbb{Z})$. In this case, let $K_{p/q}$ denote K_{pm+ql},
of m and l are chosen so that $\langle m, l\rangle = 1$. With this sign
nd l are call a standard basis of $H_1(\partial(M \setminus U);\mathbb{Z})$. (Here
ersection pairing on $H_1(\partial(M \setminus U);\mathbb{Z})$ and $\partial(M \setminus U)$ is ori-
o the "inward normal last" convention.)

1. $h_t(Q_1)$ is transverse to Q_2 at $Q_1 \cap Q_2 \cap P$ for all t.
2. $h_t|_S = \text{id}$ for all t.
3. $(h_t)_* : TR|_{S^-} \to TR|_{S^-}$ is symplectic, and hence (in view of 2 above) preserves the fibers of the normal bundle ν.

(1.20) **Proposition.** *There exists a special isotopy $\{h_t\}_{0 \le t \le 1}$ of R such that $h_1(Q_1)$ is transverse to Q_2 (i.e. their Zariski tangent spaces are transverse at each point of $Q_1 \cap Q_2$).*

Proof: Since M is a QHS, Q_1 is transverse to Q_2 at P and T_1 is transverse to T_2 in S. Choose a compactly supported isotopy of a tubular neighborhood of S^- in R which moves η_{1p} transverse to η_{2p} for each $p \in T_1^- \cap T_2^-$ and is symplectic on $TR|_{S_-}$. At this stage Q_1 is transverse to Q_2 in a neighborhood of S, and so we can find a compactly supported isotopy of R^- which moves Q_1 transverse to Q_2. □

F. Orientations.

Orientations will be important in what follows, so in this section we will establish orientation conventions. $[Y]$ will denote the orientation of the space Y. If Y is singular, this means the orientation of the top stratum. If Y is a bundle, it means an orientation of the fibers (not the total space).

First of all, orient the Heegaard surface F so that $[F]$ followed by a normal vector to F pointing into Q_2 gives $[M]$. This fixes an identification of $H^2(\pi_1(F);\mathbb{R})$ with $\mathbb{R}$, and so fixes the sign of ω.

Complex vector spaces (and almost complex manifolds) have a natural orientation: If $a_1, \ldots, a_n$ is a basis over $\mathbb{C}$, then $a_1, ia_1, \ldots, a_n, ia_n$ is an oriented basis over $\mathbb{R}$. Symplectic vector spaces and manifolds are oriented according to a compatible almost complex structure. Thus ω determines the orientations $[R]$, $[S]$, $[\nu]$ and $[\xi]$. η_j and $\hat{\theta}_j$, when lifted to $\widetilde{T_j}^-$, have J-complex structures. This determines $[\eta_j]$ and $[\hat{\theta}_j]$ (as bundles over $\widetilde{T_j}^-$).

Choose orientations of $\widetilde{T_1}$ and $\widetilde{T_2}$ so that

$$[\widetilde{S}] = [\widetilde{T_1}][\widetilde{T_2}]$$

at points of $\widetilde{T_1} \cap \widetilde{T_2}$. ($[\widetilde{S}]$ is the orientation lifted from $[S]$.)

In general, given a fibering $Y \hookrightarrow E \to B$, we choose orientations so that $[E] = [B][Y]$. Thus orientations on two of E, B or Y determine an orientation of the third. If G acts on X, there is a fibering $G \hookrightarrow$

$X \to X/G$ (at least at points where G acts freely, which is enough to determine orientations). We regard spaces of unit vectors (e.g. $\hat{\theta}_j$) as quotients of spaces of non-unit vectors (e.g. θ_j) by $\mathbf{R}^+$, the positive reals. Let $[SU(2)]$, $[S^1_0]$ and $[\mathbf{R}^+]$ be the standard orientations. The following equations determine orientations of spaces not yet oriented. (There are some cases of over determination, and it is left to the reader to check that these cases are consistent.)

$$\begin{aligned}
[R] &= [S][\xi] \\
[\xi] &= [\hat{\xi}][\mathbf{R}^+] \\
[Q_j] &= [\widetilde{T_j}][\theta_j] \\
[\eta_j] &= [\theta_j][S^1_0] \\
[\theta_j] &= [\hat{\theta}_j][\mathbf{R}^+] \\
[Q^\sharp_j] &= [Q_j][SU(2)] \\
[R^\sharp] &= [R][SU(2)]
\end{aligned}$$

(The third equation requires some comment. $[\widetilde{T_j}][\theta_j]$ determines an orientation of a double cover of a neighborhood of T_j^- in Q_j, which induces an orientation on Q_j.)

All boundaries will be oriented according to the "inward normal last" convention. (e.g. F is oriented as ∂W_2.) In particular, this fixes an orientation of ∂F^*. Hence the map

$$(1.21) \qquad \begin{aligned} \partial: \quad R^{*\sharp} &\to SU(2) \\ \rho &\mapsto \rho(\partial F^*) \end{aligned}$$

is well-defined up to conjugation by elements of $SU(2)$ (which preserves orientation). Note that $R^\sharp = \partial^{-1}(1)$. Thus, near regular points of $R^\sharp$, there is an identification of the normal fiber (in $R^{*\sharp}$) with $\mathbf{su}(2)$, well-defined up to orientation preserving linear maps (i.e. $\mathrm{Ad}\,(SU(2))$). Choose $[R^{*\sharp}]$ so that

$$(1.22) \qquad [R^{*\sharp}] = [R^\sharp][\mathbf{su}(2)].$$

($[\mathbf{su}(2)]$ is, of course, the standard orientation of $\mathbf{su}(2)$.)

Now we give an orientation convention for intersections of manifolds. Let A and B be oriented properly embedded submanifolds of an oriented manifold Y, intersecting transversely in X. Let α be the normal bundle of A in Y. Orient α so that

$$[Y] = [\alpha][A].$$

$\alpha|_X$ is the normal bundle of X in B. Orient X so that

$$[B] = [\alpha|_X][X].$$

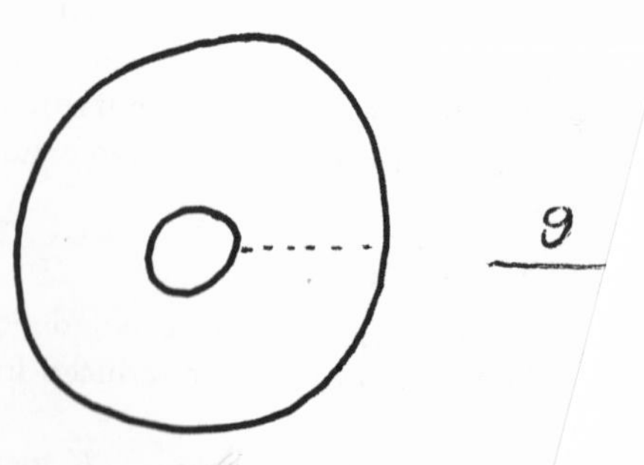

Figure 1.1: Prototypic

Note that this orientation conve
given above for boundaries in the

Note also that the induced ori
and B.

G. Dehn Twists and

Let $f : [1,2] \to \mathbf{R}$ be a
1 and $f(r) = 2\pi$ for r n
$g : A \to A$ by

for $1 \le r \le 2$ and $0 \le$
near ∂A and the the i

Let γ be a simple
be orientation preser
components of A to

The isotopy class
isotopy class of
of such maps ar

§ 2

Definition of λ

In this section we define an invariant $\lambda(M)$ of an oriented rational homology sphere M. Section A contains the definition of λ. In section B, we show that $\lambda(M)$ is independent of the various choices made in section A. Notation from Section 1 will be retained throughout.

A. Definition.

Let M be a rational homology sphere. By (1.20), Q_1 and Q_2 can be put into general position via a special isotopy. We will abuse notation and continue to denote the isotoped image of Q_1 by Q_1. The same goes for η_1, θ_1, $H^1(W_1; \mathbf{h}^\perp_{\mathrm{Ad}\rho})$, etc.

$Q_1 \cap Q_2$ now consists of a finite number of points. It would be nice if we could assign a sign to each point of $Q_1 \cap Q_2$, sum these signs, and get a well-defined invariant of M. In particular, the sum would have to be special isotopy invariant, since we began with an arbitrary special isotopy. Alas, this cannot work. For consider the case where F has genus 2.

In this case the fiber of ξ is $C(S^1 \times S^1/S^1) \cong C(S^1)$ and the fiber of θ_j is $C(\mathbb{C}P^0) \cong C(\mathrm{pt})$ (see (1.8)). ($C(X)$ denotes the cone on X.) Let $p \in T_1^- \cap T_2^-$. (Such points exist if $H_1(M;\mathbb{Z})$ has elements of order greater than two.) Then the triple $(R, Q_1, Q_2) \cap \xi_p$ is diffeomorphic to $(C(S^1), C(a_1), C(a_2))$, where $a_1, a_2 \in S^1$ are distinct. There is an isotopy $h : C(S^1) \times I \to C(S^1)$, supported away from $\partial C(S^1)$, such that $h(\,\cdot\,, 0) =$ id, $h(\,\cdot\,, 1) =$ id near the cone point, and $h(C(a_1), 1)$ intersects $C(a_2)$ once transversely (see Figure 2.1). h (or rather a suitably scaled exponential of h) extends to an isotopy h' of R supported in a neighborhood of p. Note that $h'(\,\cdot\,, 1)$ is the identity near $Q_1 \cap Q_2$. However, $h'(Q_1, 1) \cap Q_2$ has an

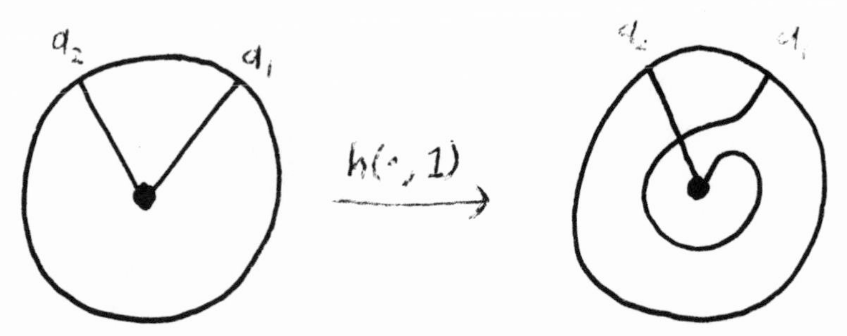

Figure 2.1: A troublesome isotopy.

extra irreducible intersection.

The above discussion makes it clear that if the irreducible points of $Q_1 \cap Q_2$ are to contribute their signs to $\lambda(M)$, the contribution of the reducible points (i.e. $T_1 \cap T_2$) must reflect global "twisting" in the normal bundle of S. In other words, we must define, for each $p \in T_1 \cap T_2$, a number $I(p)$ such that

$$\sum_{p \in Q_1^- \cap Q_2^-} \operatorname{sign}(p) + \sum_{p \in T_1 \cap T_2} I(p)$$

is independent of the choice of isotopy used to put Q_1 into general position with Q_2. Defining $I(p)$ and proving isotopy invariance will occupy most of the rest of this section.

Before proceeding with the details, we give, for motivation, a rough and oversimplified sketch of the definition of $I(p)$ in the genus 2 case. As noted above, in this case $\hat{\xi}$ is an S^1 bundle over S^- and $\hat{\theta}_j$ is a section of $\hat{\xi}$ defined over T_j^-. Assume, contrary to fact, that $\hat{\xi}$ is defined over all of S and that $\hat{\theta}_j$ is defined over all of T_j. (This allows certain technicalities to be suppressed.) Let $p \in T_1^- \cap T_2^-$. Choose arcs $\alpha_j \subset T_j$ from 1 to p. Let γ be the loop $\alpha_1 * (-\alpha_2)$. The sections $\hat{\theta}_j|_{\alpha_j}$ give a section of $\hat{\xi}|_\gamma$ with discontinuities at 1 and p. These discontinuities can be repaired by moving in the positive direction (with respect to the orientation of $\hat{\xi}$) from $\hat{\theta}_1$ to $\hat{\theta}_2$ at 1 and p. This gives a trivialization Φ of $\hat{\xi}|_\gamma$. A little thought will convince one that Φ changes by ± 1 if Q_1 is modified by the troublesome isotopy of Figure 2.1. Thus if we had a canonical trivialization Φ_0 of $\hat{\xi}|_\gamma$, we could define $I(p)$ to be the difference $\Phi - \Phi_0$.

Suppose that γ bounds a surface $E \subset S$. Then $\hat{\xi}|_E$ determines a trivialization Φ_0 of $\hat{\xi}|_\gamma$, and the difference $\Phi - \Phi_0$ is just the relative first Chern class $c_1(\hat{\xi}|_E, \Phi)$. Unfortunately, if we choose a different surface E' bounded by γ, then

$$c_1(\hat{\xi}|_E, \Phi) - c_1(\hat{\xi}|_{E'}, \Phi) = c_1(\hat{\xi}|_{E \cup (-E')}),$$

which is not in general zero. But recall (Lemma (1.15)) that the symplectic form ω represents $c_1(\hat{\xi}) = c_1(\det(\nu))$. Thus

$$I(p) \stackrel{\text{def}}{=} c_1(\hat{\xi}|_E, \Phi) - \int_E \omega$$

is independent of the choice of spanning surface E. Using the fact that T_j is lagrangian, it is not hard to show that $I(p)$ is independent of the choice the arcs α_j.

In practice, we will carry out the above construction equivariantly in the double cover $\widetilde{S}$ of S. We will also replace $\hat{\xi}$ and $\hat{\theta}_j$ with $\det^1(\nu)$ and $\det^1(\eta_j)$, since the latter bundles extend over all of $\widetilde{S}$.

Now for the details.

Let $\widetilde{S}$ be the double cover of S. Pick a metric on F, inducing an almost complex structure on ν (see (1.8)). Let $\det^1(\nu)$ be the unit determinant line bundle of the Zariski normal bundle ν of S^-, lifted to $\widetilde{S}^-$. By (1.13), $\det^1(\nu)$ extends canonically over $\widetilde{S}$. Recall from (1.13) that the Zariski normal bundle η_j of T_j^- (lifted to $\widetilde{T_j^-}$) determines a section $\det^1(\eta_j)$ of $\det^1(\nu)|_{\widetilde{T}_j^-}$ which extends smoothly and canonically over all of $\widetilde{T_j}$.

Let $p \in T_1 \cap T_2$. Let $p', p'' \in \widetilde{S}$ be the inverse images of p. (If $p \in P$, then $p' = p''$.) Choose arcs α_j' $[\alpha_j'']$ from 1 to p' $[p'']$ in $\widetilde{T_j}$ such that $\tau(\alpha_j') = \alpha_j''$, where τ is the covering involution of $\widetilde{S}$. Let $\gamma' = \alpha_1' * (-\alpha_2')$ and $\gamma'' = \alpha_1'' * (-\alpha_2'')$.

Since $\tau_* : H_1(\widetilde{S}; \mathbb{Z}) \to H_1(\widetilde{S}; \mathbb{Z})$ is equal to -1 and $\tau(\gamma') = \gamma''$, γ' is homologous to $-\gamma''$. Therefore we can choose a (possibly singular) surface E in $\widetilde{S}$ such that $\partial E = \gamma' \cup \gamma''$.

$\det^1(\eta_1)$ and $\det^1(\eta_2)$ are sections of $\det^1(\nu)$ over α_1' and α_2'. We wish to patch these sections together to get a trivialization of $\det^1(\nu)|_{\gamma'}$. This we do as follows. Let $x = 1$ or p'. $\eta_{1,x}$ and $\eta_{2,x}$ are transverse oriented lagrangian subspaces of ν_x (see (1.13)). Suppose we are given a path $P(\eta_{1,x}, \eta_{2,x})$ of oriented lagrangian subspaces from $\eta_{1,x}$ to $\eta_{2,x}$. Then $\det^1(P(\eta_{1,x}, \eta_{2,x}))$ is a path in $\det^1(\nu_x)$ connecting $\det^1(\eta_{1,x})$ to $\det^1(\eta_{2,x})$. These paths, together with $\det^1(\eta_j)|_{\alpha_j'}$, determine a map of S^1 into $\det^1(\nu)|_{\gamma'}$, namely

$$\det^1(\eta_1)|_{\alpha_1'} * \det^1(P(\eta_{1,p'}, \eta_{2,p'})) * (-\det^1(\eta_2)|_{\alpha_2'}) * (-\det^1(P(\eta_{1,1}, \eta_{2,1})).$$

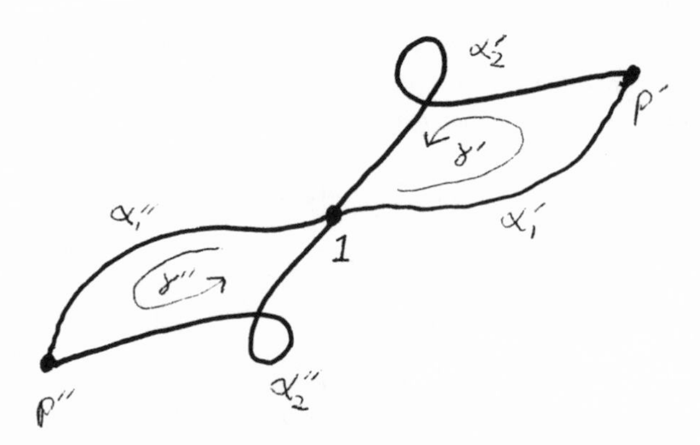

Figure 2.2: γ' and γ'' (artist's conception).

(Here the negative signs signify a reversal of path orientation.) A nonzero section of $\det^1(\nu)|_{\gamma'}$ in the same homotopy class as this map gives a trivialization of $\det^1(\nu)|_{\gamma'}$. Repeating all of the above for γ'', we get a trivialization Φ over all of ∂E.

It remains to define the path $P(\eta_{1,x}, \eta_{2,x})$. There is no unique natural choice, so instead we will define two paths, $P_+(\eta_{1,x}, \eta_{2,x})$ and $P_-(\eta_{1,x}, \eta_{2,x})$, yielding two trivializations, Φ_+ and Φ_-, and use their "average", Φ, to compute $I(p)$. (If f is an affine function from the set of trivializations to $\mathbb{R}$, then $f(\Phi) \stackrel{\text{def}}{=} (f(\Phi_+)+f(\Phi_-))/2$.) In what follows, we will usually pretend that Φ is an honest trivialization.

Let L_0, L_1 be transverse, oriented lagrangians in a symplectic vector space. Let $e_1, \ldots, e_n$ be a oriented basis of L_0. Then there is a unique basis $f_1, \ldots, f_n$ of L_1 such that $\omega(e_i, f_j) = \delta_{ij}$. (Note that if $L_{j-1} = \eta_{j,x}$ with the J-complex orientation, then this basis is oriented correctly.) Let $P_\pm(L_0, L_1)_t$ be the (oriented) span of $g_{1,t}, \ldots, g_{n,t}$, where

$$g_{i,t} = (1-t)e_i \pm (t-1)f_i.$$

(Note that $P_\pm(L_0, L_1)_t$ is transverse to L_0 and L_1 for all $0 < t < 1$. This fact will be used later.) Since the path $P_\pm(L_0, L_1)$ depends continuously on the choice of oriented basis for L_0, and the set of all such bases is connected, we see that the homotopy class of $P_\pm(L_0, L_1)$ is well defined. Note that changing the sign of ω interchanges P_+ and P_-, and so preserves Φ.

Now define

$$I(p) \stackrel{\text{def}}{=} \begin{cases} \frac{1}{2}(c_1(\det^1(\nu)|_E, \Phi) - \int_E \omega) & p \notin P \\ \frac{1}{4}(c_1(\det^1(\nu)|_E, \Phi) - \int_E \omega) & p \in P. \end{cases} \tag{2.1}$$

It must be shown that $I(p)$ does not depend on the choice of spanning surface, arcs, or the metric on F.

Suppose we choose a different surface E' such that $\partial E' = \gamma' \cup \gamma''$. Then $I(p)$ changes (up to a scalar) by

$$\begin{aligned} & c_1(\det^1(\nu)|_{E'}, \Phi) - \int_{E'} \omega - c_1(\det^1(\nu)|_E, \Phi) + \int_E \omega \\ = \; & c_1(\det^1(\nu)|_{E' \cup (-E)}) - \int_{E' \cup (-E)} \omega \\ = \; & 0. \end{aligned}$$

(Here we have used (1.15).)

Suppose we replace α_1' with $\hat{\alpha}_1'$, another arc from 1 to p' in $\widetilde{T_1}$. (We of course also replace $\alpha_1'' = \tau(\alpha_1')$ with $\hat{\alpha}_1'' = \tau(\hat{\alpha_1}')$.) Let $\hat{\gamma}' = \hat{\alpha}_1' * (-\alpha_2')$ and $\hat{\gamma}'' = \hat{\alpha}_1'' * (-\alpha_2'')$. Let δ' be the loop $\hat{\alpha}_1' * (-\alpha_1')$ and let $\delta'' = \tau\delta' = \hat{\alpha}_1'' * (-\alpha_1'')$. Let H be a surface in $\widetilde{T_1}$ such that $\partial H = \delta' \cup \delta''$. Then $\partial(E \cup H) = \hat{\gamma}' \cup \hat{\gamma}''$, and we can use $E \cup H$ in place of E to compute the new $I(p)$. Let $\hat{\Phi}$ be the trivialization of $\det^1(\nu)|_{\partial(E \cup H)}$ determined by $\det^1(\eta_j)$.

Since $H \subset \widetilde{T_1}$ and $\widetilde{T_1}$ is lagrangian with respect to ω,

$$\int_{E \cup H} \omega = \int_E \omega.$$

Since $\det^1(\eta_1)$ is defined over all of H,

$$c_1(\det^1(\nu)|_{E \cup H}, \hat{\Phi}) = c_1(\det^1(\nu)|_E, \Phi).$$

Thus $I(p)$ is independent of the choice of α_1' and α_1''. Similarly, is independent of the choice of α_2' and α_2''.

The choice of metric on F clearly does not affect the $\int_E \omega$ part of $I(p)$. The c_1 part depends continuously on the metric and takes discrete values. Since the space of metrics on F is connected, this part is also independent of the choice of metric.

Now define

$$\langle Q_1, Q_2 \rangle \stackrel{\text{def}}{=} \sum_{p \in Q_1^- \cap Q_2^-} \operatorname{sign}(p) + \sum_{p \in T_1 \cap T_2} I(p). \tag{2.2}$$

It will be useful to have a slightly different viewpoint of the above at our disposal. Define a *wiring* of Q_1 and Q_2 to be a collection of (homotopy classes of) arcs α'_j, α''_j and surfaces E, as above, for each point of $T_1 \cap T_2$. Given a wiring $\mathcal{W}$, define (for each $p \in T_1 \cap T_2$)

$$I_1(p, \mathcal{W}) \stackrel{\text{def}}{=} \begin{cases} \frac{1}{2}(c_1(\det{}^1(\nu)|_E, \Phi)) & p \notin P \\ \frac{1}{4}(c_1(\det{}^1(\nu)|_E, \Phi)) & p \in P \end{cases} \tag{2.3}$$

and

$$I_2(p, \mathcal{W}) \stackrel{\text{def}}{=} \begin{cases} \frac{1}{2}\left(-\int_E \omega\right) & p \notin P \\ \frac{1}{4}\left(-\int_E \omega\right) & p \in P. \end{cases}$$

Hence

$$I(p) = I_1(p, \mathcal{W}) + I_2(p, \mathcal{W}).$$

Define

$$A_1(Q_1, Q_2, \mathcal{W}) \stackrel{\text{def}}{=} \sum_{p \in Q_1^- \cap Q_2^-} \operatorname{sign}(p) + \sum_{p \in T_1 \cap T_2} I_1(p, \mathcal{W}) \tag{2.4}$$

$$A_2(Q_1, Q_2, \mathcal{W}) \stackrel{\text{def}}{=} \sum_{p \in T_1 \cap T_2} I_2(p, \mathcal{W})$$

Thus we have

$$\langle Q_1, Q_2 \rangle = A_1(Q_1, Q_2, \mathcal{W}) + A_2(Q_1, Q_2, \mathcal{W}). \tag{2.5}$$

Finally, define

$$\lambda(M) \stackrel{\text{def}}{=} \frac{\langle Q_1, Q_2 \rangle}{|H_1(M; \mathbb{Z})|}. \tag{2.6}$$

Note that since $I(1) = 0$, this definition of λ agrees with Casson's in the case where M is a $\mathbb{Z}$HS, up to a factor of $1/2$. (That the signs agree is not clear at this point, but will become so later (see (3.50)). In particular, $\lambda(M) = 0$ if $\pi_1(M) = 1$.

B. Well-definition.

We now show that the right hand side of (2.6) does not depend on the choice of general positioning isotopy, the choice of orientations of $\widetilde{T_1}$ and $\widetilde{T_2}$, or the choice of Heegaard splitting of M. This will show that $\lambda(M)$ is well-defined.

First we show that $\lambda(M)$ is independent of the choice of special isotopy used to put Q_1 and Q_2 into general position. This is equivalent to showing that, given a special isotopy $h : R \times I \to R$ ($h(\ \cdot\ ,0) = \mathrm{id}$) such that $Q_1^\dagger \stackrel{\mathrm{def}}{=} h(Q_1, 1)$ is in general position with Q_2, we have

$$(2.7)\quad \sum_{p \in Q_1^- \cap Q_2^-} \mathrm{sign}\,(p) + \sum_{p \in T_1 \cap T_2} I(p) = \sum_{p \in Q_1^{\dagger -} \cap Q_2^-} \mathrm{sign}\,(p) + \sum_{p \in T_1 \cap T_2} I^\dagger(p),$$

where $I^\dagger(p)$ is computed just as $I(p)$, but with Q_1, η_1, etc. replaced by $Q_1^\dagger$, $\eta_1^\dagger$, etc. (Note that we are continuing to assume that Q_1 and Q_2 have already been isotoped into general position.)

It will be useful to introduce some auxiliary spaces. Let $\hat{\xi} \subset \xi$ be the unit normal bundle of S^- in R. Consider the map

$$\begin{aligned} e : \quad \hat{\xi} \times I &\to R \\ (x, t) &\mapsto \exp(tx). \end{aligned}$$

(Here exp denotes some diffeomorphism from a neighborhood of the zero section of ξ to a neighborhood of S^- in R whose differential at the zero section is the identity.) There is a neighborhood N of $\hat{\xi} \times \{0\}$ such that e restricted to $N \setminus \hat{\xi} \times \{0\}$ is a diffeomorphism onto its image. Define

$$\overline{R^-} \stackrel{\mathrm{def}}{=} N \cup_e R^-.$$

Define $\overline{Q_j^-} \subset \overline{R^-}$ similarly. Note that $\overline{R^-}$ $[\overline{Q_j^-}]$ is a manifold with boundary $\hat{\xi} \times \{0\}$ $[\hat{\theta}_j \times \{0\}]$. h induces a map (also denoted h) $h : \overline{R^-} \times I \to \overline{R^-}$. We will identify $\overline{Q_1^-} \times I$ with its image under h.

By a general position argument, we may assume that $\overline{Q_1^-} \times I$ and $\overline{Q_2^-}$ are transverse. It follows that $(\overline{Q_1^-} \times I) \cap \overline{Q_2^-}$ is a properly embedded 1-manifold, and that $C \stackrel{\mathrm{def}}{=} \partial(\overline{Q_1^-} \times I \cap \overline{Q_2^-})$ consists of $Q_1^- \cap Q_2^-$, $Q_1^{\dagger -} \cap Q_2^-$ and $(\hat{\theta}_1 \times I) \cap \hat{\theta}_2$. (Here we have used the fact that Q_1 is transverse to Q_2 at P.) Orient $\overline{Q_1^-} \times I$ as a product. Orient $(\overline{Q_1^-} \times I) \cap \overline{Q_2^-}$ as an intersection (see (1.F)). For $c \in C$, define $s(c)$ to be $+1$ $[-1]$ if the inward pointing normal at c is positively [negatively] oriented.

With the above orientation conventions, we have

$$\begin{aligned} \sum_{p \in Q_1^- \cap Q_2^-} \mathrm{sign}\,(p) - \sum_{p \in Q_1^{\dagger -} \cap Q_2^-} \mathrm{sign}\,(p) &= \sum_{c \in Q_1^- \cap Q_2^-} s(c) + \sum_{c \in Q_1^{\dagger -} \cap Q_2^-} s(c) \\ (2.8) \qquad &= - \sum_{c \in (\hat{\theta}_1 \times I) \cap \hat{\theta}_2} s(c). \end{aligned}$$

In view of (2.8), (2.7) reduces to

$$\sum_{p\in T_1\cap T_2}(I(p)-I^\dagger(p))=\sum_{c\in(\hat\theta_1\times I)\cap\hat\theta_2}s(c). \tag{2.9}$$

Since special isotopies are fiber preserving and keep Q_1 and Q_2 Zariski transverse at P, we have

$$(\hat\theta_1\times I)\cap\hat\theta_2=\bigcup_{p\in T_1^-\cap T_2^-}(\hat\theta_{1,p}\times I)\cap\hat\theta_{2,p}.$$

Furthermore, it is easy to see that if $p\in T_1\cap T_2\cap P$, then $I(p)=I^\dagger(p)$. Thus, in order to prove (2.9) it suffices to show that

$$I(p)-I^\dagger(p)=\sum_{c\in(\hat\theta_{1,p}\times I)\cap\hat\theta_{2,p}}s(c) \tag{2.10}$$

for all $p\in T_1^-\cap T_2^-$.

First we analyze the right hand side of (2.10). Let $c\in(\hat\theta_{1,p}\times I)\cap\hat\theta_{2,p}$. Let $p'\in\widetilde{T_1}\cap\widetilde{T_2}$ be an inverse image of p. Keeping in mind the orientation conventions of (1.F), we see that

$$\begin{aligned}
s(c) &= \frac{[\partial(\overline{Q_1^-}\times I)][\partial\overline{Q_2^-}]}{[\partial\overline{R^-}]}\\
&= -\frac{[\partial\overline{Q_1^-}][I][\partial\overline{Q_2^-}]}{[\partial\overline{R^-}]}\\
&= -\frac{[\widetilde{T_1}][\hat\theta_1][I][\widetilde{T_2}][\hat\theta_2]}{[\widetilde{S}][\hat\xi]}\\
&= (-1)^{g-1}\frac{[\hat\theta_1][I][\hat\theta_2]}{[\hat\xi]}\,\frac{[\widetilde{T_1}][\widetilde{T_2}]}{[\widetilde{S}]}\\
&= (-1)^{g-1}\frac{[\hat\theta_1][I][\hat\theta_2]}{[\hat\xi]}.
\end{aligned}$$

(All juxtapositions of orientations are considered to take place at c or p'.) Therefore

$$\sum_{c\in(\hat\theta_{1,p}\times I)\cap\hat\theta_{2,p}}s(c)=(-1)^{g-1}\langle(\hat\theta_{1,p'}\times I),\hat\theta_{2,p'}\rangle_{\hat\xi_{p'}}. \tag{2.11}$$

Let $\beta : I \to \mathcal{L}^C_{p'}$ be a path of complex lagrangians, transverse to $\eta_{2,p'}$, from $\eta^\dagger_{1,p'}$ to $\eta_{1,p'}$. (This is possible, since that space of complex lagrangians transverse to a fixed one is contractable.) Let δ be the loop

$$(\eta_{1,p'} \times I) * \beta. \tag{2.12}$$

Since $(\hat{\theta}_{1,p'} \times I) = \mathbb{C}P(\eta_{1,p'} \times I)$ and $\mathbb{C}P(\beta)$ is disjoint from $\hat{\theta}_{2,p'} = \mathbb{C}P(\eta_{2,p'})$, (2.11) becomes

$$\begin{aligned} \sum_{c \in (\hat{\theta}_{1,p} \times I) \cap \hat{\theta}_{2,p}} s(c) &= (-1)^{g-1} \langle \mathbb{C}P(\delta), \mathbb{C}P(\eta_{2,p'}) \rangle_{\hat{\xi}_{p'}} \\ &= (-1)^{g-1} \psi([\delta]). \end{aligned} \tag{2.13}$$

(Recall that $\psi : \pi_1(\mathcal{L}^C_{p'}) \to \mathbb{Z}$ was defined in (1.10).)

Now for the left hand side of (2.10). First of all, note that since special isotopies are fixed on S, we can use the same wiring $\mathcal{W}$ for computing $I(p)$ and $I^\dagger(p)$. It follows that

$$I(p) - I^\dagger(p) = I_1(p, \mathcal{W}) - I_1^\dagger(p, \mathcal{W}) = -\frac{1}{2}(\Phi - \Phi^\dagger). \tag{2.14}$$

(Φ is computed using $\eta_1 \times \{0\} = \eta_1$, $\Phi^\dagger$ is computed using $\eta_1 \times \{1\} = \eta_1^\dagger$.)

Note that

$$\Phi - \Phi^\dagger = (\Phi|_{\gamma'} - \Phi^\dagger|_{\gamma'}) + (\Phi|_{\gamma''} - \Phi^\dagger|_{\gamma''}).$$

Let $\Psi : \det^1(\nu)|_{\gamma'} \to S^1$ be a trivialization. Then

$$\Phi|_{\gamma'} - \Phi^\dagger|_{\gamma'} = \deg(\Psi(a)) - \deg(\Psi(a^\dagger)),$$

where

$$\begin{aligned} a = (-\det{}^1(\eta_2)|_{\alpha'_2}) * (-\det{}^1(P(\eta_{1,1}, \eta_{2,1}))) * \\ (\det{}^1(\eta_1)|_{\alpha'_1}) * (\det{}^1(P(\eta_{1,p'}, \eta_{2,p'}))) \\ a^\dagger = (-\det{}^1(\eta_2)|_{\alpha'_2}) * (-\det{}^1(P(\eta^\dagger_{1,1}, \eta_{2,1}))) * \\ (\det{}^1(\eta_1^\dagger)|_{\alpha'_1}) * (\det{}^1(P(\eta^\dagger_{1,p'}, \eta_{2,p'}))). \end{aligned}$$

(Here deg denotes the degree of a map from S^1 to S^1, and the negative signs indicate that a path should be traversed in the reverse direction.) Consider the 1-parameter family of paths

$$(-\det{}^1(\eta_2)|_{\alpha'_2}) * (-\det{}^1(P(\eta_{1,1} \times \{t\}, \eta_{2,1}))) * (\det{}^1(\eta_1 \times \{t\})|_{\alpha'_1})$$

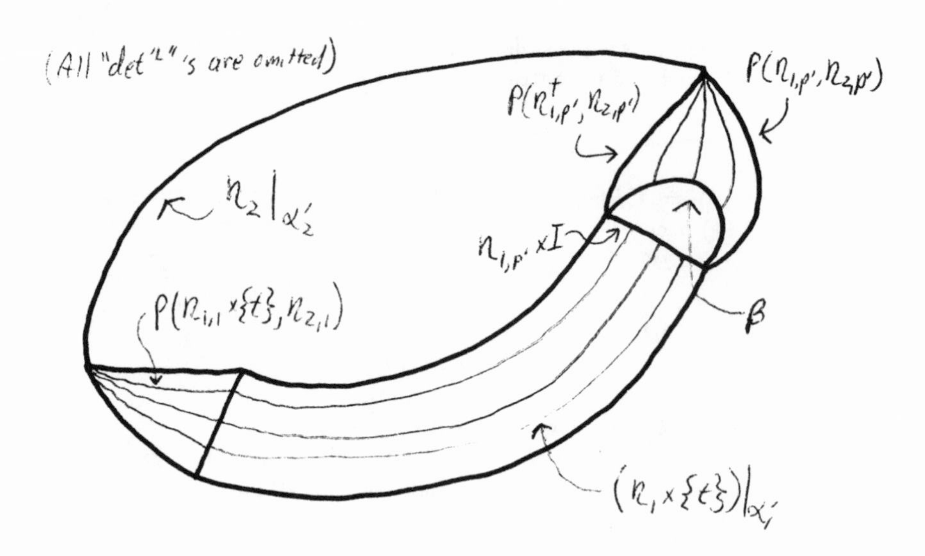

Figure 2.3: Various paths in $\det^1(\nu)|_{\gamma'}$.

$(0 \le t \le 1)$. Adding the boundary of the above square to $a \cup a^\dagger$, we see that

$$\Phi|_{\gamma'} - \Phi^\dagger|_{\gamma'} = \deg(\Psi(b)),$$

where

$$b = (-\det^1(\eta_{1,p'} \times I)) * (\det^1(P(\eta_{1,p'}, \eta_{2,p'}))) * (-\det^1(P(\eta^\dagger_{1,p'}, \eta^\dagger_{2,p'}))).$$

(Here $\det^1(\eta_{1,p'} \times I)$ is viewed as a path from $\det^1(\eta_{1,p'})$ to $\det^1(\eta^\dagger_{1,p'})$ in $\det^1(\nu_{p'})$.)

Recall the path β of complex lagrangians transverse to $\eta_{2,p'}$ from $\eta^\dagger_{1,p'}$ to $\eta_{1,p'}$. The 1-parameter family of paths $P(\beta_t, \eta_{2,p'})$ can be viewed as a homotopy (rel end points) from

$$(\det^1(P(\eta_{1,p'}, \eta_{2,p'}))) * (-\det^1(P(\eta^\dagger_{1,p'}, \eta_{2,p'})))$$

to $-\det^1(\beta)$. Hence

$$\Phi|_{\gamma'} - \Phi^\dagger|_{\gamma'} = \deg(\det^1(-\delta)) = -\varphi([\delta]),$$

where δ is as defined in (2.12) and φ is the homomorphism defined in (1.10). Note that Ψ can be (and has been) omitted, since everything is taking place in the single fiber $\det^1(\nu_{p'})$.

By equivariance, we also have

$$\Phi|_{\gamma''} - \Phi^\dagger|_{\gamma''} = -\varphi([\delta]),$$

and hence

(2.15) $$I(p) - I^\dagger(p) = \varphi([\delta]).$$

(2.10) now follows from (2.13), (2.15) and (1.12), and the proof of isotopy invariance is complete.

In the above argument, the fact that the isotopy was fixed on S was needed only to show that the A_2 part of $\langle Q_1, Q_2\rangle$ was invariant. The A_1 part of $\langle Q_1, Q_2\rangle$ is invariant under a less restricted class of isotopies, and this fact will be used in the proof of (3.36).

More precisely, define a *not-so-special isotopy* $\{h_t\}_{0\leq t\leq 1}$ of R to be one which satisfies

1. $h_t(Q_1)$ is transverse to Q_2 at $Q_1 \cap Q_2 \cap P$ for all t.
2. $h_t(T_1)$ is transverse to T_2 in S for all t.
3. h_t induces a symplectic bundle map of ν for all t.

Let $\mathcal{W}$ be a wiring of Q_1 and Q_2 and $\{h_t\}$ be a not-so-special isotopy. Then condition 2 above allows us to choose wirings $h_t(\mathcal{W})$ of $h_t(Q_1)$ and Q_2 so that the arcs and surfaces of $h_t(\mathcal{W})$ vary continuously with t. An argument similar to the one used to establish the special isotopy invariance of $\langle Q_1, Q_2\rangle$ shows

(2.16) *Let $\mathcal{W}$ be a wiring of Q_1 and Q_2 and $\{h_t\}$ be a not-so-special isotopy. Then $A_1(h_1(Q_1), Q_2, h_1(\mathcal{W})) = A_1(Q_1, Q_2, \mathcal{W})$.*

Suppose that the orientations of $\widetilde{T_1}$ and $\widetilde{T_2}$ are both changed. Then, according to the orientation conventions of (1.F), the orientations of Q_1 and Q_2 also change, while all other orientations remain the same. It is now easy to see that these changes do not affect (2.6).

Finally, it must be shown that (2.6) does not depend on the choice of Heegaard splitting for M. First we recall the notion of a stabilization of a Heegaard splitting. Let (V_1, V_2, Σ) be the standard genus one Heegaard splitting of S^3. Let

$$(W_1^\natural, W_2^\natural, F^\natural) \stackrel{\text{def}}{=} (W_1, W_2, F)\#(V_1, V_2, \Sigma).$$

$(W_1^\natural, W_2^\natural, F^\natural)$ is called a stabilization of (W_1, W_2, F). We can make the identifications

$$\begin{aligned} F^\natural &= F\#\Sigma \\ W_j^\natural &= W_j \#_\partial V_j \end{aligned}$$

(where $\#_\partial$ denotes the boundary connected sum). See [AM] for more details.

By [S] (see also [AM]), any two Heegaard splittings of M are equivalent after a sufficient number of stabilizations. Thus it suffices to show that (2.6) does not change when (W_1, W_2, F) is stabilized to $(W_1^\natural, W_2^\natural, F^\natural)$.

Use the superscript $\natural$ to denote objects associated to the stabilized Heegaard splitting. (e.g. $R^\natural = \mathrm{Hom}(\pi_1(F^\natural), SU(2))/SU(2)$.) Let F^- be F minus a disk. Define Σ^- similarly. Make the identifications

$$\begin{aligned} F^\natural &= F^- \cup \Sigma^- \\ F &= F^- \cup D^2 \\ \Sigma &= \Sigma^- \cup D^2. \end{aligned}$$

$\pi_1(F^\natural)$ is generated by $\pi_1(F^-)$ and $\pi_1(\Sigma^-)$. We can identify

$$R = \{[\rho] \in R^\natural \mid \rho(\pi_1(\Sigma^-)) = \{1\}\}.$$

Similarly,

$$Q_j = \{[\rho] \in Q_j^\natural \mid \rho(\pi_1(V_j)) = \{1\}\}.$$

It is easy to see that

$$Q_1^\natural \cap Q_2^\natural = Q_1 \cap Q_2. \tag{2.17}$$

Our first task is to show that the special isotopies of R used to put Q_1 and Q_2 into general position can be extended to $R^\natural$ so that (2.17) remains true. Toward this end, we analyze the normal bundles of R in $R^\natural$ and Q_j in $Q_j^\natural$.

Let $\gamma \overset{\text{def}}{=} \partial F^- = \partial\Sigma^-$. Let $[\rho] \in R$. By examining the Mayer-Vietoris sequence for the triple (F, F^-, D^2) with coefficients in the flat vector bundle $\mathbf{g}_{\mathrm{Ad}\rho}$, one finds that

$$T_{[\rho]}R = H^1(F; \mathbf{g}_{\mathrm{Ad}\rho}) = \ker(H^1(F^-; \mathbf{g}_{\mathrm{Ad}\rho}) \to H^1(\gamma; \mathbf{g}_{\mathrm{Ad}\rho})).$$

The Mayer-Vietoris sequence for $(F^\natural, F^-, \Sigma^-)$ yields

$$0 \to H^1(F^\natural; \mathbf{g}_{\mathrm{Ad}\rho}) \to H^1(F^-; \mathbf{g}_{\mathrm{Ad}\rho}) \oplus H^1(\Sigma^-; \mathbf{g}_{\mathrm{Ad}\rho}) \to H^1(\gamma; \mathbf{g}_{\mathrm{Ad}\rho}).$$

Since $\mathbf{g}_{\mathrm{Ad}\rho}$ is trivial over Σ^-, the map $H^1(\Sigma^-;\mathbf{g}_{\mathrm{Ad}\rho}) \to H^1(\gamma;\mathbf{g}_{\mathrm{Ad}\rho})$ is zero and

$$\begin{aligned} T_{[\rho]}R^\natural &= H^1(F^\natural;\mathbf{g}_{\mathrm{Ad}\rho}) \\ &= H^1(F;\mathbf{g}_{\mathrm{Ad}\rho}) \oplus H^1(\Sigma^-;\mathbf{g}_{\mathrm{Ad}\rho}) \\ &= T_{[\rho]}R \oplus H^1(\Sigma^-;\mathbf{g}). \end{aligned}$$

Therefore the Zariski normal bundle of R in $R^\natural$ can be identified with the trivial bundle $R \times H^1(\Sigma^-;\mathbf{g})$. It is not hard to see that the splitting $T_{[\rho]}R \oplus H^1(\Sigma^-;\mathbf{g})$ is an orthogonal direct sum of symplectic vector spaces. Similarly, the Zariski normal bundle of Q_j in $Q_j^\natural$ can be identified with the trivial bundle $Q_j \times H^1(V_j;\mathbf{g})$. Since $H^1(V_1;\mathbf{g})$ and $H^1(V_2;\mathbf{g})$ are transverse inside $H^1(\Sigma^-;\mathbf{g})$, the normal fibers of Q_1 and Q_2 are transverse inside the normal fiber of R at all points of $Q_1 \cap Q_2$.

It is now easy to see that a special isotopy of Q_1 in R can be extended to a special isotopy of $Q_1^\natural$ in $R^\natural$ so that (2.17) and the product structure of the normal bundle of R in $R^\natural$ are preserved.

Assume now that $Q_1^\natural$ and $Q_2^\natural$ have been put into general position as above. If it can be shown that for each $p \in T_1 \cap T_2$, $|I^\natural(p)| = |I(p)|$, and that for each $p \in Q_1^- \cap Q_2^-$,

$$\operatorname{sign}^\natural(p) = \operatorname{sign}(p), \tag{2.18}$$

then, using the isotopy invariance of $\langle Q_1^\natural, Q_2^\natural\rangle$, it will follow that

$$\langle Q_1^\natural, Q_2^\natural\rangle = \langle Q_1, Q_2\rangle.$$

First we show that $|I^\natural(p)| = |I(p)|$ for $p \in T_1 \cap T_2 = T_1^\natural \cap T_2^\natural$. We can use the same arcs α_j', α_j'' and surface E to compute $I^\natural(p)$ and $I(p)$. Since $\omega^\natural|_{\widetilde{S}} = \omega$, $\int_E \omega^\natural = \int_E \omega$. Let $\bar\nu$ be the bundle $\widetilde{S} \times H^1(\Sigma^-;\mathbf{h}^\perp_{\mathrm{Ad}\rho})$. Then $\nu^\natural|_{\widetilde{S}} = \nu \oplus \bar\nu$. Hence $\det(\nu^\natural)|_{\widetilde{S}} = \det(\nu) \otimes \det(\bar\nu)$. Similarly, $\det(\eta_j^\natural)|_{\widetilde{T_j}} = \det(\eta_j) \otimes \det(\bar\eta_j)$, where $\bar\eta_j \stackrel{\mathrm{def}}{=} \widetilde{T_j} \times H^1(W_j;\mathbf{h}^\perp)$.

Let Y_1 and Y_2 be symplectic vector spaces. Let $K_i, L_i \subset Y_i$ be oriented transverse lagrangian subspaces. It is easy to see from the definition of the path $P_\pm$ that

$$P_\pm(K_1 \oplus K_2, L_1 \oplus L_2) = P_\pm(K_1, L_1) \oplus P_\pm(K_2, L_2).$$

Using the canonical trivialization of $\det(\bar\nu)$, we can identify $\det^1(\nu^\natural)$ with $\det^1(\nu)$. Using $P_\pm$ to patch together the sections $\det^1(\bar\eta_j)$ of $\det^1(\bar\nu)$, one gets a trivialization $\bar\Phi_\pm$ of $\det^1(\bar\nu)|_{\partial E}$. It follows from the above remarks that $\Phi_{\pm}^{\natural} = \Phi_{\pm} \otimes \bar{\Phi}_{\pm}$. So it suffices to show that $\bar{\Phi}_{\pm}$ is in the same homotopy class as the canonical trivialization of $\det^1(\bar{\nu})|_{\partial E}$. This is obvious, since $\det^1(\bar{\eta}_j)$ is constant with respect to this trivialization.

Next we establish (2.18). Let $p \in Q_1^- \cap Q_2^-$. Let $K_j \stackrel{\text{def}}{=} H^1(V_j; \mathbf{g})$, $L \stackrel{\text{def}}{=} H^1(\Sigma^-; \mathbf{g})$. Orient L as a symplectic vector space and choose orientations of K_1 and K_2 so that

$$[L] = (-1)^{g-1}[K_1][K_2].$$

It follows from the orientation conventions of (1.F) (though not without some effort) that

$$[Q_j^{\natural}] = [Q_j][K_j].$$

Hence

$$\begin{aligned} \operatorname{sign}^{\natural}(p) &= \frac{[Q_1^{\natural}][Q_2^{\natural}]}{[R^{\natural}]} \\ &= \frac{[Q_1][K_1][Q_2][K_2]}{[R][L]} \\ &= (-1)^{3(3g-3)} \frac{[Q_1][Q_2]}{[R]} \frac{[K_1][K_2]}{[L]} \\ &= \frac{[Q_1][Q_2]}{[R]} \\ &= \operatorname{sign}(p). \end{aligned}$$

This completes the proof that λ is well-defined.

§ 3

Various Properties of λ

In this section we prove various technical lemmas needed for the proof of the general Dehn surgery formula of the next section. Many of the proofs are easy generalizations of Casson's work in the $\mathbb{Z}$HS case, and this has been indicated by the words "Casson for $\mathbb{Z}$HS case".

Throughout this section, retain all notation from the previous two sections.

(3.1) **Lemma.** *Let M be a $\mathbb{Q}$HS. Let $-M$ denote M with the opposite orientation. Then*

$$\lambda(-M) = \lambda(M).$$

Proof: (Casson for $\mathbb{Z}$HS case.) The only parts of the constructions used for defining $\lambda(M)$ which are sensitive to the orientation of M are the orientations of various spaces and the sign of ω. (In particular, the "patching procedure" used over points of $T_1 \cap T_2$ does not depend on the sign of ω.) Hence, for $p \in T_1 \cap T_2$, the absolute value of $I(p)$ does not depend on the orientation of M. It now follows from (2.2) and the isotopy invariance of $\langle Q_1, Q_2 \rangle$ that absolute value of $\langle Q_1, Q_2 \rangle$ is independent of the orientation of M.

All that is left to do is check signs. It suffices to do this at a single irreducible point of $Q_1 \cap Q_2$. Changing the orientation of M changes the orientation of F which changes the sign of ω. Changing the sign of ω changes $[S]$ by $(-1)^g$ and $[R]$ by $(-1)^{3g-3}$. This means that $[\widetilde{T_1}]$ (say), and hence $[Q_1]$, change by $(-1)^g$. Thus sign(p) changes by $(-1)^{g+3g-3} = -1$.

□

The following lemma is the prototype for most of the others in this section.

(3.2) **Lemma.** *Let γ be a separating simple closed curve on F. Let $h : F \to F$ be a left-handed Dehn twist along γ. Let $M_n = W_1 \cup_{h^n} W_2$. (That is, the gluing map $\partial W_1 \to \partial W_2$ is composed with h^n.) Then*

$$\lambda(M_n) = \lambda(M) + n\lambda'(\gamma)$$

for some $\lambda'(\gamma) \in \mathbb{Q}$, independent of n.

Proof: (Casson for $\mathbb{Z}$HS case.) h induces maps of $\pi_1(F)$, R and $R^\sharp$, which will also be denoted by h. It is clear that

$$\lambda(M_n) = \frac{\langle h^n(Q_1), Q_2\rangle}{|H_1(M;\mathbb{Z})|}.$$

(Here we have used the fact that $H_1(M_n;\mathbb{Z}) \cong H_1(M;\mathbb{Z})$ for all n.) So what we must show is that

$$\lambda(M_{n+1}) - \lambda(M_n) = \frac{(\langle h^{n+1}(Q_1), Q_2\rangle - \langle h^n(Q_1), Q_2\rangle)}{|H_1(M;\mathbb{Z})|} = \lambda'(\gamma)$$

for some $\lambda'(\gamma) \in \mathbb{Q}$, independent of n.

The idea of the proof is to represent the difference $h(Q_1) - Q_1$ as a cycle $D \subset R^-$. It will follow that

$$\langle h^{n+1}(Q_1), Q_2\rangle - \langle h^n(Q_1), Q_2\rangle = \langle h^n(D), Q_2\rangle, \tag{3.3}$$

where the right hand side denotes a homological intersection number. The proof will be complete upon showing that $h(D)$ is homologous to D.

Our first task is to construct the difference cycle D. Let F_1 and F_2 be the components of F cut along γ. Then $\pi_1(F)$ is the free product of $\pi_1(F_1)$ and $\pi_1(F_2)$ amalgamated over $\pi_1(\gamma)$. If the location of the base point is chosen properly, then h is given by

$$h(\alpha) = \begin{cases} \alpha, & \alpha \in \pi_1(F_1) \\ \gamma\alpha\gamma^{-1}, & \alpha \in \pi_1(F_2). \end{cases}$$

Correspondingly, for $\rho \in R^\sharp$ we have

$$h(\rho)\alpha = \begin{cases} \rho(\alpha), & \alpha \in \pi_1(F_1) \\ \rho(\gamma)\rho(\alpha)\rho(\gamma)^{-1}, & \alpha \in \pi_1(F_2). \end{cases}$$

This map descends to one on R.

There is a open ball $U \subset \mathbf{su}(2)$ such that $\exp : U \to SU(2) \setminus \{-1\}$ is a diffeomorphism. For $A = \exp(a) \in \exp(U)$ and $t \in [0,1]$, define

$$A^t \stackrel{\text{def}}{=} \exp(ta). \tag{3.4}$$

Also define $(-1)^0 \stackrel{\text{def}}{=} 1$. Let $V_\epsilon \subset SU(2)$ be the closed ball of radius ϵ (with respect to some Ad-invariant metric) centered at $-1 \in SU(2)$. Let

$$\begin{aligned} N^\sharp &= \{\rho \in R^\sharp \mid \rho(\gamma) = -1\} \\ N^\sharp_\epsilon &= \{\rho \in R^\sharp \mid \rho(\gamma) \in V_\epsilon\}. \end{aligned}$$

Let $f : SU(2) \to [0,1]$ be a smooth, Ad-invariant, monotonic (on $SU(2)/\text{Ad}$) function such that $f = 1$ outside V_ϵ and $f(-1) = 0$. For $t \in [0,1]$, define $h_t : R^\sharp \to R^\sharp$ by

$$h_t(\rho)\alpha = \begin{cases} \rho(\alpha), & \alpha \in \pi_1(F_1) \\ \rho(\gamma)^{f(\rho(\gamma))t}\rho(\alpha)\rho(\gamma)^{-f(\rho(\gamma))t}, & \alpha \in \pi_1(F_2). \end{cases}$$

$\{h_t\}$ descends to an isotopy of R, also denoted $\{h_t\}$. Note that $h_0 = \text{id}$ and $h_1 = h$ on $R \setminus N_\epsilon$.

We now show that $\{h_t\}$ is a special isotopy (see (1.19)). Since $[\gamma]$ lies in the commutator subgroup of $\pi_1(F)$, the differential of the map

$$\begin{aligned} R^\sharp &\to SU(2) \\ \rho &\mapsto \rho(\gamma) \end{aligned}$$

is 0 at $P^\sharp$. It follows that $h_{t*} : T_pR \to T_pR$ is the identity for $p \in P$. Since $\rho(\gamma) = 1$ for $\rho \in S^\sharp$, $h_t|_S = \text{id}$. It follows from Theorem 4.5 of [G3] that h_t is a symplectic map on all of R, a fortiori on $TR|_{S^-}$. Since $h_{t*} : TS^- \to TS^-$ is the identity, h_{t*} must preserve the fibers of ν.

By the special isotopy invariance proved in (2.C),

$$\langle Q_1, Q_2 \rangle = \langle h_1(Q_1), Q_2 \rangle.$$

We may assume without loss of generality that Q_1 in transverse to ∂N_ϵ. Let D be the cycle

$$(-h_1(Q_1 \cap N_\epsilon)) \cup h(Q_1 \cap N_\epsilon).$$

Note that $h_1(Q_1)$ coincides with $h(Q_1)$ on $R \setminus N_\epsilon$. Thus it follows from (2.2) that

$$\langle h(Q_1), Q_2 \rangle - \langle Q_1, Q_2 \rangle = \langle D, Q_2 \rangle, \tag{3.5}$$

where the right hand side is to be regarded as the homological intersection number of $[D] \in H_{3g-3}(R^-;\mathbf{Z})$ and $[Q_2] \in H_{3g-3}(R,S;\mathbf{Z})$. (3.3) clearly follows from (3.5).

Now we must show that $h_*([D]) = [D] \in H_{3g-3}(R^-;\mathbf{Z})$. This follows from the facts that $D \subset N_\epsilon$, $h|_N = \mathrm{id}$, and N_ϵ deformation retracts onto N. It will be useful in what follows, however, to have an explicit representation of $[D]$ as a cycle in N. This construction occupies the remainder of the proof.

Let

$$R_i^* = \mathrm{Hom}(\pi_1(F_i), SU(2))/SU(2)$$

$(i = 1,2)$. There is a natural map

$$\pi : R \to R_1^* \times R_2^*$$

which is a fibering over the image of N. I claim that

$$[D] = 2[\pi^{-1}(\pi(Q_1 \cap N))] \in H_{3g-3}(R^-;\mathbf{Z}). \tag{3.6}$$

(Note that the sign of the right hand side is ambiguous, since no orientation of N has been specified.) The idea of the proof is to let ϵ approach zero.

Define

$$\begin{aligned} k : \mathrm{int}(V_\epsilon) &\to SU(2) \\ X &\mapsto X^{f(X)}. \end{aligned}$$

Note that

$$h_1(\rho)\alpha = \begin{cases} \rho(\alpha), & \alpha \in \pi_1(F_1) \\ k(\rho(\gamma))\rho(\alpha)k(\rho(\gamma))^{-1}, & \alpha \in \pi_1(F_2). \end{cases}$$

f can be chosen so that k is a diffeomorphism onto $SU(2) \setminus V_\epsilon$. Since $Q_1^\sharp$ is (without loss of generality) transverse to $N^\sharp$, there is a unique Ad-invariant diffeomorphism

$$l : (Q_1^\sharp \cap N^\sharp) \times V_\epsilon \to Q_1^\sharp \cap N_\epsilon^\sharp$$

such that for all $\rho \in Q_1^\sharp \cap N^\sharp$ and $X \in V_\epsilon$, $l(\rho, X)(\gamma) = X$.

It follows that as $\epsilon \to 0$, $h_1(Q_1^\sharp \cap N_\epsilon^\sharp)$ tends to the image of the map

$$\begin{aligned} m : (Q_1^\sharp \cap N^\sharp) \times (SU(2) \setminus \{-1\}) &\to N^\sharp \\ m(\rho, X) &= \begin{cases} \rho(\alpha), & \alpha \in \pi_1(F_1) \\ X\rho(\alpha)X^{-1}, & \alpha \in \pi_1(F_2), \end{cases} \end{aligned}$$

while $Q_1^\sharp \cap N_\epsilon^\sharp$ tends to $Q_1^\sharp \cap N^\sharp = m(Q_1^\sharp \cap N^\sharp, -1)$. This implies (3.6). (The factor of 2 arises because $\mathrm{Ad} : SU(2) \to \mathrm{Aut}\,(SU(2))$ has degree 2.) □

(3.7) **Corollary.** *Let $K \subset M$ be a null-homologous knot in a* $\mathbb{Q}$*HS. Let $K_{1/n}$ denote $1/n$ Dehn surgery on K. Then there is a number $\lambda'(K) \in \mathbb{Q}$ such that*

$$\lambda(K_{1/n}) = \lambda(M) + n\lambda'(K)$$

for all n.

Proof: (Casson for $\mathbb{Z}$HS case.) Let U be a regular neighborhood of K. Let $A \subset \partial U$ be a regular neighborhood of a simple closed curve representing the longitude of K. Then $1/n$ Dehn surgery on K is equivalent to cutting M along ∂U and regluing after performing n left-handed Dehn twists on A. The important thing to notice here is that this operation depends only on the annulus A and not on the cutting surface which contains it. Thus, by (3.2), it suffices to find a Heegaard surface for M which contains A as a separating annulus.

Let E be a Seifert surface for K. Fix an identification of $E \times I$ with a bicollar of E in M. Let C be a collar of ∂E in E. Then we can take U to be $C \times I$ and A to be $\partial U \times I$.

Pick a relative handle decomposition of $(M, E \times I)$. That is,

$$M = E \times I \cup \{\text{1-handles}\} \cup \{\text{2-handles}\} \cup \{\text{3-handles}\}.$$

After some sliding, we may assume that the attaching regions of the 1-handles lie in $E \times \{1\}$. It is now easy to see that $\partial(E \times I \cup \{1 - \text{handles}\})$ is the desired Heegaard surface. □

(3.8) **Corollary.** *Let $K, L \subset M$ be a pair of null-homologous knots with linking number 0 in a* $\mathbb{Q}$*HS M. Let $(K_{1/m}, L_{1/n})$ denote the* $\mathbb{Q}$*HS obtained from $1/m$ Dehn surgery on K and $1/n$ Dehn surgery on L. Then there is a number $\lambda''(K, L) \in \mathbb{Q}$ such that*

$$\lambda(K_{1/m}, L_{1/n}) = \lambda(M) + m\lambda'(K) + n\lambda'(L) + mn\lambda''(K, L).$$

Proof: (Casson for $\mathbb{Z}$HS case.) Note that since $\mathrm{lk}\,(K, L) = 0$, $1/n$ Dehn surgery on L does not affect the longitude of K, and vice-versa. The result now follows from (3.7). □

Note that

$$\begin{aligned}(3.9)\ \lambda''(K, L) &= \lambda(K_{1/m}, L_{1/n}) - (K_{1/(m-1)}, L_{1/n}) \\ &\quad -(K_{1/m}, L_{1/(n-1)}) + (K_{1/(m-1)}, L_{1/(n-1)})\end{aligned}$$

for any $m, n \in \mathbb{Z}$.

(3.10) **Lemma.** *Let α and γ be disjoint separating simple closed curves on F. (It follows that α and γ are null-homologous in M and have linking number 0.) Then $\lambda''(\alpha,\gamma) = 0$.*

Proof: (Casson for $\mathbb{Z}$HS case.) Let h_α $[\, h_\gamma]: F \to F$ be a Dehn twist along α $[\gamma]$. Let h_α and h_γ also denote the induced maps on $\pi_1(F)$ and R. Let

$$M_{m,n} \overset{\text{def}}{=} W_1 \cup_{h_\alpha^m h_\gamma^n} W_2.$$

This is the same as doing $1/m$ Dehn surgery on α and $1/n$ Dehn surgery on γ. By (3.9), it suffices to show that

$$\lambda(M_{1,1}) - \lambda(M_{1,0}) - \lambda(M_{0,1}) + \lambda(M_{0,0}) = 0$$

or, equivalently, that

$$\langle h_\alpha h_\gamma(Q_1), Q_2\rangle - \langle h_\alpha(Q_1), Q_2\rangle - \langle h_\gamma(Q_1), Q_2\rangle + \langle Q_1, Q_2\rangle = 0.$$

Let

$$N = \{[\rho] \in R \,|\, \rho(\gamma) = -1\} \subset R^-.$$

Note that h_α preserves N. Let $D \subset N$ be the difference cycle, as defined in the proof of (3.2), representing $h_\gamma(Q_1) - Q_1$. Then

$$\begin{aligned} \langle h_\gamma(Q_1), Q_2\rangle - \langle Q_1, Q_2\rangle &= \langle D, Q_2\rangle \\ \langle h_\alpha h_\gamma(Q_1), Q_2\rangle - \langle h_\alpha(Q_1), Q_2\rangle &= \langle h_\alpha(D), Q_2\rangle. \end{aligned}$$

h_α acts trivially on $H_*(F;\mathbb{Z})$ and so, by (1.16) and the universal coefficient theorem, also acts trivially on $H_{3g-3}(N;\mathbb{Q})$. Therefore

$$\langle h_\alpha(D), Q_2\rangle = \langle D, Q_2\rangle.$$

□

(3.11) **Corollary.** *Let $K, L \subset M$ be null-homologous knots bounding disjoint Seifert surfaces. Then $\lambda''(K, L) = 0$.*

Proof: (Casson for $\mathbb{Z}$HS case.) By (3.10), it suffices to situate K and L as disjoint separating curves on a Heegaard surface for M. This can be done as in the proof of (3.7). □

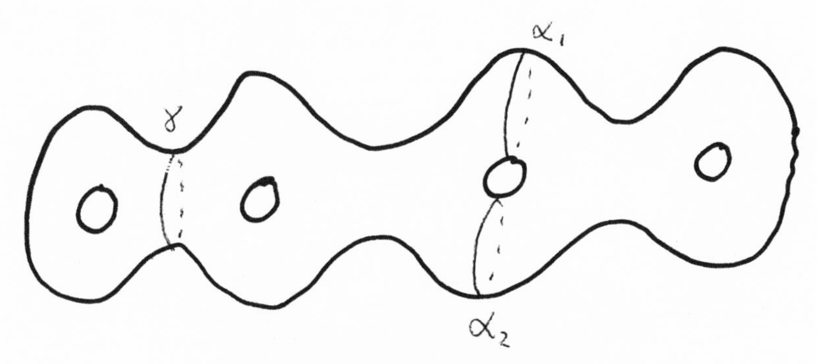

Figure 3.1: The setup for (3.12).

(3.12) **Lemma.** *Let $\alpha_1, \alpha_2, \gamma \subset F$ be disjoint simple closed curves such that γ separates F and α_1 and α_2 jointly separate F (see Figure 3.1). Let h_α : $F \to F$ be a right-handed Dehn twist along α_1 composed with a left-handed Dehn twist on α_2. Let $M_n = W_1 \cup_{h_\alpha^n} W_2$. Then $\lambda'(\gamma; M_n) = \lambda'(\gamma; M_0)$ for all n, where $\lambda'(\gamma; M_n)$ denotes $\lambda'(\gamma)$ with γ thought of as a knot in M_n.*

Proof: The proof is identical to that of (3.10), since the only properties of h_α used were that it preserved N_γ and acted trivially on $H^*(F;\mathbb{Z})$. □

(3.13) **Corollary.** *Let $K, L_1, L_2 \subset M$ be knots such that K bounds an embedded surface E, L_1 and L_2 cobound an embedded surface H, and E and H are disjoint. Let M_n be the result of performing $1/n$ Dehn surgery on L_1 and $-1/n$ Dehn surgery on L_2, where the surgery coefficients are given with respect to the framings determined by H. Then $\lambda'(K; M_n) = \lambda'(K; M_0)$ for all n.*

Proof: This follows from (3.12) and an argument similar to the one used in the proof of (3.7). □

The next lemma is most conveniently stated in terms of

$$\bar{\lambda}(M) \stackrel{\text{def}}{=} |H_1(M;\mathbb{Z})|\lambda(M) = \langle Q_1, Q_2 \rangle. \tag{3.14}$$

The above results also hold with λ replaced with $\bar{\lambda}$. So we have, for example, an invariant $\bar{\lambda}'(K)$ for null-homologous knots K, and $\bar{\lambda}''(K, L) = 0$ if K and L are a boundary link.

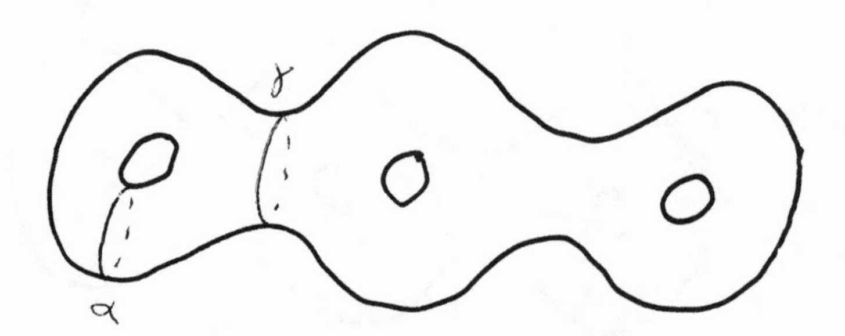

Figure 3.2: More curves called γ and α.

Let γ be a separating curve on F such that one component of $F \setminus \gamma$ has genus one. Let α be a non-separating curve in this component. Let h_α be a Dehn twist along α and let $M_n = W_1 \cup_{h_\alpha^n} W_2$ (see Figure 3.2). We wish to determine the effect of this operation on $\bar{\lambda}'(\gamma)$.

Modifying the gluing map of the Heegaard splitting by n Dehn twists along α is, of course, equivalent to performing $1/n$ Dehn surgery on α, with respect to the basis of $H_1(\partial(\text{nbd}(\alpha)); \mathbb{Z})$ determined by the framing of α given by the normal bundle of α in F. Let p/q be the slope of the longitude of α with respect to this basis. After possibly changing the sign of one of the basis vectors and the sign of the Dehn twist, we may assume that $p/q < 0$.

(3.15) **Lemma.** *$\bar{\lambda}'(\gamma; M_n) = \bar{\lambda}'(\gamma; M_0)$ for all $n > q/p$.*

(If $n < q/p$, the statement remains true up to sign.)

Proof: Let h_γ be a Dehn twist along γ. Let h_α and h_γ also denote the induced maps on $\pi_1(F)$, R, etc. Let $N_\gamma = \{[\rho] \in R \mid \rho(\gamma) = -1\}$ and define N_α similarly. Let D be the difference cycle, as defined in (3.2), representing $h_\gamma(Q_1) - Q_1$. Thus we have

$$\bar{\lambda}'(\gamma; M_0) = \langle D, Q_2 \rangle.$$

Similarly,

$$\bar{\lambda}'(\gamma; M_n) = \langle h_\alpha^n(D), Q_2 \rangle.$$

Consider the action of the maps h_α^n ($n \in \mathbb{Z}$) on $T_1 R$. These maps are colinear in the vector space $\text{End}(T_1 R)$ and so extend naturally to a one (real) parameter family of maps. It is easy to check that $h_\alpha^n(T_1 Q_1)$ is not transverse to $T_1 Q_2$ precisely when $n = q/p$.

Thus, for $n > q/p$ ($n \in \mathbb{Z}$), we can assign orientations (according to the conventions of (1.F)) so that $h_\alpha^n : Q_1 \to h_\alpha^n(Q_1)$ is orientation preserving . So, in order to show that $\bar{\lambda}'(\gamma; M_n) = \bar{\lambda}'(\gamma; M_0)$ (for $n > q/p$), it must be shown that

$$\langle h_\alpha^n(D), Q_2 \rangle = \langle D, Q_2 \rangle.$$

In fact, we will show that $h_\alpha^n(D)$ is isotopic to D.

Let F' be F cut along α. Let $\beta \subset F$ be a curve intersecting α once transversely. Then $\pi_1(F)$ is generated by $\pi_1(F')$ and β. If base points are chosen properly, the action of h_α on $\pi_1(F)$ is given by

$$\begin{aligned} h_\alpha(x) &= x, & x \in \pi_1(F') \\ h_\alpha(\beta) &= \alpha\beta. \end{aligned}$$

It follows that the action of h_α on $R^\sharp$ is given by

$$\begin{aligned} h_\alpha(\rho)x &= \rho(x), & x \in \pi_1(F') \\ h_\alpha(\rho)\beta &= \rho(\alpha)\rho(\beta). \end{aligned}$$

Let $N_\alpha \stackrel{\text{def}}{=} \{[\rho] \in R \mid \rho(\alpha) = -1\}$. For $[\rho] \in R \setminus N_\alpha$ and $t \in \mathbb{R}$, define

$$\begin{aligned} h_t(\rho)x &= \rho(x), & x \in \pi_1(F') \\ h_t(\rho)\beta &= \rho(\alpha)^t\rho(\beta). \end{aligned}$$

(cf. (3.4).) For $n \in \mathbb{Z}$, $h_\alpha^n = h_n$. Hence h_α^n is isotopic to the identity on $R \setminus N_\alpha$.

Since γ bounds a genus one surface containing α as a non-separating curve, $\rho(\alpha) = -1$ implies $\rho(\gamma) = 1$. Since $D \subset N_\gamma = \{[\rho] \in R \mid \rho(\gamma) = -1\}$, this implies that $D \subset R^- \setminus N_\alpha$. Hence $h_\alpha^n(D)$ is isotopic to D. □

(3.16) **Corollary.** *Let $K \subset M$ be a knot in a $\mathbb{Q}HS$ bounding a genus one Seifert surface $E \subset M$. Let $L \subset E$ be a simple closed curve. Let $L_{1/n}$ denote $1/n$ Dehn surgery on L, with respect to the basis of $H_1(\partial(\text{nbd}(L)); \mathbb{Z})$ determined by E (see the discussion preceding (3.15)). Let p/q be the slope of the longitude of L. Assume, without loss of generality, that $p/q < 0$. Then*

$$\bar{\lambda}'(K; L_{1/n}) = \bar{\lambda}'(K; M)$$

for all $n > q/p$.

Proof: Similar to the proof of (3.7). □

Let $K \subset M$ be a fibered knot in a $\mathbb{Q}$HS. Let E be a fiber and let $g : E \to E$ be the monodromy of the fibering. Let

$$\begin{aligned} Z &\stackrel{\text{def}}{=} \operatorname{Hom}(\pi_1(E), SU(2))/SU(2) \\ Z_{-1} &\stackrel{\text{def}}{=} \{[\rho] \in Z \mid \rho(\partial E) = -1\}. \end{aligned}$$

Note that Z_{-1} is a manifold.

Let $g^* : Z_{-1} \to Z_{-1}$ be the induced map. Let $\operatorname{Lef}(g^*)$ denote the Lefschetz number of g^*. Recall that since Z_{-1} is a manifold, $\operatorname{Lef}(g^*)$ is equal to the intersection number of the graph of g^* with the diagonal in $Z_{-1} \times Z_{-1}$.

(3.17) **Lemma.** *Let K and g^* be as above. Then $\bar{\lambda}'(K) = \pm 2 \operatorname{Lef}(g^*)$.*

Proof: (Casson for $\mathbb{Z}$HS case.) Let $p : M \setminus K \to S^1 = [0, 2\pi]/(0 \sim 2\pi)$ be the fibering projection. Let

$$\begin{aligned} E_0 &\stackrel{\text{def}}{=} p^{-1}(0) \cup K \\ E_\pi &\stackrel{\text{def}}{=} p^{-1}(\pi) \cup K \\ F &\stackrel{\text{def}}{=} E_0 \cup E_\pi \\ W_1 &\stackrel{\text{def}}{=} p^{-1}([0, \pi]) \cup K \\ W_2 &\stackrel{\text{def}}{=} p^{-1}([\pi, 2\pi]) \cup K. \end{aligned}$$

(F, W_1, W_2) is a Heegaard splitting of M. For the sake of notational consistency, let γ denote the image of K in F. We can take the fiber E above to be E_0. Identify E_0 and E_π via the product structure of W_1. With these identifications in mind, we have

$$\begin{aligned} R^\sharp &= \{(\rho, \rho') \in Z^\sharp \times Z^\sharp \mid \rho(\gamma) = \rho'(\gamma)\} \\ Q_1^\sharp &= \{(\rho, \rho') \in Z^\sharp \times Z^\sharp \mid \rho = \rho'\} \\ Q_2^\sharp &= \{(\rho, \rho') \in Z^\sharp \times Z^\sharp \mid \rho = g^* \rho'\} \\ N^\sharp &= \{(\rho, \rho') \in Z^\sharp \times Z^\sharp \mid \rho(\gamma) = \rho'(\gamma) = -1\}, \end{aligned}$$

where N is as in the proof of (3.2).

There is a natural fibration

$$SO(3) \to N \xrightarrow{\pi} Z_{-1} \times Z_{-1}.$$

Let $\Delta \subset Z_{-1} \times Z_{-1}$ be the diagonal. Let $\widetilde{\Delta} \stackrel{\text{def}}{=} \pi^{-1}(\Delta)$. Let D be the difference cycle, as defined in the proof of (3.2). Note that $\pi(Q_1 \cap N) = \Delta$. It follows from (3.6) that D is homologous to $2\widetilde{\Delta}$.

The proof can now be completed using some standard intersection number yoga. We have (up to sign, with $\langle \cdot, \cdot \rangle_X$ denoting the intersection number in X),

$$\begin{aligned}
\bar{\lambda}'(K) &= \langle D, Q_2 \rangle_R \\
&= 2\langle \widetilde{\Delta}, Q_2 \rangle_R \\
&= 2\langle \widetilde{\Delta}, Q_2 \cap N \rangle_N \\
&= 2\langle \Delta, \pi(Q_2 \cap N) \rangle_{Z_{-1} \times Z_{-1}} \\
&= 2\langle \Delta, \text{graph of } g^* \rangle_{Z_{-1} \times Z_{-1}} \\
&= 2\,\mathrm{Lef}(g^*).
\end{aligned}$$

□

We are now in a position to do some calculations which will be needed in the proof of (3.36). Let $L_{p,q}$ denote the p, q lens space (i.e. p/q-surgery on the unknot in S^3). It is easy to see that, as oriented manifolds,

$$L_{p,q} = -L_{p,p-q}.$$

In particular, $L_{2,1} = -L_{2,1}$. Therefore, by (3.1),

$$\bar{\lambda}(L_{2,1}) = 0. \tag{3.18}$$

Let K be a knot in S^3. Let $K^\dagger \subset S^3$ be a knot which differs from K by a crossing change (i.e. K and $K^\dagger$ coincide except in a 3-ball, where they differ as shown in Figure 3.3). Let $K_{p/q}$ denote p/q-surgery on K. Let $J \subset S^3 \setminus K$ be an unknot surrounding the crossing, as shown in Figure 3.4. Note that -1-surgery on J transforms K into $K^\dagger$. J bounds a genus one Seifert surface E consisting of an unknotting disk D for J with a neighborhood of $K \cap D$ removed and a tube surrounding half of K glued in. (See Figure 3.5.) Let $A \subset E$ be a curve isotopic in $S^3 \setminus K$ to a meridian on K. Note that in $K_{p/q}$, the longitude on A is a q/p curve.

Let $K_{p/q}J_{a/b}A_{c/d}$ denote the manifold obtained by doing p/q-surgery on K, a/b-surgery on J and c/d-surgery on A. By (3.16), -1-surgery on A in $K_{p/q}$ does not affect $\bar{\lambda}'(J)$, so long as $p/q > 0$. In other words,

$$\bar{\lambda}(K_{p/q}J_{-1/1}A_{-1/1}) - \bar{\lambda}(K_{p/q}J_{1/0}A_{-1/1}) = \\ \bar{\lambda}(K_{p/q}J_{-1/1}A_{1/0}) - \bar{\lambda}(K_{p/q}J_{1/0}A_{1/0}).$$

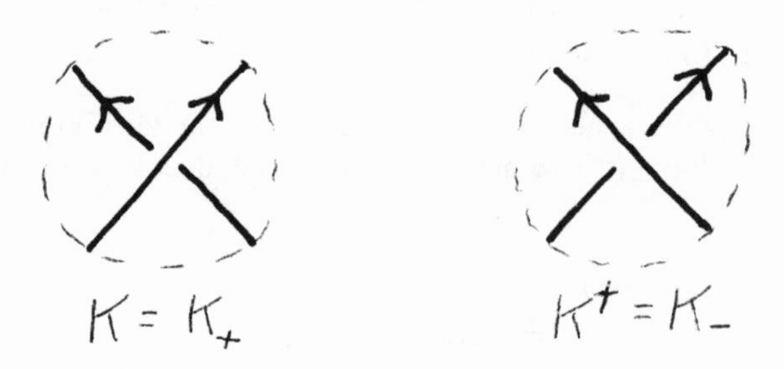

Figure 3.3: The non-coinciding parts of K and $K^\dagger$.

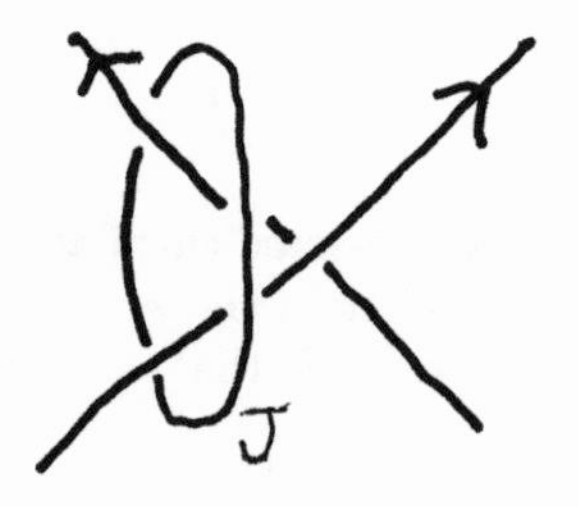

Figure 3.4: J, surgery upon which effects a crossing change

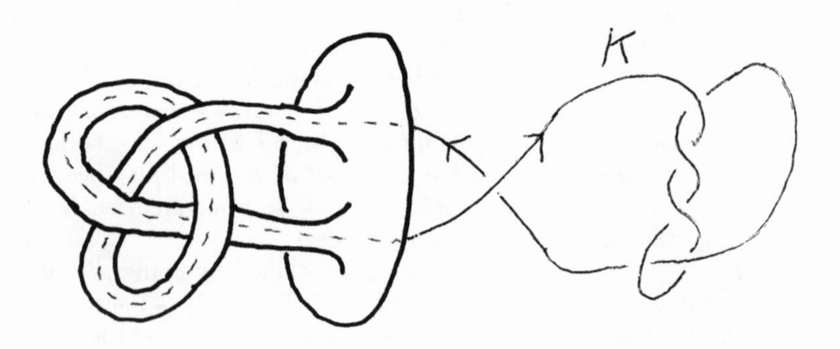

Figure 3.5: The genus one Seifert surface for J

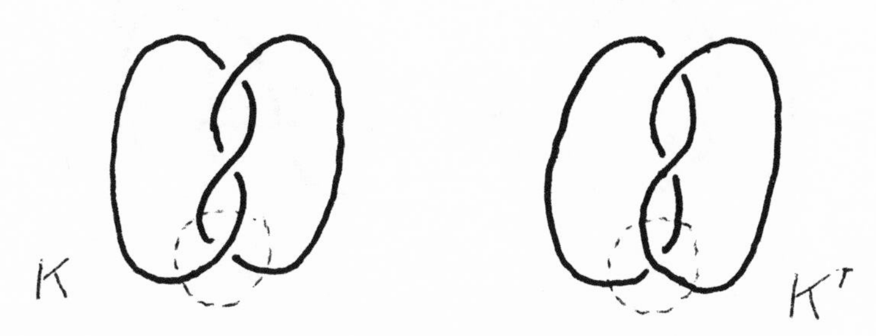

Figure 3.6: K and $K^\dagger$ (special case).

for all $p/q > 0$. This is equivalent to

$$\bar{\lambda}(K^\dagger_{(p+q)/q}) - \bar{\lambda}(K^\dagger_{p/q}) = \bar{\lambda}(K_{(p+q)/q}) - \bar{\lambda}(K_{p/q}). \tag{3.19}$$

Now we specialize to the case where K is the right-handed trefoil knot and $K^\dagger$ is the unknot (see Figure 3.6). Four applications of (3.19) yield

$$\bar{\lambda}(L_{5,1}) - \bar{\lambda}(L_{1,1}) = \bar{\lambda}(K_{5/1}) - \bar{\lambda}(K_{1/1}). \tag{3.20}$$

K is a fibered knot with genus one fiber (see, e.g., [Ro]). For the genus one case, it is not hard to see that Z_{-1} (defined in the proof of (3.17)) consists of a single point. Hence $\mathrm{Lef}(g^*) = \pm 1$ and, by (3.17), $\bar{\lambda}(K_{1/1}) = \bar{\lambda}'(K) = \lambda'(K) = \pm 2$. (It will follow from the proof of (3.36) that $\lambda'(K) = 2$.)

$L_{1,1} \cong S^3$, so $\bar{\lambda}(L_{1,1}) = 0$. It follows from [Mo] that $K_{5/1} \cong -L_{5,1}$ (see also Figure 3.7). Therefore, by (3.1) and (3.20),

$$2\bar{\lambda}(L_{5,1}) = \mp 2$$

or

$$\bar{\lambda}(L_{5,1}) = \mp 1. \tag{3.21}$$

(3.22) **Lemma.** *Let M be a $\mathbb{Q}HS$ and Σ be a $\mathbb{Z}HS$. Let K be a knot in Σ and let $K^\dagger$ be the corresponding knot in $\Sigma \# M$. Then*

$$\lambda'(K^\dagger) = \lambda'(K).$$

Proof: Let (W_{1a}, W_{2a}, F_a) $[(W_{1b}, W_{2b}, F_b)]$ be a Heegaard splitting of Σ $[M]$. Without loss of generality, K coincides a separating curve $\gamma \subset F_a$

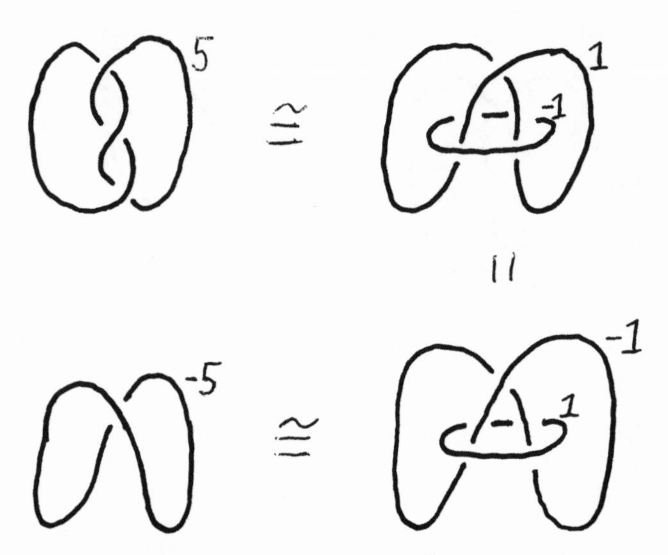

Figure 3.7: The proof that $K_{5/1} = -L_{5,1}$.

(see (3.7)). Let (W_1, W_2, F) be the connected sum of (W_{1a}, W_{2a}, F_a) and (W_{1b}, W_{2b}, F_b). (So that F is the connected sum of F_a and F_b, and W_j is the boundary connected sum of W_{ja} and W_{jb}. See II.2.c of [AM].) Let R, Q_j, S, T_j, etc. be as usual. Let E_x ($x = a$ or b) be the part of F_x coming from F. (So $E_x = F_x \setminus \text{disk}$.) Then $F = E_a \cup E_b$ and $\alpha \stackrel{\text{def}}{=} \partial E_a = \partial E_b = E_a \cap E_b$.

Let

$$\begin{aligned} R_x^{*\sharp} &= \operatorname{Hom}(\pi_1(E_x), SU(2)) \\ R_x^{\sharp} &= \{\rho \in R_x^{*\sharp} \mid \rho(\alpha) = 1\} \\ Q_{jx}^{\sharp} &= \operatorname{Hom}(\pi_1(W_{jx}), SU(2)) \end{aligned}$$

($x = a$ or b). Then we can make the identification

$$R^{\sharp} = \{(\rho_a, \rho_b) \in R_a^{*\sharp} \times R_b^{*\sharp} \mid \rho_a(\alpha) = \rho_b(\alpha)\}.$$

With respect to this identification,

$$Q_j^{\sharp} = Q_{ja}^{\sharp} \times Q_{jb}^{\sharp}.$$

Let $\{h_t^{\sharp}\}$ $[\{h_{at}^{\sharp}\}]$ be the isotopy of (3.2) for $\gamma \subset F$ $[\gamma \subset F_a]$, and let D $[D_a]$ be the associated difference cycle. Hence

$$\lambda'(K^{\dagger}) = \frac{\langle D, Q_2 \rangle}{|H_1(M; \mathbb{Z})|}$$

$$\lambda'(K) \quad = \quad \langle D_a, Q_{2a} \rangle .$$

To prove the lemma it suffices to show that

$$(3.23) \qquad \langle D, Q_2 \rangle = \pm |H_1(M; \mathbb{Z})| \langle D_a, Q_{2a} \rangle$$

and to check the sign.

It is easy to see that

$$h_t^\sharp = h_{at}^\sharp \times \mathrm{id}.$$

It follows that

$$(3.24) \qquad D^\sharp = D_a^\sharp \times Q_{1b}^\sharp .$$

By choosing compatible isotopies on R_a^- and R^-, we may assume that D_a is in general position with respect to Q_{2a} and that (3.24) still holds.

Choose a neighborhood V_a of $D_a \cap Q_{2a}$ in R_a^- over which there is a smooth section $f : V_a \to R_a^\sharp$ of the quotient map $R_a^\sharp \to R_a$. Let U be a neighborhood of 1 in $SU(2)$. Choose an immersion

$$\bar{f} : V_a \times U \to R_a^{*\sharp}$$

so that $\bar{f}(\,\cdot\,, 1) = f$ and $\bar{f}(p, u)(\alpha) = u$. Let $N_b^\sharp$ be a neighborhood of $R_b^\sharp$ in $R_b^{*\sharp}$.

Define

$$\begin{array}{rcl} g : V_a \times V_b^\sharp & \to & R^- \\ (p_a, \rho_b) & \mapsto & \pi(\bar{f}(p_a, \rho_b(\alpha)), \rho_b), \end{array}$$

where π is the quotient map from $R^\sharp \subset R_a^{*\sharp} \times R_b^{*\sharp}$ to R. It is easy to see that g is a diffeomorphism onto its image. Furthermore,

$$\begin{array}{rcl} D \cap Q_2 & \subset & g(V_a \times V_b^\sharp) \\ g^{-1}(D) & = & (V_a \cap D_a) \times Q_{1b}^\sharp \\ g^{-1}(Q_2) & = & (V_a \cap Q_{2a}) \times Q_{2b}^\sharp . \end{array}$$

It follows that

$$\langle D, Q_2 \rangle = \pm \langle D_a, Q_{2a} \rangle \langle Q_{1b}^\sharp, Q_{2b}^\sharp \rangle_{R_b^{*\sharp}} .$$

But by Proposition III.1.1 of [AM],

$$\langle Q_{1b}^\sharp, Q_{2b}^\sharp \rangle_{R_b^{*\sharp}} = \pm |H_1(M; \mathbb{Z})|.$$

This proves (3.23).

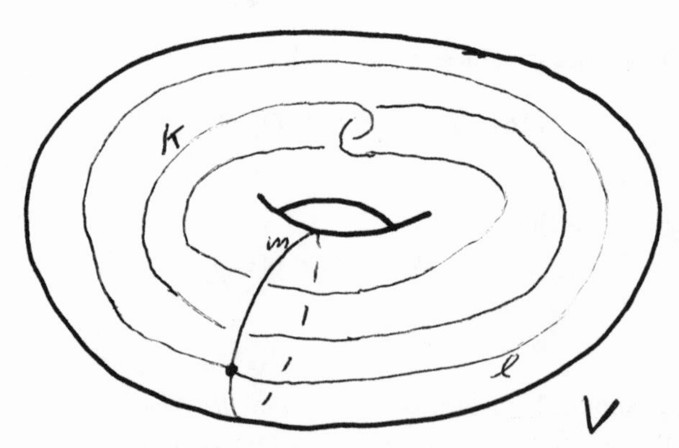

Figure 3.8: A pre-untwisted double.

To determine the sign, one need only consider orientations near the trivial representation. Since a neighborhood of the trivial representation is independent (up to diffeomorphism) of the manifolds involved, it suffices to check a single example, e.g. $M = S^3$. The lemma is clearly true in this case, so it is true in general. □

In the proof of the next lemma we will need the concept of a clean intersection. Let A and B be oriented submanifolds of an oriented manifold M which intersect in an (oriented) submanifold Z. Assume that $\dim(A) + \dim(B) = \dim(M)$. (Note that A and B are not transverse if $\dim(Z) > 0$.) A and B are said to intersect *cleanly* if

$$TZ = TA|_Z \cap TB|_Z.$$

Let $\nu(Z)$ be the intersection of the normal bundles of A and B in M, restricted to Z. $\nu(Z)$ is an orientable, $\dim(Z)$ dimensional vector bundle over Z, and it is not hard to see that

$$\langle A, B\rangle = \pm e(\nu(Z)), \tag{3.25}$$

where $e(\nu(Z))$ denotes the Euler number of $\nu(Z)$.

Consider the solid torus V containing the knot K, as shown in Figure 3.8. If V is embedded is a $\mathbb{Q}$HS M in such a way that the curve $l \in \partial V$ is parallel to a longitude of $\overline{M \setminus V}$, then we say that K (considered as a knot in M) is an *untwisted double* (with negative clasp).

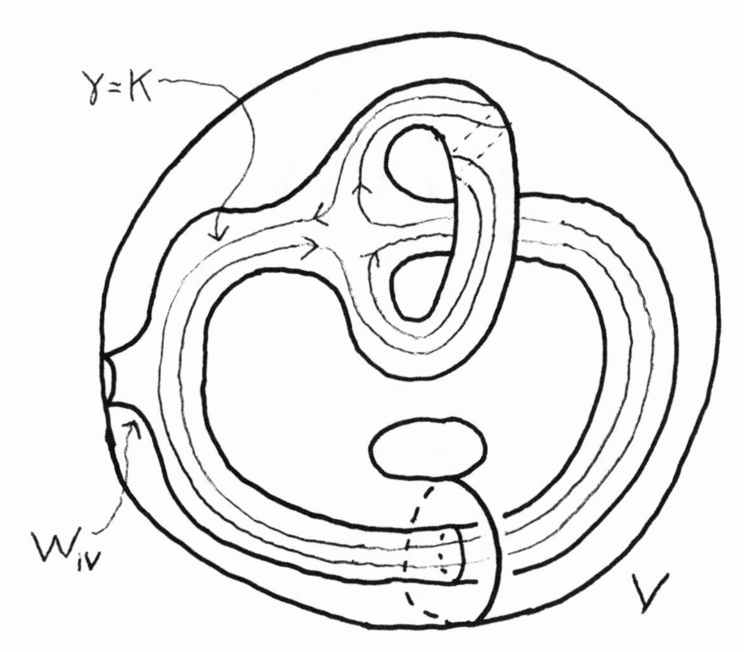

Figure 3.9: $W_{1V} \subset V$

(3.26) **Lemma.** *Let $K \subset M$ be an untwisted double. Then $\lambda'(K) = 0$.*

Proof: As usual, we begin by constructing the appropriate Heegaard splitting. Fix some embedding of V in M which realizes K as an untwisted double. Identify V with its image under this embedding. K lies on the surface of a genus two handlebody $W_{1V} \subset V$, as shown in Figure 3.9. (W_{1V} can be thought of as $\langle$a regular neighborhood of the core of $V\rangle$ union $\langle$an unknotted circle which intersects the core once$\rangle$ union $\langle$an arc connecting the core to $\partial V\rangle$.) Let W_{2V} be the closure in V of $V \setminus W_{1V}$. Let $\Lambda \stackrel{\text{def}}{=} \overline{M \setminus V}$. Choose a collar $\partial V \times I \subset \Lambda$ of ∂V in Λ. Let $W_{2\Lambda} \subset \Lambda$ be $(W_{2V} \cap \partial V) \times I$ union some 1-handles in Λ. Let $W_{1\Lambda}$ be the closure in Λ of $\Lambda \setminus W_{2\Lambda}$. If the 1-handles are chosen properly, then $W_{1\Lambda}$ will be a handlebody, and $W_1 \stackrel{\text{def}}{=} W_{1V} \cup W_{1\Lambda}$ and $W_2 \stackrel{\text{def}}{=} W_{2V} \cup W_{2\Lambda}$ will form a Heegaard splitting of M. Let $F \stackrel{\text{def}}{=} \partial W_1 = \partial W_2$. Let $F_V \stackrel{\text{def}}{=} F \cap V$ and $F_\Lambda \stackrel{\text{def}}{=} F \cap \Lambda$. Note that the natural maps $\pi_1(F_V) \to \pi_1(W_{jV})$ and $\pi_1(F_\Lambda) \to \pi_1(W_{j\Lambda})$ $(j = 1, 2)$ are all surjective. For the sake of notational consistency, let γ denote K thought of as a curve in F.

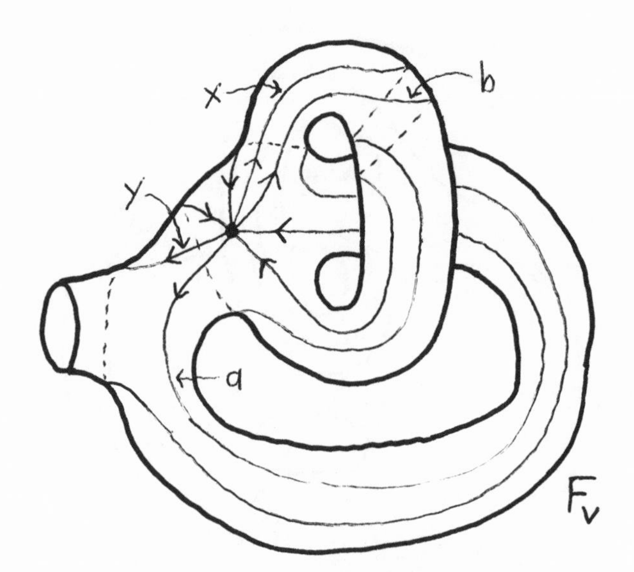

Figure 3.10: a, b, x and y.

Let R, Q_j, S, etc., be as usual. Let

$$R_V^{*\sharp} \stackrel{\text{def}}{=} \operatorname{Hom}(\pi_1(F_V), SU(2))$$
$$Q_{jV}^{\sharp} \stackrel{\text{def}}{=} \operatorname{Hom}(\pi_1(W_{jV}), SU(2)).$$

Define $R_\Lambda^{*\sharp}$ and $Q_{j\Lambda}^{\sharp}$ similarly. $R^\sharp$ can be identified with a subset of $R_V^{*\sharp} \times R_\Lambda^{*\sharp}$. It is easy to see that with respect to this identification, $Q_1^\sharp = Q_{1V}^\sharp \times Q_{1\Lambda}^\sharp$ and $Q_2^\sharp \subset Q_{2V}^\sharp \times Q_{2\Lambda}^\sharp$. ($Q_2^\sharp$ is equal to the set of all $(\rho_V, \rho_\Lambda) \subset Q_{2V}^\sharp \times Q_{2\Lambda}^\sharp$ which agree on $\pi_1(W_2 \cap \partial V)$.)

Choose a set of generators $a, b, x, y \in \pi_1(F_V)$ as shown in Figure 3.10. For $\rho \in R_V^{*\sharp}$, let $A \stackrel{\text{def}}{=} \rho(a)$, $B \stackrel{\text{def}}{=} \rho(b)$, $X \stackrel{\text{def}}{=} \rho(x)$, $Y \stackrel{\text{def}}{=} \rho(y)$. The functions A, B, X and Y give an identification of $R_V^{*\sharp}$ with $SU(2)^4$.

Let $c, d, e, \partial \in \pi_1(F_V)$ be as shown in Figure 3.11. We have

$$\begin{aligned} c &= xy^{-1}x^{-1}bab^{-1} \\ d &= x^{-1}b \\ e &= ba^{-1}b^{-1}xyx^{-1}b \\ \partial &= aba^{-1}b^{-1}xyx^{-1}y^{-1} \\ \gamma &= aba^{-1}b^{-1}. \end{aligned}$$

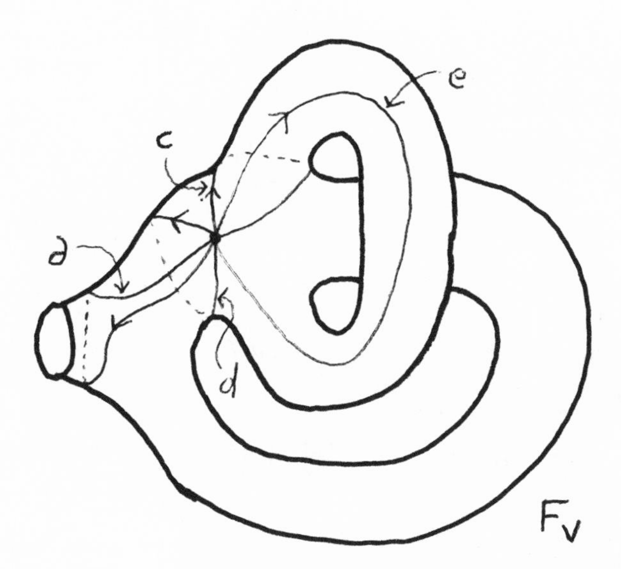

Figure 3.11: c, d, e and ∂

(In the last equation, γ is confused with an element of $\pi_1(F_V)$ in its conjugacy class.) It follows that

$$\begin{aligned} Q^\sharp_{1V} &= \{(A,B,X,Y) \in R^{*\sharp}_V \mid XY^{-1}X^{-1}BAB^{-1} = 1, X^{-1}B = 1\} \\ &= \{(A,B,X,Y) \in R^{*\sharp}_V \mid X = B, Y = A\} \\ Q^\sharp_{2V} &= \{(A,B,X,Y) \in R^{*\sharp}_V \mid BA^{-1}B^{-1}XYX^{-1}B = 1\}. \end{aligned}$$

Let $D \subset R$ be the difference cycle associated to γ, as described in the proof of (3.2). The lemma will be proved if it can be shown that $\langle D, Q_2 \rangle = 0$.

Our next goal is to describe D more explicitly. Let F_1 and F_2 be the components of F cut along γ. $R^\sharp$ can be identified with a subset of $\mathrm{Hom}(\pi_1(F_1), SU(2)) \times \mathrm{Hom}(\pi_1(F_2), SU(2))$. Let $N^\sharp \stackrel{\mathrm{def}}{=} \{\rho \in R^\sharp \mid \rho(\gamma) = -1\}$. It is not hard to show that Q_1 intersects N transversely. By (3.6) we may assume

(3.27) $D^\sharp = 2\{(\rho_1, g\rho_2) \in R^\sharp \mid g \in \mathrm{Aut}\,(SU(2)), (\rho_1,\rho_2) \in N^\sharp \cap Q^\sharp_1\}$,

where $(g\rho_2)(z) \stackrel{\mathrm{def}}{=} g(\rho_2(z))$.

In the present situation,

$$N^\sharp \cap Q^\sharp_1 = \{(A,B,X,Y) \in R^\sharp_V \mid ABA^{-1}B^{-1} = -1,\ X = B,\ Y = A\} \times Q^\sharp_{1\Lambda}.$$

(Here we are thinking of $R^\sharp$ as a subset of $R_V^{*\sharp} \times R_\Lambda^{*\sharp}$ again.) γ is isotopic to the boundary of a regular neighborhood of curves representing a and b. It follows from (3.27) that we take $D^\sharp$ to be $D_V^\sharp \times Q_{1\Lambda}^\sharp$, where

$$\begin{aligned} D_V^\sharp &= 2\{(gA, gB, X, Y) \in R_V^\sharp \,|\, g \in \mathrm{Aut}\,(SU(2)), \\ &\qquad\qquad ABA^{-1}B^{-1} = -1, X = B, Y = A\} \\ &= 2\{(gA, gB, X, Y) \in R_V^\sharp \,|\, ABA^{-1}B^{-1} = -1, XYX^{-1}Y^{-1} = -1\}. \end{aligned}$$

Let $U \in \mathbf{su}(2)$ be a ball around 0 such that $\exp : U \to SU(2) \setminus \{-1\}$ is a diffeomorphism. Let $L \subset U$ be the 2-sphere which maps to the elements of trace 0. Scale the metric on $\mathbf{su}(2)$ so that L is the sphere of radius $\pi/2$. Then $ABA^{-1}B^{-1} = -1$ if and only if $A = \exp(u)$ and $B = \exp(v)$ where $u, v \in L$ and u is orthogonal to v.

Let $R_V^\sharp \stackrel{\text{def}}{=} \{\rho \in R_V^{*\sharp} \,|\, \rho(\partial) = 1\}$. Let $S_V^\sharp$ be the abelian representations in $R_V^\sharp$. Define $R_\Lambda^\sharp$ and $S_\Lambda^\sharp$ similarly. Let $Z^\sharp \stackrel{\text{def}}{=} Z_V^\sharp \times R_\Lambda^\sharp$, where

$$Z_V^\sharp \stackrel{\text{def}}{=} \{\rho \in R_V^\sharp \,|\, \rho(y) \neq \pm 1\}.$$

For $\rho \in Z_V^\sharp$, let $u(\rho) \in U$ be the unit vector such that

$$\exp(tu(\rho)) = \rho(y)$$

for some $0 < t < \pi$. There is an S^1-action on $Z_V^\sharp$ defined by

$$(\theta\rho)(z) = \exp(\theta u(\rho))\rho(z)\exp(-\theta u(\rho))$$

($\rho \in Z_V^\sharp$, $z \in \pi_1(F_V)$, $\theta \in [0, \pi]/(0 \sim \pi) = S^1$). By acting trivially on the second factor, this induces an action of S^1 on $Z_V^\sharp \times R_\Lambda^\sharp = Z^\sharp$. This action clearly descends to one on $Z = Z^\sharp/SU(2)$. S^1 acts freely on

$$(3.28)\quad Z_- \stackrel{\text{def}}{=} Z \setminus \left((Z_V^\sharp \times S_\Lambda^\sharp)/SU(2) \cup ((S_V^\sharp \cap Z_V^\sharp) \times R_\Lambda^\sharp)/SU(2)\right).$$

Note that both $D = D \cap Z$ and $Q_2 \cap Z$ are invariant under this action. Thus if we can equivariantly isotope D in Z so that it intersects Q_2 in isolated principal orbits and these intersections are clean, then it will follow from (3.25) that $\langle D, Q_2 \rangle$ is equal to the Euler number of an oriented vector bundle over a closed 1-manifold, and hence is zero.

The first step will be to isotope D in Z so that $D \cap Z \subset Z_-$.

Let $\rho_V = (A, B, X, Y) \in D_V^\sharp$. We will define a deformation $\rho_{V,t} = (A_t, B_t, X_t, Y_t)$ ($t \in [0, \epsilon]$, ϵ near 0). Define

$$\begin{aligned} X_t &= X \\ Y_t &= \exp(tu(\rho_V))Y. \end{aligned} \tag{3.29}$$

It follows that

$$X_tY_tX_t^{-1}Y_t^{-1} = -\exp(-2tu(\rho_V)).$$

Assume for the moment that we can find A_t and B_t (depending smoothly on ρ_V and t, and such that $A_0 = A$ and $B_0 = B$) which satisfy

$$A_tB_tA_t^{-1}B_t^{-1} = -\exp(2tu(\rho_V)) \tag{3.30}$$

and

$$(\mathrm{Ad}(H)\rho_V)_t = (\mathrm{Ad}(H)A_t, \mathrm{Ad}(H)B_t, \mathrm{Ad}(H)X_t, \mathrm{Ad}(H)Y_t) \tag{3.31}$$

for all $H \in SU(2)$. It follows that

$$\rho_{V,t}(\partial) = A_tB_tA_t^{-1}B_t^{-1}X_tY_tX_t^{-1}Y_t^{-1} = 1,$$

and hence that $\rho_{V,t} \in Z_V^\sharp$.

Let $\{D_{V,t}^\sharp\}$ denote the resulting deformation of $D_V^\sharp$. Define a deformation $\{D_t^\sharp\}$ of $D^\sharp$ by $(\rho_V, \rho_\Lambda)_t = (\rho_{V,t}, \rho_\Lambda)$ for $(\rho_V, \rho_\Lambda) \in D_V^\sharp \times Q_{1\Lambda}^\sharp = D^\sharp$. It follows from (3.31) that this deformation descends to a deformation $\{D_t\}$ of D in Z which commutes with the action of S^1 on Z.

Now to define A_t and B_t. Consider the map

$$\begin{aligned} c : SU(2) \times SU(2) &\rightarrow SU(2) \\ (G,H) &\mapsto GHG^{-1}H^{-1}. \end{aligned}$$

Let $W \stackrel{\mathrm{def}}{=} SU(2) \setminus \{1\}$. c is a submersion, and hence a fibering, over W (with fiber isomorphic to $c^{-1}(-1) \cong SO(3)$). Suppose that there is a smooth bundle map

$$f : W \times c^{-1}(-1) \rightarrow c^{-1}(W)$$

such that $f(-1, x) = x$ and such that f commutes with the actions of SU(2) on $W \times c^{-1}(-1)$ and $c^{-1}(W)$. Then

$$(A_t, B_t) \stackrel{\mathrm{def}}{=} f(-\exp(2tu(\rho_V)), (A, B))$$

clearly satisfies (3.30) and (3.31).

To define f, choose an $SU(2)$-invariant connection on the fibering

$$c^{-1}(W) \rightarrow W.$$

Let $f(w, x)$ (for $w \neq -1$) be the parallel transport of x along the arc connecting -1 to w in the abelian subgroup containing w. It is easy to see that this works.

Next we show that for some (in fact almost all) $t \in [0, \epsilon]$, $D_t \cap Q_2 \subset Z_-$. Clearly $D^\sharp_{V,t} \cap S^\sharp_V = \emptyset$ for all t. So, by (3.28), what remains to be shown is that

$$(D^\sharp_t \cap Q^\sharp_2) \cap (Z^\sharp_V \times S^\sharp_\Lambda) = \emptyset. \tag{3.32}$$

Note that

$$\begin{aligned} D^\sharp_t \cap Q^\sharp_2 \cap (Z^\sharp_V \times S^\sharp_\Lambda) &\subset ((D^\sharp_{V,t} \times Q^\sharp_{1\Lambda}) \cap (Q^\sharp_{2V} \times Q^\sharp_{2\Lambda})) \cap (Z^\sharp_V \times S^\sharp_\Lambda) \\ &= (D^\sharp_{V,t} \cap Q^\sharp_{2V}) \times (Q^\sharp_{1\Lambda} \cap Q^\sharp_{2\Lambda} \cap S^\sharp_\Lambda). \end{aligned}$$

$Q^\sharp_{1\Lambda} \cap Q^\sharp_{2\Lambda} \cap S^\sharp_\Lambda$ is just the space of abelian representations $\pi_1(\Lambda) \to SU(2)$. Since K is an *untwisted* double, the curve l in Figure 3.8 must have finite order (say $d \in \mathbb{Z}$) in $H_1(\Lambda; \mathbb{Z})$. Thus if $\bar{l} \in \pi_1(F_V)$ is freely homotopic in Λ to l and $\rho_\Lambda \in Q^\sharp_{1\Lambda} \cap Q^\sharp_{2\Lambda} \cap S^\sharp_\Lambda$, then $\rho_\Lambda(\bar{l})^d = 1$. On the other hand, l is freely homotopic in W_{2V} to y (see Figures 3.9 and 3.10). Therefore if

$$\rho \in ((D^\sharp_{V,t} \cap Q^\sharp_{2V}) \times (Q^\sharp_{1\Lambda} \cap Q^\sharp_{2\Lambda} \cap S^\sharp_\Lambda)) \cap Q^\sharp_2,$$

then $Y_t^d = \rho(y)^d = 1$. It is clear from (3.29) that this is possible for only a finite number of t. Thus (3.32) holds for all but finitely many t. Let t_0 be one of these and define $D' \overset{\text{def}}{=} D_{t_0}$. D' is the desired deformation of D which contains only principal orbits.

Now for the clean intersection property. First we show that near $D' \cap Q_2$, $Q_2 \cap Z$ is a manifold of dimension $3g-5$ (where g is the genus of F). There is a free basis $\bar{m}, \bar{l}, x_3, \ldots, x_g$ of $\pi_1(W_2)$ such that $\bar{m}$ $[\bar{l}]$ is freely homotopic to m $[l]$ and such that ∂ is freely homotopic to $\bar{m}\bar{l}\bar{m}^{-1}\bar{l}^{-1}$. Hence $Q_2 \cap Z$ is diffeomorphic to

$$\begin{aligned} \{(\rho(\bar{m}), \rho(\bar{l}), \rho(x_3), \ldots, \rho(x_g)) \in SU(2)^g \mid \\ \rho(\bar{m})\rho(\bar{l})\rho(\bar{m})^{-1}\rho(\bar{l})^{-1} = 1,\ \rho(\bar{l}) \neq \pm 1\}/SU(2). \end{aligned}$$

It is now easy to see that the irreducible part of $Q_2 \cap Z$ is a manifold of dimension $3g-5$. Since D' consists entirely of irreducible representations, we are done.

Let $\nu(Q_2 \cap Z; Q_2)$ denote the normal bundle of $Q_2 \cap Z$ in Q_2. Let $\nu(Z; R)$ denote the normal bundle of Z in R. I claim that near $D' \cap Q_2$

$$\nu(Q_2 \cap Z; Q_2) = TQ_2 \cap \nu(Z; R). \tag{3.33}$$

Let $f : R^\sharp \to SU(2)$ be given by $f(\rho) \overset{\text{def}}{=} \rho(\partial)$. (3.33) follows from the facts that $Z^\sharp$ is an open submanifold of $f^{-1}(1)$, f is a submersion near $D'^\sharp \cap Q^\sharp_2$, and $f|_{Q^\sharp_2}$ has rank 2.

Now isotope D'/S^1 inside Z/S^1 so that it is in general position with respect to $(Q_2 \cap Z)/S^1$ and so that $(D'/S^1) \cap ((Q_2 \cap Z)/S^1) \subset Z_-/S^1$. Cover this isotopy with an isotopy of D'. $D' \cap Q_2$ is now a closed 1-manifold. (3.33) and the fact that D'/S^1 is in general position with respect to $(Q_2 \cap Z)/S^1$ imply that D' and Q_2 intersect cleanly. Hence $\langle D', Q_2 \rangle$ is equal to the Euler number of an oriented line bundle over a closed 1-manifold, i.e. $\langle D', Q_2 \rangle = 0$. □

(3.34) **Lemma.** *Let K be a knot in a $\mathbb{Q}HS$ M with self-linking zero. That is, there exists $x \in H_1(\partial(\mathrm{nbd}(K)); \mathbb{Z})$ such that $\langle m, x \rangle = 1$ and $l = \langle m, l \rangle x$, where m is the meridian and l is the longitude of K. Let $K_{1/n}$ denote $1/n$-surgery on K, with respect to the basis m, x. Then there exists $\lambda'(K) \in \mathbb{Q}$ such that*

$$\lambda(K_{1/n}) = \lambda(M) + n\lambda'(K).$$

Note that if $\langle m, x \rangle = \pm 1$, this is (3.7). Note also that since $|H_1(K_{1/n}; \mathbb{Z})| = |H_1(M; \mathbb{Z})|$, $\bar{\lambda}(K_{1/n}) = \bar{\lambda}(M) + n\bar{\lambda}'(K)$, where $\bar{\lambda}'(K) \overset{\text{def}}{=} |H_1(M; \mathbb{Z})| \lambda'(K)$.

Proof: First we construct the appropriate Heegaard splitting of M. Let U be a closed regular neighborhood of K. Let $d \overset{\text{def}}{=} |\langle m, l \rangle|$. Assume that $d > 1$. Let $E \subset \overline{M \setminus U}$ be a Seifert surface for K. That is, $[\partial E] = l \in H_1(\partial U; \mathbb{Z})$ and ∂E consists of d disjoint parallel curves. Let $E \times [-1, 1] \subset M$ be a bicollar of E. As in the proof of (3.7), we can add 1-handles to $E \times \{-1\}$ so that $W_2 \overset{\text{def}}{=} E \times [-1, 1] \cup \{1 - \text{handles}\}$ and $W_1 \overset{\text{def}}{=} \overline{M \setminus W_2}$ are a Heegaard splitting of M. As usual, let $F \overset{\text{def}}{=} \partial W_1 = \partial W_2$.

Let $\gamma_1, \ldots, \gamma_d$ be the boundary components of $E = E \times \{0\}$, thought of as curves on F. Let h be a left-handed Dehn twist along γ_1. Then

$$K_{1/n} = W_1 \cup_{h^n} W_2,$$

so what must be shown is that

$$\langle h^n(Q_1), Q_2 \rangle - \langle h^{n-1}(Q_1), Q_2 \rangle$$

is independent of n. The proof of this is similar to that of (3.2). Extra complications arise because γ_1 is non-separating.

Let $a \subset E$ be an arc connecting γ_1 to, say, γ_2. Let $\beta \subset F$ be the loop $\partial a \times [-1, 1] \cup a \times \{1\} \cup a \times \{-1\}$. Note that β and γ_1 intersect once transversely. Let $F' \subset F$ be F cut open along γ_1. Then the action of h on $\pi_1(F)$ and $R^\sharp$ is as described in the proof of (3.15) (with h replacing h_α and γ_1 replacing α).

Let $V_\epsilon \subset SU(2)$ be the closed ball of radius ϵ (with respect to some Ad-invariant metric) centered at $-1 \in SU(2)$. Let

$$\begin{aligned} N^\sharp &= \{\rho \in R^\sharp \mid \rho(\gamma_1) = -1\} \\ N^\sharp_\epsilon &= \{\rho \in R^\sharp \mid \rho(\gamma_1) \in V_\epsilon\}. \end{aligned}$$

Let $f : SU(2) \to [0,1]$ be a smooth, Ad-invariant, monotonic (on $SU(2)/\mathrm{Ad}$) function such that $f = 1$ outside V_ϵ and $f(-1) = 0$. For $t \in [0,1]$, define $h^\sharp_t : R^\sharp \to R^\sharp$ by

$$\begin{aligned} h^\sharp_t(\rho)x &= \rho(x), & x \in \pi_1(F') \\ h^\sharp_t(\rho)\beta &= \rho(\gamma_1)^{f(\rho(\gamma_1))t}\rho(\beta). \end{aligned}$$

$\{h^\sharp_t\}$ descends to an isotopy $\{h_t\}$ of R. Note that $h_0 = \mathrm{id}$ and $h_1 = h$ on $R \setminus N_\epsilon$.

Let $[\rho] \in T_1$ (i.e. ρ is an abelian representation of $\pi_1(W_1)$). Since the γ_i's are all homologous in W_1, $\rho(\gamma_1) = \rho(\gamma_2) = \cdots = \rho(\gamma_d)$. Since $\gamma_1 \cup \cdots \cup \gamma_d$ is a boundary in W_1, $\rho(\gamma_1)^d = 1$. Since T_1 is connected and $\rho(\gamma_1) = 1$ if ρ is the trivial representation, $\rho(\gamma_1) = 1$ for all $[\rho] \in T_1$. Therefore $h_t|_{T_1} = \mathrm{id}$.

It follows from Theorem 4.7 of [G3] that h_t is a symplectic map of R. Hence h_t acts symplectically on ν. It is easy to see that $h_t(Q_1)$ is transverse to Q_2 at P. $h_t|_S$ is not the identity, so $\{h_t\}$ is not quite a special isotopy. However, since $h_t|_{T_1} = \mathrm{id}$, the proof of special isotopy invariance in Section 2 goes through anyway and we have

$$\langle Q_1, Q_2 \rangle = \langle h_1(Q_1), Q_2 \rangle.$$

Let

$$D \overset{\text{def}}{=} h(Q_1 \cap N_\epsilon) \cup (-h_1(Q_1) \cap N_\epsilon).$$

Since $T_1 \cap N = \emptyset$, $D \subset R^-$. Therefore

$$\langle h(Q_1), Q_2 \rangle - \langle Q_1, Q_2 \rangle = \langle D, Q_2 \rangle,$$

where the right hand side denotes a homological intersection number.

As in the proof of (3.2), it now suffices to show that $h(D)$ is homologous to D. This will follow if it can be shown that h acts trivially on $H_*(N; \mathbb{Q})$, since D can be homotoped into N.

Let $F^\dagger \subset F$ be the complement of a regular neighborhood of $\gamma_1 \cup \beta$. Let

$$R^{\dagger\sharp} \overset{\text{def}}{=} \{\rho \in \mathrm{Hom}(\pi_1(F^\dagger), SU(2)) \mid \rho(\partial F^\dagger) = 1\}.$$

Define $S^{\dagger\sharp}$ and $P^{\dagger\sharp}$ similarly. $N^\sharp$ can be identified with $R^{\dagger\sharp} \times SU(2)$, where the second factor is identified with $\rho(\beta)$. With respect to this identification,

$$h(\tau, X) = (\tau, -X).$$

Note that $h^2 = \mathrm{id}$.

There is a natural map

$$\pi : N \to R^{\dagger}.$$

$\pi^{-1}(p)$ is diffeomorphic to $SU(2)$, D^2 or I, according to whether p is in $R^{\dagger -}$, $S^{\dagger -}$ or $P^{\dagger}$, respectively.

We will need the following fact, the proof of which is left to the reader.

(3.35) *Let f be an endomorphism of a long exact sequence of vector spaces over $\mathbb{Q}$*

$$\cdots \to A_i \to B_i \to C_i \to A_{i-1} \to \cdots$$

such that $f^2 = \mathrm{id}$, $f|_{A_i} = \mathrm{id}$ and $f|_{B_i} = \mathrm{id}$ (for all i). Then $f|_{C_i} = \mathrm{id}$ for all i.

h descends to the identity on $R^{\dagger}$. Since $\pi : \pi^{-1}(P^{\dagger}) \to P^{\dagger}$ is a homotopy equivalence, h acts trivially on $H_*(\pi^{-1}(P^{\dagger}); \mathbb{Q})$. Similarly, h acts trivially on $H_*(\pi^{-1}(S^{\dagger}_{-}); \mathbb{Q})$. Applying (3.35) to the Gysin sequence of the fibering $\pi^{-1}(R^{\dagger}_{-}) \to R^{\dagger}_{-}$ we see that h acts trivially on $H_*(\pi^{-1}(R^{\dagger}_{-}); \mathbb{Q})$. Let U be regular neighborhood of $P^{\dagger}$ in $S^{\dagger}$. Applying (3.35) to the Mayer-Vietoris sequence of the triple $(\pi^{-1}(S^{\dagger}), \pi^{-1}(S^{\dagger}_{-}), \pi^{-1}(U))$ yields that h acts trivially on $H_*(\pi^{-1}(S^{\dagger}); \mathbb{Q})$. (This uses the easily proved fact that h acts trivially on $H_*(\pi^{-1}(S^{\dagger}_{-} \cap U); \mathbb{Q})$.) Similarly, h acts trivially on $H_*(\pi^{-1}(R^{\dagger}); \mathbb{Q}) = H_*(N; \mathbb{Q})$. □

(3.36) **Lemma.** *Let $K \subset M$ be a knot in a $\mathbb{Q}HS$ and let $N \overset{\mathrm{def}}{=} M \setminus \mathrm{nbd}(K)$. Let $l \in H_1(\partial N; \mathbb{Z})$ be a longitude of K and let $a, b \in H_1(\partial N; \mathbb{Z})$ be primitive homology classes such that $\langle a, l\rangle\langle b, l\rangle > 0$ and $\langle a, b\rangle = 1$. Let $c, d \in H_1(\partial N; \mathbb{Z})$ be primitive homology classes such that $\langle a, c\rangle \geq 0$, $\langle a, d\rangle \geq 0$, $\langle c, b\rangle \geq 0$, $\langle d, b\rangle \geq 0$ and $\langle c, d\rangle = 1$ (see Figure 3.12). For primitive $x \in H_1(\partial N; \mathbb{Z})$, let N_x denote Dehn surgery along x. Then there is a real number $r(K, a, b)$ (depending on K, a and b, but not on c and d) such that*

$$\bar{\lambda}(N_{c+d}) - \bar{\lambda}(N_c) - \bar{\lambda}(N_d) - \frac{1}{6}(|H_1(N_c; \mathbb{Z})| - |H_1(N_d; \mathbb{Z})|) = r(K, a, b).$$

Remarks: It will be shown in the next section that $r(K, a, b) = 0$ for all K, a and b. Note that (in view of the lemma) knowledge of $\bar{\lambda}(N_a)$, $\bar{\lambda}(N_b)$ and $r(K, a, b)$ determines $\bar{\lambda}(N_c)$ for all primitive $c \in H_1(\partial N; \mathbb{Z})$ lying between a and b (i.e. $\langle a, c\rangle \geq 0$, $\langle c, b\rangle \geq 0$).

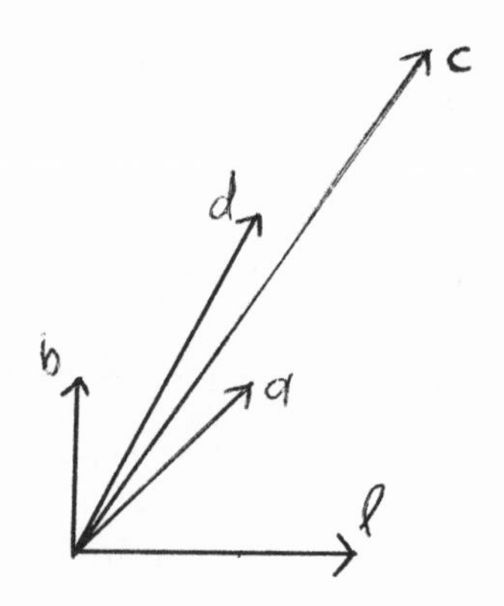

Figure 3.12: a, b, c, d and l.

Proof: The proof is long and technical.

For a, b as in the statement of the lemma, define

$$r(a,b) \stackrel{\text{def}}{=} \bar{\lambda}(N_{a+b}) - \bar{\lambda}(N_a) - \bar{\lambda}(N_b) - \frac{1}{6}(|H_1(N_a;\mathbb{Z})| - |H_1(N_b;\mathbb{Z})|).$$

The lemma will be proved if it can be shown that

$$\begin{aligned} r(a, a+b) &= r(a,b) \\ r(a+b, b) &= r(a,b), \end{aligned}$$

for any pair (c,d) (as in the statement of the lemma) is related to (a,b) by a sequence of the elementary transformations $a \mapsto a$, $b \mapsto a+b$ and $a \mapsto a+b$, $b \mapsto b$.

First we construct the appropriate Heegaard splitting. Let $U \stackrel{\text{def}}{=} \text{nbd}(K)$. Let α and β be simple closed curves on ∂U $(= \partial N)$ which represent a and b and which intersect once transversely. Attach 1-handles to U, disjoint from α and β, so that $W_1 \stackrel{\text{def}}{=} U \cup \{\text{1-handles}\}$ and $W_2 \stackrel{\text{def}}{=} \overline{M \setminus W_1}$ form a Heegaard splitting of M. Let F, R, Q_j, etc. be as usual. Let $\gamma \subset F$ be the boundary of a regular neighborhood of $\alpha \cup \beta$ in F. Let $\delta_1, \ldots, \delta_{g-1} \subset F$ be the boundaries of disks transverse to the cores of the 1-handles. (Note that $\rho(\delta_i) = 1$ for $[\rho] \in Q_1$.)

Let $z \in H_1(\partial U;\mathbb{Z})$ be primitive and let $\xi \subset \partial U$ be a simple closed curve representing z and disjoint from γ. (Disjointness from γ guarantees that $\xi \subset F$.) Define

$$\begin{aligned} Q_1^z &\stackrel{\text{def}}{=} \{[\rho] \in R \,|\, \rho(\xi) = 1, \rho(\delta_i) = 1, 1 \le i \le g-1\} \\ Q_1^{-z} &\stackrel{\text{def}}{=} \{[\rho] \in R \,|\, \rho(\xi) = -1, \rho(\delta_i) = 1, 1 \le i \le g-1\}. \end{aligned}$$

Note that Q_1^z and Q_2 are representation spaces corresponding to a Heegaard splitting which represents z-surgery on K. Hence

$$\bar{\lambda}(N_z) = \langle Q_1^z, Q_2 \rangle.$$

Assume that Q_2 has been isotoped into general position with respect to $Q_1^{\pm z}$ for all primitive $z \in H_1(\partial U; \mathbf{Z})$, $\langle z, l \rangle \neq 0$.

Let h be a left-handed Dehn twist along β. As usual, let h also denote the induced maps on homology groups, homotopy groups and representation spaces. One can easily verify that $h(a) = a + b$, and hence that

$$h(Q_1^{a+b}) = Q_1^a.$$

Following the pattern of previous lemmas, we will, to the extent possible, represent $h(Q_1^{a+b}) - Q_1^{a+b}$ as a "difference cycle". This will be more delicate than in previous cases, since $h|_{T_1} \neq \text{id}$.

Let F' be F cut along β. Then $\pi_1(F)$ is generated by $\pi_1(F')$ and α. If base points are chosen properly, the action of h on $R^\sharp$ is given by

$$\begin{aligned} h(\rho)x &= \rho(x)\,, & x \in \pi_1(F') \\ h(\rho)\alpha &= \rho(\beta)\rho(\alpha)\,. \end{aligned}$$

Let $f : SU(2) \times [0,1) \to [0,1]$ be a smooth, Ad-invariant function such that $f(X,t) = 1$ if $\operatorname{tr}(X) \geq -1 - t$, $f(-1,t) = 0$ for all t, and $f(\,\cdot\,,t)$ is monotonic (on $SU(2)/\text{Ad}$) for all t. For $t \in [0,1)$, define $h_t^\sharp : R^\sharp \to R^\sharp$ by

$$\begin{aligned} h_t^\sharp(\rho)x &= \rho(x)\,, & x \in \pi_1(F') \\ h_t^\sharp(\rho)\alpha &= \rho(\beta)^{f(\rho(\beta),t)t}\rho(\alpha)\,. \end{aligned}$$

$\{h_t^\sharp\}$ descends to an isotopy $\{h_t\}$ of R.

Arguing as in the proof of (3.2), we see that as $t \to 1$, $h_t(Q_1^{a+b}) \to Q_1^a \cup Q_1^{-b}$. For notational convenience, let $h_1(Q_1^{a+b}) \stackrel{\text{def}}{=} Q_1^a \cup Q_1^{-b}$. Note also that $h_0 = \text{id}$ and that as $t \to 1$, $h_t \to h$ on $\{[\rho] \in R \,|\, \rho(\beta) \neq -1\}$. Hence Q_1^{-b} will play the role of the difference cycle. To make this precise will require closer examination of the singularities of R.

First we show that $\{h_t\}$ is a not-so-special isotopy (cf (2.16)). The conditions on a, b and l guarantee that $h_t(Q_1^{a+b})$ is transverse to Q_2 at P. By Theorem 4.7 of [G3], h_t is a symplectic map of R to itself, and so h_t acts symplectically on ν. All that remains to be shown is that $h_t(\widetilde{T_1})$ is transverse to $\widetilde{T_2}$.

Pick curves κ_i, $1 \leq i \leq g-1$, so that $\{\delta_i, \kappa_i, \alpha, \beta\}$ is a standard symplectic set of curves on F. Let

$$\begin{aligned} X &\stackrel{\text{def}}{=} \{\rho \in \widetilde{S} \,|\, \rho(\delta_i) = 1, 1 \leq i \leq g-1\} \cong (S_0^1)^{g+1} \\ V &\stackrel{\text{def}}{=} \{\rho \in \widetilde{S} \,|\, \rho(\delta_i) = \rho(\kappa_i) = 1, 1 \leq i \leq g-1\} \cong (S_0^1)^2. \end{aligned}$$

Let $\pi : X \to V$ be the projection (with respect to the coordinates induced from $\{\delta_i, \kappa_i, \alpha, \beta\}$). Use $A \stackrel{\text{def}}{=} \rho(\alpha)$ and $B \stackrel{\text{def}}{=} \rho(\beta)^{-1}$ as coordinates on V. Then

$$\begin{aligned}
\pi(\widetilde{T_1}^a) &= \widetilde{T_1}^a \cap V = \{(A,B) \in V \mid A = 1\} \\
\pi(\widetilde{T_1}^{\pm b}) &= \widetilde{T_1}^{\pm b} \cap V = \{(A,B) \in V \mid B = \pm 1\} \\
\pi(\widetilde{T_1}^{a+b}) &= \widetilde{T_1}^{a+b} \cap V = \{(A,B) \in V \mid A - B = 1\} \\
\pi(\widetilde{T_2} \cap X) &= \{(A,B) \in V \mid uA + vB = 1\}, \qquad (3.37)
\end{aligned}$$

for some integers $u, v > 0$. (Note that we are writing group composition in S_0^1 additively.) $\pi : \widetilde{T_2} \cap X \to \pi(\widetilde{T_2} \cap X)$ is a covering. Let n be its degree. (If we ignore the fact that we are dealing with tori rather than vector spaces, then $\pi(\widetilde{T_2} \cap X)$ is the symplectic reduction of the lagrangian torus $\widetilde{T_2}$ with respect to the coisotropic torus X (see [We]).) Note that

$$|H_1(N_a; \mathbb{Z})| = |\widetilde{T_1}^a \cap \widetilde{T_2}| = nv \qquad (3.38)$$

$$|H_1(N_b; \mathbb{Z})| = |\widetilde{T_1}^b \cap \widetilde{T_2}| = nu \qquad (3.39)$$

$h_t(\widetilde{T_1}^{a+b}) \subset X$ for all t, and $\pi(h_t(\widetilde{T_1}^{a+b})) = h_t(\widetilde{T_1}^{a+b}) \cap V$. Thus $h_t(\widetilde{T_1}^{a+b})$ is transverse to $\widetilde{T_2}$ if and only if $h_t(\widetilde{T_1}^{a+b}) \cap V$ is transverse to $\pi(\widetilde{T_2} \cap X)$. This follows from the monotonicity of the function f used in defining h_t. This completes the proof that h_t is a not-so-special isotopy.

Let $\mathcal{W}_0$ be a wiring of Q_1^{a+b} and Q_2. Extend this to a continuous, 1-parameter family $\mathcal{W}_t \stackrel{\text{def}}{=} h_t(\mathcal{W}_0)$ of wirings of $h_t(Q_1^{a+b})$ and Q_2, $0 \le t < 1$. By (2.16),

$$A_1(h_t(Q_1^{a+b}), Q_2, \mathcal{W}_t) = A_1(Q_1^{a+b}, Q_2, \mathcal{W}_0) \qquad (3.40)$$

for all $0 \le t < 1$. Our next goal is to make sense of (3.40) when $t = 1$.

Recall that as $t \to 1$, $h_t(Q_1^{a+b})$ tends to $h_1(Q_1^{a+b}) = Q_1^a \cup Q_1^{-b}$. The paths and surfaces of $\mathcal{W}_t$ can be chosen so that they have limits as $t \to 1$ and so that the $\widetilde{T_1}$ paths consist of an initial segment which lies in $\pi^{-1}(1)$ and a final segment which projects 1 to 1 into V and on which $\rho(\delta_i)$ is constant, $1 \le i \le g-1$ (see Figure 3.15). Define $\mathcal{W}_1$ to be the limit of $\mathcal{W}_t$ as $t \to 1$. Then $\mathcal{W}_1$ restricts to wirings $\mathcal{W}_1^a$ of Q_1^a and Q_2 and $\mathcal{W}_1^{-b}$ of Q_1^{-b} and Q_2. (If $p \in \widetilde{T_1}^a \cap \widetilde{T_1}^{-b} \cap \widetilde{T_2}$, then there are two points of $h_t(\widetilde{T_1}^{a+b}) \cap \widetilde{T_2}$ which tend to p, and correspondingly two configurations of arcs and surface in $\mathcal{W}_1$. Arbitrarily assign one of these to $\mathcal{W}_1^a$ and the other to $\mathcal{W}_1^{-b}$.)

We will now define $A_1(Q_1^{-b}, Q_2, \mathcal{W}_1^{-b})$ in such a way that

$$(3.41)\quad A_1(Q_1^a, Q_2, \mathcal{W}_1^a) + A_1(Q_1^{-b}, Q_2, \mathcal{W}_1^{-b}) = A_1(Q_1^{a+b}, Q_2, \mathcal{W}_0).$$

Let $p \in T_1^{-b} \cap T_2$ and let $p', p'' \in \widetilde{T_1}^{-b} \cap \widetilde{T_2}$ be the inverse images of p in $\widetilde{S}$. (If $p \in P$, then $p' = p''$.) Let $\alpha_1' \in \mathcal{W}_1^{-b}$ be the arc from 1 to p' in $\widetilde{T_1}^a \cup \widetilde{T_1}^{-b}$. α_1' consists of an initial segment in $\widetilde{T_1}^a$ and a final segment in $\widetilde{T_1}^{-b}$. (If $p' \in \widetilde{T_1}^a \cap \widetilde{T_1}^{-b}$, then the final segment consists only of p'.) The normal bundles of $\widetilde{T_1}^a$ and $\widetilde{T_1}^{-b}$ determine sections of $\det^1(\nu)$ over these two segments, but the sections do not coincide in the middle. Patch them together in the middle so that the relative homotopy class of the resulting section agrees with the relative homotopy class of the limit as $t \to 1$ of the corresponding sections over the appropriate arcs in $\mathcal{W}_t$. Define a section of $\det^1(\nu)$ over α_1'' similarly. This done, we can now proceed as in (2.A) to define the trivializations $\Phi_\pm$. Now define $I_1(p, \mathcal{W}_1^{-b})$ by (2.3) and $A_1(Q_1^{-b}, Q_2, \mathcal{W}_1^{-b})$ by (2.4).

We now establish (3.41). Since Q_2 is in general position with respect to $h_1(Q_1^{a+b})$, Q_2 is in general position with respect to $h_t(Q_1^{a+b})$ for all t sufficiently close to 1. Hence

$$(3.42)\quad \sum_{p \in h_t(Q_1^{a+b})^- \cap Q_2^-} \operatorname{sign}(p) = \sum_{p \in (Q_1^a)^- \cap Q_2^-} \operatorname{sign}(p) + \sum_{p \in (Q_1^{-b})^- \cap Q_2^-} \operatorname{sign}(p)$$

for t near 1. If it can be shown that

$$(3.43)\quad \sum_{p \in h_t(T_1^{a+b}) \cap T_2} I_1(p, \mathcal{W}_t) = \sum_{p \in T_1^a \cap T_2} I_1(p, \mathcal{W}_1^a) + \sum_{p \in T_1^{-b} \cap T_2} I_1(p, \mathcal{W}_1^{-b})$$

for t near 1, then (3.41) will follow from (3.42), (3.43), (3.40) and (2.4).

Let $p' \in h_t(\widetilde{T_1}^{a+b}) \cap \widetilde{T_2} \subset \widetilde{S}$ and let $p \in S$ be the corresponding point downstairs. Define

$$\widetilde{I}_1(p', \mathcal{W}_t) = \begin{cases} \frac{1}{2} I_1(p, \mathcal{W}_t), & p \notin P \\ I_1(p, \mathcal{W}_t), & p \in P. \end{cases}$$

Make similar definitions for $p' \in \widetilde{T_1}^a \cap \widetilde{T_2}$ and $p' \in \widetilde{T_1}^{-b} \cap \widetilde{T_2}$. Then (3.43) is equivalent to

$$(3.44)\quad \sum_{p' \in h_t(\widetilde{T_1}^{a+b}) \cap \widetilde{T_2}} \widetilde{I}_1(p', \mathcal{W}_t) = \sum_{p' \in \widetilde{T_1}^a \cap \widetilde{T_2}} \widetilde{I}_1(p', \mathcal{W}_1^a) + \sum_{p' \in \widetilde{T_1}^{-b} \cap \widetilde{T_2}} \widetilde{I}_1(p', \mathcal{W}_1^{-b})$$

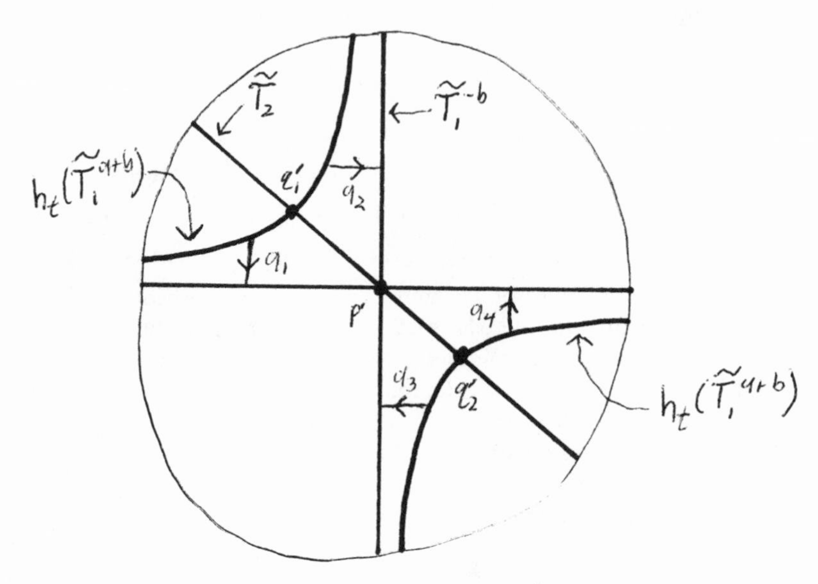

Figure 3.13: A neighborhood of $p' \in \widetilde{T_1}^a \cap \widetilde{T_1}^{-b} \cap \widetilde{T_2}$, projected onto V.

Let $p' \in (\widetilde{T_1}^{-b} \cap \widetilde{T_2}) \setminus \widetilde{T_1}^a$ and let q' be the corresponding point in $h_t(\widetilde{T_1}^{a+b}) \cap \widetilde{T_2}$. It is clear from the definition of $I_1(p, \mathcal{W}_1^{-b})$ that

$$\widetilde{I}_1(p', \mathcal{W}_1^{-b}) = \widetilde{I}_1(q', \mathcal{W}_t)$$

for t near 1. If $p' \in (\widetilde{T_1}^a \cap \widetilde{T_2}) \setminus \widetilde{T_1}^{-b}$, the corresponding statement is even clearer. Now let $p' \in \widetilde{T_1}^a \cap \widetilde{T_1}^{-b} \cap \widetilde{T_2}$. In this case there are two points of $h_t(\widetilde{T_1}^{a+b}) \cap \widetilde{T_2}$, q_1' and q_2', which tend to p' as $t \to 1$ (see Figure 3.13). (3.44) will be established if we can show that

$$\widetilde{I}_1(p', \mathcal{W}_1^a) + \widetilde{I}_1(p', \mathcal{W}_1^{-b}) = \widetilde{I}_1(q_1', \mathcal{W}_t) + \widetilde{I}_1(q_2', \mathcal{W}_t) \tag{3.45}$$

for t near 1.

Let $U \subset \widetilde{S}$ be a neighborhood of p' containing q_1' and q_2'. Let

$$\Psi : \det^1(\nu)|_U \to S^1$$

be a trivialization. Let the arcs a_1, a_2, a_3 and $a_4 \subset U$ be as shown in Figure 3.13. $h_t(\eta_1^{a+b})$, for various t near 1, determines a section of $\det^1(\nu)$ over a_i, $1 \le i \le 4$. Thus all of the subspaces pictured in Figure 3.13 have

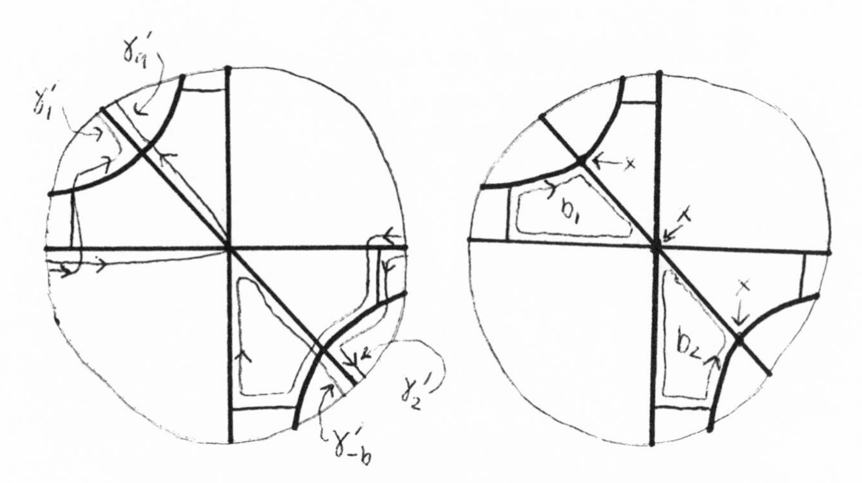

Figure 3.14: Non-coinciding parts of the homotoped γ's.

sections of $\det^1(\nu)$ defined over them. These sections do not agree at p', q'_1 and q'_2, but they can be patched together using the paths $P_\pm$ in $\mathcal{L}(\nu)$ (see (2.A)).

Without loss of generality, the limit of the part of $\mathcal{W}_t$ associated to q'_1 is in $\mathcal{W}_1^a$ and the limit of the part of $\mathcal{W}_t$ associated to q'_2 is in $\mathcal{W}_1^{-b}$. Let γ'_1, γ'_2, γ'_a and γ'_{-b} be the loops in $\mathcal{W}_t$, $\mathcal{W}_t$, $\mathcal{W}_1^a$ and $\mathcal{W}_1^{-b}$, respectively, through q'_1, q'_2, p' and p', respectively. These loops can be homotoped, together with their sections of $\det^1(\nu)$, so that γ'_1 (and its section) coincides with γ'_a and γ'_2 coincides with γ'_{-b}, except in U, where they differ as shown in Figure 3.14. We may assume that the sections over the (homotoped) γ's in Figure 3.14 agree with the sections defined in the previous paragraph.

Let b_1 $[b_2]$ be the loop in U representing $\gamma'_1 - \gamma'_a$ $[\gamma'_2 - \gamma'_{-b}]$, as shown in Figure 3.14. Using the paths $P_\pm$ (over the points marked 'x' in the figure), we get sections of $\det^1(\nu)$ over b_1 and b_2. Let $e_1^\pm$ and $e_2^\pm$ denote the degree of these sections with respect to the trivialization Ψ. Let $\Phi'_{1\pm}$, $\Phi'_{2\pm}$, $\Phi'_{a\pm}$ and $\Phi'_{-b\pm}$ be the sections of $\det^1(\nu)$ over γ'_1, γ'_2, γ'_a and γ'_{-b}. Then we have

$$\begin{aligned} \Phi'_{1\pm} - \Phi'_{a\pm} &= e_1^\pm \\ \Phi'_{2\pm} - \Phi'_{-b\pm} &= e_2^\pm . \end{aligned}$$

(The trivialization Ψ of $\det^1(\nu)$ over U is used to compare the sections.) Therefore (3.45) follows from

$$e_1^+ + e_1^- + e_2^+ + e_2^- = 0. \tag{3.46}$$

This will be proved by a symmetry argument.

Note that $e_i^{\pm}$ depends only on the intersections with U of $h_t(\widetilde{T_1}^{a+b})$, $\widetilde{T_1}^{a}$, $\widetilde{T_1}^{-b}$ and $\widetilde{T_2}$, and their normal bundles. Note also that it does not change $e_i^{\pm}$ to vary $\widetilde{T_2}$ and its normal bundle, so long as they are kept in general position with respect to $h_t(\widetilde{T_1}^{a+b})$, $\widetilde{T_1}^{a}$, $\widetilde{T_1}^{-b}$ and their normal bundles.

Let $H^1 \stackrel{\text{def}}{=} H_1(F;\mathbf{R})$. It follows from (1.13) that $\nu|_U$ can be identified with a subset of $(H^1 \otimes \mathbf{h}) \times (H^1 \otimes \mathbf{h}^{\perp})$. (If $p' \notin P$, then p' will not correspond to $0 \in H^1 \otimes \mathbf{h}$.) Let $\bar{\alpha}$, $\bar{\beta}$, $\bar{\delta}_i$, $\bar{\kappa}_i$ be the coordinates on H^1 induced by the basis α, β, δ_i, κ_i of H^1. So, for example, $\widetilde{T_1}^{-b} \cap U$ corresponds to $\{(\bar{\alpha}, \bar{\beta}, \bar{\delta}_i, \bar{\kappa}_i) \in H^1 \,|\, \bar{\alpha} = 0, \bar{\delta}_i = 0, 1 \le i \le g-1\} \otimes \mathbf{h}$.

Consider the involution σ of H^1 given by

$$\begin{aligned} \bar{\alpha} &\mapsto \bar{\beta} \\ \bar{\beta} &\mapsto \bar{\alpha} \\ \bar{\delta}_i &\mapsto \bar{\delta}_i \\ \bar{\kappa}_i &\mapsto -\bar{\kappa}_i \end{aligned}$$

σ preserves the cup product pairing on H^1 up to sign, and so induces an involution (also denoted σ) of $\nu|_U$. σ interchanges $\widetilde{T_1}^{a} \cap U$ and $\widetilde{T_1}^{-b} \cap U$ (and their normal bundles). $\widetilde{T_2} \cap U$ and its normal bundle can be isotoped (keeping things in general position) so that they are preserved by σ. h_t can be chosen so that $h_t(\widetilde{T_1}^{a+b}) \cap U$ and its normal bundle are preserved by σ. σ interchanges the loops b_1 and b_2. Since $\sigma^*\omega = -\omega$, σ interchanges the paths $\det^1(P_+)$ and $\det^1(P_-)$ (in the fibers of $\det^1(\nu)$ over p', q_1' and q_2'), and also reverses the orientation of $\det^1(\nu)|_U$. Therefore

$$e_1^+ + e_1^- = -(e_2^+ + e_2^-),$$

This completes the proof of (3.41).

Our next goal is to show that

$$\begin{aligned} (3.47)\ A_2(Q_1^{a+b}, Q_2, \mathcal{W}_0) &= A_2(Q_1^{a}, Q_2, \mathcal{W}_1^{a}) + A_2(Q_1^{-b}, Q_2, \mathcal{W}_1^{-b}) \\ &\quad + \frac{1}{6}(|H_1(N_a;\mathbf{Z})| - |H_1(N_b;\mathbf{Z})|) + \frac{1}{2}|H_1(N_b;\mathbf{Z})|. \end{aligned}$$

This is just a matter of computing the change in $\int_E \omega$ for various surfaces E in $\mathcal{W}_0$ and $\mathcal{W}_1$. We will do this for the case where u is odd and v is even, the other three cases being similar.

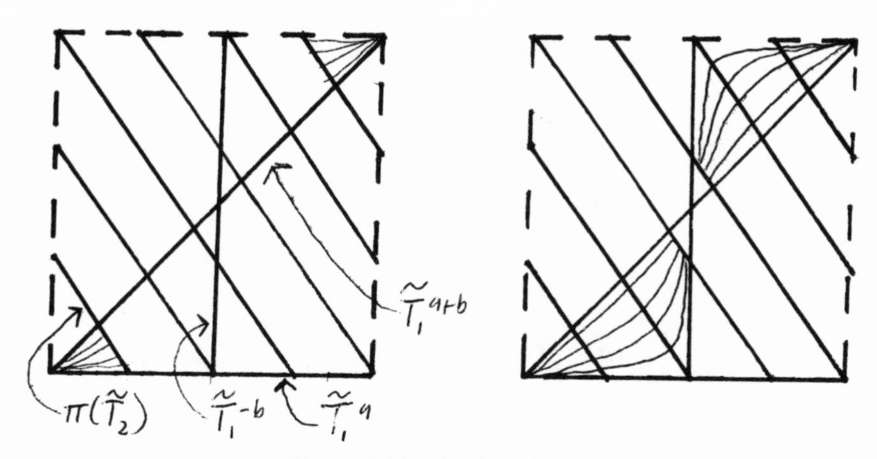

Figure 3.15: Tracks of arcs.

Note that if E is any surface in X, then

$$\int_E \omega = \int_{\pi(E)} \omega .$$

If E_0 is a surface in $\mathcal{W}_0$ and E_1 is the corresponding surface in $\mathcal{W}_1$, then we may assume that E_0 and E_1 differ by the track of the arcs α_1' and α_1'' under the isotopy $\{h_t\}$. So we must add up the ω-areas of these tracks after projecting them onto V. Two typical examples are shown in Figure 3.15.

Identify V with $I \times I/((0,x) \sim (1,x),(x,0) \sim (x,1))$. Then

$$\pi(\widetilde{T_2} \cap X) = \{(A,B) \in I \times I \mid uA + vB = k, k \in \mathbb{Z}, 0 \leq k < u+v\}.$$

Thus to each point of $\widetilde{T_1}^{a+b} \cap \widetilde{T_2}$ there is associated an integer k, $0 \leq k < u+v$. Since $\pi : \widetilde{T_2} \cap X \to \pi(\widetilde{T_2} \cap X)$ is an n-fold covering, each k is associated to n points of $\widetilde{T_1}^{a+b} \cap \widetilde{T_2}$.

Let $p' \in \widetilde{T_1}^{a+b} \cap \widetilde{T_2}$ and let k be the associated integer. Assume $0 \leq k \leq (u+v-1)/2$. Let α_1' be the path of $\mathcal{W}_0$ from 1 to p'. Let ΔE be the track of $h_t(\alpha_1')$. Keeping in mind that $\int_V \omega = -8$ (see (1.15)), it is not hard to see that

$$\int_{\pi(\Delta E)} \omega = \begin{cases} \frac{4k^2}{v(u+v)} , & 0 \leq k \leq v/2 \\ 1 - \frac{4}{u(u+v)}(\frac{u+v}{2} - k)^2 , & v/2 \leq k \leq (u+v-1)/2. \end{cases}$$

Therefore

$$A_2(Q_1^{a+b}, Q_2, \mathcal{W}_0) - A_2(Q_1^{a}, Q_2, \mathcal{W}_1^{a}) - A_2(Q_1^{-b}, Q_2, \mathcal{W}_0^{-b})$$

$$
\begin{aligned}
&= n\left(\sum_{k=0}^{v/2}\frac{4k^2}{v(u+v)}+\sum_{k=v/2+1}^{(u+v-1)/2}\left(1-\frac{4}{u(u+v)}\left(\frac{u+v}{2}-k\right)^2\right)\right)\\
&= n\left(\frac{4}{v(u+v)}\frac{v(v+2)(v+1)}{24}+\frac{u-1}{2}\right.\\
&\qquad \left.-\frac{4}{u(u+v)}\left(\frac{(u^2-1)u}{24}-\frac{(u^2-1)}{8}+\frac{(u-1)}{8}\right)\right)\\
&= n\left(\frac{1}{6}(v-u)+\frac{1}{2}u\right)\\
&= \frac{1}{6}(|H_1(N_a;\mathbb{Z})|-|H_1(N_b;\mathbb{Z})|)+\frac{1}{2}|H_1(N_b;\mathbb{Z})|.
\end{aligned}
$$

This completes the proof of (3.47).

Define

$$
\langle Q_1^{-b},Q_2\rangle \stackrel{\text{def}}{=} A_1(Q_1^{-b},Q_2,\mathcal{W}_1^{-b})+A_2(Q_1^{-b},Q_2,\mathcal{W}_1^{-b}).
$$

It follows from (3.41) and (3.47) that

$$
\begin{aligned}
r(a,b) &= \bar\lambda(N_{a+b})-\bar\lambda(N_a)-\bar\lambda(N_b)-\frac{1}{6}(|H_1(N_a;\mathbb{Z})|-|H_1(N_b;\mathbb{Z})|)\\
&= (A_1(Q_1^{a+b},Q_2,\mathcal{W}_0)+A_2(Q_1^{a+b},Q_2,\mathcal{W}_0)-A_1(Q_1^{a},Q_2,\mathcal{W}_1^{a})\\
&\quad -A_2(Q_1^{a},Q_2,\mathcal{W}_1^{a})-\langle Q_1^{b},Q_2\rangle)-\frac{1}{6}(|H_1(N_a;\mathbb{Z})|-|H_1(N_b;\mathbb{Z})|)\\
&= (\langle Q_1^{-b},Q_2\rangle-\langle Q_1^{b},Q_2\rangle)+\frac{1}{2}|H_1(N_b;\mathbb{Z})|. \qquad (3.48)
\end{aligned}
$$

Now, finally, we show that $r(a+b,b)=r(a,b)$. The transformation $a\mapsto a+b$, $b\mapsto b$ is effected by the Dehn twist h. Applying (3.48) and keeping in mind that $h(Q_1^{\pm b})=Q_1^{\pm b}$, we see that

$$
\begin{aligned}
r(a+b,b)-r(a,b) &= \langle h(Q_1^{-b}),Q_2\rangle-\langle Q_1^{-b},Q_2\rangle\\
&= A_1(Q_1^{-b},Q_2,\mathcal{V}_1^{-b})-A_1(Q_1^{-b},Q_2,\mathcal{W}_1^{-b})\\
&\qquad +A_2(Q_1^{-b},Q_2,\mathcal{V}_1^{-b})-A_2(Q_1^{-b},Q_2,\mathcal{W}_1^{-b}),
\end{aligned}
$$

where $\mathcal{V}_1^{-b}$ is induced from some wiring $\mathcal{V}_0$ of $Q_1^{a+2b}=h(Q_1^{a+b})$ and Q_2, as above.

Let $p'\in\widetilde{T_1}^{-b}\cap\widetilde{T_2}$. Let α_1' and β_1' be the paths of $\mathcal{W}_1^{-b}$ and $\mathcal{V}_1^{-b}$ connecting p' to 1 through $\widetilde{T_1}^{-b}$. We may assume that α_1' and β_1' differ

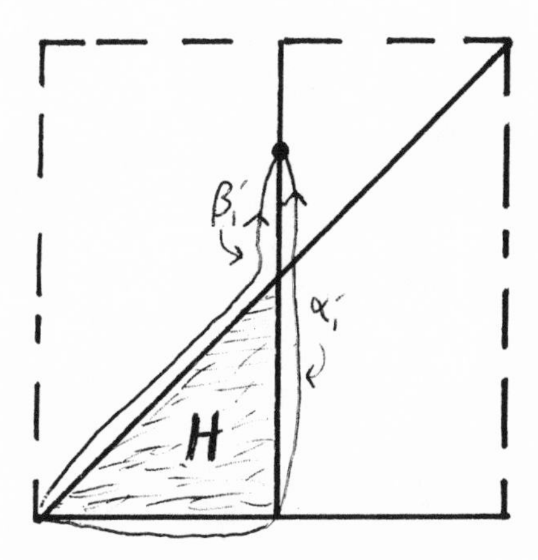

Figure 3.16: $\partial^{-1}(\alpha_1' - \beta_1')$.

by the boundary of a triangle $H \subset X$ parallel to V as shown in Figure 3.16. Recall that τ is the covering involution of $\widetilde{S}$. We may assume that $E^\dagger = E \cup H \cup \tau(H)$, where E and $E^\dagger$ are the surfaces in $\mathcal{W}_1^{-b}$ and $\mathcal{V}_1^{-b}$ associated to p' (and $p'' = \tau(p')$).

Using the path $P_\pm$ to patch things up at the trivial representation, we get a trivialization (induced by $\det^1(\eta_1)|_{\alpha_1'}$ and $\det^1(h(\eta_1))|_{\beta_1'}$) of $\det^1(\nu)|_{\partial H}$. Let $e_\pm \in \mathbb{Z}$ be difference between this trivialization and the one induced from $\det^1(\nu)|_H$. It is easy to see that

$$\widetilde{I}_1(p', \mathcal{V}_1^{-b}) - \widetilde{I}_1(p', \mathcal{W}_1^{-b}) = \frac{e_+ + e_-}{4}.$$

Since $\widetilde{T_1}^{-b} \cap \widetilde{T_2}$ consists of $|H_1(N_b; \mathbb{Z})|$ points,

$$A_1(Q_1^{-b}, Q_2, \mathcal{V}_1^{-b}) - A_1(Q_1^{-b}, Q_2, \mathcal{W}_1^{-b}) = \frac{e_+ + e_-}{4} |H_1(N_b; \mathbb{Z})|.$$

Similarly,

$$\widetilde{I}_2(p', \mathcal{V}_1^{-b}) - \widetilde{I}_2(p', \mathcal{W}_1^{-b}) = \frac{1}{2} \int_H \omega = \frac{1}{2}$$

and

$$A_2(Q_1^{-b}, Q_2, \mathcal{V}_1^{-b}) - A_2(Q_1^{-b}, Q_2, \mathcal{W}_1^{-b}) = \frac{1}{2} |H_1(N_b; \mathbb{Z})|.$$

Therefore

$$r(a + b, b) - r(a, b) = |H_1(N_b; \mathbb{Z})| \left(\frac{e_+ + e_-}{4} + \frac{1}{2}\right). \tag{3.49}$$

So what must be shown is that

$$x \stackrel{\text{def}}{=} \frac{e_+ + e_-}{4} + \frac{1}{2} = 0.$$

Note that x is independent of Q_2, and hence of K, a and b. It therefore suffices to show that x is zero for some particular knot K and homology classes a, b.

Let K be the unknot in S^3. Let $N = S^3 \setminus \text{nbd}(K)$. Let $b \in H_1(\partial(N); \mathbb{Z})$ be a meridian and let $l \in H_1(\partial(N); \mathbb{Z})$ be a longitude, oriented so that $\langle l, b \rangle = 1$. Then

$$N_b \cong S^3$$

and, for $p \in \mathbb{Z}$,

$$N_{pb+l} \cong -L_{p,1},$$

where $L_{p,1}$ denotes the $p, 1$ lens space, i.e. $p/1$ surgery on K.

Let

$$r_0 \stackrel{\text{def}}{=} r(b + l, b).$$

Then, by (3.49),

$$r(pb + l, b) = r_0 + x(p - 1)$$

for $p \geq 2$. Hence

$$\begin{aligned}
\bar{\lambda}(L_{p,1}) &= -\bar{\lambda}(N_{pb+l}) \\
&= -\bar{\lambda}(N_{(p-1)b+l}) - \bar{\lambda}(N_b) - \frac{1}{6}(|H_1(N_{(p-1)b+l}; \mathbb{Z})| - |H_1(N_b; \mathbb{Z})|) \\
&\qquad - r((p-1)b + l, b) \\
&= \bar{\lambda}(L_{p-1,1}) - \frac{1}{6}(p - 2) - r_0 - x(p - 2) \\
&= \bar{\lambda}(L_{p-1,1}) + (-\frac{1}{6} - x)p + (\frac{1}{3} - r_0 + 2x).
\end{aligned}$$

Therefore, using the fact that $\bar{\lambda}(L_{1,1}) = \bar{\lambda}(S^3) = 0$,

$$\bar{\lambda}(L_{p,1}) = (-\frac{1}{12} - \frac{x}{2})p^2 + (-r_0 + \frac{1}{4} + \frac{3x}{2})p + (r_0 - \frac{1}{6} - x).$$

But, by (3.18) and (3.21),

$$\begin{aligned}
\bar{\lambda}(L_{2,1}) &= 0 \\
\bar{\lambda}(L_{5,1}) &= \mp 1.
\end{aligned}$$

Solving for r_0 and x, we get

$$\begin{aligned}
r_0 &= 0 \\
x &= \frac{\pm 1 - 1}{6}.
\end{aligned}$$

Since $4x \in \mathbb{Z}$, we must have $\bar{\lambda}(L_{5,1}) = -1$ and $x = 0$. This completes the proof that $r(a+b,b) = r(a,b)$. The proof that $r(a,a+b) = r(a,b)$ is similar. The proof of (3.36) is complete. □

As a byproduct of the previous proof we proved that $\bar{\lambda}(L_{5,1}) = -1$. The proof of (3.21) shows that this is equivalent to

(3.50) **Lemma.** *Let K be the right-handed trefoil knot in S^3. Then $\lambda'(K) = 2$.* □

(3.51) **Lemma.** *Let $K \subset M$ be an unknot in a $\mathbb{Q}HS$. Then $r(K,a,b) = 0$ for all appropriate $a,b \in H_1(\partial(\mathrm{nbd}(K));\mathbb{Z})$ (see (3.36)).*

Proof: As usual, let $N \stackrel{\mathrm{def}}{=} M \setminus \mathrm{nbd}(K)$. Let $m, l \in H_1(\partial N;\mathbb{Z})$ be a standard meridian-longitude basis of $H_1(\partial N;\mathbb{Z})$. It suffices to show that $r(K, m+il, m+(i+1)l) = 0$ for all $i \in \mathbb{Z}$. Since $N_{pm+ql} \cong N_{pm+(ip+q)l}$ for all $i \in \mathbb{Z}$ and relatively prime $p,q \in \mathbb{Z}$, it suffices to show that $r(K,m,m+l) = 0$. But $N_m \cong N_{m+l} \cong M$, while $N_{2m+l} \cong M\#\mathbb{R}P^2$. Therefore (3.51) is implied by

(3.52) $\bar{\lambda}(M\#\mathbb{R}P^3) = 2\bar{\lambda}(M)$ *for any $\mathbb{Q}HS$ M.*

Let (W_{1a}, W_{2a}, F_a) be a Heegaard splitting for M and let (W_{1b}, W_{2b}, F_b) be the genus one Heegaard splitting of $\mathbb{R}P^3$. Let (W_1, W_2, F) be the connected sum of these two Heegaard splittings. Appropriate appropriate notation from the proof of (3.22) (e.g. R_a, R_b, Z, $q: Z \to R_a \times R_b$, etc.).

As in the proof of (3.22), we may assume that

$$Q_j = q^{-1}(Q_{aj} \times Q_{bj})$$

holds with Q_{1a} transverse to Q_{2a} in R_a. (Note that Q_{1b} is automatically transverse to Q_{2b} in R_b; no isotoping is required.) Therefore

$$Q_1 \cap Q_2 = \bigcup_{(p_a,p_b)} q^{-1}(p_a,p_b),$$

where (p_a,p_b) runs through $(Q_{1a} \cap Q_{2a}) \times (Q_{1b} \cap Q_{2b})$. $(Q_{1b} \cap Q_{2b})$ consists of two points: the trivial representation 1_b and a non-trivial representation $\sigma \in P_b$. It follows that, for each (p_a,p_b), $q^{-1}(p_a,p_b)$ consists of a single point, and that Q_1 and Q_2 are transverse. Note that $q^{-1}(p_a,p_b)$ lies in R^-,

S^- or P exactly when p_a lies in R_a^-, S_a^- or P_a, respectively. For notational simplicity, we will henceforth denote $q^{-1}(p_a, p_b)$ by (p_a, p_b).

We have

$$\bar{\lambda}(M \# \mathbb{R}P^3) = \sum_{p \in Q_{1a}^- \cap Q_{2a}^-} (\operatorname{sign}(p, 1_b) + \operatorname{sign}(p, \sigma)) + \sum_{p \in T_{1a} \cap T_{2a}} (I(p, 1_b) + I(p, \sigma)).$$

The proof will be complete if it can be shown that

$$\bar{\lambda}(M) = \sum_{p \in Q_{1a}^- \cap Q_{2a}^-} \operatorname{sign}(p, 1_b) + \sum_{p \in T_{1a} \cap T_{2a}} I(p, 1_b) \tag{3.53}$$

and that

$$\operatorname{sign}(p, 1_b) = \operatorname{sign}(p, \sigma) \tag{3.54}$$

$$I(r, 1_b) = I(r, \sigma) \tag{3.55}$$

for all $p \in Q_{1a}^- \cap Q_{2a}^-$ and $r \in T_{1a} \cap T_{2a}$.

The proof of (3.54) is similar to the proof of the corresponding fact in the proof of (3.22), and so is omitted.

$R_a \times \{1_b\}$ and $R_a \times \{\sigma\}$ can be regarded as subspaces of R. Arguing as in the proof (in (2.B)) of the stabilization invariance of λ, we see that $R_a \times \{p\}$ has a trivial normal bundle in R ($p = 1_b$ or σ). Similarly, $Q_{ja} \times \{p\}$ has a trivial normal bundle in Q_j. This implies (3.53). It also implies that a neighborhood of the triple $(S_a, T_{1a}, T_{2a}) \times \{1_b\}$ in (R, Q_1, Q_2) is diffeomorphic to a neighborhood of the triple $(S_a, T_{1a}, T_{2a}) \times \{\sigma\}$ in (R, Q_1, Q_2).

For $r \in T_{1a} \cap T_{2a}$, let $I((1_a, \sigma), (r, \sigma))$ be the same as $I(r, \sigma)$, but with $(1_a, \sigma)$ playing the role of $1 = (1_a, 1_b)$ (i.e. the paths of the wiring connect (r, σ) to $(1_a, \sigma)$). Clearly

$$I(r, 1_b) = I((1_a, \sigma), (r, \sigma)). \tag{3.56}$$

On the other hand, it is easy to see that

$$I(r, \sigma) = xI(1_a, \sigma) + I((1_a, \sigma), (r, \sigma)), \tag{3.57}$$

where $x = 1$ if $r \in P_a$ and $x = 2$ if $r \notin P_a$. Since $\{1_a\} \times (R_b, Q_{1b}, Q_{2b})$ has trivial normal bundle in (R, Q_1, Q_2),

$$\lambda(\mathbb{R}P^3) = \frac{I(1_a, 1_b) + I(1_a, \sigma)}{2}.$$

But $\lambda(\mathbb{R}P^3) = 0$ (by (3.1)) and $I(1_a, 1_b) = 0$. Hence

$$I(1_a, \sigma) = 0. \tag{3.58}$$

(3.55) now follows from (3.56), (3.57) and (3.58). □

(3.59) **Lemma.** *Let K and L be knots in a $\mathbb{Q}HS$ M such that L is null-homologous and does not link K (i.e. L bounds a Seifert surface disjoint from K). Let M_n denote $1/n$ Dehn surgery along L. Let $r(K, M_n;\ \cdot\ ,\ \cdot\ )$ denote the function introduced in (3.36) (defined on a certain subset of $H_1(\partial(\mathrm{nbd}(K)); \mathbb{Z}) \times H_1(\partial(\mathrm{nbd}(K)); \mathbb{Z})$), with K thought of as a knot in M_n. Then $r(K, M_n;\ \cdot\ ,\ \cdot\ )$ is independent of n.*

Proof: Retain all notation from the proof of (3.36).

Note that since $\mathrm{lk}\,(K, L) = 0$, surgery on L does not affect the longitude of K, and so the domain of $r(K, M_n;\ \cdot\ ,\ \cdot\ )$ is independent of n. Let $(a, b) \in H_1(\partial(\mathrm{nbd}(K)); \mathbb{Z}) \times H_1(\partial(\mathrm{nbd}(K)); \mathbb{Z})$ be in the domain (i.e. $\langle a, b\rangle = 1$ and the longitude of K is not a nonnegative linear combination of a and b).

Choose the Heegaard splitting for M so that, in addition to satisfying the conditions specified in the proof of (3.36) (i.e. W_1 contains $\mathrm{nbd}(K)$ as a boundary connected summand), L coincides with a separating curve $\gamma \subset F$, disjoint from α and β.

Let $h : F \to F$ be a left-handed Dehn twist along γ. As usual, let h also denote the various maps which it induces. Let

$$N \stackrel{\mathrm{def}}{=} \{[\rho] \in R \,|\, \rho(\gamma) = -1\}.$$

Let $D \subset N$ be the difference cycle representing $h(Q_2) - Q_2$, as constructed in (3.2) (with the roles of Q_1 and Q_2 interchanged).

By (3.48),

$$r(K, M_n; a, b) = \langle Q_1^{-b}, h^n(Q_2)\rangle - \langle Q_1^b, h^n(Q_2)\rangle + \frac{1}{2}|H_1(M_{nb}; \mathbb{Z})|,$$

where M_{nb} denotes $1/n$ surgery on L and b surgery on K. Since

$$|H_1(M_{nb}; \mathbb{Z})| = |H_1(M_{0b}; \mathbb{Z})|$$

for all n,

$$\begin{aligned} r(K, M_{n+1}; a, b) - r(K, M_n; a, b) &= \langle Q_1^{-b}, h^{n+1}(Q_2)\rangle - \langle Q_1^b, h^{n+1}(Q_2)\rangle \\ &\quad - \langle Q_1^{-b}, h^n(Q_2)\rangle + \langle Q_1^b, h^n(Q_2)\rangle \\ &= \langle Q_1^{-b}, h^n(D)\rangle - \langle Q_1^b, h^n(D)\rangle. \end{aligned}$$

So the lemma will be proved if it can be shown that

$$\langle Q_1^{-b}, h^n(D)\rangle = \langle Q_1^b, h^n(D)\rangle$$

for all $n \in \mathbb{Z}$.

Let $x_1, y_1, \ldots, x_g, y_g$ be a symplectic basis of $\pi_1(F)$ such that α represents x_1, β represents y_1, and γ represents $\prod_1^k [x_i, y_i]$ for some k. This gives rise to an identification

$$R^\sharp = \{(X_1, Y_1, \ldots, X_g, Y_g) \in (SU(2))^{2g} \mid \prod_1^g [X_i, Y_i] = 1\}.$$

Define $\tau^\sharp : R^\sharp \to R^\sharp$ by

$$\tau^\sharp(X_1, Y_1, \ldots, X_g, Y_g) = (X_1, -Y_1, X_2, Y_2, \ldots, X_g, Y_g).$$

$\tau^\sharp$ descends to a map $\tau : R \to R$.

It is not hard to see that

$$\tau(Q_1^{-b}) = Q_1^b$$

(preserving orientations). By (1.18), τ maps N to itself and

$$\tau_* : H_{3g-3}(N; \mathbb{Z}) \to H_{3g-3}(N; \mathbb{Z})$$

is the identity. In particular, $\tau(h^n(D))$ is homologous to $h^n(D)$ for all n. Therefore

$$\begin{aligned} \langle Q_1^{-b}, h^n(D)\rangle &= \langle \tau(Q_1^{-b}), \tau(h^n(D))\rangle \\ &= \langle Q_1^b, h^n(D)\rangle. \end{aligned}$$

□

§ 4

The Dehn Surgery Formula

In this section we state and prove a formula for how λ transforms under a general Dehn surgery (i.e. a Dehn surgery on a $\mathbb{Q}$HS which yields another $\mathbb{Q}$HS). Notation from previous sections is *not* retained.

Before stating the Dehn surgery formula, we need some preliminary definitions. Let N be the complement of a knot in a $\mathbb{Q}$HS. Let $\Delta_N(t^{1/2})$ be the Alexander polynomial of N, normalized so that it is symmetric under $\Delta_N(1) = 1$ and $t^{1/2} \leftrightarrow t^{-1/2}$. Define

$$\Gamma(N) \stackrel{\text{def}}{=} \frac{d^2}{dt^2}\Delta_N(1).$$

Section B contains a more detailed definition of Γ, as well as the proofs of various properties of Γ used in this section.

Let $a, b, l \in H_1(T^2;\mathbb{Z})$ be such that a and b are primitive (i.e. represented by simple closed curves), $\langle a, l\rangle \neq 0$ and $\langle b, l\rangle \neq 0$. Choose a basis x, y of $H_1(T^2;\mathbb{Z})$ such that $\langle x, y\rangle = 1$ and $l = dy$ for some $d \in \mathbb{Z}$. Define

$$(4.1)\, \tau(a, b; l) \stackrel{\text{def}}{=} -s(\langle x, a\rangle, \langle y, a\rangle) + s(\langle x, b\rangle, \langle y, b\rangle) + \frac{d^2-1}{12}\frac{\langle a, b\rangle}{\langle a, l\rangle\langle b, l\rangle},$$

where $s(q, p)$ denotes the Dedekind sum

$$\begin{aligned} s(q,p) &\stackrel{\text{def}}{=} (\operatorname{sign}(p))\sum_{k=1}^{|p|}((k/p))((kq/p)) \\ ((x)) &\stackrel{\text{def}}{=} \begin{cases} 0, & x \in \mathbb{Z} \\ x - [x] - 1/2, & \text{otherwise} \end{cases} \end{aligned}$$

Note that $\tau(a,b;l)$ depends only on a, b, l and $\langle\cdot,\cdot\rangle$, not on x or y. Section A contains the proofs of various properties of Γ used in this section.

(4.2) Theorem. *Let $N = M\setminus\text{nbd}(K)$ be the complement of a knot K in a $\mathbb{Q}HS$ M. Let $l \in H_1(\partial N;\mathbb{Z})$ be the longitude. Let $a,b \in H_1(\partial N;\mathbb{Z})$ be primitive and such that $\langle a,l\rangle \neq 0$, $\langle b,l\rangle \neq 0$. Then*

$$\lambda(N_b) = \lambda(N_a) + \tau(a,b;l) + \frac{\langle a,b\rangle}{\langle a,l\rangle\langle b,l\rangle}\Gamma(N).$$

Note: Casson proved that for K a knot in a $\mathbb{Z}$HS, $\lambda'(K) = \Gamma(N)$ (cf (3.7)). The proof of (4.2) is in the same spirit as Casson's, but is much more involved.

Proof: By (5.1), there is some invariant of $\mathbb{Q}$HS's, call it λ_c, for which (4.2) holds. ((4.2) could be proved using a result weaker than (5.1), but (5.1) needs to be proved anyway, so why bother proving a redundant weaker result?) Let

$$\delta \stackrel{\text{def}}{=} \lambda - \lambda_c$$

and

$$\overline{\delta} \stackrel{\text{def}}{=} |H_1(\,\cdot\,;\mathbb{Z})|\,\delta = \bar{\lambda} - \bar{\lambda}_c.$$

Then (4.2) is equivalent to

(4.3) *Let $N = M\setminus\text{nbd}(K)$ be the complement of a knot K in a $\mathbb{Q}HS$ M. Then $\delta(N_a)$ is independent of the primitive homology class $a \in H_1(\partial N;\mathbb{Z})$.*

Let $l = dk$, where $k \in H_1(\partial N;\mathbb{Z})$ is primitive and d is a positive integer. Choose $m \in H_1(\partial N;\mathbb{Z})$ such that $\langle m,k\rangle = 1$. For $(p,q) \in \mathbb{Z}^2$ primitive and $p \neq 0$, define

$$N(p,q) \stackrel{\text{def}}{=} N_{pm+qk}.$$

By (3.34),

$$\bar{\lambda}(N(1,q)) = \bar{\lambda}(N(1,0)) + \bar{\lambda}'(N)q$$

for some $\bar{\lambda}'(N) \in \mathbb{Q}$ and all $q \in \mathbb{Z}$. By (4.1) and elementary linear algebra,

$$\bar{\lambda}_c(N(1,q)) = \bar{\lambda}_c(N(1,0)) + |H_1(N(1,0);\mathbb{Z})|\left(\frac{\Gamma(N)}{d^2} + \frac{d^2-1}{12d^2}\right)q.$$

Therefore

(4.4)
$$\overline{\delta}(N(1,q)) = \overline{\delta}_0 + \overline{\delta}'q,$$

where

$$\begin{aligned}\overline{\delta}_0 &\stackrel{\text{def}}{=} \bar{\lambda}(N(1,0)) - \bar{\lambda}_c(N(1,0))\\ \overline{\delta}' &\stackrel{\text{def}}{=} \bar{\lambda}'(N) - |H_1(N(1,0);\mathbb{Z})|\left(\frac{\Gamma(N)}{d^2} + \frac{d^2-1}{12d^2}\right).\end{aligned}$$

Let $a = p_1 m + q_1 k$, $b = p_2 m + q_2 k \in H_1(\partial N;\mathbb{Z})$ be primitive and such that $\langle a, b\rangle = 1$ and $\langle a, l\rangle, \langle b, l\rangle > 0$ (i.e. $p_1, p_2 > 0$). Let $s = [q_1/p_1] \in \mathbb{Z}$. ($[x]$ denotes the greatest integer $\leq x$.) Then a and b are non-negative linear combinations of $m + sk$ and $m + (s+1)k$. By (3.36),

$$\begin{aligned}(4.5)\quad \bar{\lambda}(N(p_1+p_2, q_1+q_2)) &= \bar{\lambda}(N(p_1,q_1)) + \bar{\lambda}(N(p_2,q_2))\\ &\quad + \frac{1}{6}(|H_1(N(p_1,q_1);\mathbb{Z})| - |H_1(N(p_2,q_2);\mathbb{Z})|) + r(s),\end{aligned}$$

where

$$r(s) \stackrel{\text{def}}{=} r(m+sk, m+(s+1)k).$$

We now prove an analogous statement for $\bar{\lambda}_c$. Let $h \stackrel{\text{def}}{=} |\mathrm{Tor}(H_1(N;\mathbb{Z}))|$. Then

$$|H_1(N_x;\mathbb{Z})| = |\langle x, l\rangle| h$$

for all primitive $x \in H_1(\partial N;\mathbb{Z})$ not colinear with l. Let a and b be as above. Applying (4.2), we have

$$\begin{aligned}\bar{\lambda}_c(N_a) &= \langle a,l\rangle h\left(\lambda_c(N_{a+b}) + \tau(a+b,a;l) + \frac{\langle a+b,a\rangle}{\langle a+b,l\rangle\langle a,l\rangle}\Gamma(N)\right)\\ &= \frac{\langle a,l\rangle}{\langle a+b,l\rangle}\bar{\lambda}_c(N_{a+b}) + h\langle a,l\rangle\tau(a+b,a;l) + \frac{h\langle b,a\rangle}{\langle a+b,l\rangle}\Gamma(N).\end{aligned}$$

Similarly,

$$\bar{\lambda}_c(N_b) = \frac{\langle b,l\rangle}{\langle a+b,l\rangle}\bar{\lambda}_c(N_{a+b}) + h\langle b,l\rangle\tau(a+b,b;l) + \frac{h\langle a,b\rangle}{\langle a+b,l\rangle}\Gamma(N).$$

Adding these two equations and applying (A.17) yields

$$\bar{\lambda}_c(N_a) + \bar{\lambda}_c(N_b) = \bar{\lambda}_c(N_{a+b}) - \frac{h}{6}(\langle a,l\rangle - \langle b,l\rangle),$$

or

$$\begin{aligned}(4.6)\quad \bar{\lambda}_c(N(p_1+p_2, q_1+q_2)) &= \bar{\lambda}_c(N(p_1,q_1)) + \bar{\lambda}_c(N(p_2,q_2))\\ &\quad + \frac{1}{6}(|H_1(N(p_1,q_1);\mathbb{Z})| - |H_1(N(p_2,q_2);\mathbb{Z})|).\end{aligned}$$

Subtracting (4.6) from (4.5) gives

$$(4.7)\quad \overline{\delta}(N(p_1+p_2,q_1+q_2)) = \overline{\delta}(N(p_1,q_1)) + \overline{\delta}(N(p_2,q_2)) + r([\frac{q_1}{p_1}]).$$

(4.4) and repeated applications of (4.7) now give, for primitive $(p,q) \in \mathbb{Z}^2$ with $p > 0$,

$$\overline{\delta}(N(p,q)) = \overline{\delta}_0 + (\overline{\delta}_0 + r([\frac{q}{p}]))(p-1) + q\overline{\delta}'.$$

Dividing by $|H_1(N(p,q);\mathbb{Z})| = dhp$ yields

$$(4.8)\qquad \delta(N(p,q)) = \delta_0 + \frac{(p-1)r([\frac{q}{p}])}{dhp} + \frac{q\delta'}{p}.$$

So what must be shown is that $\delta' = 0$ and that $r(s) = 0$ for all $s \in \mathbb{Z}$. (It follows from (4.2), (5.1) and the fact that $\lambda(S^3) = 0$ that $\delta_0 = 0$ also.)

To proceed further, we need the following special case of (4.3).

(4.9) **Lemma.** *Let $K \subset M$ be a knot in a* $\mathbb{Q}$*HS such that*

$$[K] \in [\pi_1(M), [\pi_1(M), \pi_1(M)]].$$

Then (4.3) holds for $N = M \setminus \mathrm{nbd}(K)$. □

The proof will be given later.

Let $M = N(1,0)$ and let $K \subset M$ be the core of the surgery solid torus (so that $N = M \setminus \mathrm{nbd}(K)$). Let $G = \pi_1(M)$. Since M is a $\mathbb{Q}$HS, $H_1(M;\mathbb{Z}) = G/[G,G]$ and $[G,G]/[G,[G,G]]$ are both finite groups. Therefore there exists positive $n \in \mathbb{Z}$ such that $n[K] \in [G,[G,G]]$

Choose positive $j, p \in \mathbb{Z}$ such that p and jn are relatively prime. Let $L \subset \partial N = \partial(\mathrm{nbd}(K))$ be a simple closed curve representing the homology class $(p)m + (jn)k$. Let $A \subset \partial(N)$ be an annulus neighborhood of L. Consider the manifold M' obtained via i left-handed Dehn twists along A. Since $A \subset \partial N$, M' is equivalent to some Dehn surgery on N, namely $M' = N(1+ijnp, ij^2n^2)$. On the other hand, after an isotopy A is contained in $\partial(\mathrm{nbd}(L))$, and so is equivalent to some Dehn surgery on L. Since $[L] = jn[K]$, $[L] \in [G,[G,G]]$. Therefore, by (4.9),

$$\delta(1+ijnp, ij^2n^2) = \delta_0$$

for all i. Comparing this with (4.8), we get

$$\frac{ijnp}{dh(1+ijnp)} r([\frac{ij^2n^2}{1+ijnp}]) + \frac{ij^2n^2}{1+ijnp}\delta' = 0$$

(for $i \geq 0$).

Letting i, j and p take all possible values provides enough independent equations to prove that $\delta' = 0$ and $r(s) = 0$ for all $s \in \mathbb{Z}$. This completes the proof of (4.2), modulo that of (4.9). □

To prove (4.9), we will need the following lemma.

(4.10) **Lemma.** *Let $K^-, K^+ \subset M$ be two null-homologous knots in a $\mathbb{Q}HS$ which differ by a crossing change. (That is, K^- and K^+ coincide except in a 3-ball, where they differ as shown in Figure 3.3.) Let N^- and N^+ be the corresponding knot complements. Via the standard bases, we can identify $H_1(N^-;\mathbb{Z})$ with $H_1(N^+;\mathbb{Z})$ (see (1.G)). Then for any primitive $a \in H_1(N^-;\mathbb{Z}) = H_1(N^+;\mathbb{Z})$ not colinear with the longitude,*

$$\delta(N_a^-) = \delta(N_a^+).$$

□

The proof will be given later.

Proof of (4.9): By (4.10), it suffices to establish (4.9) for some knot in the homotopy class of K. That is, we are free to change crossings of K whenever it is convenient.

Since $[K] \in [\pi_1(M), [\pi_1(M), \pi_1(M)]]$, we may assume (after possibly changing some crossings of K) that K bounds an embedded surface $E \subset M$ which has a standard symplectic set of curves $X_1, Y_1, \ldots, X_k, Y_k \subset E$ such that each Y_i is null-homologous (i.e. $[Y_i] \in [\pi_1(M), \pi_1(M)]$. E can be thought of as a disk with bands attached, one band for each X_i or Y_i (see Figure 4.1). Let b_1 be the band associated to Y_1. Via further crossing changes, it can be arranged that the self-linking of b_1 is zero (see Figure 4.2).

Let $A \subset M$ be a curve parallel to Y_1, disjoint from E, and such that $\mathrm{lk}(Y_1, A) = 1$. Further require that A is close to Y_1, in the sense that an annulus bounded by A and Y_1 intersects E in two arcs contained in b_1 (see Figure 4.3). Let B be an unknotted curve linking Y_1 and disjoint from E, as shown in Figure 4.3.

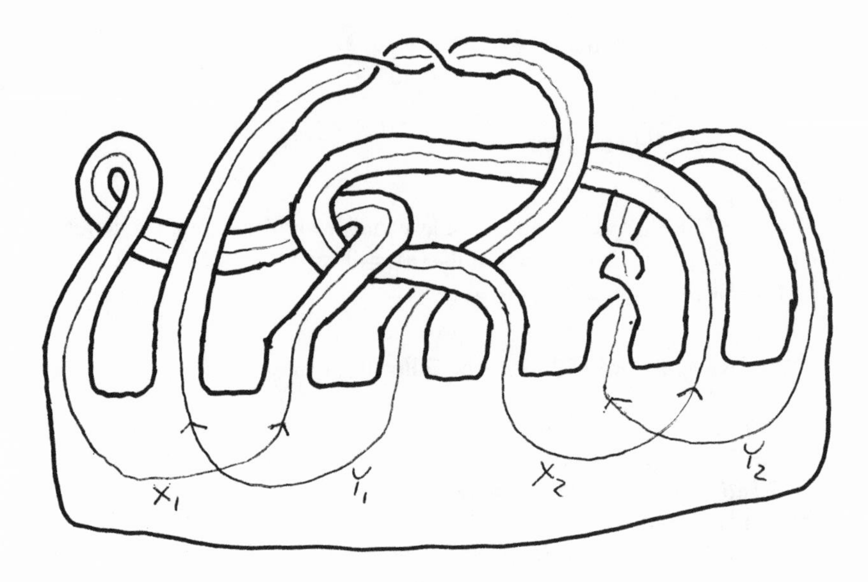

Figure 4.1: The surface E.

Figure 4.2: Changing the self linking of b_1.

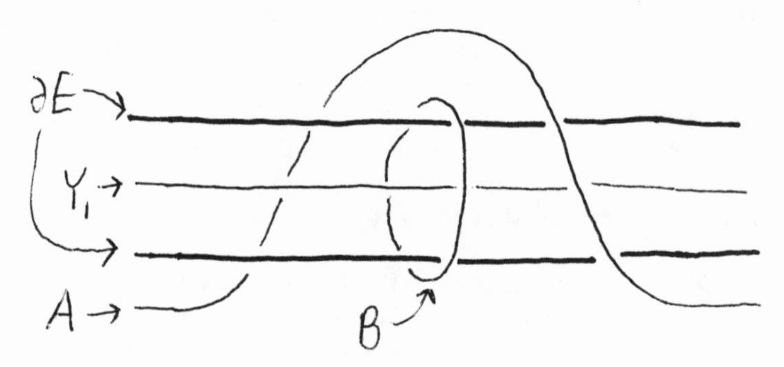

Figure 4.3: A, B, Y_1 and E.

Figure 4.4: L, surgery along which effects a crossing change.

A and B together bound a surface F which is disjoint from b_1. (This uses the fact that Y_1 is null-homologous and b_1 has self-linking number zero.) Without loss of generality, all intersections of F with E consist of bands of E passing though F. After still more crossing changes on K (i.e. passing bands through b_1), we may assume that F is disjoint from E and that A is still close to Y_1 in the above sense.

We are now in a position to apply (3.13) and (3.59). Let $M^\dagger$ denote M after $+1$ Dehn surgery on A and -1 Dehn surgery on B. Let $K^\dagger \subset M^\dagger$ denote the knot corresponding to K. By (3.13), $\lambda'(K^\dagger) = \lambda'(K)$. By (3.59), the function r (of (3.36)) also does not change. The double surgery along A and B does not change the linking numbers of any curves disjoint from F. Therefore, by (B.10), $\lambda'_c(K^\dagger) = \lambda'_c(K)$ (i.e. $\Gamma(K^\dagger) = \Gamma(K)$). So by (4.8), it suffices to prove (4.3) for $K^\dagger$.

Note, however, that in $M^\dagger$ E can be compressed along Y_1, reducing its genus by one. (The compressing disk passes once through each of the surgery solid tori of A and B.) Repeating the above procedure reduces our task to proving (4.3) for a genus zero knot (i.e. an unknot) in a $\mathbb{Q}$HS. This follows from (4.8), (3.51), and the obvious facts that $\lambda'(U) = \lambda'_c(U) = 0$ for an unknot U. □

Proof of (4.10): Let L be an unknot surrounding the crossing (see Figure 4.4), so that K_+ is equivalent to K_- in the manifold obtained by $+1$ surgery on L. (This manifold is, of course, diffeomorphic to M.) Note that $\mathrm{lk}(K_-, L) = 0$, so, by (3.59), surgery on L does not change the function r (of (3.36)). Therefore, by (4.8), what must be shown is that

$$\begin{aligned} 0 &= \delta'(K_+) - \delta'(K_-) \\ &= \lambda'(K_+) - \lambda'(K_-) - \lambda'_c(K_+) + \lambda'_c(K_-). \end{aligned}$$

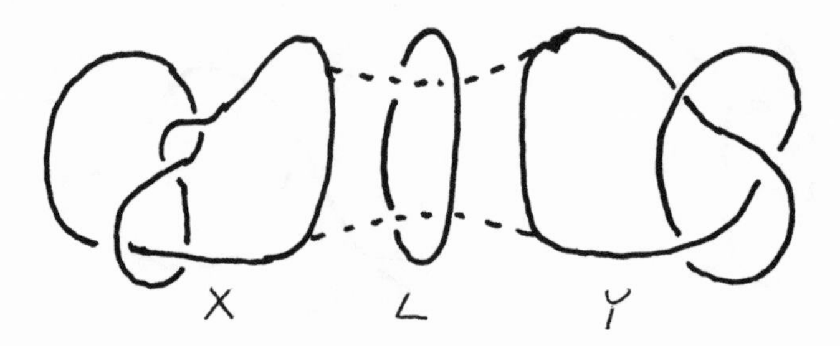

Figure 4.5: Smoothing the crossing, yielding X and Y.

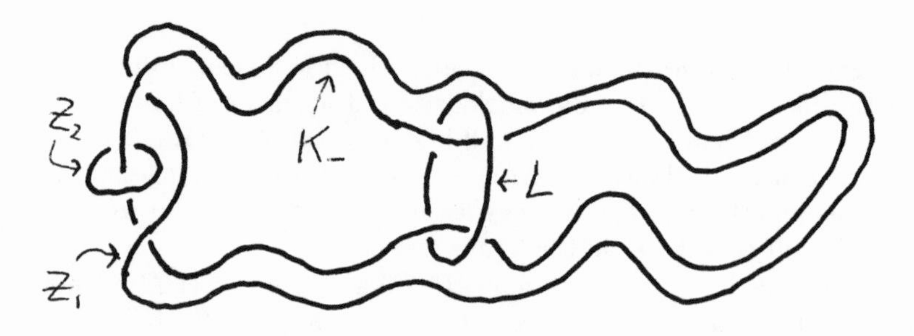

Figure 4.6: Z_1, Z_2, K_- and L

Let $X, Y \subset M$ be the curves obtained from "smoothing" the crossing, as shown in Figure 4.5. Note that K_- is a band connected sum of X and Y.

Using (B.10) it is an elementary exercise, left to the reader, to show that

$$\lambda_c'(K_+) - \lambda_c'(K_-) = \Gamma(K_+) - \Gamma(K_-) = -2\,\mathrm{lk}\,(X, Y).$$

(Hint: Select a Seifert surface for K_- which is disjoint from L, and select as part of its symplectic basis a curve which passes through L once and a curve which is parallel to X (or Y).) So the proof will be complete if it can be shown that

$$\lambda''(K_-, L) = \lambda'(K_+) - \lambda'(K_-) = -2\mathrm{lk}\,(X, Y). \tag{4.11}$$

Let D be an unknotting disk for L which intersects K_- in two points. Let Z_1 be parallel to K_-, have linking number 1 with K_-, and be disjoint from D (see Figure 4.6). Let Z_2 be an unknot linking K_- once, as shown in Figure 4.6. Z_1 and Z_2 together bound a surface F which is disjoint from K_- and D. (F can be constructed by taking a Seifert surface for K_- which

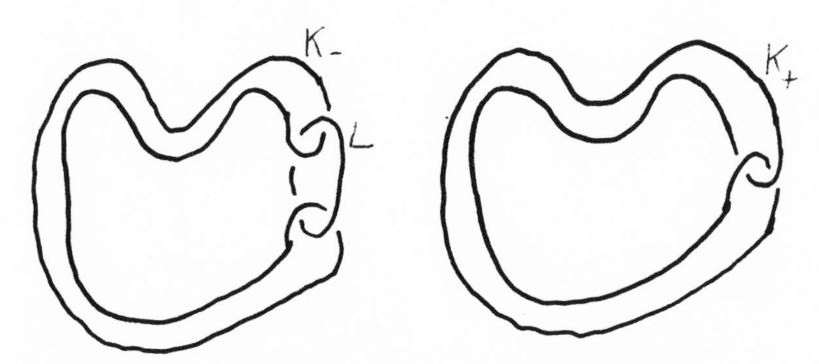

Figure 4.7: K_- and K_+ in $M^\dagger$

is disjoint from L, isotoping it so that it is disjoint from D and intersects K_- once transversely, and removing a neighborhood of the intersection point.) L bounds a genus one surface consisting of D with a neighborhood of $D \cap K_-$ excised and a tube surrounding half of K_-. This surface is disjoint from K_- and F.

We now apply (3.13). Let $M^\dagger$ be the manifold obtained from $+1$ surgery on Z_1 and -1 surgery on Z_2. By (3.13), this surgery does not change $\lambda'(L)$ (which is clearly zero anyway, since L is an unknot in both M and $M^\dagger$). The same is true (except for the parenthetical remark) if $+1$ surgery is done on K_- in both M and $M^\dagger$. Therefore the surgery on Z_1 and Z_2 does not change $\lambda''(K_-, L)$ (cf. (3.9)). It is easy to check that the surgery also does not change $\mathrm{lk}(X, Y)$. Thus it suffices to show that (4.11) holds in $M^\dagger$.

Note that in $M^\dagger$, X (say) can be isotoped through the surgery solid tori of Z_2 and Z_1 so that it ends up parallel to Y. It follows that in $M^\dagger$ K_- is unknotted and K_+ is a double, with negative clasp, of Y (see Figure 4.7). Hence $\lambda''(K_-, L) = \lambda'(K_+)$.

To each doubled knot K there is associated the self linking number of the core of K. In the above case this linking number is equal to $\mathrm{lk}(X, Y)$. So what must be shown is

(4.12) *Let $K \subset M$ be a doubled knot with negative clasp. Let $\mathrm{lk}(K) \in \mathbb{Q}$ denote the self linking number of the core of K. Then $\lambda'(K) = -2\mathrm{lk}(K)$.*

To prove this we will need

(4.13) **Lemma.** *Let V be a solid torus in a 3-manifold M. Let $(L, K) \subset V \subset M$ be the link shown in Figure 4.8. Then*

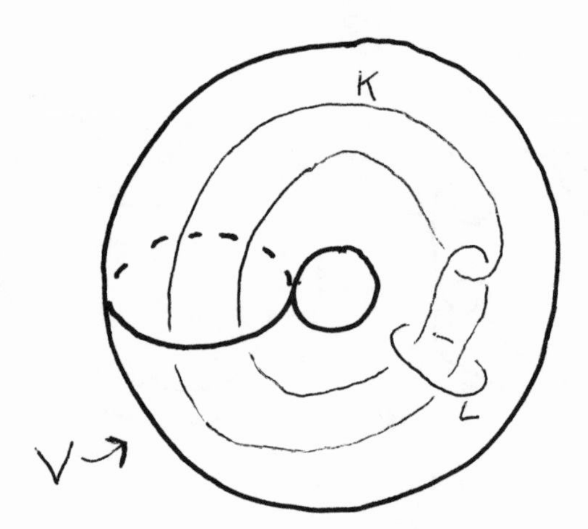

Figure 4.8: A two component link.

$$\lambda''(K, L) = 2.$$

Proof: Let $(Z_1, Z_2) \subset V$ be the link shown in Figure 4.9. L and the pair (Z_1, Z_2) bound disjoint Seifert surfaces in the complement of K (see Figure 4.10). Therefore, by (3.13), performing $(1, -1)$-surgery on (Z_1, Z_2) does not change $\lambda''(K, L)$. After this surgery is performed, we can slide K over Z_1, obtaining a two component link contained in a 3-ball, as shown in Figure 4.11. It now follows from Figure 4.12, (3.22) and (3.50) that $\lambda''(K, L) = 2$. □

The strategy for proving (4.12) is to reduce to the case of an untwisted double ($\text{lk}(K) = 0$) and apply (3.26).

K bounds a genus one Seifert surface consisting of an annulus B parallel to the core of K and an "overpass" with one twist (see Figure 4.13). Let V be a solid torus neighborhood of the core of K such that K is contained in the interior of V. Let β be the core of the annulus B. Without loss of generality, ∂V contains an annulus neighborhood of β in B. Let $\mu \subset \partial V$ be a meridian of V. Let $m = [\mu] \in H_1(\partial V; \mathbb{Z})$ and $b = [\beta] \in H_1(\partial V; \mathbb{Z})$. Choose orientations of μ and β so that $\langle m, b\rangle = 1$. Let $l = pm + qb \in H_1(\partial V; \mathbb{Z})$ be a longitude of $\overline{M \setminus V}$. Clearly,

$$\text{lk}(K) = -p/q.$$

Consider the effect of a left-handed Dehn twist along μ. Let $M^\dagger$ denote the new $\mathbb{Q}$HS so obtained, and let $K^\dagger \subset M^\dagger$ correspond to K. ($M^\dagger$ is, of

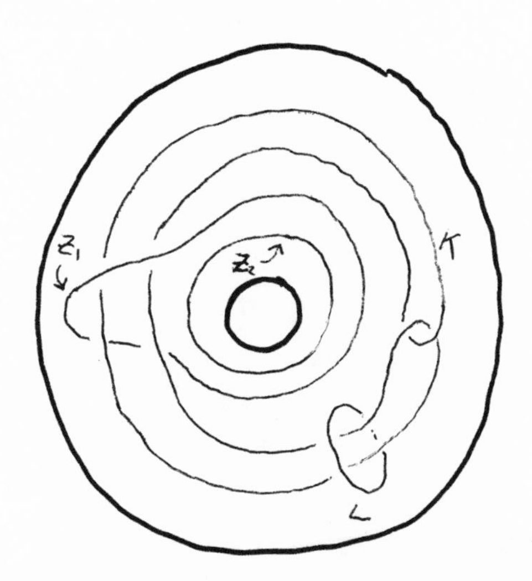

Figure 4.9: Another two component link.

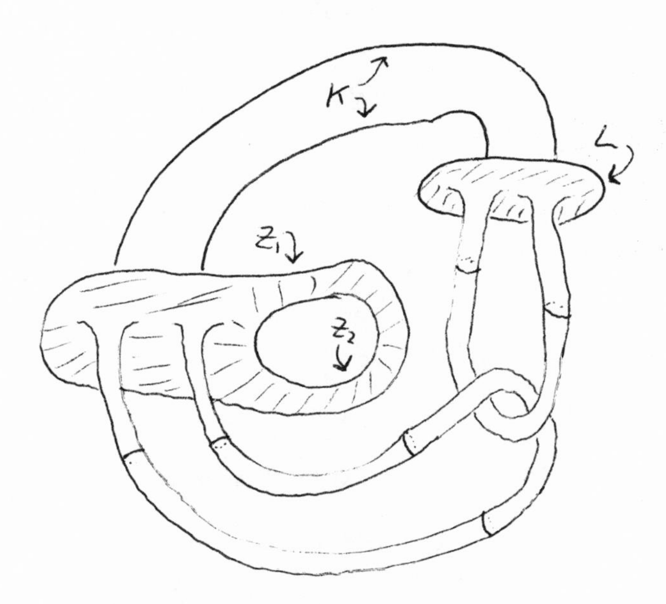

Figure 4.10: Seifert surfaces for L and (Z_1, Z_2).

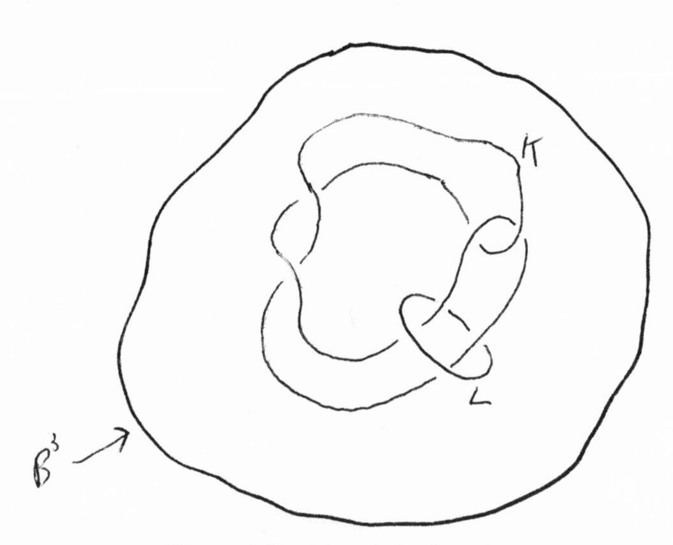

Figure 4.11: K and L after the surgery.

course, diffeomorphic to M.) $M^\dagger$ can also be thought of as the result of $+1$ Dehn surgery on μ. By (4.13),

$$\begin{aligned} \lambda'(K^\dagger) - \lambda'(K) &= \lambda''(K,\mu) \\ &= 2. \end{aligned} \tag{4.14}$$

Let $l^\dagger = p^\dagger m + q^\dagger b$ be the longitude of $\overline{M^\dagger \setminus V}$. It is easy to see that

$$\begin{cases} p^\dagger = p + q \\ q^\dagger = q. \end{cases} \tag{4.15}$$

Hence $\mathrm{lk}\,(K^\dagger) = -(p+q)/q$, and so

$$-\,2\mathrm{lk}\,(K^\dagger) + 2\mathrm{lk}\,(K) = 2. \tag{4.16}$$

It follows from (4.14) and (4.16) that (4.12) holds for K if and only if it holds for $K^\dagger$.

Now consider the effect of a right-handed Dehn twist along β. Let $M^\dagger$ denote the resulting 3-manifold, $K^\dagger \subset M^\dagger$ denote the knot corresponding to K, and $l^\dagger = p^\dagger m + q^\dagger b$ be the longitude of $\overline{M^\dagger \setminus V}$. It is easy to see that

$$\begin{cases} p^\dagger = p \\ q^\dagger = q - p. \end{cases} \tag{4.17}$$

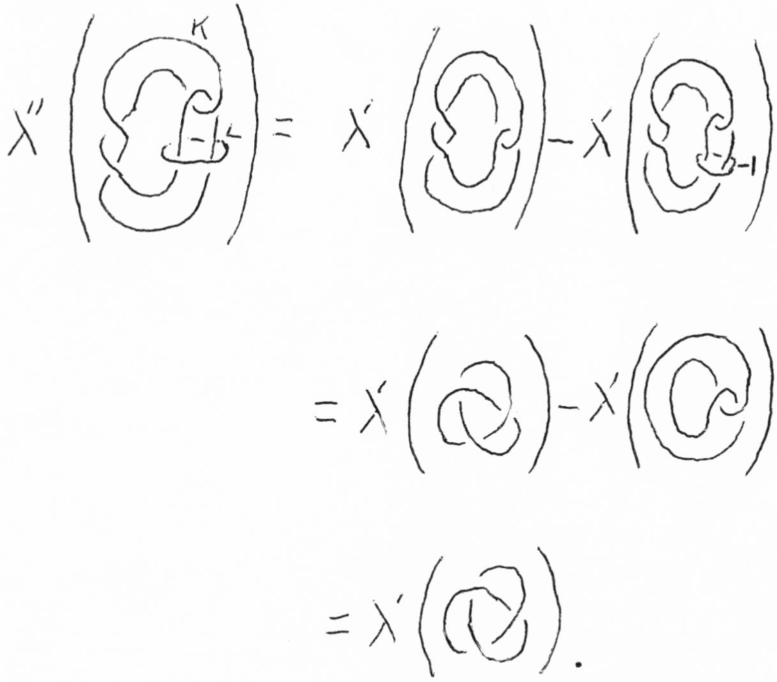

Figure 4.12: The end of the proof.

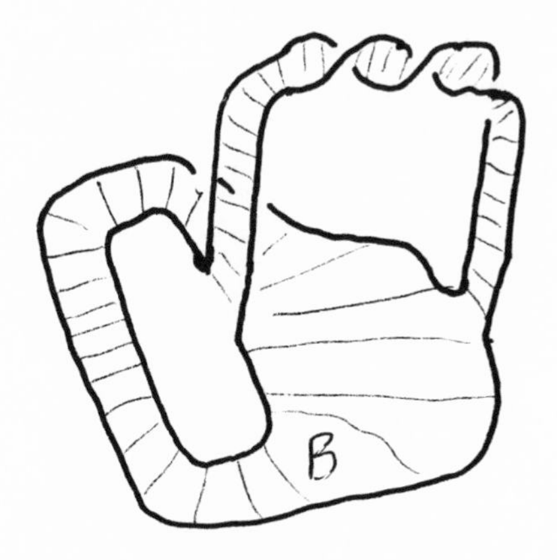

Figure 4.13: The Seifert surface for K

Assume that $q^\dagger$ has the same sign as q. In this case we can apply (3.16) and conclude

$$\bar{\lambda}'(K^\dagger) = \bar{\lambda}'(K).$$

Let

$$h = |\mathrm{Tor}(H_1(M \setminus V; \mathbf{Z}))|.$$

Then $|H_1(M;\mathbf{Z})| = h\langle m, l\rangle$ and $|H_1(M^\dagger;\mathbf{Z})| = h\langle m, l^\dagger\rangle$. Therefore

$$\begin{aligned} \lambda'(K^\dagger) &= \frac{\bar{\lambda}'(K^\dagger)}{|H_1(M^\dagger;\mathbf{Z})|} \\ &= \frac{|H_1(M;\mathbf{Z})|}{|H_1(M^\dagger;\mathbf{Z})|}\frac{\bar{\lambda}'(K)}{|H_1(M;\mathbf{Z})|} \\ &= \frac{q}{q-p}\lambda'(K). \end{aligned} \tag{4.18}$$

On the other hand,

$$\begin{aligned} \mathrm{lk}\,(K^\dagger) &= -\frac{p}{q-p} \\ &= -\frac{q}{q-p}\frac{p}{q} \\ &= \frac{q}{q-p}\mathrm{lk}\,(K). \end{aligned} \tag{4.19}$$

It follows from (4.18) and (4.19) that (4.12) holds for K if and only if it hold for $K^\dagger$.

Using the transformations (4.15) and (4.17) and their inverses (with the restriction that $q^\dagger$ and q have the same sign in (4.17)), one can reduce (4.12) to the case where $p = 0$. That is, $\mathrm{lk}\,(K) = 0$ and, by (3.26), $\lambda'(K) = 0$. This completes the proof of (4.12), and also that of (4.10). □

§ 5

Combinatorial Definition of λ

In this section we show that the Dehn surgery formula of (4.2) (and the fact that $\lambda(S^3) = 0$) not only uniquely determine λ, but also can be used to give an alternative definition of λ.

For the sake of self-containedness, recall that if N is a 3-manifold, $\partial N \cong T^2$, and $a \in H_1(\partial N; \mathbb{Z})$ is primitive (represented by a simple closed curve), then N_a denotes the closed 3-manifold obtained from Dehn surgering N along a. (See also (1.G).) If N is the complement of a knot in a $\mathbb{Q}$HS, a *longitude* of N is defined to be a generator of $\ker(H_1(\partial N; \mathbb{Z}) \to H_1(N; \mathbb{Z}))$. This is unique up to sign.

All boundaries of manifolds are oriented according to the "inward normal last" convention.

For the definitions of τ and Γ, see the beginning of Section 4 or Sections A and B.

(5.1) **Theorem.** *There is a unique function*

$$\lambda : \{\text{rational homology spheres}\} \to \mathbb{Q}$$

such that

1. $\lambda(S^3) = 0$
2. *For N a knot complement is a $\mathbb{Q}HS$ and l a longitude of N*

$$\lambda(N_b) = \lambda(N_a) + \tau(a, b; l) + \frac{\langle a, b\rangle}{\langle a, l\rangle\langle b, l\rangle}\Gamma(N)$$

for all primitive $a, b \in H_1(\partial N; \mathbb{Z}), \langle a, l\rangle \neq 0, \langle b, l\rangle \neq 0$.

(Note: A version of this theorem, applying to the case where both N_a and N_b are homology lens spaces, has been proved independently by Boyer and Lines [BL].)

The proof of (5.1) occupies the rest of this section. Roughly, the idea of the proof is to represent a rational homology sphere M as surgery on a link in S^3, compute $\lambda(M)$ by surgering one component at a time and applying (5.1.2) (and (5.1.1)), and then show that the result is independent of the choice of surgery representation.

(5.2) *Definition.* A *permissible surgery sequence* (or PSS) $\mathcal{N} = (N^i, m_i, s_i, h_i)$ consists of 3-manifolds N^i with $\partial N_i \cong T^2$ $(1 \leq i \leq n)$, primitive elements $m_i, s_i \in H_1(\partial N^i; \mathbb{Z})$ $(1 \leq i \leq n)$, and homeomorphisms $h_i : N^i_{s_i} \to N^{i+1}_{m_{i+1}}$ $(1 \leq i \leq n-1)$ such that $N^1_{m_1} \cong S^3$ and each $N^i_{s_i}$ is a $\mathbb{Q}$HS. We say $\mathcal{N}$ *represents* $N^n_{s_n}$. Two PSS's $\mathcal{N} = (N^i, m_i, s_i, h_i)$ and $\mathcal{N}' = (N'^i, m'_i, s'_i, h'_i)$ are considered the same if there is a commutative ladder

$$\begin{array}{ccccccccc}
\cdots & \hookleftarrow & N^i & \hookrightarrow & N^i_{s_i} & \xrightarrow{h_i} & N^{i+1}_{m_{i+1}} & \hookleftarrow & N^{i+1} & \hookrightarrow & \cdots \\
 & & \downarrow & & \downarrow & & \downarrow & & \downarrow & & \\
\cdots & \hookleftarrow & N'^i & \hookrightarrow & N'^i_{s'_i} & \xrightarrow{h'_i} & N'^{i+1}_{m'_{i+1}} & \hookleftarrow & N'^{i+1} & \hookrightarrow & \cdots
\end{array}$$

which takes m_i to m'_i and s_i to s'_i.

Given a PSS $\mathcal{N}$, define $\lambda(\mathcal{N})$ to be the value of $\lambda(N^n_{s_n})$ computed by applying (5.1.1) and (5.1.2) to the sequence of surgeries in $\mathcal{N}$. That is,

$$\text{(5.3)} \qquad \lambda(\mathcal{N}) \stackrel{\text{def}}{=} \sum_{i=1}^{n} \left(\tau(m_i, s_i; l_i) + \frac{\langle m_i, s_i \rangle}{\langle m_i, l_i \rangle \langle s_i, l_i \rangle} \Gamma(N^i) \right),$$

where $l_i \in H_1(\partial N^i; \mathbb{Z})$ is a longitude of N^i.

The unicity of λ follows from

(5.4) **Lemma.** *Every $\mathbb{Q}$HS is represented by a PSS.* □

The proof will be given later.

(Note that if the requirement of rationality is dropped, this is equivalent to the fact ([L]) that every closed 3-manifold can be obtained from Dehn surgery on a link in S^3.)

The proof will be complete if we can show that $\lambda(\mathcal{N}) = \lambda(\mathcal{N}')$ for any two PSS's representing the same $\mathbb{Q}$HS M, for then we can (well-) define

$\lambda(M)$ to be $\lambda(\mathcal{N})$ (for any $\mathcal{N}$ representing M), and $\lambda(M)$ so defined clearly satisfies (5.1.1) and (5.1.2).

(5.5) We now introduce four "moves" for modifying PSS's. Throughout, $\mathcal{N}$ will denote (N^i, m_i, s_i, h_i), $1 \leq i \leq n$.

1. *Isotopy.* Change the homeomorphisms h_i by isotopies.

2. *Stabilization.* Fix $0 \leq j \leq n$ and let $K \subset N$ be an unknotted solid torus in $N^j_{s_j}$. (If $j = 0$, define $N^j_{s_j}$ to be $N^1_{m_1}$.) Let $N' = N^j_{s_j} \backslash K$, m' be a meridian of K, and $s' = m' + l'$, where l' is a longitude of N'. Let $h' : N'_{s'} \to N^{j+1}_{s_{j+1}}$ be a homeomorphism which coincides with h_j outside a regular neighborhood of K union its unknotting disk. Insert (N', m', s', h') into $\mathcal{N}$ after (N^j, m_j, s_j, h_j).

3. *Prolongation.* Fix j and let r be a primitive element of $H_1(\partial N^j; \mathbb{Z})$ such that $\langle r, l\rangle \neq 0$ (l a longitude of N^j). Replace (N^j, m_j, s_j, h_j) with $(N^j, m_j, r, \mathrm{id})$, (N^j, r, s_j, h_j).

4. *Permissible transposition.* Fix $1 \leq j \leq n-1$. Suppose that $\overline{N^j_{s_j} \backslash N^j}$ is disjoint from $h_j^{-1}(\overline{N^{j+1}_{m_{j+1}} \backslash N^{j+1}})$, that is, the jth and $j+1^{\text{st}}$ surgery tori are disjoint. Let $R \overset{\text{def}}{=} N^j \cap h_j^{-1}(N^{j+1})$. We can identify ∂R with $\partial N^j \cup \partial N^{j+1}$. We have the following commutative diagram :

$$\begin{array}{ccccc} R_{m_j m_{j+1}} & \leftarrow & R_{m_{j+1}} & \rightarrow & R_{s_j m_{j+1}} \\ \uparrow & & \uparrow & & \uparrow \\ R_{m_j} & \leftarrow & R & \rightarrow & R_{s_j} \\ \downarrow & & \downarrow & & \downarrow \\ R_{m_j s_{j+1}} & \leftarrow & R_{s_{j+1}} & \rightarrow & R_{s_j s_{j+1}}. \end{array}$$

All maps are inclusions. Note that $R_{m_{j+1}} = N^j$, $R_{s_j} = N^{j+1}$, $R_{s_j m_{j+1}} = N^j_{s_j} = N^{j+1}_{m_{j+1}}$, etc. Suppose that the closed 3-manifold $R_{m_j s_{j+1}}$ is a $\mathbb{Q}$HS. Then we can form a new PSS $\mathcal{N}'$ by replacing N^j with R_{m_j}, N^{j+1} with $R_{s_{j+1}}$ and interchanging m_j, s_j and m_{j+1}, s_{j+1}. In other words, we interchange the order of the jth and $j+1^{\text{st}}$ surgeries.

(5.6) **Lemma.** *Any two PSS's which represent the same $\mathbb{Q}$HS are related by a finite sequence of isotopies, stabilizations (and inverse stabilizations), prolongations and permissible transpositions.* □

The proof will be given later.

(If the requirement of rationality is dropped, this follows easily from [K1].)

Clearly $\lambda(\mathcal{N})$ does not change if we modify $\mathcal{N}$ by an isotopy.

If $\mathcal{N}$ is changed by a stabilization, then (using the notation from the definition of stabilization) $\lambda(\mathcal{N})$ changes by

$$\tau(m', m'+l'; l') + \frac{\langle m', m'+l' \rangle}{\langle m', l' \rangle \langle m'+l', l' \rangle} \Gamma(N').$$

The first term is zero by (A.15) and (A.13). The second term is zero because N' is the complement of an unknot, so $\Gamma(N') = 0$. Thus stabilizations do not change $\lambda(\mathcal{N})$.

If $\mathcal{N}$ is modified by a prolongation, (retaining the above notation and dropping the "j" 's) $\lambda(\mathcal{N})$ changes by

$$\tau(m, r; l) + \tau(r, s; l) - \tau(m, s; l)$$
$$+ \frac{\langle m, r \rangle}{\langle m, l \rangle \langle r, l \rangle} \Gamma(N) + \frac{\langle r, s \rangle}{\langle r, l \rangle \langle s, l \rangle} \Gamma(N) - \frac{\langle m, s \rangle}{\langle m, l \rangle \langle s, l \rangle} \Gamma(N).$$

The τ terms cancel by (A.14). The coefficients of $\Gamma(N)$ add up to zero by elementary linear algebra. Hence prolongations do not change $\lambda(\mathcal{N})$.

Thus we have reduced (5.1) to

(5.7) **Lemma.** *Let $\mathcal{N}$ and $\mathcal{N}'$ be two PSS's which differ by a permissible transposition. Then $\lambda(\mathcal{N}) = \lambda(\mathcal{N}')$.*

Proof: Recall the notation from the definition of permissible transposition. Let $a = m_j$, $b = s_j$, $x = m_{j+1}$ and $y = s_{j+1}$. Let $\partial_1 = \partial R_x = \partial R_y \subset \partial R$ and $\partial_2 = \partial R_a = \partial R_b \subset \partial R$. Thus $a, b \in H_1(\partial_1; \mathbf{Z})$ and $x, y \in H_1(\partial_2; \mathbf{Z})$. Let l_1, l_1', l_2 and l_2' be longitudes of R_x, R_y, R_a and R_b, respectively. In view of (5.3), what we must show is

$$\text{(5.8)}\quad \tau(a, b; l_1) + \tau(x, y; l_2') + \frac{\langle a, b \rangle}{\langle a, l_1 \rangle \langle b, l_1 \rangle} \Gamma(R_x) + \frac{\langle x, y \rangle}{\langle x, l_2' \rangle \langle y, l_2' \rangle} \Gamma(R_b) =$$
$$\tau(x, y; l_2) + \tau(a, b; l_1') + \frac{\langle x, y \rangle}{\langle x, l_2 \rangle \langle y, l_2 \rangle} \Gamma(R_a) + \frac{\langle a, b \rangle}{\langle a, l_1' \rangle \langle b, l_1' \rangle} \Gamma(R_y).$$

Consider the map $i_* : H_1(\partial_1; \mathbf{Q}) \to H_1(R; \mathbf{Q})$. i_* must either be an isomorphism or have rank 1.

Case (i). i_* an isomorphism. In this case the inclusions induce isomorphisms

$$H_1(\partial_1; \mathbf{Q}) \cong H_1(R; \mathbf{Q}) \cong H_1(\partial_2; \mathbf{Q}).$$

We will identify these spaces via the above isomorphisms and call them collectively V. Let $\langle \cdot,\cdot\rangle_i$ denote the intersection pairing on $H_1(\partial_i;\mathbb{Q})$. Note that under this identification $\langle \cdot,\cdot\rangle_1 = -\langle \cdot,\cdot\rangle_2$. Let $\langle \cdot,\cdot\rangle \stackrel{\text{def}}{=} \langle \cdot,\cdot\rangle_1$. This inner product allows us to identify V and V^*. V has four distinguished lattices: $H_1(\partial_1;\mathbb{Z})$, $H_1(\partial_2;\mathbb{Z})$, $\mathrm{Free}(H_1(R;\mathbb{Z}))$ and $\mathrm{Hom}(H_1(R;\mathbb{Z}),\mathbb{Z}) \cong H_2(R,\partial R;\mathbb{Z})$. For $r \in V$, define $\nu(r)$ to be the minimal positive integer such that $\nu(r)r \in H_2(R,\partial R;\mathbb{Z})$.

We will need the following two lemmas.

(5.9) **Lemma.** *There is a symmetric bilinear form B defined on V such that*

$$\Gamma(R_{r_i}) = \nu(r_i)^2 B(r_i,r_i) - \frac{1}{12}\nu(r_i)^2 + \frac{1}{12}$$

for all primitive $r_i \in H_1(\partial_i;\mathbb{Z})$, $i = 1$ or 2.

Proof: Let $\widetilde{R_{r_i}}$ be the universal free abelian cover of R_{r_i}. Let $\widetilde{R}$ be the induced cover of $R \subset R_{r_i}$. Let $\widetilde{R}'$ be the universal free abelian cover of R. We have the following commutative diagram

$$\begin{array}{ccc} \widetilde{R}' & & \\ \downarrow & & \\ \widetilde{R} & \hookrightarrow & \widetilde{R_{r_i}} \\ \downarrow & & \downarrow \\ R & \hookrightarrow & R_{r_i}. \end{array}$$

Let

$$\begin{aligned} \Delta(\widetilde{R_{r_i}}), \Delta(\widetilde{R}) &\in \mathbb{Q}[\mathrm{Free}\,(H_1(R_{r_i};\mathbb{Z}))] \\ \Delta(\widetilde{R}') &\in \mathbb{Q}[\mathrm{Free}\,(H_1(R;\mathbb{Z}))] \end{aligned}$$

be the Alexander polynomials. We will express $\Delta(\widetilde{R})$ in terms of $\Delta(\widetilde{R}')$ and then $\Delta(\widetilde{R_{r_i}})$ in terms of $\Delta(\widetilde{R})$.

The inclusion $i : R \hookrightarrow R_{r_i}$ induces a map

$$i_* : \mathbb{Q}[\mathrm{Free}\,(H_1(R;\mathbb{Z}))] \to \mathbb{Q}[\mathrm{Free}\,(H_1(R_{r_i};\mathbb{Z}))].$$

Let t be a generator of $\mathrm{Free}\,(H_1(R_{r_i};\mathbb{Z}))$, written multiplicatively. By (B.4),

$$\Delta(\widetilde{R}) = (1-t)i_*(\Delta(\widetilde{R}')).$$

More concretely, recall that Free $(H_1(R;\mathbb{Z}))$ can be identified with a lattice in V. Then for $x \in$ Free $(H_1(R;\mathbb{Z}))$,

$$i_*(x) = t^{\langle x,\nu(r_i)r_i\rangle}$$

(after possibly replacing t with t^{-1}). Thus, if $\Delta(\widetilde{R}') = \sum_j c_j[x_j]$ ($c_j \in \mathbb{Q}$, $x_j \in$ Free $(H_1(R;\mathbb{Z}))$), then

$$\Delta(\widetilde{R}) = (1-t)\sum_j c_j t^{\langle x_j,\nu(r_i)r_i\rangle}. \tag{5.10}$$

We now relate $\Delta(\widetilde{R_{r_i}})$ to $\Delta(\widetilde{R})$. The map

$$i_* : H_1(\widetilde{R};\mathbb{Q}) \to H_1(\widetilde{R_{r_i}};\mathbb{Q})$$

is a surjective map of $\mathbb{Q}$[Free $(H_1(R_{r_i};\mathbb{Z}))$] modules. Therefore, by (B.1),

$$\Delta(\widetilde{R}) = \Delta(\widetilde{R_{r_i}})\Delta(\ker(i_*)). \tag{5.11}$$

Let $\partial_k = \partial R \backslash \partial R_{r_i}$. The inverse image $\widetilde{\partial_k}$ of ∂_k in $\widetilde{R'}$ consists of $\nu(r_i)$ copies of $S^1 \times \mathbb{R}$. It is not hard to see that the inclusion induced map $H_1(\widetilde{\partial_k};\mathbb{Q}) \to H_1(\widetilde{R};\mathbb{Q})$ is an isomorphism onto $\ker(i_*)$. Free $(H_1(R_{r_i};\mathbb{Z}))$ (acting as deck translations) cyclically permutes the components of $\widetilde{\partial_k}$. It follows that

$$\Delta(\ker(i_*)) = t^{\nu(r_i)} - 1.$$

Combining this with (5.10) and (5.11) yields

$$\Delta(\widetilde{R_{r_i}}) = \frac{\sum c_j t^{\langle x_j,\nu(r_i)r_i\rangle}}{1+t+\cdots+t^{\nu(r_i)-1}}. \tag{5.12}$$

In order to compute $\Gamma(R_{r_i})$, we must normalize and symmetrize (5.12). It follows from the results of [M2] that we may assume that $\Delta(\widetilde{R}') = \sum c_j[x_j]$ is symmetric under $x_j \leftrightarrow -x_j$, though possibly at the cost of $x_j \in \frac{1}{2}$Free $(H_1(R;\mathbb{Z})) \subset V$. This implies that the numerator of (5.12) is symmetric under $t \leftrightarrow t^{-1}$. We may further assume that $\sum c_j = 1$. Let

$$E_l(t) \stackrel{\text{def}}{=} l^{-1} \sum_{j=-\frac{l-1}{2}}^{j=\frac{l-1}{2}} t^j.$$

$E_{\nu(r_i)}$ is the normal, symmetric form of the denominator of (5.12). Hence the normal, symmetric form of the Alexander polynomial of R_{r_i} is

$$\Delta_{\text{SN}}(R_{r_i}) \stackrel{\text{def}}{=} \frac{\sum c_j t^{\langle x_j,\nu(r_i)r_i\rangle}}{E_{\nu(r_i)}(t)}.$$

Taking the formal second derivative with respect to t and evaluating at $t = 1$, we have

$$\Gamma(R_{r_i}) = \frac{d^2}{dt^2}\Delta_{\mathrm{SN}}(R_{r_i})|_{t=1} = \nu(r_i)^2 B(r_i, r_i) - \frac{1}{12}\nu(r_i)^2 + \frac{1}{12}$$

where

$$B(a,b) \stackrel{\text{def}}{=} \sum c_j \langle x_j, a\rangle \langle x_j, b\rangle.$$

□

(5.13) **Lemma.** *Let C be any symmetric bilinear form on V, and let a_1, b_1, a_2, $b_2 \in V$ be pairwise linearly independent. Then*

$$\frac{\langle a_2, b_2\rangle}{\langle a_2, a_1\rangle\langle b_2, a_1\rangle} C(a_1, a_1) - \frac{\langle a_2, b_2\rangle}{\langle a_2, b_1\rangle\langle b_2, b_1\rangle} C(b_1, b_1) + \frac{\langle a_1, b_1\rangle}{\langle a_1, a_2\rangle\langle b_1, a_2\rangle} C(a_2, a_2) - \frac{\langle a_1, b_1\rangle}{\langle a_1, b_2\rangle\langle b_1, b_2\rangle} C(b_2, b_2) = 0.$$

Proof: There exist non-zero $\alpha, \beta, \gamma, \delta \in \mathbb{Q}$ such that $a_1 = \alpha a_2 + \beta b_2$ and $b_1 = \gamma a_2 + \delta b_2$. Hence

$$\begin{aligned} C(a_1, a_1) &= \alpha^2 C(a_2, a_2) + \alpha\beta C(a_2, b_2) + \beta^2 C(b_2, b_2) \\ C(b_1, b_1) &= \gamma^2 C(a_2, a_2) + \gamma\delta C(a_2, b_2) + \delta^2 C(b_2, b_2). \end{aligned}$$

Eliminating $C(a_2, b_2)$ gives

$$\frac{1}{\alpha\beta} C(a_1, a_1) - \frac{1}{\gamma\delta} C(b_1, b_1) = \left(\frac{\alpha}{\beta} - \frac{\gamma}{\delta}\right) C(a_2, a_2) + \left(\frac{\beta}{\alpha} - \frac{\delta}{\gamma}\right) C(b_2, b_2).$$

Substituting $\alpha = \langle a_1, b_2\rangle / \langle a_2, b_2\rangle$, $\beta = \langle a_1, a_2\rangle / \langle b_2, a_2\rangle$, etc., and noting that $\alpha\delta - \beta\gamma = \langle a_1, b_1\rangle / \langle a_2, b_2\rangle$, we get

$$\frac{\langle a_2, b_2\rangle\langle b_2, a_2\rangle}{\langle a_1, b_2\rangle\langle a_1, a_2\rangle} C(a_1, a_1) - \frac{\langle a_2, b_2\rangle\langle b_2, a_2\rangle}{\langle b_1, b_2\rangle\langle b_1, a_2\rangle} C(b_1, b_1) = \frac{\langle a_2, b_2\rangle\langle a_1, b_1\rangle}{\langle a_1, a_2\rangle\langle b_1, a_2\rangle} C(a_2, a_2) - \frac{\langle a_2, b_2\rangle\langle a_1, b_1\rangle}{\langle a_1, b_2\rangle\langle b_1, b_2\rangle} C(b_2, b_2).$$

Dividing by $\langle a_2, b_2\rangle$ completes the proof. □

We now establish (5.7) for case (i). First observe that $l_1 = \nu(x)x$, $l_1' = \nu(y)y$, $l_2 = \nu(a)a$ and $l_2' = \nu(b)b$. Also, since we have identified $H_1(\partial_2; \mathbb{Z})$ with a lattice in V and the pairing on V has the opposite sign from the intersection pairing on $H_1(\partial_2; \mathbf{Z})$, the signs of terms in (5.8) involving elements elements of $H_1(\partial_2; \mathbf{Z})$ must be changed. (5.8) now becomes

$$(5.14)\ \tau(a,b;\nu(x)x) - \tau(a,b;\nu(y)y) + \tau(x,y;\nu(a)a) - \tau(x,y;\nu(b)b) =$$
$$-\frac{\langle a,b\rangle}{\langle a,\nu(x)x\rangle\langle b,\nu(x)x\rangle}\Gamma(R_x) + \frac{\langle a,b\rangle}{\langle a,\nu(y)y\rangle\langle b,\nu(y)y\rangle}\Gamma(R_y)$$
$$-\frac{\langle x,y\rangle}{\langle x,\nu(a)a\rangle\langle y,\nu(a)a\rangle}\Gamma(R_a) + \frac{\langle x,y\rangle}{\langle x,\nu(b)b\rangle\langle y,\nu(b)b\rangle}\Gamma(R_b).$$

By (5.9) and (5.13), the right hand side is equal to

$$\frac{1}{12}\Bigg(\frac{\langle a,b\rangle}{\langle a,x\rangle\langle b,x\rangle} - \frac{\langle a,b\rangle}{\langle a,y\rangle\langle b,y\rangle} + \frac{\langle x,y\rangle}{\langle x,a\rangle\langle y,a\rangle} - \frac{\langle x,y\rangle}{\langle x,b\rangle\langle y,b\rangle}$$
$$-\frac{\langle a,b\rangle}{\langle a,\nu(x)x\rangle\langle b,\nu(x)x\rangle} + \frac{\langle a,b\rangle}{\langle a,\nu(y)y\rangle\langle b,\nu(y)y\rangle}$$
$$-\frac{\langle x,y\rangle}{\langle x,\nu(a)a\rangle\langle y,\nu(a)a\rangle} + \frac{\langle x,y\rangle}{\langle x,\nu(b)b\rangle\langle y,\nu(b)b\rangle}\Bigg).$$

By (A.16), this differs from the left hand side by

$$\frac{1}{4}\Bigg(\operatorname{sign}\frac{\langle a,b\rangle}{\langle a,x\rangle\langle b,x\rangle} - \operatorname{sign}\frac{\langle a,b\rangle}{\langle a,y\rangle\langle b,y\rangle}$$
$$+\operatorname{sign}\frac{\langle x,y\rangle}{\langle x,a\rangle\langle y,a\rangle} - \operatorname{sign}\frac{\langle x,y\rangle}{\langle x,b\rangle\langle y,b\rangle}\Bigg).$$

This is zero, and the proof of (5.7) is complete in case (i).

Case (ii). i_* has rank one. In this case, $l_1 = l_1'$ and $l_2 = l_2'$, so (5.8) reduces to

$$(5.15)\quad \frac{\langle a,b\rangle}{\langle a,l_1\rangle\langle b,l_1\rangle}(\Gamma(R_x) - \Gamma(R_y)) = \frac{\langle x,y\rangle}{\langle x,l_2\rangle\langle y,l_2\rangle}(\Gamma(R_a) - \Gamma(R_b)).$$

This follows immediately from

(5.16) **Lemma.** *There is a $\beta = \beta(R) \in \mathbf{Q}$ such that*

$$\begin{aligned}\Gamma(R_a) - \Gamma(R_b) &= 2\frac{\langle a,b\rangle}{\langle a,l_1\rangle\langle b,l_1\rangle}\beta \\ \Gamma(R_x) - \Gamma(R_y) &= 2\frac{\langle x,y\rangle}{\langle x,l_2\rangle\langle y,l_2\rangle}\beta.\end{aligned}$$

(If R is the complement of a two component link in S^3 with linking number zero, then β is the Sato-Levine invariant.)

Proof: We will use (B.10) to compute Γ. There are connected surfaces E_1, $E_2 \in R$ such that $[\partial E_i] = l_i \in H_1(\partial_i; \mathbb{Z})$. Put E_1 and E_2 in general position. Then $E_1 \cap E_2$ is a closed 1-manifold. Orient $E_1 \cap E_2$ so that its orientation followed by the positive normal of E_1 followed by the positive normal of E_2 gives the orientation of R. Let $(E_1 \cap E_2)^+$ denote the push-off of $E_1 \cap E_2$ in E_1 in the positive normal direction of E_2. Define

$$\beta(R) \stackrel{\text{def}}{=} \text{lk}\,(E_1 \cap E_2, (E_1 \cap E_2)^+).$$

It is not hard to show that this does not depend on the choice of E_1 and E_2.

Let $\alpha_1, \ldots, \alpha_{2g}$ be curves representing a symplectic basis of $H_1(E_2; \mathbb{Z})$ and such that $[E_1 \cap E_2] = k[\alpha_2]$ for some $k \in \mathbb{Z}$. Since $\langle [\alpha_i], [E_1 \cap E_2] \rangle = 0$ for $i \geq 2$, $\alpha_2, \ldots, \alpha_{2g}$ rationally bound surfaces in R. Thus any linking number involving them is independent of how ∂_1 is surgered. It follows from this and (B.10) that

$$\Gamma(R_a) - \Gamma(R_b) = 2(\text{lk}_a(\alpha_1^-, \alpha_1) - \text{lk}_b(\alpha_1^-, \alpha_1))\text{lk}\,(\alpha_2^+, \alpha_2),$$

where lk_a $[\text{lk}_b]$ denotes the linking number in R_a $[R_b]$.

Let $N(\alpha_1)$ be a closed tubular neighborhood of α_1 in R and let $R^- \stackrel{\text{def}}{=} R \setminus N(\alpha_1)$. As in case (i) above, the inclusions into R^- induce an isomorphism

$$H_1(\partial_1; \mathbb{Q}) \cong H_1(\partial N(\alpha_1); \mathbb{Q}),$$

and we identify these two spaces. Let $p = [\alpha_1^-] \in H_1(\partial N(\alpha_1); \mathbb{Q})$, and let $m \in H_1(\partial N(\alpha_1); \mathbb{Q})$ be a meridian of $N(\alpha_1)$ such that $\langle m, p \rangle = 1$. Then

$$\text{lk}_c(\alpha_1^-, \alpha_1) = -\frac{\langle p, c \rangle}{\langle m, c \rangle},$$

where $c = a$ or b. Note also that $km = l_1$, up to sign. Therefore

$$\begin{aligned}
\Gamma(R_a) - \Gamma(R_b) &= 2\left(-\frac{\langle p, a \rangle}{\langle m, a \rangle} + \frac{\langle p, b \rangle}{\langle m, b \rangle}\right) \text{lk}\,(\alpha_2^+, \alpha_2) \\
&= 2\left(\frac{\langle p, a \rangle \langle m, b \rangle - \langle m, a \rangle \langle p, b \rangle}{\langle m, a \rangle \langle m, b \rangle}\right)\left(\frac{\beta(R)}{k^2}\right) \\
&= 2\frac{\langle a, b \rangle}{\langle a, l_1 \rangle \langle b, l_1 \rangle}\beta(R).
\end{aligned}$$

This completes the proof of the first half of (5.16). The proof of the other half is similar. □

The proof of (5.7) is also complete. □

It remains to prove (5.4) and (5.6). The first step is to reformulate them in terms of framed link diagrams.

Definitions. An *ordered framed link* (or OFL) $(L, s) = ((L_1, s_1), \ldots, (L_n, s_n))$ consists of an ordered link $L = (L_1, \ldots, L_n) \subset S^3$ and primitive homology classes $s_i \in H_1(\partial(\mathrm{nbd}(L_i)); \mathbb{Z})$. Let m_i be a meridian of $\mathrm{nbd}(L_i)$ and l_i be a longitude of $S^3 \setminus \mathrm{nbd}(L_i)$, oriented so that $\langle m_i, l_i \rangle = 1$. Then $s_i = \alpha_i m_i + \beta_i l_i$ (α_i, $\beta_i \in \mathbb{Z}$), and we identify s_i with $\alpha_i/\beta_i \in \mathbb{Q} \cup \{\infty\}$. The *result* of (L, s), $M(L, s)$, is $S^3 \setminus \mathrm{nbd}(L)$ surgered along $s_1, \ldots, s_n$. (L, s) is *permissible* if $M((L_1, s_1), \ldots, (L_k, s_k))$ is a QHS for each $k \leq n$. (L, s) is *integral* if $\langle m_i, s_i \rangle$ is an integer for each i, or, equivalently, if the rational number corresponding to s_i is an integer. An integral PSS is defined similarly.

Let (L, s) be an integral OFL. Choose orientations of the components of L and represent each s_i by a simple closed curve $S_i \subset \partial(\mathrm{nbd}(L_i))$ which is (oriented) parallel to L_i. The matrix $[\mathrm{lk}\,(L_i, S_j)]$ is called the *linking matrix* of (L, s). ($\mathrm{lk}\,(\cdot, \cdot)$ denotes the linking number in S^3.) It is not hard to show that the linking matrix is a presentation matrix for $H_1(M(L, s); \mathbb{Z})$. This implies

(5.17) **Lemma.** *An integral OFL (L, s) is permissible if and only if*

$$\det_k(L, s) \stackrel{\text{def}}{=} \det([\mathrm{lk}\,(L_i, S_j)]_{1 \leq i,j \leq k})$$

is non-zero for each $k \leq n$. □

A permissible OFL determines a PSS in the obvious way: In the notation of (5.2), $N^i = M((L_1, s_1), \ldots, (L_{i-1}, s_{i-1})) \setminus \mathrm{nbd}(L_i)$. Modulo isotopies of PSS's, every PSS corresponds to some OFL.

(5.18) Recall ([K1]) that any two integral OFL's representing the same 3-manifold are related by a finite sequence of the following three moves.

[Inverse] stabilization. Add [subtract] an unknotted and unlinked component K to L with framing ± 1. K may be inserted anywhere in the ordering.

Handle slides. Fix i and j. Let S_i be a simple closed curve in $\partial\,\mathrm{nbd}(L_i)$ representing $\pm s_i$. Replace L_j with a band connected sum of L_j and S_i (use any band). If L_i and L_j have framings a_i and $a_j \in \mathbb{Z}$, the framing for the new L_j is $a_i + a_j \pm \mathrm{lk}\,(L_i, L_j)$. (The sign depends on whether the band preserves or reverses the orientations of L_i and L_j.) In terms of the linking matrix, this corresponds to adding or subtracting the ith row and column to the jth row and column.

Transpositions. Exchange (L_i, s_i) and (L_{i+1}, s_{i+1}).

As a final preliminary, we need the following lemma, the proof of which is left to the reader.

(5.19) **Lemma.** *Let $P_1, \ldots, P_m$ be polynomials defined on $\mathbb{Z}^k$. If, for each $1 \le i \le m$, there is a $v_i \in \mathbb{Z}^k$ such that $P_i(v_i) \ne 0$, then there is a $v \in \mathbb{Z}^k$ such that $P_i(v) \ne 0$ for all $1 \le i \le m$.* □

We are now ready to prove (5.4). In view of the above remarks, this follows from

(5.20) **Lemma.** *Every $\mathbb{Q}HS$ M is represented by a permissible (and also integral) OFL.*

Proof: By [L], every 3-manifold is represented by an integral OFL. Let (L, s) represent M. We will modify (L, s) by stabilizations and handle slides to obtain a permissible integral OFL.

First add unknots $K_1, \ldots, K_n$ with framing 1 to (L, s) at the end of the ordering. Next, choose integers $b_1, \ldots, b_n$ and slide each L_i over K_i b_i times. Call the resulting OFL (L', s'). Let A be the linking matrix of (L, s) and B be the diagonal matrix with diagonal $(b_1, \ldots, b_n)$. Then the linking matrix of (L', s') is

$$A' = \begin{bmatrix} A + B^2 & B \\ B & I \end{bmatrix}.$$

We must choose the b_i's so that $\det_k(L', s') \ne 0$ for all $k \le 2n$ (see (5.17)). $\det_k(L', s')$ is a polynomial function of the b_i's. In view of (5.19), it suffices to show that each of these polynomials individually has a nonsolution. If we choose the b_i's sufficiently large, the matrix $A + B^2$ (thought of as a bilinear form) will be positive definite, and hence $\det_k(L', s')$ will be positive for $k \leq n$. If the b_i's are all zero, then, for $n+1 \leq k \leq 2n$, $\det_k(L', s') = \det(A) \neq 0$. □

Next, as a step toward (5.6), we prove

(5.21) **Lemma.** *Any PSS can be prolonged (in the sense of* (5.5.3)*) to an integral PSS.*

Proof: Recall the notation of (5.2). It suffices to find, for each i, primitive homology classes $c_{i0} = m_i$, $c_{i1}, \ldots, c_{ik_i} = s_i \in H_1(\partial N^i; \mathbb{Z})$ such that $\langle c_{ij}, c_{ij+1} \rangle = \pm 1$ and $\langle c_{ij}, l_i \rangle \neq 0$ for all j. This is possible, by elementary number theory. □

We now introduce moves analogous to those in (5.6) for modifying permissible integral OFL's.

Ordered handle slides. Slide L_i over L_j, *where* $i \geq j$. The corresponding PSS changes by an isotopy.

[Inverse] stabilizations. Same as in (5.18). The corresponding PSS changes by a stabilization (as defined in (5.5)).

Permissible transpositions. Interchange L_i and L_{i+1} in the ordering, provided that the resulting OFL is also permissible. The corresponding PSS changes by a permissible transposition (as defined in (5.5)).

The proof of (5.6) will be complete upon the proof of

(5.22) **Lemma.** *Any two permissible integral OFL's,* (L, s) *and* (L', s')*, which represent the same rational homology sphere are related by a finite sequence of ordered handle slides, [inverse] stabilizations and permissible transpositions.*

Proof: After stabilizing (L, s) and (L', s'), we may assume, by Kirby's theorem, that (L, s) and (L', s') are related by a finite sequence of ordered handle slides and transpositions. Unfortunately, the transpositions are not necessarily permissible. In order to correct this, we do the following.

Step 1. Add n (= the number of components of L and L') unknots $K_1, \ldots, K_n$ with framing 1 to (L, s) at the beginning of the ordering.

Step 2. Choose integers $b_1, \ldots, b_n$, satisfying conditions specified below, and slide L_i over K_i b_i times.

Step 3. Via transpositions, move $K_1, \ldots, K_n$ to the end of the ordering.

Step 4. Perform a sequence of transpositions and ordered handle slides on the L_i's as before.

Step 5. Move the K_i's (via transpositions) back to the beginning of the ordering.

Step 6. Perform appropriate handle slides of the L_i's over the K_i's, yielding (L', s') plus n unknots with framing 1 (i.e. $K_1, \ldots, K_n$).

Step 7. Remove $K_1, \ldots, K_n$.

We must choose $b_1, \ldots, b_n$ so that the transpositions in steps 3, 4 and 5 are permissible. By (5.17) and (5.19) it suffices to find, for each OFL occuring in steps 3–5 and each $1 \leq k \leq 2n$, a choice of the b_i's such that $\det_k$ of that OFL is nonzero. In steps 3 and 5, $b_1 = \cdots = b_n = 0$ is such a choice. In step 4 with $k \leq n$, choose the b_i's large enough that the linking matrix restricted to the L_i's is positive definite. In step 4 with $k \geq n+1$, choose $b_1 = \cdots = b_n = 0$. □

§ 6

Consequences of the Dehn Surgery Formula

In this section we use the Dehn surgery formula (4.2) to prove various things about λ.

(6.1) **Proposition.** *Let M_1 and M_2 be $\mathbb{Q}HS$s. Let $M_1 \# M_2$ denote the connected sum of M_1 and M_2. Then*

$$\lambda(M_1 \# M_2) = \lambda(M_1) + \lambda(M_2).$$

Proof: Let L_i be a permissible ordered framed link representing M_i, $i = 1, 2$ (see Section 5). Let $L_1 \cup L_2$ denote the permissible ordered framed link obtained by situating L_1 and L_2 in disjoint 3-balls in S^3 and putting L_2 after L_1 in the ordering. Then $L_1 \cup L_2$ represents $M_1 \# M_2$. The presence of the L_1 part of $L_1 \cup L_2$ does not affect homological information or Alexander polynomials of the L_2 part, so the proposition now follows easily from (4.2). □

The following is a special case of (4.2).

(6.2) **Proposition.** *Let K be a null-homologous knot in a $\mathbb{Q}HS$ M. Let $K_{p/q}$ denote p/q-surgery on K. Let $L_{p/q}$ denote the p, q-lens space (i.e. p/q-surgery on the unknot in S^3). Then*

$$\lambda(K_{p/q}) = \lambda(M) + \lambda(L_{p/q}) + \frac{q}{p}\Gamma(K).$$

□

(4.1) and (4.2) imply

(6.3) **Proposition.** *Let $L_{p/q}$ denote the p, q-lens space. Then*

$$\lambda(L_{p/q}) = -s(q,p).$$

□

It is perhaps worth noting that (6.3) implies that $4p\lambda(L_{p/q})$ is equal to the signature defect of $L_{p/q}$ (see [HZ]).

(6.4) **Proposition.** $6|H_1(M;\mathbb{Z})|\lambda(M) = 6\bar{\lambda}(M) \in \mathbb{Z}$ *for all* $\mathbb{Q}$*HSs* M.

Proof: This can, in principle, be proved using (4.2), but it will be easier to use (3.34) and (3.36) (which are key ingredients in proving (4.2)).

By (5.4) and the fact that (6.4) is true for S^3, it suffices to show that $6\bar{\lambda}(K_a) - 6\bar{\lambda}(K_b) \in \mathbb{Z}$ for any knot K in a $\mathbb{Q}$HS and any appropriate homology classes $a, b \in H_1(\partial(\text{nbd}K);\mathbb{Z})$. This follows from (3.34), (3.36), the fact that $\bar{\lambda}'(K) \in \mathbb{Z}$ (which is clear from the proof of (3.34)), and the fact that $r(K, \cdot, \cdot) = 0$ (which was proved in Section 4). □

The next result relates λ to the μ-invariant of a $\mathbb{Z}_2$-homology sphere ($\mathbb{Z}_2$HS). All facts concerning μ-invariants which are stated without proof can be found in [K2], or can be proved easily using the techniques of [K2] and will be proved in [W2].

Let M be an oriented 3-manifold and let σ be a spin structure on M. Let W be an oriented 4-manifold with spin structure ω such that $\partial W = M$ and ω restricts to σ. Define

$$\mu(M,\sigma) = \text{sign}\,(W) \mod 16,$$

where sign (W) denotes the signature of the intersection pairing on $H_2(W;\mathbb{Z})$. This is well defined by Rochlin's theorem, which states that the signature of a closed spin 4-manifold is congruent to 0 mod 16. If M is a $\mathbb{Z}_2$HS then then it has a unique spin structure and we can speak unambiguously of $\mu(M)$.

(6.5) **Proposition.** *Let* M *be a* $\mathbb{Z}_2$*HS. Then*

$$4|H_1(M;\mathbb{Z})|^2\lambda(M) \equiv_{16} \mu(M)$$

(where $\equiv_{16}$ *denotes congruence mod 16).*

(Note: The case where M is an odd homology lens space has been proved independently in [BL].)

Proof: Define an *odd Dehn surgery* to be a Dehn surgery on a $\mathbb{Z}_2$HS which yields a $\mathbb{Z}_2$HS.

(6.6) **Lemma.** *Any $\mathbb{Z}_2HS$ can be obtained from S^3 via a finite sequence of odd Dehn surgeries.*

Proof: Let M be a $\mathbb{Z}_2$HS. Represent M as surgery on an integral framed link $L = (L_1, \ldots, L_n)$ in S^3 (see Section 5). Let $A = [A_{ij}]$ be the linking matrix of L. Let d_k be the determinant of the $k \times k$ submatrix of A corresponding to the first k components of L. Then L represents a sequence of odd surgeries if and only if d_k is odd for $1 \leq k \leq n$. Thus it suffices to show that A can be transformed by row and column operations and stabilizations so that it has this property.

Let $\overline{A} = [\overline{A}_{ij}]$ be A reduced mod 2. After possibly stabilizing, we may assume that $\overline{A}$ has a 1 on the diagonal. Then arguing as in the proof of Theorem 4.3 of [MH], we see that $\overline{A}$ can be transformed via row and column operations into the identity matrix (over $\mathbb{Z}_2$). (The crucial ingredient in the argument is to see that any symmetric inner product space over $\mathbb{Z}_2$ has an element of square 0. But if $\langle a, a\rangle = \langle b, b\rangle = 1$ and $a \neq b$, then $\langle a+b, a+b\rangle = 0$.) Performing the corresponding operations on A produces a matrix with odd partial determinants. □

In view of (6.6), it suffices to compute, modulo 16, how the left and right hand sides of (6.5) change when M is modified by an odd Dehn surgery, and to check that these changes are equal. We first consider the right hand side.

Let M be a $\mathbb{Z}_2$HS, let K be a knot in M, and let $N = M \setminus \mathrm{nbd}(K)$. Let $k \in H_1(\partial N; \mathbb{Z})$ be such that $l = dk$, where l is the longitude of N and $d \in \mathbb{Z}$. Then $H^1(N_k; \mathbb{Z}_2) \cong \mathbb{Z}_2$, so N_k has two spin structures, σ_1 and σ_2. $\mu(N_k, \sigma_1)$ differs from $\mu(N_k, \sigma_2)$ by a multiple of 8. Define the *arf invariant* of N to be

$$\mathrm{arf}(N) \stackrel{\mathrm{def}}{=} \frac{\mu(N_k, \sigma_1) - \mu(N_k, \sigma_2)}{8} \in \mathbb{Z}_2.$$

Let $m \in H_1(\partial N; \mathbb{Z})$ be such that $\langle m, k\rangle = 1$. Let $N_{p,q} \stackrel{\mathrm{def}}{=} N_{pm+qk}$. Let $L_{p/q}$ denote, as usual, the p, q lens space.

(6.7) **Lemma.** *For all relatively prime p, q such that $N_{p,q}$ is a $\mathbb{Z}_2HS$ (i.e. such that p is odd),*

$$\mu(N_{p,q}) \equiv_{16} \mu(N_{1,0}) + \mu(L_{p/q}) + 8q\,\mathrm{arf}(N).$$

□

The case where M is a $\mathbb{Z}$HS is proved in [Go]. The proof of the general case is similar and will be given in [W2].

(6.8) **Lemma.**

$$\operatorname{arf}(N) \equiv_2 \frac{1}{2}|\operatorname{Tor}(H_1(N;\mathbb{Z}))|^2 \left(\Gamma(N) - \frac{d^2-1}{12}\right),$$

where $|\operatorname{Tor}(H_1(N;\mathbb{Z}))|$ *denotes the order of the the torsion subgroup of* $H_1(N;\mathbb{Z})$. □

Again, this is well known for the case that M is a $\mathbb{Z}$HS, and the general case will be proved in [W2].

The next lemma is equivalent to (6.5) in the case where M is an odd lens space.

(6.9) **Lemma.** *Let* p *and* q *be relatively prime with* p *odd. Then*

$$\mu(L_{p/q}) \equiv_{16} -4p^2 s(q,p).$$

□

By the remark following (6.3), proving (6.9) amounts to establishing a relation between the μ-invariant of an odd lens space and its signature defect. As these two invariants are both defined in terms of bounded 4-manifolds, it is not surprising the such a relation exists. Details can be found in [BL], Lemma 4.4.

(6.10) **Lemma.** *If* M *is a* $\mathbb{Z}_2$*HS, then* $\mu(M)$ *is even.*

Proof: This follows from (6.6), (6.7), (6.9), and the fact that $-4p^2s(q,p)$ is even if p is odd. □

(6.11) **Lemma.** *Let* N *and* d *be as above. Then*

$$\frac{2|\operatorname{Tor}(H_1(N;\mathbb{Z}))|(d^2-1)}{3} \equiv_{16} 0.$$

Proof: If d is not divisible by 3, then d^2-1 is divisible by 24. If d is divisible by 3, then d^2-1 is divisible by 8 and $|\operatorname{Tor}(H_1(N;\mathbb{Z}))|$ is divisible by 3. (The longitude represents an element of order d in $\operatorname{Tor}(H_1(N;\mathbb{Z}))$.) □

Finally, note that

(6.12) **Lemma.** *Let $e, o \in \mathbb{Z}$, e even, o odd. Then*

$$o^2 e \equiv_{16} e.$$

□

We are now ready to prove (6.5). Let N be as above. By (6.6), it suffices to show that (6.5) holds for $N_{p,q}$ (p odd, $p > 0$) assuming that it holds for, say, $N_{1,0}$. Let $h = |H_1(N_{1,0}; \mathbb{Z})|$. Note that $|H_1(N_{p,q}; \mathbb{Z})| = ph$ and $|\mathrm{Tor}(H_1(N; \mathbb{Z}))| = h/d$. We have

$$\begin{aligned}
\mu(N_{p,q}) &\equiv_{16} \mu(N_{1,0}) + \mu(L_{p/q}) + 8q\,\mathrm{arf}(N) \\
&\qquad \text{(by (6.7))} \\
&\equiv_{16} 4p^2h^2\lambda(N_{1,0}) - 4p^2h^2 s(q,p) + 4pq\left(\frac{h}{d}\right)^2\left(\Gamma(N) - \frac{d^2-1}{12}\right) \\
&\qquad \text{(by the inductive assumption, (6.12), (6.9) and (6.8))} \\
&= 4p^2h^2\Bigg(\lambda(N_{1,0}) + \tau(m, pm+qk, dk) + \\
&\qquad \frac{\langle m, pm+qk\rangle}{\langle m, dk\rangle\langle pm+qk, dk\rangle}\Gamma(N)\Bigg) - 8qp\left(\frac{h^2}{d^2}\right)\frac{d^2-1}{12} \\
&\qquad \text{(by (4.1))} \\
&\equiv_{16} 4|H_1(N_{p,q}; \mathbb{Z})|^2\lambda(N_{p,q}) \\
&\qquad \text{(by (4.2) and (6.11)).}
\end{aligned}$$

□

This completes the proof of (6.5).

§ A

Dedekind Sums

In this section we prove various properties of the function τ. We give an alternative definition of τ ((A.4), below), show that this is equivalent to (4.1), and use the alternative definition to derive the properties of τ used in Sections 4 and 5. (Thus it is not necessary to ever mention Dedekind sums.) At the end, we show that the two definitions agree. The alternative definition is inspired by Theorem 1 of [H].

The treatment in this section is somewhat crude, but it has the advantage of being self-contained.

The alternative definition will require a few preliminary definitions.

Let $\widetilde{SL}_2(\mathbb{R})$ be the universal cover of $SL_2(\mathbb{R})$. $\widetilde{SL}_2(\mathbb{R})$ acts on $\widetilde{\mathbb{R}^2_*}$, the universal cover of $\mathbb{R}^2_* \stackrel{\text{def}}{=} \mathbb{R}^2 \backslash \{0\}$. This action commutes with the action of $\mathbb{R}_+$ on $\widetilde{\mathbb{R}^2_*}$, so there is an induced action of $\widetilde{SL}_2(\mathbb{R})$ on $\mathbb{R}_P \stackrel{\text{def}}{=} \widetilde{\mathbb{R}^2_*}/\mathbb{R}_+ \cong \mathbb{R}$. $\widetilde{SL}_2(\mathbb{R})$ acts on $\mathbb{R}_P$ via periodic diffeomorphisms, that is, diffeomorphisms which commute with the deck transformations of the covering

$$\mathbb{R}_P \to (\mathbb{R}^2_*)/\mathbb{R}_+ = S^1,$$

and we parameterize $\mathbb{R}_P$ so that a generator of the deck transformations is $x \mapsto x+1$. (Note that $\widetilde{SL}_2(\mathbb{R})$ acts via diffeomorphisms of period 1/2.) Give $\mathbb{R}^2$ its standard orientation, orient $\mathbb{R}^2_*$ and $\widetilde{\mathbb{R}^2_*}$ compatibly, and orient $\mathbb{R}_P$ so that its orientation preceded by the standard one on $\mathbb{R}_+$ gives the orientation on $\widetilde{\mathbb{R}^2\backslash\{0\}}$.

Let G be the group of diffeomorphisms of $\mathbb{R}$ with period 1. For $g \in G$, define

$$\operatorname{rot}(g) \stackrel{\text{def}}{=} \lim_{n\to\infty} \frac{g^n(x) - x}{n}$$

where $x \in \mathbb{R}$. By a well known folk theorem, this limit exists and is independent of x. Furthermore,

$$\text{rot}\,(hgh^{-1}) = \text{rot}\,(g) \tag{A.1}$$

for all $g, h \in G$. Since we have an action of $\widetilde{SL}_2(\mathbb{R})$ on $\mathbb{R}_P$ and an identification of $\mathbb{R}_P$ with $\mathbb{R}$ up to periodic diffeomorphisms, we get a map

$$\text{rot} : \widetilde{SL}_2(\mathbb{R}) \to \mathbb{R}.$$

Let $\widetilde{SL}_2(\mathbb{Z}) = \pi^{-1}(SL_2(\mathbb{Z})) \subset \widetilde{SL}_2(\mathbb{R})$. There is an exact sequence

$$1 \to K \to \widetilde{SL}_2(\mathbb{Z}) \to SL_2(\mathbb{Z}) \to 1,$$

where $K \cong \mathbb{Z}$. It is easy to verify that rot restricts to an isomorphism from K to $\mathbb{Z}$.

For H any Group, let H_{ab} denote $H/[H,H]$. Let $\varphi : \widetilde{SL}_2(\mathbb{Z}) \to \widetilde{SL}_2(\mathbb{Z})_{\text{ab}} \cong \mathbb{Z}$ be the projection. Clearly

$$\varphi(hgh^{-1}) = \varphi(g) \tag{A.2}$$

Since $SL_2(\mathbb{Z})_{\text{ab}} \cong \mathbb{Z}_{12}$, $\varphi(K)$ has index 12 in $\widetilde{SL}_2(\mathbb{Z})_{\text{ab}}$. Fix an identification of $\widetilde{SL}_2(\mathbb{Z})_{\text{ab}}$ with $\mathbb{Z}$ by requiring that

$$\varphi(k) = 12\,\text{rot}\,(k) \tag{A.3}$$

for all $k \in K$.

More concretely, let

$$\begin{aligned} U &= \begin{pmatrix} 1 & 1 \\ 0 & 1 \end{pmatrix} \in SL_2(\mathbb{Z}) \\ L &= \begin{pmatrix} 1 & 0 \\ 1 & 1 \end{pmatrix} \in SL_2(\mathbb{Z}). \end{aligned}$$

There are elements $\widetilde{U}$ and $\widetilde{L}$ of $\widetilde{SL}_2(\mathbb{Z})$ such that $\pi(\widetilde{U}) = U$, $\pi(\widetilde{L}) = L$, and such that $\widetilde{SL}_2(\mathbb{Z})$ is generated by $\widetilde{U}$ and $\widetilde{L}$ subject to the single relation

$$\widetilde{L}\widetilde{U}^{-1}\widetilde{L}\widetilde{U}\widetilde{L}^{-1}\widetilde{U} = 1.$$

In terms of these generators,

$$\begin{aligned} \varphi(\widetilde{U}) &= -1 \\ \varphi(\widetilde{L}) &= 1. \end{aligned}$$

Let V be a two dimensional integral lattice equiped with an antisymmetric inner product $\langle\cdot,\cdot\rangle$ such that $\langle a,b\rangle=\pm1$ if and only if a,b is a basis of V. (For example, $V=H_1(T^2;\mathbb{Z})$ with the intersection pairing.) Choose an identification of V with $\mathbb{Z}^2$ (equiped with its standard pairing $\langle(a_1,a_2),(b_1,b_2)\rangle=a_1b_2-a_2b_1$). This induces an identification of $\widetilde{\mathrm{Aut}}\,(V)$ with $\widetilde{SL}_2(\mathbb{Z})$, and hence functions rot $:\widetilde{\mathrm{Aut}}\,(V)\to\mathbb{R}$ and $\varphi:\widetilde{\mathrm{Aut}}\,(V)\to\mathbb{Z}$. It follows from (A.1) and (A.2) that these functions are independent of the choice of identification of V with $\mathbb{Z}^2$.

For $a,b,l\in V$ define

$$\delta(a,b;l)\stackrel{\text{def}}{=}\begin{cases}0 & \text{if } \langle a,l\rangle\langle b,l\rangle>0\\ 1 & \text{otherwise}\end{cases}$$

and

$$\epsilon(a,b;l)\stackrel{\text{def}}{=}\begin{cases}0 & \text{if } \langle a,b\rangle>0\\ 1 & \text{if } \langle a,b\rangle<0 \text{ and } \delta(a,b;l)=0\\ -1 & \text{if } \langle a,b\rangle<0 \text{ and } \delta(a,b;l)=1\\ 0 & \text{if } \langle a,b\rangle=0 \text{ and } \delta(a,b;l)=0\\ -1 & \text{if } \langle a,b\rangle=0 \text{ and } \delta(a,b;l)=1.\end{cases}$$

Let V_{pr} denote the primitive elements of V, i.e. those that are not non-trivial multiples of other elements. Let $a,b\in V_{\mathrm{pr}}$ and $l\in V$. Choose $\hat{a},\hat{b}\in V_{\mathrm{pr}}$ and $\hat{l}\in V\otimes\mathbb{Q}$ so that $\langle a,\hat{a}\rangle=\langle b,\hat{b}\rangle=\langle l,\hat{l}\rangle=1$. Define $X\in\mathrm{Aut}\,(V)$ by

$$\begin{aligned}X(a)&=b\\ X(\hat{a})&=\hat{b}.\end{aligned}$$

Choose $\widetilde{X}\in\widetilde{\mathrm{Aut}}\,(V)$ such that $\pi(\widetilde{X})=X$. Finally, define

$$\text{(A.4)}\quad \tau(a,b;l)=\frac{1}{12}\left(\frac{\langle a,\hat{l}\rangle}{\langle a,l\rangle}-\frac{\langle\hat{a},l\rangle}{\langle a,l\rangle}-\frac{\langle b,\hat{l}\rangle}{\langle b,l\rangle}+\frac{\langle\hat{b},l\rangle}{\langle b,l\rangle}\right)+\frac{1}{12}\varphi(\widetilde{X})-\left[\mathrm{rot}\,(\widetilde{X})+\frac{1}{2}\epsilon(a,b;l)\right]-\frac{1}{2}\delta(a,b;l).$$

($[x]$ denotes that greatest integer $\leq x$.)

(Note: Steve Boyer has pointed out to me that

$$\frac{1}{12}\varphi(\widetilde{X})-\left[\mathrm{rot}\,(\widetilde{X})+\frac{1}{2}\epsilon(a,b;l)\right]-\frac{1}{2}\delta(a,b;l)=$$
$$-\frac{1}{12}\Phi\left(\begin{bmatrix}\langle b,\hat{a}\rangle & \langle\hat{b},\hat{a}\rangle\\ -\langle b,a\rangle & -\langle\hat{b},a\rangle\end{bmatrix}\right)+\frac{1}{4}\mathrm{sign}\left(\frac{\langle a,b\rangle}{\langle a,l\rangle\langle b,l\rangle}\right),$$

where Φ is Radamacher's Φ-function

$$\begin{aligned} \Phi : SL_2(\mathbb{Z}) &\rightarrow \mathbb{Z} \\ \Phi\left(\begin{bmatrix} w & x \\ y & z \end{bmatrix}\right) &= \begin{cases} \frac{x}{z}, & y = 0 \\ -12s(z,y) + \frac{w+z}{y}, & y \neq 0 \end{cases} \end{aligned}$$

(see [RG]).)

Before showing that τ is well defined and proving that it has the desired properties, we need the following six lemmas. The proofs are elementary and are left to the reader.

(A.5) **Lemma.** *Let $\widetilde{X} \in \widetilde{SL}_2(\mathbb{R})$. Then $\pi(\widetilde{X}) \in SL_2(\mathbb{R})$ has a positive [negative] eigenvalue if and only if* $\mathrm{rot}\,(\widetilde{X}) \in \mathbb{Z}$ $[\mathrm{rot}\,(\widetilde{X}) \in \mathbb{Z} + \frac{1}{2}]$. □

(A.6) **Lemma.** *Let $\widetilde{X} \in \widetilde{SL}_2(\mathbb{R})$. Then* $\mathrm{tr}(\pi(\widetilde{X})) \geq 2$ $[\mathrm{tr}(\pi(\widetilde{X})) \leq -2]$ *if and only if $\pi(\widetilde{X})$ has a positive [negative] eigenvalue.* □

(A.7) **Lemma.** *Let $\widetilde{X}$, $\widetilde{Y} \in \widetilde{SL}_2(\mathbb{Z})$. Then*

$$|\mathrm{rot}\,(\widetilde{X}) + \mathrm{rot}\,(\widetilde{Y}) - \mathrm{rot}\,(\widetilde{X}\widetilde{Y})| < 1.$$

□

(For (A.7), keep in mind that elements of $\widetilde{SL}_2(\mathbb{Z})$ act on $\mathbb{R}_P$ with period 1/2, not merely 1.)

(A.8) **Lemma.** *Let $\widetilde{X}$, $\widetilde{Y} \in \widetilde{SL}_2(\mathbb{Z})$. If $[\widetilde{X}, \widetilde{Y}] = 1$, then*

$$\mathrm{rot}\,(\widetilde{X}) + \mathrm{rot}\,(\widetilde{Y}) - \mathrm{rot}\,(\widetilde{X}\widetilde{Y}) = 0.$$

□

For $a \in \mathbb{R}^2$ and $t \in \mathbb{R}$, define $S(a,t) \in SL_2(\mathbb{R})$ by $S(a,t)(b) = b + t\langle a, b\rangle a$. The mapping $t \mapsto S(a,t)$ is a homomorphism. Let $\widetilde{S}(a,t)$ denote the lift to $\widetilde{SL}_2(\mathbb{R})$. If $a \in \mathbb{Z}^2$ and $t \in \mathbb{Z}$, then $S(a,t) \in SL_2(\mathbb{Z})$ and $\widetilde{S}(a,t) \in \widetilde{SL}_2(\mathbb{Z})$. Let e_1 and e_2 be the standard basis of $\mathbb{Z}^2$. Note that $\widetilde{U} = \widetilde{S}(e_1, 1)$ and $\widetilde{L} = \widetilde{S}(e_2, -1)$, where $\widetilde{U}$ and $\widetilde{L}$ are as defined above.

(A.9) **Lemma.** *Let $\widetilde{X} \in \widetilde{SL}_2(\mathbb{R})$ and $a \in \mathbb{R}^2$. Then there is an integer n such that*

$$\mathrm{rot}\,(\widetilde{X}\widetilde{S}(a,t)) \subset n + (\mathrm{sign}\,\langle a, \pi(\widetilde{X})(a)\rangle)\,[0, \frac{1}{2}]$$

for all $t \in \mathbb{R}$. □

(A.10) **Lemma.** *Let* $a \in \mathbb{Z}^2_{pr}$, $n \in \mathbb{Z}$. *Then*

$$\varphi(\widetilde{S}(a,n)) = -n.$$

□

We now show that $\tau(a,b;l)$ is independent of the choices of $\widetilde{X}$, $\hat{a}$, $\hat{b}$ and $\hat{l}$. Denote the terms on the right hand side of (A.4) by 1 ($\frac{1}{12}\frac{\langle a,\hat{l}\rangle}{\langle a,l\rangle}$) through 7 ($-\frac{1}{2}\delta(a,b;l)$).

If we replace $\widetilde{X}$ with $k\widetilde{X}$, $k \in K$, then term 5 changes by $\operatorname{rot}(k)$ (by (A.3)), term 6 changes by $-\operatorname{rot}(k)$ (by (A.8) and the fact that K is contained in the center of $\widetilde{SL}_2(\mathbb{Z})$), and the other terms are not affected. Thus $\tau(a,b;l)$ does not depend on the choice of $\widetilde{X}$.

If we replace $\hat{a}$ with $\hat{a}+na$ ($n \in \mathbb{Z}$), we must replace X with $XS(a,-n)$ and $\widetilde{X}$ with $\widetilde{X}\widetilde{S}(a,-n)$. Terms 1, 3, 4 and 7 are not affected. By (A.9), term 6 does not change, while term 2 changes by $-\frac{n}{12}$ and (by (A.10)) term 5 changes by $\frac{n}{12}$. Thus $\tau(a,b;l)$ does not depend on the choice of $\hat{a}$.

Similar arguments show that $\tau(a,b;l)$ does not depend on the choices of $\hat{b}$ and $\hat{l}$. Therefore $\tau(a,b;l)$ is well-defined.

Now we show that τ has the properties required in sections 4 and 5. In the statements of the following lemmas, all variables (e.g. a, b, l) should be considered to be quantified by "for all", subject to the constraint that they lie in the domain of τ.

(A.11) **Lemma.** $\tau(-a,b;l) = \tau(a,-b;l) = \tau(a,b;-l) = \tau(a,b;l)$.

Proof: Changing the sign of a, b or l does not affect terms 1–4. Changing the sign of l does not affect terms 5–7.

There is a unique $\widetilde{R} \in \widetilde{SL}_2(\mathbb{Z})$ such that $\pi(\widetilde{R}) = \begin{bmatrix} -1 & 0 \\ 0 & -1 \end{bmatrix}$, $\operatorname{rot}(\widetilde{R}) = \frac{1}{2}$ and $\varphi(\widetilde{R}) = 6$. Furthermore, $\widetilde{R}$ is contained in the center of $\widetilde{SL}_2(\mathbb{Z})$. If we change the sign of a or b, we must replace $\widetilde{X}$ with $\widetilde{R}\widetilde{X}$. Suppose $\delta(a,b;l) = 0$. Then $\delta(a,b;l)$ changes by 1, $\epsilon(a,b;l)$ changes by -1, and $\operatorname{rot}(\widetilde{R}\widetilde{X}) = \operatorname{rot}(\widetilde{R}) + \operatorname{rot}(\widetilde{X}) = \operatorname{rot}(\widetilde{X}) + \frac{1}{2}$ (by (A.8)). Therefore term 5 changes by $\frac{1}{2}$, term 6 does not change, and term 7 changes by $-\frac{1}{2}$. The case $\delta(a,b;l) = 1$ is similar. □

(A.12) **Lemma.** *τ changes sign if the orientation on V changes.*

Proof: If we change the orientation on V, we must also change the signs of $\hat{a}$, $\hat{b}$ and $\hat{l}$. Thus terms 1–4 change sign. $\widetilde{X}$ and $\delta(a,b;l)$ remain the same, but rot and φ (and hence term 5) change sign. After possibly changing the sign of (say) a, we may assume (by (A.11)) that $\delta(a,b;l) = 0$. By choosing $\hat{a}$ and $\hat{b}$ properly, we may assume that $\operatorname{rot}(\widetilde{X}) \in \mathbb{Z}$. Hence term 6 and (trivially) term 7 also change sign. □

(A.13) **Lemma.** $\tau(b,a;l) = -\tau(a,b;l)$.

Proof: This is similar to 3 above. □

(A.14) **Lemma.** $\tau(a,b;l) + \tau(b,c;l) + \tau(c,a;l) = 0$.

Proof: After possibly changing the signs of a, b or c, we may assume (by (A.11)) that $\langle a,l\rangle$, $\langle b,l\rangle$, $\langle c,l\rangle > 0$. After possibly reordering a, b and c, we may assume (by (A.13)) that $\langle a,b\rangle$, $\langle b,c\rangle > 0$, $\langle c,a\rangle < 0$. (If, say, $\langle a,b\rangle = 0$ then we are done, by (A.13) and (A.11).) We can choose $\widetilde{X_{ab}}$, $\widetilde{X_{bc}}$ and $\widetilde{X_{ca}}$ so that $\widetilde{X_{ab}}\widetilde{X_{bc}}\widetilde{X_{ca}} = 1$. By choosing $\hat{a}$, $\hat{b}$ and $\hat{c}$ properly, we may assume that $\operatorname{rot}(\widetilde{X_{ab}})$, $\operatorname{rot}(\widetilde{X_{bc}})$ and $\operatorname{rot}(\widetilde{X_{ca}})$ are integers. By (A.7), this implies that $\operatorname{rot}(\widetilde{X_{ab}}) + \operatorname{rot}(\widetilde{X_{bc}}) + \operatorname{rot}(\widetilde{X_{ca}}) = 0$.

The contributions of terms 1–4 to $\tau(a,b;l)+\tau(b,c;l)+\tau(c,a;l)$ clearly cancel. The contribution of term 5 cancels because φ is a homomorphism. Since $\epsilon(a,b;l)$, $\epsilon(b,c;l)$ and $\epsilon(c,a;l)$ are either 0 or 1, the contribution from term 6 cancels. The contribution from term 7 cancels because $\delta(a,b;l) = \delta(b,c;l) = \delta(c,a;l) = 0$. □

(A.15) **Lemma.** $\tau(a + n\langle l,a\rangle l, b + m\langle l,b\rangle l; l) = \tau(a,b;l)$ *for all* $m,n \in \mathbb{Z}$ *if* $l \in V_{pr}$.

Proof: Suppose we replace a with $a + n\langle l,a\rangle l = S(l,n)(a)$. Then we must also replace $\hat{a}$ with $\hat{a} + n\langle l,\hat{a}\rangle l$ and $\widetilde{X}$ with $\widetilde{X}\widetilde{S}(l,-n)$. Thus term 1 changes by $-\frac{n}{12}$, term 5 changes by $\frac{n}{12}$ (by (A.10)), and terms 2, 3, 4 and 7 remain the same.

Suppose that $n < 0$. After possibly changing the sign of a and the orientation on V, we may assume (by (A.11) and (A.12)) that $\delta(a,b;l) = 0$ and $\langle a,b\rangle > 0$ or $a = b$. Under these circumstances the proper choice of $\hat{a}$ and $\hat{b}$ will guarantee that

$$\begin{aligned} \operatorname{tr}(\pi(\widetilde{X}\widetilde{S}(l,t))) &= -\langle \hat{a} + t\langle l,\hat{a}\rangle l, b\rangle + \langle a + t\langle l,a\rangle l, \hat{b}\rangle \\ &\geq 2 \end{aligned}$$

for all real $0 \leq t \leq -n$. Therefore (by (A.5) and (A.6)) $\operatorname{rot}(\widetilde{X}\widetilde{S}(l,-n)) = \operatorname{rot}(\widetilde{X}) \in \mathbb{Z}$. Since $\epsilon(a,b;l) = 0$ or 1 and $\epsilon(a + n\langle l,a\rangle l, b; l) = 0$ or 1 (because

$\delta(a,b;l)=0$ and $\delta(a+n\langle l,a\rangle l,b;l)=0$), term 6 does not change. The case $n>0$ is similar. □

(A.16) **Lemma.** $\tau(a,b;l_1)-\tau(a,b;l_2)=$

$$-\frac{1}{12}\left(\frac{\langle a,b\rangle}{\langle a,l_1\rangle\langle b,l_1\rangle}-\frac{\langle a,b\rangle}{\langle a,l_2\rangle\langle b,l_2\rangle}-\frac{\langle l_1,l_2\rangle}{\langle l_1,a\rangle\langle l_2,a\rangle}+\frac{\langle l_1,l_2\rangle}{\langle l_1,b\rangle\langle l_2,b\rangle}\right)$$
$$+\frac{1}{4}\left(\operatorname{sign}\frac{\langle a,b\rangle}{\langle a,l_1\rangle\langle b,l_1\rangle}-\operatorname{sign}\frac{\langle a,b\rangle}{\langle a,l_2\rangle\langle b,l_2\rangle}\right).$$

Proof: Since both sides of the identity asserted in (A.16) are antisymmetric with respect to $a\leftrightarrow b$, it suffices to consider the case $\langle a,b\rangle>0$. (The case $\langle a,b\rangle=0$ follows from (A.11) and (A.13).) The contribution of terms 1–4 to $\tau(a,b;l_1)-\tau(a,b;l_2)$ is

$$\frac{1}{12}\left(\frac{\langle a,\hat{l}_1\rangle\langle b,l_1\rangle-\langle\hat{a},l_1\rangle\langle b,l_1\rangle-\langle b,\hat{l}_1\rangle\langle a,l_1\rangle+\langle\hat{b},l_1\rangle\langle a,l_1\rangle}{\langle a,l_1\rangle\langle b,l_1\rangle}\right.$$
$$\left.-\frac{\langle a,\hat{l}_2\rangle\langle b,l_2\rangle-\langle\hat{a},l_2\rangle\langle b,l_2\rangle-\langle b,\hat{l}_2\rangle\langle a,l_2\rangle+\langle\hat{b},l_2\rangle\langle a,l_2\rangle}{\langle a,l_2\rangle\langle b,l_2\rangle}\right).$$

Upon repeated applications of the identity

$$\langle x,y\rangle=\langle x,\hat{z}\rangle\langle z,y\rangle-\langle x,z\rangle\langle\hat{z},y\rangle$$

(where $\langle z,\hat{z}\rangle=1$), this becomes

$$-\frac{1}{12}\left(\frac{\langle a,b\rangle}{\langle a,l_1\rangle\langle b,l_1\rangle}-\frac{\langle a,b\rangle}{\langle a,l_2\rangle\langle b,l_2\rangle}-\frac{\langle l_1,l_2\rangle}{\langle l_1,a\rangle\langle l_2,a\rangle}+\frac{\langle l_1,l_2\rangle}{\langle l_1,b\rangle\langle l_2,b\rangle}\right).$$

The contributions from term 5 clearly cancel. Since $\langle a,b\rangle>0$, $\epsilon(a,b;l_i)=0$ and hence the contribution from term 6 is zero. Keeping in mind that $\langle a,b\rangle>0$, we see that the contribution from term 7 is

$$-\frac{1}{2}\delta(a,b;l_1)+\frac{1}{2}\delta(a,b;l_2)=\frac{1}{4}\left(\operatorname{sign}\frac{\langle a,b\rangle}{\langle a,l_1\rangle\langle b,l_1\rangle}-\operatorname{sign}\frac{\langle a,b\rangle}{\langle a,l_2\rangle\langle b,l_2\rangle}\right)$$

□

(A.17) **Lemma.** *Let a, b and l be such that $\langle a,b\rangle=1$ and $\langle a,l\rangle\langle b,l\rangle>0$. Then*

$$\langle a,l\rangle\tau(a+b,a;l)+\langle b,l\rangle\tau(a+b,b;l)=\frac{1}{6}(\langle b,l\rangle-\langle a,l\rangle).$$

Proof: By (A.15) and (A.13)

$$\tau(a+b,a;b)=\tau(a,a;b)=0.$$

Therefore, by (A.16),

$$\begin{aligned}\langle a,l\rangle\tau(a+b,a;l) &= -\frac{\langle a,l\rangle}{12}\left(\frac{\langle b,a\rangle}{\langle a+b,l\rangle\langle a,l\rangle}-\frac{\langle b,a\rangle}{\langle a,b\rangle\langle a,b\rangle}\right.\\ &\qquad\left.-\frac{\langle l,b\rangle}{\langle l,a+b\rangle\langle b,a+b\rangle}+\frac{\langle l,b\rangle}{\langle l,a\rangle\langle b,a\rangle}\right).\end{aligned}$$

Similarly,

$$\begin{aligned}\langle b,l\rangle\tau(a+b,b;l) &= -\frac{\langle b,l\rangle}{12}\left(\frac{\langle a,b\rangle}{\langle a+b,l\rangle\langle b,l\rangle}-\frac{\langle a,b\rangle}{\langle b,a\rangle\langle b,a\rangle}\right.\\ &\qquad\left.-\frac{\langle l,a\rangle}{\langle l,a+b\rangle\langle a,a+b\rangle}+\frac{\langle l,a\rangle}{\langle l,b\rangle\langle a,b\rangle}\right).\end{aligned}$$

Adding these two equations establishes the lemma. □

It is perhaps worth noting that τ is uniquely determined by (A.14), (A.15) and (A.16).

Finally we show that the above definition of τ, (A.4), agrees with (4.1).

Choose an identification of V with $\mathbb{Z}^2$ so that $l=(d,0)$ for some $d\in\mathbb{Z}$. Let $a=(u,v)$ and $b=(p,q)$. Without loss of generality (see (A.11)), $q,v>0$.

First we calculate $\tau((0,1),(p,q);(1,0))$. Choose a continued fraction expansion $p/q=\langle a_0,a_1,\ldots,a_r\rangle$ of p/q. We may assume that $a_i\in\mathbb{Z}$, r is even, and $a_1,\ldots,a_r>0$. (See [NZ] for these and other elementary facts about continued fractions.) Let

$$X=U^{a_0}L^{a_1}U^{a_2}\cdots L^{a_{r-1}}U^{a_r}.$$

Note that $X(0,1)=(p,q)$. Let $(\hat{p},\hat{q})=X(-1,0)$. Then $\langle(p,q),(\hat{p},\hat{q})\rangle=1$ and

$$\frac{\hat{p}}{\hat{q}}=-\langle a_0,\ldots,a_{r-1}\rangle.$$

Note also that

$$\frac{\hat{q}}{q}=-\langle 0,a_r,\ldots,a_1\rangle.$$

Let $\widetilde{X} = \widetilde{U}^{a_0}\widetilde{L}^{a_1}\widetilde{U}^{a_2}\ldots\widetilde{L}^{a_{r-1}}\widetilde{U}^{a_r}$. It follows that

$$\varphi(\widetilde{X}) = -a_0 + a_1 - a_2 + \cdots + a_{r-1} - a_r.$$

Since $a_1, \ldots, a_r > 0$, $\operatorname{rot}(\widetilde{X}) \in [0, 1/2]$ and $\operatorname{rot}(\widetilde{X}) = 0$ if $p \geq 0$. Since

$$\delta((0,1),(p,q);(1,0)) = 0,$$

$\epsilon((0,1),(p,q);(1,0)) = 0$ if $p \leq 0$ and $\epsilon((0,1),(p,q);(1,0)) = 1$ if $p > 0$. Hence

$$\left[\operatorname{rot}(\widetilde{X}) + \frac{1}{2}\epsilon((0,1),(p,q);(1,0))\right] = 0.$$

By Theorem 1 of [H],

$$\begin{aligned} s(p,q) &= \frac{1}{12}\big(\langle a_0, a_1, \ldots, a_r\rangle - \langle 0, a_r, \ldots, a_1\rangle \\ &\qquad - a_0 + a_1 - a_2 + \cdots + a_{r-1} - a_r\big). \end{aligned}$$

Putting the above facts together we have

$$\begin{aligned} \tau((0,1),(p,q);(1,0)) &= \frac{1}{12}\left(\frac{0}{1} - \frac{0}{1} - \frac{p}{-q} + \frac{-\hat{q}}{q}\right) + \frac{1}{12}\varphi(\widetilde{X}) \\ &= \frac{1}{12}\big(\langle a_0, a_1, \ldots, a_r\rangle - \langle 0, a_r, \ldots, a_1\rangle \\ &\qquad -a_0 + a_1 - a_2 + \cdots + a_{r-1} - a_r\big) \\ &= s(p,q). \end{aligned}$$

Similarly,

$$\tau((u,v)(0,1);(1,0)) = -s(u,v).$$

Therefore, by (A.14) and (A.16),

$$\tau((u,v),(p,q);(d,0)) = -s(u,v) + s(p,q) + \frac{(d^2-1)}{12}\frac{(uq - vp)}{(dv)(dq)}.$$

This is equivalent to (4.1).

§ B

Alexander Polynomials

This section contains facts about Alexander polynomials (and second derivatives evaluated at 1 thereof) needed elsewhere in the paper. First we define Alexander polynomials. (For more details, see [M1].)

Let R be a commutative ring and M be a finitely generated R-module. Let M have generators $x_1, \ldots, x_n$ subject to the relations $\sum a_{ij}x_j = 0$, $1 \leq i \leq m$. The ideal of R generated by all $n \times n$ determinants of the matrix $[a_{ij}]$ is called the *order ideal* on M and denoted $o(M)$. (If $m < n$, $o(M) \stackrel{\text{def}}{=} 0$.) This is independent of the choice of presentation of M. o is multiplicative in the following sense.

(B.1) **Proposition.** *If $M_1 \subset M_2$ are R-modules, then*

$$o(M_2) = o(M_1)o(M_2/M_1).$$

□

If $o(M)$ is principal, let $\Delta(M)$ denote a generator of $o(M)$. $\Delta(M)$ is well-defined up to units in R.

If $\widetilde{X} \to X$ is a covering with group of deck transformations D, then $H_1(\widetilde{X}; \mathbb{Q})$ is a module over $\mathbb{Q}[D]$, and we define $o(\widetilde{X})$ to be $o(H_1(\widetilde{X}; \mathbb{Q}))$. If $D \cong \mathbb{Z}$, then $\mathbb{Q}[D]$ is a principal ideal domain, and we define $\Delta(\widetilde{X}) \in \mathbb{Q}[D]$ to be $\Delta(H_1(\widetilde{X}; \mathbb{Q}))$. This is well-defined up to units in $\mathbb{Q}[D]$, that is, elements of the form ct^k, where $c \in \mathbb{Q}$, t is a generator of D, and $k \in \mathbb{Z}$.

We are particularly interested in the case where X is the complement of a tubular neighborhood of a knot in a $\mathbb{Q}$HS, or, equivalently, X is a compact 3-manifold with $H_*(X; \mathbb{Q}) \cong H_*(S^1; \mathbb{Q})$. In this case the *universal* *free abelian cover* $\widetilde{X}_{FA}$ of X (determined by $\pi_1(X) \to \text{Free}(H_1(X;\mathbb{Z})) \cong \mathbb{Z}$) has deck transformations isomorphic to $\mathbb{Z}$ and we define $\Delta(X)$ (also denoted Δ_X), the *Alexander polynomial* of X, to be $\Delta(\widetilde{X}_{FA})$. Another case of interest is when X is the complement of a two component link in a $\mathbb{Q}$HS. $\widetilde{X}_{FA}$ has deck translations $\text{Free}(H_1(X;\mathbb{Z})) \cong \mathbb{Z} \oplus \mathbb{Z}$.

(B.2) **Lemma.** *Let X be the complement of a two component link in a $\mathbb{Q}$HS. Then $H_1(\widetilde{X}_{FA};\mathbb{Q})$ has a square presentation matrix over $\mathbb{Q}[\mathbb{Z} \oplus \mathbb{Z}]$.*

It follows that $o(\widetilde{X}_{FA})$ is principal, and we define $\Delta(X)$ to be $\Delta(\widetilde{X}_{FA})$.

Proof: Choose a basis x, y of $\text{Free}(H_1(X;\mathbb{Z})) \cong \mathbb{Z}^2$. There is a 2-complex $Y \subset X$ onto which X deformation retracts. For notational convenience, we now redefine X to be Y. Without loss of generality, X contains a single 0-cell p. Let $e_1, \ldots, e_n$ be the 1-cells of X and $f_1, \ldots, f_m$ be the 2-cells of X. Since the Euler characteristic of X is zero, $m = n - 1$. Without loss of generality

$$\begin{aligned} [e_1] &= x \in \text{Free}(H_1(X;\mathbb{Z})) \\ [e_2] &= y \in \text{Free}(H_1(X;\mathbb{Z})). \end{aligned}$$

Let $\tilde{p}$ be a fixed lift of p in $\widetilde{X}_{FA}$. Choose lifts $\tilde{e}_1, \ldots, \tilde{e}_n$ of $e_1, \ldots, e_n$ so that for all i and some $a_i, b_i \in \mathbb{Z}$

$$\partial(\tilde{e}_i) = (a_i x + b_i y)\tilde{p} - \tilde{p}.$$

(It follows that $a_1 = 1$, $b_1 = 0$, $a_2 = 0$, $b_2 = 1$.) Define the following 1-cycles in $\widetilde{X}_{FA}$:

$$\begin{aligned} l_1 &\stackrel{\text{def}}{=} \tilde{e}_1 + (x)\tilde{e}_2 - (y)\tilde{e}_1 - \tilde{e}_2 \\ l_{i-1} &\stackrel{\text{def}}{=} \tilde{e}_i - \text{sign}(a_i)(\sum_{k=0}^{a_i} kx + b_i y)\tilde{e}_1 - \text{sign}(b_i)(\sum_{k=0}^{b_i} ky)\tilde{e}_2, \end{aligned}$$

$3 \leq i \leq n$ (see Figure B.1).

Let $\widetilde{X}^1_{FA}$ denote the 1-skeleton of $\widetilde{X}_{FA}$. It is not hard to see that the classes $[l_1], \ldots, [l_{n-1}]$ freely generate $H_1(\widetilde{X}_{FA};\mathbb{Q})$ (over $\mathbb{Q}[\text{Free}(H_1(X;\mathbb{Z}))]$). There is an exact sequence

$$\text{(B.3)} \qquad H_2(\widetilde{X}_{FA}, \widetilde{X}^1_{FA};\mathbb{Q}) \xrightarrow{\partial} H_1(\widetilde{X}^1_{FA};\mathbb{Q}) \to H_1(\widetilde{X}_{FA};\mathbb{Q}) \to 0.$$

$H_2(\widetilde{X}_{FA}, \widetilde{X}^1_{FA};\mathbb{Q})$ is freely generated by $n-1$ lifts of the 2-cells $f_1, \ldots, l_{n-1}$ (one lift for each 2-cell). Hence (B.3) is equivalent to an $(n-1)$ by $(n-1)$

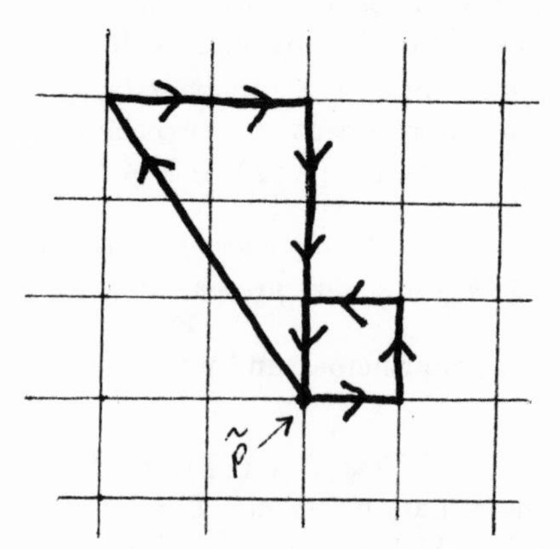

Figure B.1: Cycles in $\widetilde{X}_{FA}$.

presentation matrix for $H_1(\widetilde{X}_{FA}; \mathbb{Q})$. (Note that if X is the complement of a two component link of linking number zero, then this matrix will have rank $n-2$.) $\square$

Let X be the complement of a two component link in a $\mathbb{Q}$HS. Let

$$\varphi : \text{Free}\,(H_1(X;\mathbb{Z})) \to \mathbb{Z}$$

be a surjective homomorphism. Let t be a generator of $\mathbb{Z}$, with group composition written multiplicatively. Let $\varphi_* : \mathbb{Q}[\text{Free}\,(H_1(X;\mathbb{Z}))] \to \mathbb{Q}[\mathbb{Z}] = \mathbb{Q}[t,t^{-1}]$ be the induced map. Let $\widetilde{X}$ be the cover of X defined by the homomorphism $\pi_1(X) \to \text{Free}\,(H_1(X;\mathbb{Z})) \xrightarrow{\varphi} \mathbb{Z}$.

The next lemma is used in Section 5.

(B.4) **Lemma.** $\Delta(\widetilde{X}) = (1-t)\varphi_*(\Delta(\widetilde{X}_{FA}))$.

Proof: Retain notation from the proof of (B.2). The 2-complex Y (alias X) can be chosen so that $\varphi([e_1]) = t$ and $[e_2]$ generates the kernel of φ. Let $\widetilde{X}^1$ denote the 1-skeleton of $\widetilde{X}$. Let $\pi : \widetilde{X}_{FA} \to \widetilde{X}$ be the natural projection, and let π also denote its restriction to 1-skeletons.

Let k be the 1-cycle in $\widetilde{X}$ consisting of the lift of e_2 which contains $\pi(\tilde{p})$. $H_1(\widetilde{X};\mathbb{Q})$ is freely generated (over $\mathbb{Q}[t,t^{-1}]$) by $[k]$ and $[\pi(l_i)]$, $2 \le i \le n-1$. Note that

$$[\pi(l_1)] = (1-t)[k].$$

Let f be a 2-cell of X and $\tilde{f}$ be a lift of f to $\widetilde{X}_{FA}$ Let

$$[\partial(\tilde{f})] = \sum_{1}^{n-1} x_i[l_i],$$

where $x_i \in \mathbb{Q}[\mathrm{Free}\,(H_1(X;\mathbb{Z}))]$. Then

$$[\partial(\pi(\tilde{f}))] = \sum_{1}^{n-1} \varphi_*(x_i)[\pi(l_i)].$$

It follows that a square presentation matrix for $H_1(\widetilde{X};\mathbb{Q})$ is obtained by taking the square presentation matrix for $H_1(\widetilde{X}_{FA};\mathbb{Q})$ of (B.2), applying φ_* to the entries, and multiplying the first column by $1-t$. □

Let $\sum c_j[x_j] \in \mathbb{Q}[\mathbb{Z}]$. Let $\langle c\rangle \stackrel{\text{def}}{=} \sum c_j$ and $\langle cx\rangle \stackrel{\text{def}}{=} \sum c_j x_j$. If $\langle c\rangle \neq 0$, define

$$\Gamma(\sum c_j[x_j]) \stackrel{\text{def}}{=} \frac{\sum c_j\left(x_j - \frac{\langle cx\rangle}{\langle c\rangle}\right)^2}{\langle c\rangle}.$$

(If we think of $\sum c_j[x_j]$ as representing masses c_j at positions $x_j \in \mathbb{Z} \subset \mathbb{R}$, then Γ is the moment of inertia about the center of mass, divided by the total mass.) Note that Γ does not change if we multiply by units of $\mathbb{Q}[\mathbb{Z}]$. Thus $\Gamma(\Delta(M))$ is well-defined, so long as the sum of the coefficients of $\Delta(M)$ is not zero. If N is a knot complement (in a $\mathbb{Q}$HS), it is well known that $\Delta(N)$ has the form

$$\Delta(N) = \sum_j c_j(t^j + t^{-j}),$$

where t is a generator of $\mathbb{Z}$ written multiplicatively and j ranges over a finite set of half integers. (We are implicitly extending $\mathbb{Q}[\mathbb{Z}]$ to $\mathbb{Q}[\frac{1}{2}\mathbb{Z}]$. The case of half odd integers arises when the multiplicity of the longitude of the knot is even. See the proof of (B.10).) If we normalize so that $\Delta(N)|_{t=1} = 1$ (which is possible), then

(B.5) $$\Gamma(N) \stackrel{\text{def}}{=} \Gamma(\Delta(N)) = 2\sum_{j=0}^{n} c_j j^2 = \frac{d^2}{dt^2}\Delta(N)|_{t=1}.$$

We now relate Γ to Seifert matrices. (What follows is standard material (see, for example, [Ro]), except for complications arising from the fact that the longitude is not necessarily a primitive element of $H_1(\partial N;\mathbb{Z})$.)

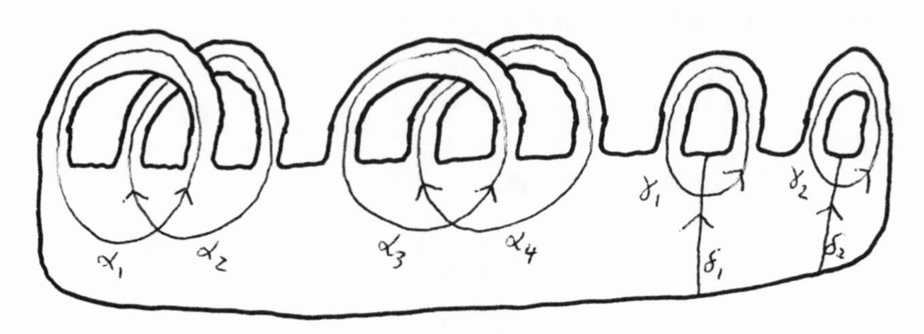

Figure B.2: Curves on E.

Let $E \subset N$ be a Seifert surface, that is, a connected, properly embedded surface representing a generator of $H_2(N, \partial N; \mathbb{Z}) \cong \mathbb{Z}$. Further assume that ∂E consists of d parallel (as opposed to antiparallel) curves on ∂N. (This can always be arranged.) Let $\alpha_1, \ldots, \alpha_{2g}, \gamma_1, \ldots, \gamma_{d-1}$ be simple closed curves representing a basis of $H_1(E; \mathbb{Q})$, as shown in Figure B.2. Orient the α_is so that

$$\langle \alpha_i, \alpha_j \rangle = \begin{cases} 1, & i \text{ odd}, j = i + 1 \\ -1, & i \text{ even}, j = i - 1 \\ 0, & \text{otherwise.} \end{cases}$$

Let Y be the complement of an open regular neighborhood of E in N. ∂Y contains two copies of E; call them E^+ and E^-. Let $\alpha_i^\pm \subset E^\pm$ be the curve corresponding to α_i. Let $a_i^\pm = [\alpha_i^\pm] \in H_1(Y; \mathbb{Q})$. Define $\gamma_i^\pm \subset E^\pm$ and $c_i^\pm \in H_1(Y; \mathbb{Q})$ similarly. $\widetilde{N}_{\mathrm{FA}}$ can be constructed by taking $\mathbb{Z}$ copies of Y and identifying E^+ of the ith copy with E^- of the $i+1^{\text{st}}$ copy. Identify Y with the zeroth copy of Y. Let $e_1, \ldots, e_n$ be a basis of $H_1(Y; \mathbb{Q})$. $H_1(\widetilde{N}_{\mathrm{FA}}; \mathbb{Q})$ is generated over $\mathbb{Q}[\mathbb{Z}] \cong \mathbb{Q}[t, t^{-1}]$ by $e_1, \ldots, e_n$ subject to the relations

$$\text{(B.6)} \qquad \begin{cases} a_j^+ = t a_j^-, & 1 \leq j \leq 2g \\ c_j^+ = t c_j^-, & 1 \leq j \leq d - 1 \end{cases}$$

In order to get an explicit presentation matrix for $H_1(\widetilde{N}_{\mathrm{FA}}; \mathbb{Q})$, we need to be more explicit about the basis $e_1, \ldots, e_n$. Let M be a $\mathbb{Q}$HS which contains N as the complement of an open solid torus U. Let $E^* \stackrel{\text{def}}{=} E \cup U$. By Alexander duality, $H_1(Y; \mathbb{Q}) \cong H_1(E^*; \mathbb{Q})$. More concretely, let $\epsilon_1, \ldots, \epsilon_n$ be curves representing a basis of $H_1(E^*; \mathbb{Q})$. Let V be the $\mathbb{Q}$-vector space with basis $\epsilon_1, \ldots, \epsilon_n$. Let $\eta \subset Y$ be a curve. Then the correspondence

$$[\eta] \mapsto \sum_i \mathrm{lk}(\eta, \epsilon_i) \epsilon_i$$

gives rise to to an isomorphism from $H_1(Y;\mathbb{Q})$ to V. For the ϵ_is we can take $\alpha_1, \ldots, \alpha_{2g}$, $\delta_1, \ldots, \delta_{d-1}$, where δ_i consists of an arc in E joining the 0th and i^{th} boundary components together with an arc in U (see Figure B.2). Orient the γ_is and δ_is so that

$$\langle \gamma_i, \delta_j \rangle = \begin{cases} 1, & i = j \\ 0, & \text{otherwise.} \end{cases}$$

With the above remarks in mind, it is easy to see that (B.6) gives rise to a presentation matrix

$$\begin{pmatrix} A & B \\ C & D \end{pmatrix},$$

where

$$\begin{aligned} A_{ij} &= \operatorname{lk}(\alpha_i^+, \alpha_j) - t \operatorname{lk}(\alpha_i^-, \alpha_j) \\ B_{ij} &= \operatorname{lk}(\alpha_i^+, \delta_j) - t \operatorname{lk}(\alpha_i^-, \delta_j) \\ C_{ij} &= \operatorname{lk}(\gamma_i^+, \alpha_j) - t \operatorname{lk}(\gamma_i^-, \alpha_j) \\ D_{ij} &= \operatorname{lk}(\gamma_i^+, \delta_j) - t \operatorname{lk}(\gamma_i^-, \delta_j). \end{aligned}$$

If $E^{\pm}$ is deformed slightly near its boundary, then $\partial E^{\pm}$ consists of d copies of $\gamma_i^{\pm}$. Since α_j is disjoint from $E^{\pm}$, $\operatorname{lk}(\gamma_i^{\pm}, \alpha_j) = \frac{1}{d}\langle \alpha_j, E^{\pm} \rangle = 0$, and hence $C = 0$. Therefore

$$\Delta(N) = \det(A)\det(D). \tag{B.7}$$

If every entry of A and D is multiplied by $t^{-1/2}$, that is

$$\begin{aligned} A_{ij} &= t^{-1/2}\operatorname{lk}(\alpha_i^+, \alpha_j) - t^{1/2}\operatorname{lk}(\alpha_i^-, \alpha_j) \\ D_{ij} &= t^{-1/2}\operatorname{lk}(\gamma_i^+, \delta_j) - t^{1/2}\operatorname{lk}(\gamma_i^-, \delta_j), \end{aligned}$$

then (B.7) changes by a unit (of $\mathbb{Q}[t^{1/2}, t^{-1/2}]$) to become symmetric in $t^{1/2}$ and $t^{-1/2}$. Since this is a desirable property for $\Delta(N)$ to have, consider this change to have been effected in what follows.

If $\beta_1, \beta_2 \subset E$ are simple closed curves, then

$$\operatorname{lk}(\beta_1^+, \beta_2) - \operatorname{lk}(\beta_1^-, \beta_2) = \langle \beta_1, \beta_2 \rangle.$$

It follows that

$$\begin{aligned} A_{ij}|_{t=1} &= \langle \alpha_i, \alpha_j \rangle \\ D_{ij}|_{t=1} &= \langle \gamma_i, \delta_j \rangle, \end{aligned}$$

and hence that $\det(A)|_{t=1} = \det(D)|_{t=1} = 1$. Thus (B.7) is the symmetric, normal form of $\Delta(N)$, and

$$\Gamma(N) = \left.\frac{d^2}{dt^2}\right|_{t=1} (\det(A)\det(D)). \tag{B.8}$$

Using the fact that

$$\left.\frac{d}{dt}\right|_{t=1} (\det(A)) = \left.\frac{d}{dt}\right|_{t=1} (\det(D)) = 0,$$

(B.8) becomes

$$\Gamma(N) = \left.\frac{d^2}{dt^2}\right|_{t=1} (\det(A)) + \left.\frac{d^2}{dt^2}\right|_{t=1} (\det(D)). \tag{B.9}$$

It is not hard to verify that

$$D_{ij} = \begin{cases} \frac{j}{d}(t^{1/2} - t^{-1/2}), & j < i \\ \frac{j}{d}(t^{1/2} - t^{-1/2}) + t^{-1/2}, & j = i \\ (\frac{j}{d} - 1)(t^{1/2} - t^{-1/2}), & j > i. \end{cases}$$

A fun-filled exercise in algebraic manipulation (left to the reader, of course) now yields

$$\Gamma(N) = 2 \sum_{1 \le i,j \le g} \det \begin{bmatrix} \mathrm{lk}\,(\alpha_{2i-1}^-, \alpha_{2j-1}) & \mathrm{lk}\,(\alpha_{2i-1}^+, \alpha_{2j}) \\ \mathrm{lk}\,(\alpha_{2i}^-, \alpha_{2j-1}) & \mathrm{lk}\,(\alpha_{2i}^+, \alpha_{2j}) \end{bmatrix} + \frac{d^2 - 1}{12}. \tag{B.10}$$

Bibliography

[AM] S. Akbulut and J. McCarthy, Casson's invariant for oriented homology 3-spheres – an exposition, Princeton mathematical notes 36, Princeton university press, 1990.

[AS4] M. F. Atiyah and I. M. Singer, The index of elliptic operators: IV, Ann. Math. **93** (1971), 119-138.

[AS] M. F. Atiyah and I. M. Singer, Dirac operators coupled to vector potentials, Proc. Nat. Acad. Sci. **81** (1984), 2597-2600.

[BL] S. Boyer and D. Lines, Surgery formulae for Casson's invariant and extensions to homology lens spaces, UQAM Rapport de recherche 66 (1988).

[BN] S. Boyer and A. Nicas, Varieties of group representations and Casson's invariant for rational homology 3-spheres, preprint (1986).

[G1] W. Goldman, The symplectic nature of the fundamental groups of surfaces, Adv. Math. **54** (1984), 200-225.

[G2] W. Goldman, Representations of fundamental groups of surfaces, in Proceedings of Special Year in Topology, Maryland 1983-1984, Lecture notes in mathematics 1167, Springer-Verlag, 1985.

[G3] W. Goldman, Invariant functions on Lie groups and hamiltonian flows on surface group representations, Invent. math. **85** (1986), 263-302.

[GM] W. Goldman and J. Millson, The deformation theory of representations of fundamental groups of compact Kähler manifolds, Publ. Math. I.H.E.S. **67** (1988), 43-96.

[Go] C. Gordon, Knots, homology spheres, and contractable 4-manifolds, Topology **14** (1975), 151-172.

[H] D. Hickerson, Continued fractions and density results for Dedekind sums, J. Reine Angew. Math. **290** (1977), 113-116.

[HZ] F. Hirzebruch and D. Zagier, The Atiyah-Singer index theorem and elementary number theory, Publish or perish, 1974.

[K1] R. Kirby, A calculus for framed links in S^3, Invent. math. **45** (1978), 35-56.

[K2] R. Kirby, The topology of 4-manifolds, Lecture notes in mathematics 1374, Springer-Verlag, 1989.

[L] W. B. R. Lickorish, A representation of orientable combinatorial 3-manifolds, Ann. Math. **76** (1962), 531-538.

[M1] J. W. Milnor, Infinite cyclic coverings, in Conference on the topology of manifolds (Michigan State Univ., East Lansing Mich., 1967), Prindle, Weber and Schmidt, 1968.

[M2] J. W. Milnor, A duality theorem for Reidemeister torsion, Ann. Math. **76** (1962), 137-147.

[MH] J. Milnor and D. Husemoller, Symmetric Bilinear forms, Ergebnisse der mathematik und ihrer grenzegebiete 73, Springer-Verlag, 1973.

[Mo] L. Moser, Elementary surgery on a torus knot, Pacific J. Math. **38** (1971), 737-745.

[N1] P. E. Newstead, Topological properties of some spaces of stable bundles, Topology **6** (1967), 241-262.

[N2] P. E. Newstead, Characteristic classes of stable bundles of rank 2 over an algebraic curve, Trans. A. M. S. **169**, 337-345 (1972).

[NZ] I. Niven and H. Zuckerman, An introduction to the theory of numbers, 3rd ed., John Wiley and sons, 1972.

[Q] D. Quillen, Determinants of Cauchy-Riemann operators over a Riemann surface, Funct. anal. appl. **19** (1985), 31-34.

[RG] H. Rademacher and E. Grosswald, Dedekind sums, Carus mathematical monographs 16, M.A.A., 1972.

[Ro] D. Rolfsen, Knots and links, Publish or perish, 1976.

[S] J. Singer, Three dimensional manifolds and their Heegaard diagrams, Trans. AMS **35** (1933), 88-111.

[W1] K. Walker, An extension of Casson's invariant to rational homology spheres, Bull. AMS **22** (1990) 261-268.

[W2] K. Walker, The μ-invariant as a topological quantum field theory, in preparation.

[We] A. Weinstein, Lectures on symplectic manifolds, C.B.M.S. regional conference series in math., no. 29, A.M.S., 1979.

Happy Birth
You are my Shero!

June 2019

"As the father of Title IX, Birch Bayh has left a lasting impact on our country. In *Birch Bayh: Making a Difference*, it is clear his influence and his contributions will continue to affect all Americans for generations to come in many ways."

—Billie Jean King, founder of the Billie Jean King Leadership Initiative

"Robert Blaemire's celebration of Birch Bayh's career evokes a different and better time in American politics, when leaders really did think about making a difference. There are many vivid portraits of Bayh's work on landmark civil rights bills and other legislation. My favorite passage cites Bayh's question for his staff when facing a tough choice: 'Just tell me what you feel is the right thing to do.' Read this memoir and remember a time when people like Bayh, and the decent, compassionate politics of the heartland, were truly the American way."

—David Ignatius, columnist, *The Washington Post*

"Robert Blaemire's *Birch Bayh* is a marvelous biography of Bayh, a dynamo in Indiana politics and the national scene throughout the 20th century. My takeaway, after reading, was that Bayh, a consummate public servant, would have made an excellent president. Highly recommended!"

—Douglas Brinkley, author of *Rightful Heritage: Franklin D. Roosevelt and the Land of America*

"My friend Birch Bayh has led a life of remarkable public service dedicated, always, to making a difference. His supporters and opponents will long remember his skill as a campaigner combined with his ability to reach across the political aisle and achieve constitutional amendments and timely legislation that strengthened our nation."

—Richard Lugar, former United States Senator from Indiana

"Birch Bayh was one of the most consequential lawmakers of the 20th century, responsible for constitutional amendments and a long list of legislative accomplishments that changed and improved America. Robert Blaemire has given us a biography that does justice to a great American, a vivid portrait of the man and the Senate at a time when Bayh could work with allies and adversaries alike."

—Norman Ornstein, Resident Scholar, American Enterprise Institute

"The story of Birch Bayh's political career is completely inspiring, especially in an era that has lost touch with bipartisanship and civility. A must read for Hoosiers and for anyone interested in how democracy worked, when it really worked."

—Ted Widmer, historian and former presidential speechwriter

"In Indiana's, and the nation's, political history, perhaps no elected official has produced the legislative achievements crafted by US Senator Birch Bayh. In addition to authoring two constitutional amendments—the Twenty-Fifth and Twenty-Sixth—Bayh produced the landmark Title IX legislation, providing women with equal opportunities in public education. Bayh has long needed a comprehensive biography, and Robert Blaemire has provided an insider's account of Bayh's life and career and places him among Indiana's leading political figures."

—Ray E. Boomhower, author of *Robert F. Kennedy and the 1968 Indiana Primary*

"Highlights the life of one of our most remarkable United State Senators, not just in Indiana but in the nation. *Birch Bayh* shows the dedication of a man to his state and country through more than 25 years of elected office."

—Geoffrey Paddock, author of *Indiana Political Heroes*

Birch Bayh

MAKING A DIFFERENCE

ROBERT BLAEMIRE

INDIANA UNIVERSITY PRESS

This book is a publication of

Indiana University Press
Office of Scholarly Publishing
Herman B Wells Library 350
1320 East 10th Street
Bloomington, Indiana 47405 USA

iupress.indiana.edu

 The paper used in this publication meets the minimum requirements of the American National Standard for Information Sciences—Permanence of Paper for Printed Library Materials, ANSI Z39.48-1992.

Manufactured in the United States of America

Cataloging information is available from
the Library of Congress.

ISBN 978-0-253-03917-0 (hardback)
ISBN 978-0-253-03918-7 (ebook)

1 2 3 4 5 24 23 22 21 20 19

To

NICK AND DAN,
WHO ARE MAKING A DIFFERENCE IN
THEIR OWN WORLDS,
AND OF WHOM I AM SO PROUD.

Contents

Acknowledgments

BIRCH BAYH PASSED AWAY ON MARCH 14, 2019, AT THE AGE OF 91. Up until the time of his death, he had been living in Easton, Maryland, with his wife, Kitty. Age had slowed him down physically but not mentally, and all evidence seemed to indicate that he enjoyed his retirement on Maryland's Eastern Shore. Kitty took great care of him, and he often talked about the huge and important role she played in his life.

This biography was a labor of love. My long relationship with Birch Bayh and the admiration I have for his career fueled my enthusiasm for this project. This book wouldn't have been possible without the enormous support and cooperation of Birch and Kitty Bayh. I conducted twelve video interviews with Birch over a period of over four years; the material and photographs Birch and Kitty provided to me were invaluable.

But that is only part of the story: The assistance of former staff colleagues helped me fill in many of the details, both of politics and the legislative process. This work would have been incomplete without the contributions from these people.

Video interviews were conducted with former staffers Gordon Alexander, Jay Berman, David Bochnowski, Bob Boxell and his wife Peggy, Tom Connaughton, Terry Crone, Jim Freidman, Mary Grabianowski, Bob Hinshaw, Bob Keefe, Pat Long, P.A. Mack, Louis Mahern, Ann Moreau, Bill Moreau, Fred Nation, Allan Rachles, David Rubenstein, Diane Meyer Simon, Joe Smith, Jeff Smulyan, Darry Sragow, and Trish Whitcomb.

Video interviews were also conducted with former Congressman Lee Hamilton, former senator Richard Lugar, and Senators Patrick Leahy and Orrin Hatch. The video interview and conversations with former senator Evan Bayh were important additions to this work.

Additional nonvideo interviews and conversations, both verbal and written, that helped advance my research were with former staffers Nels Ackerson, Chris Aldridge, Gail Alexander, Joe Allen, Lew Borman, Tom Buis, Mary Jane Checchi, Ann Church, Patty Dewey, John Dibble, Barbara Dixon, Jim English, Mathea Falco, Kevin Faley, Ed Grimmer, Mary Jolly, Ron Klain, Gary Kornell, Ann Latscha, Barbara Leeth, Tim Leeth, Eve Lubalin, Susan McCarthy, Lynne Mann, Tim Minor, Jay Myerson, Carol Ann Nix, Nancy Papas, John Rector, Joe Rees, John Reuther, Steve Richardson, Abby Saffold, Jerry Udell, and Mark Wagner. Conversations with former mayor Richard Gordon Hatcher were also very valuable. I am also indebted to Herb Simon for our conversation about his relationship with Birch as well as his other efforts to help make this book possible.

I am indebted to my brother-in-law, Lincoln Caplan, for his sage advice about writing a book and getting it published. He connected me to David Korzenik, who provided me with valuable legal advice. My gratitude to both.

A particular debt of gratitude is owed to those who read the text and provided valuable additions, corrections and editing improvements. First on that list is Joanna Caplan, but also Jay Berman, Terry Crone, Kate Cruikshank, Pat Long, Lynne Mann, Diane Meyer, Bill Moreau, Nancy Papas, and Joe Rees. This book is considerably improved because of their contributions.

Thanks is also due to Chris Bayh for his careful reading of the text and the suggestions that he made.

To Bill Moreau, my friend and consigliere, only you know the full story of your contributions that made this possible.

To those I have worked with at Indiana University Press, professionals all, my gratitude goes out to Ashley Runyon and Gary Dunham.

Finally, like so many acknowledgements in books published year in and year out, I need to thank my family. They have been told Birch Bayh stories for the entire time they have known me and despite that, their support and encouragement were total.

To all these people I say thank you. If you had told me as an 18-year-old entering college that I would spend the next thirteen years working for Senator Birch Bayh, I would have been incredulous. I certainly never thought I would become his biographer. But words cannot adequately describe the affection I feel for him or the impact our relationship has had on my life. He entrusted me with a great many things in his career while I was very young, and my gratitude can never be fully expressed. It is my hope that telling his story will begin to pay that debt.

Birch Bayh

Introduction

UNITED STATES SENATOR BIRCH BAYH, A DEMOCRAT FROM Indiana, served in the Senate for eighteen years during a tumultuous time in American history. A campaign slogan of his was "One Man Can Make a Difference," and it reflected his belief and his motivation to seek higher office. His story is out of Horatio Alger: a man from America's heartland without wealth, elected Speaker of the House in the Indiana legislature at age thirty and US senator at thirty-four, author of two constitutional amendments and of landmark legislation with effects lasting through the ages.

His story is not only one of substantial personal accomplishment, however. It is also a story about an era when things worked in American government, an era when Democrats worked with Republicans, when giants walked the halls of Congress and when the public interest was served in ways that are still felt today. It was an era of change, with social movements seeking civil rights, women's rights, gay rights, and environmental protections. It was an era of assassinations, scandal, and a presidential resignation. America's attention was focused on the United States Senate, and Birch Bayh was very much in the midst of that attention.

Three highlights of Birch Bayh's history illustrate his impact.

* * *

The twentieth century was nearing its fourth quarter as the United States faced the largest political scandal in its history. The Watergate revelations had been headline news from the time of the break-in at the Democratic National Committee's Watergate office in June 1972 until President Richard Nixon resigned from the presidency two years and two months later.

Before the Nixon resignation, Vice President Spiro Agnew resigned his office in disgrace.

What became clear as the Watergate scandal unfolded and Nixon's involvement became evident was that if the vice presidency remained vacant, Speaker of the House Carl Albert would be next in line. Instead of handing the office to a traditional successor, Nixon would be turning over the entire administration to the Democratic Party. For the first time in American history, the Constitution provided a process to fill the vacancy. The Twenty-Fifth Amendment, ratified in 1967, provided a smooth process, and Michigan congressman Gerald Ford became vice president.

Under intense pressure with his congressional support dwindling, President Nixon resigned. Without the Twenty-Fifth Amendment, the smooth transition from Nixon to Ford might not have taken place after the Nixon resignation. The amendment would be invoked a second time when Ford became president, and the vacancy in the office of vice president would be filled once again, this time by former New York governor Nelson Rockefeller. The Twenty-Fifth Amendment, ratified only seven years before, was invoked twice in 1974, changing American history in the process.

* * *

In June 2012, President Barack Obama's White House hosted a fortieth anniversary celebration of a groundbreaking legislative measure called Title IX, a celebration to highlight the dramatic impact the measure has had on American life. Passed into law in 1972, Title IX was an amendment added during the reauthorization of the Higher Education Act of 1965, taking language out of the proposed Equal Rights Amendment:

"No person in the United States shall, on the basis of sex, be excluded from participation in, be denied the benefits of, or be subjected to discrimination under any education program or activity receiving federal financial assistance."

Title IX has had a tremendous impact on American life, changing women's level of participation in all walks of life because of the educational opportunities created for women by its passage. Most attention has been on the role it played in creating women's sports, but that was more of a side effect than a central intention. The culture of the country has been changed by Title IX. One example of that change is a quotation from that White House ceremony in June 2012. Laurel Ritchie, president of the Women's National Basketball Association (WNBA), spoke of taking her young niece to many WNBA games but said that when she took her to see her first NBA game, her niece said to her, "Auntie, I didn't know boys played basketball."

* * *

In 1980, most new innovations and inventions emerging from universities and small businesses were not being brought to market because of government patent policy. The policy said, in effect, that if taxpayers are helping to pay for these innovations, the right to bring them to the public belongs to the public. But the result was that they were not being brought to the public at all. Because Congress finally acted to meet this crisis, things look much different in the decades since. University inventions have spurred on the creation of thousands of new American companies; university patent licensing has brought more than 4,350 new products onto the market. There are now thousands of university–industry licensing partnerships in effect; most of those partnerships are with small companies, leading to the commercialization from this federally funded research of more than 200 new drugs, vaccines, or in vitro devices.

It has been estimated that between 1996 and 2013, university patent licensing contributed $1.18 trillion to the US economy and supported nearly four million good-paying jobs. No other system in the world even approximates these economic and public health benefits derived for American citizens from publicly supported research made possible by this act of Congress, the Bayh–Dole Act. One company that grew out of a federal–university partnership that could not have happened prior to Bayh–Dole is Google.

One person, Birch Bayh, wrote all three of these measures: the Twenty-Fifth Amendment to the Constitution, Title IX, and the Bayh–Dole Act. He also guided to passage the Twenty-Sixth Amendment to the Constitution that lowered the voting age to eighteen, was the Senate sponsor of the unsuccessful Equal Rights Amendment, led the successful Senate opposition to two Nixon Supreme Court nominations, and was a leader in Senate passage of legislation dealing with juvenile delinquency, gun control, the Foreign Intelligence Surveillance Act (FISA), and the original funding of the Washington, DC subway system. He developed a mastery of inside and outside power: the ability to accomplish goals within an institution while motivating influencers outside to help make those accomplishments happen. He is the only person since the Founding Fathers to have steered two constitutional amendments to passage. It can be argued that his greatest contribution was in preventing certain proposed constitutional amendments from becoming the law of the land.

His life demonstrates that a person of ability and ambition can reach the heights of a chosen career and make a difference. This is his story.

1

Farmer, Soldier, Legislator

Democracy is the one form of society which guarantees
to every new generation of men the right to imagine
and to attempt to bring to pass a better world.

FRANKLIN D. ROOSEVELT[1]

THE FUTURE STATE OF INDIANA WAS POPULATED BY NATIVE Americans as early as 8000 BCE. It was explored and then claimed by the French in the 1670s and transferred to British control after the French and Indian War.

Established from a portion of the Northwest Territory, Indiana grew in population until it became the nineteenth state in 1816. Many settlers moved there for the trapping, to farm, or simply to seek a better life. The first governor of the Indiana Territory was William Henry Harrison, who served from 1800 to 1813. As a general in the Indian wars, inexorably forcing the Native Americans off the land they had possessed for centuries, he made his reputation in Indiana and began a political career that would result in his election as president in 1840. Many of Indiana's ninety-two counties would be named for his officers in those Indian wars.[2]

As America expanded westward, the National Road was built across the territories, ultimately reaching Indianapolis in 1829. President Lincoln's Homestead Act encouraged further migration into states and territories like Indiana with the promise of free land to those hardy enough to get there and homestead the land. Indiana's population grew as rivers and waterways generated commercial activity and steel mills were built in the northwest, beckoning skilled and unskilled laborers. As the railroad was built, it would play a large role in populating the young state.[3]

The lure of railroad jobs provided the incentive for Christopher Bayh (pronounced Bye) to leave Germany at age twenty and migrate to Indiana in 1858, part of a wave of German immigration that brought to America almost 900,000 in the previous eight years.[4] He arrived in the United States with only a paper bag containing some belongings and wearing a pair of overalls with a note pinned to them: "Send this man out to work on the railroad." He came from a long line of Bayhs in Germany and, like many of that era, saw America as the land of opportunity and hope. The railroad would be his vehicle to realize his dreams. Eventually, he found himself in southern Indiana, near the town of Spencer in Owen County, slightly north of where Abraham Lincoln resided as a boy.

Christopher Bayh (1836–1915) made a life for himself in Indiana, marrying Christina Crauf, another German immigrant, in 1867. Christopher lived seventy-nine years. His wife, Christina (1840–1895), preceded him in death at the age of fifty-five. They were the parents of two sons, John (1867–1966) and Frederick (1871–1947).

Fred grew up in nearby Spencer, where he and his brother owned Bayh Brothers Wagon and Buggy Shop, eventually also owning a hardware store. He walked to and from his hardware store reliably at the same time every day, which his neighbors said allowed them to "set their watches" by Fred Bayh. Fred married Nettie Evans (1872–1934), and they had three children: Ruth, Bernard, and Birch. The image of Fred and Nettie Bayh sitting in a swing on the porch of their brick house is indelible. Fred's grandson remembered a nice wagon his grandfather made for him. Nettie was the daughter of Jesse and Elizabeth Evans, and her death would be the first one her grandson ever remembered.

Ruth, later Ruth Bayh Bourne, kept the house her parents owned and lived in it throughout her life. Her father, Fred, died in that house as well. Bernard was remembered as a fine left-handed pitcher and would be the father of two sons, Fred and Bill. Birch Evans Bayh, born in Quincy, Indiana, on September 29, 1893, would have a distinguished career as a teacher, a coach of high school and college basketball and baseball, a high school basketball referee, and a military officer. Once asked about the genesis of his first name, he said that his mother had been reading a romantic novel before his birth. The hero of the novel was named Birch, and she was so enamored of him that she took the name for her son.

John Hollingsworth (1865–1953) and Mary Katherine Ward (1865–1951) came from families that migrated from Great Britain and settled in the Shenandoah Valley before moving west and settling in southern Indiana. They married and had one child, Leah Ward Hollingsworth, born in 1897 at Union Hospital in Terre Haute. Leah grew up to be a teacher at Fayette Township High School in Terre Haute, where she would meet a young coach at Indiana State Normal School, later known as Indiana State Teachers College and now as Indiana State University. The young coach was Birch Evans Bayh. They would marry and have two children: Birch Evans Jr. (born in 1928) and Mary Alice (born in 1930). Birch and Mary Alice were about eighteen months apart in age and would graduate from the same Fayette Township High School where their mother had taught.

Born on January 22, 1928, Birch Evans Bayh Jr. arrived at an interesting time in American history. America was prospering in the decade after the end of World War I. Calvin Coolidge was president. The year before Birch's birth, Charles Lindbergh flew across the Atlantic Ocean and Babe Ruth hit sixty home runs.

In 1928, the world stage saw Leon Trotsky arrested in Moscow and exiled and saw Stalin launch his first five-year plan. The Kellogg–Briand Pact was signed in Paris, outlawing aggressive war forever. At home, there was the first regular schedule of television programming in Schenectady, New York, and Walt Disney released a cartoon in which Mickey Mouse first appeared. The Democratic Party nominated the first Catholic for president, Governor Al Smith of New York, who was defeated by Herbert Hoover. The convention was simulcast on radio and television.[5]

Birch Evans Bayh Sr., father of Birch and Mary Alice Bayh, served as an example for his children. His passion was physical education and fitness. He served as an NCAA head basketball coach and was basketball and baseball coach at Indiana State from 1918 to 1923, also serving as its athletic director and a professor of physical education. He made quite a reputation for himself in Terre Haute as a coach and statewide as a basketball referee. During his political career, his son would travel around

Indiana, regularly running into people with flattering stories about his father, including memories of watching him referee basketball games.

As basketball coach at Indiana State, Birch Sr. led the Sycamores to their first fifteen-win season in 1920–21 and two years later to a 20-and-5 season. His .640 winning percentage ranks him sixth in school history. He became director of physical education for the Terre Haute Public Schools and later held the same job in Washington, DC. He also received an exercise certificate (associate degree) from the North American Gymnastics Union of Indianapolis and coauthored one of the first physical education manuals for American schools. He still holds the record for refereeing ten Indiana high school basketball championship games. His was a hard act to follow but one that brought pride to his children.

Birch Evans Bayh Jr., hereinafter referred to simply as Birch, would be called Bud within the family. His earliest memory was when he was only four or five years old, when the train arrived near the fairgrounds to unload the circus that would soon appear. He remembered the sound of the calliope and recalled watching the men putting up the circus tent and unloading the animals. His mother, Leah, was one hundred percent mother and housewife, a good cook who made her own clothes. His father had courted her in his Model A Ford, and after they married, they made their home in what Birch would describe as a "little bungalow" at 242 Barton Avenue in Terre Haute.[6]

Terre Haute when Birch arrived was not much of a metropolis. It has been described as the northern center of the Ku Klux Klan in the 1920s, "full of demented kooks—including many of the city's most prominent citizens—prancing around in their sheets and pillowcases."[7]

When Birch was in the third grade, the Washington, DC public school system hired his father to head its department of physical education. The young family moved to the Lux Manor area in Rockville, Maryland, to a house at 7 Sedgwick Lane near Old Georgetown Road, five or six miles from DC. The house cost just $17,000. Birch attended Bethesda Elementary School and Alta Vista Elementary School. The latter school was experimenting with classes that included students from two grades at the same time, which may help to explain why he would graduate from high school at age seventeen. Starting in 1939, he attended Leland Jr. High for grades seven through nine. With a January birthdate, his parents could choose whether to hold him back or start him early. They did the latter, and he was always among the youngest in his classes.

When Birch was twelve years old, Leah was hospitalized with uterine cancer, now detectable with a Pap smear. She was at Columbia Hospital for Women in Washington, DC, where her husband would spend lots of time. Every day after school, Birch called the hospital to talk with his mother or to ask his father's permission to do whatever he had plans for that afternoon or evening. He remembered being in bed with his father during this period, listening to him describe his mother's condition and asking him, "Dad, do you mean Mom might die?" His father replied, "We hope that won't be the case."

One afternoon he called the hospital and the operator said, "There's no Mrs. Bayh here." He replied, "I talked to her this morning." She put him on hold while she checked and on returning told him, "Mrs. Bayh died. I'm sorry." The phone went dead.

Mary Alice remembered Birch screaming, which he had blanked from his memory.

In 1940 the family moved back to Indiana, to the 360-acre Hollingsworth farm just south of Shirkieville, near Terre Haute. Birch had completed the first year of high school at Bethesda-Chevy Chase in Bethesda, Maryland, and transferred to Fayette Township High School in Vigo County, where his mother had taught. Birch felt his grandparents had saved his life; being raised by a single father in the DC area never would have brought him the benefits of living on a farm with his grandparents. He remembered them as pioneer types, Granddad John Hollingsworth having crossed the Shenandoah Mountains in a covered wagon. Mary Katherine "Kate" Hollingsworth was remembered as five feet tall with knee-length hair that she usually kept wrapped in a bun behind her head. He remembered that they showed no emotion, but he did recall sitting at the top of the stairs with his sister, listening to his grandparents talking with his father and bringing him to task for "taking their girl to Washington," partially blaming him for her death. Birch remembered no other women ever being present in his father's life.

Birch Bayh Sr. had been a captain in World War I; he was a major by the age of twenty-one, and he reentered the military in May 1941. He joined the army and toured the United States, visiting bases where he taught physical fitness to young pilots. Within a month after Pearl Harbor, he received his commission and shipped off to Kunming, China, to work with the Flying Tigers. "Flying Tigers" was the nickname for the First American Volunteer Group of the Chinese air force, recruited under presidential authority to assist in the defense of China against the Japanese. His ranking meant he would always be addressed as Colonel Bayh. Birch and Mary Alice continued to live in Shirkieville with their grandparents.

Birch always regretted being too young to fight in World War II. When they lived in Rockville, their father would sometimes drive them to Rehoboth, Delaware, right after the war in Europe began. He recalled seeing oil washed up on the beaches because of the damage the German navy was doing to Allied shipping. Later Birch seriously considered enlisting for the Korean War, but the five-year commitment was daunting when the popular expectation was that the war would only last a matter of months.

Birch grew close to his grandparents. His affection for the farm was great, and he grew to love agriculture. He was given calves to tend and learned to grow tomatoes. He found that hard work could be joyous. His chores included lighting his grandmother's coal stove, feeding the chickens, putting down hay for the cattle and corn for the hogs, collecting eggs, milking cows, waxing the linoleum kitchen floor, combining the wheat, putting up bales of hay in the barn, making and mending fences, churning milk for butter, plowing fields, planting crops (corn, soybean, wheat, hay), and tending to a vegetable garden. Eventually he would proudly show his 4-H club calf at the Vigo County fair.

Mary Alice had no interest in household chores and spent her time reading movie magazines, something Kate Hollingsworth could not appreciate. Birch greatly loved his grandfather and became his principal farm assistant. He never saw him take a drop of liquor and assumed he never did, until he found a fifth of whiskey in the woodshed, something he never told a soul. After supper, his grandfather always smoked a pipe or a five-cent La Fendrich cigar. He was a dyed-in-the-wool Democrat who would move close to his great big cabinet radio to listen to Franklin Roosevelt's fireside chats.

In the summer of 1944, Birch won the Campbell Soup Tomato Growing Contest, a statewide competition that awarded for the largest tomato with the highest quality. He was elected president of his 4-H club and also placed third in the NJVGA (National Junior Vegetable Growers Association) Demonstration Contest, winning its regional scholarship in 1945. While busy with his agricultural activities, he found himself to be a comfortable and successful public speaker. And in 1945 he went on to win the county and district Rural Youth Speech Contests.

Around this time, Campbell Soup officials visited the Hollingsworth farm to discuss the possibility of growing tomatoes there for the soup company. Birch could hear the discussion one night from the top of the stairs and grew excited about the prospect of the farm growing a crop he felt he knew so well, plus the prospect of a relationship with such a successful franchise.

But John Hollingsworth demurred, saying that he had done well with the crops he had and saw no need to add tomatoes to the mix. Birch was deflated when he heard this and the polite exit by the representatives of Campbell Soup, but then he heard his grandmother admonish her husband. She had seen the expression on her grandson's face.

"John," she said, "if the boy wants tomatoes, he gets tomatoes."

"Yes, Kate," was the reply. They began growing tomatoes on the farm. Birch later said that it was the GI Bill and tomatoes (the latter earning $1,600), which helped him go to Purdue and paid all his bills.

Birch was given a six-acre plot of land and a team of mismatched horses in order to cultivate his tomatoes, yielding one hundred tons by the end of the year. At harvest time, while still in school, he rose at dawn twice a week, drove to Terre Haute, and picked up twelve to fifteen migrant workers. After school, he worked with them in the field. In the evening, they loaded seven tons of tomatoes into three-foot hampers on a flatbed truck for the twelve-mile drive to Campbell's processing plant. His record for one day's work was loading seventy-two hampers. The work ethic became a part of his life, something he rarely thought about. Farming responsibilities meant a job with no days off. He was physically fit and outdoors daily.

While in high school, Birch had his first romantic relationship, with a girl named Billie Marie Hatcher. Billie's uncle Oren was the Fayette Township trustee. Her father owned a car that he allowed Birch to drive: his first experience with an automatic transmission. Billie's mother demonstrated her interest in fostering the Bayh–Hatcher relationship. In the waning days of their romance, Billie's mother sent Birch a note telling him that Billie had eyes for another young man and if Birch was serious about her, he had to do something. He never did.

In May 1945, Birch graduated from Fayette Township High School in a class of twenty-six students. He was both vice president and salutatorian of his senior class. As a member of 4-H, he had been the president of Vigo County Rural Youth. A year later he would be president of Indiana Rural Youth, an organization sponsored by the Farm Bureau. He wanted to enlist in the war, which would end that August, but he was too young. During that summer, his father's good friend Jake Maehling, principal of Woodrow Wilson Junior High School in Terre Haute, offered to take Birch to his alma mater, Purdue University, and show him around. They made the trip to West Lafayette and stayed in the Alpha Tau Omega fraternity house. During that visit, Birch made up his mind that he was going to become a Purdue Boilermaker and that he would rush the ATO fraternity. Jake Maehling became one of the most important people in Birch's life, almost a surrogate father. As Birch once mused, "He must have seen something in me he liked." He always suspected that his father had written to Jake, asking him to look out for his son.

Birch recalled his first car, a red Mercury convertible with a white top. He told his grandparents that he wanted a motorcycle, leading his grandfather to take him shopping for an automobile instead. When his grandfather asked him which one he wanted, Birch pointed out the Mercury convertible, feeling that its $800 price tag was too expensive. "If that's what you want," was the response, and it became his.

Not long after he was given the car, he took a curve too fast and went off the road. The car flipped over, and Birch had to crawl out of the upside-down car through its trunk, escaping with only a bloody nose. He couldn't believe what he had done to his very first car. It wouldn't be the last time he escaped serious injury in an accident.

Cars weren't his only form of transportation. Birch and a friend pooled their resources to buy an Aeronca single-engine airplane that cost them very little. They paid an instructor $50 to teach them the basics, and they used it to fly back and forth between Purdue and Shirkieville. As convenient as it seemed and as much as Birch loved the flying experience, the potential dangers were obvious. His friend once landed the plane in a strange field and almost turned it over. Once when Birch landed it, he realized he was flying with less than a gallon of gas left in the tank. Soon the plane was sold, and the boys made $125 apiece.

Mary Alice started acting in high school plays and pursued a degree in theater at Indiana State. Knowing their father would not approve, she and Birch led him to think she was getting a degree in education, and he never learned the truth. Coincidentally, Mary Alice and Birch graduated college on the same day. Birch's graduation came late because he enlisted in the army between his freshman and sophomore years at Purdue, an eighteen-month enlistment. After graduation, Mary Alice followed her passion and moved to New York City, where she appeared in a number of off-Broadway plays, mostly in character roles. She later moved to Baltimore, where she also acted in plays.

While in the army, Birch trained at Fort Lee, Virginia, and went on to spend a year in Germany with the 529 Military Police Company, serving from October 4, 1946, until February 12, 1948. While he was proud to win his battalion's marksmanship award, what made him even prouder was becoming captain of the baseball team. His marksmanship later served him well: as a supporter of gun control, he shot guns in front of suspicious journalists, surprising them with his expertise. But his first glimpse of fame was related to his agricultural background and the circumstances of postwar Germany.

One of his superior officers, Corporal George Rademacher, had a German girlfriend and became particularly interested in helping those in the area who were struggling. Birch suggested that they learn how to grow their own food in their own gardens and sent a letter to the Vigo County agricultural agent, Mildred Schlosser: "Please send at once $4 worth of vegetable garden seeds. Be sure to put in some sweet corn. I enclose check." The letter was signed Birch Bayh, Private First Class. She mailed seeds to Birch at the army barracks. A superior officer saw the seed packets in Birch's footlocker and admonished him, and he ended up carrying them around in his pockets until he and Rademacher could meet with the Germans who were interested in growing their own food. In the town of Hungen, they created thirty-six plots of vegetables that helped many residents avoid starvation during those difficult times.

Birch was particularly worried about the children in the area, and he enlisted their help. He helped ninety children from forty-five families design and plant vegetable gardens. When the harvest came, "each family received 30 pounds of cabbage, seven pounds of beans, a peck of spinach, eight pounds of turnips, six of rutabagas, a peck and a half of tomatoes. The children took home parsley, cucumbers, peppers, beets, lettuce, kale, chard and herbs."[8]

Mildred Schlosser, a friend of the Bayh family, was friendly with author Karl Detzer, who wrote an article about farming being done in Germany with the assistance of a private from Indiana. The article, entitled "GI Ambassador," appeared in the November 1948 *Reader's Digest*. It described Birch as Miss Schlosser's "star pupil in agricultural extension work." Birch had been a 4-H president for two terms and his tomato patch had won the A&P Tea Company's $200 prize as the "best teenager's garden in the state."[9]

One of Birch's army buddies was Ralph from Oregon. Most soldiers had photos of women in their lockers, usually a girlfriend from home; Birch's locker contained a picture of his girlfriend, Billie Marie. Ralph, on the other hand, had posted pictures of Mount Hood in Oregon. He was even less a man of the world than Birch was. On one of their furloughs, Birch, Ralph, and a few other army friends made an excursion to Paris, the most exotic of European cities and a visit he would never forget.

Birch also had a German girlfriend. At first Lila Limbach, a telephone operator, was just a pleasant voice on the phone and one who spoke perfect English. During their courtship, Birch discovered that she was an epileptic, having witnessed one of her fits. He remained committed to her until she started telling him he needed to spend less time playing baseball and more time with her. He didn't take well to that attitude, calling it "unpardonable."

While in Germany, Birch jogged on a large track near an athletic facility. Inside the facility was a boxing ring where weekly fights were held. He decided to try boxing, something he had done at Purdue. Every Sunday the Red Cross chartered a boat and on it American GIs competed with Polish soldiers. Birch won his first fight against a Polish boxer. Once he mastered how to deflect a right cross while reacting with a strong blow to the opponent's midsection, his next fight lasted only one minute and thirteen seconds. He started winning, and he yearned for more.

After finishing his commitment to the army, Birch resumed his studies at Purdue. While browsing the campus bookstore, he saw a poster that said "Read about Purdue's own GI Ambassador" but did not realize it was about him. His fraternity brothers held a celebration because of the article, something he would never forget. Later, they elected Birch fraternity president, a position he held for his final two years at Purdue, which he considered one of the great honors in his life. He was also elected president of the senior class and chosen as Purdue's Outstanding Agricultural Student, having been awarded Ralston Purina's Danforth Fellowship, given to one student at each land grant college in the country. For his senior class presidential campaign, he organized the campus into precincts and assigned his fraternity brothers to talk to students in their designated precincts. Birch spoke to every student in the senior class and sent postcards to all of them. He even launched "Vote for Bayh" balloons around the campus. Birch spoke through a loudspeaker while riding in a bright red farm wagon hitched to a car driven around the Purdue campus, which proved to be excellent training for the future.

He took up boxing again, eventually becoming Purdue's champion in the light heavyweight division. On a boxing trip to Gary, Indiana, he loaned a teammate his mouthpiece to help keep the teeth where they belong. His teammate got knocked out, and Bayh was called up to fight a boxer described as "the next Joe Louis." Entering the ring with a fever he had been nursing all weekend and still without his mouthpiece, Birch was quickly punched in the mouth and thrown out of the ring. He came back with a fury and knocked his opponent down seven times but lacked the energy to put him away. Nonetheless, he won the fight against an opponent who was hardly the next Joe Louis.

The next day, during a fight with a boxer from Notre Dame University, Birch knocked his opponent down right away. The fighter got back on his feet and was soon back on the mat. This time the referee stopped the fight, making up for what he had failed to do the night before. That day Birch was introduced to one of the most famous boxers to come out of Indiana, Tony Zale of Gary. During the introduction, Zale just said, "Hi, bud," never taking his attention away from the bout taking place.

Whenever Birch later thought of fighters getting "punch drunk," he remembered Tony Zale, who may have fought one fight too many.

Though Birch won a Golden Glove as a light heavyweight boxer, boxing wasn't the only sport in which he excelled. He was passionate about baseball, his first love in sports. He played third base or shortstop and was quite a hitter. He was once told that he looked like his father out there at third base, where Birch Sr. had also played, the only difference being his father's chaw of tobacco; he had never known that his father even chewed tobacco. He fondly remembered playing shortstop for Purdue against Notre Dame and recalled gazing up at the famous golden dome. A fellow Purdue teammate was Bill Skowron. Known as Moose, Skowron went on to become a New York Yankee and play fourteen major league seasons. Hoping to play baseball professionally, Birch attended a tryout for the Brooklyn Dodgers but had difficulty hitting a major league curve ball.

During his college days at Purdue, Birch made what would be two lifelong friendships, one with P. A. Mack, who would be the best man at his wedding and later join the Senate staff, and the other with Wayne Townsend, whose Indiana political career paralleled Birch's in a number of ways. Townsend remembered a college event when Birch wanted desperately to date a woman who was already pinned to another student. Nonetheless, she agreed to go out with him. He planned a romantic canoe ride on a nearby lake. They ended up capsizing and walking back to her sorority house sopping wet, where they encountered her pin-mate waiting for her. To avoid a violent confrontation, Birch successfully used the persuasive powers he would later find to be so useful in politics.[10]

After graduating with a degree in agricultural economics, Birch took part in a Rural Youth Speech Contest at the Congress Hotel in Chicago. The subject of his oration was public service. He called it "We Grow by Serving Others," a topic that would resonate throughout his life. It was there on December 3, 1951, that he met Marvella Hern of Enid, Oklahoma. Marvella was representing the state of Oklahoma. Daughter of a wheat farmer and named after a Norwegian aunt, Lillian Morvilla, Marvella was a freshman at Oklahoma A&M, now known as Oklahoma State University, and had been the governor of Girls State and president of Girls' Nation. Birch noticed her in a cafeteria with a group of other contestants. He joined the group and, noting she was from Oklahoma, introduced himself saying, "Sit over here, Oklahoma, and get to meet some folks from Indiana." Many times after this meeting and in light of the fact that Marvella won the contest, Birch would say, "She won the contest, and I won the girl."

Both Marvella and Birch were farm kids and smart, ambitious, and intensely interested in public affairs. He prided himself in his public speaking ability and was strongly attracted to this pretty young woman who could best him.

They started dating right away and, because of the considerable distance between their homes, burned up the long-distance phone lines, especially on Sunday evenings. The phones were generally unreliable, especially after a rainstorm. They were known as "party lines," as they were shared by four or five families. Birch knew fairly quickly that Marvella was the girl for him, and soon he hoped to marry her. Marvella must have felt the same way. After returning home from Chicago, both Birch and Marvella told their families they had met the person they would marry. Birch later learned that Marvella's mother, Bernett, chewed out her husband, saying, "Delbert, I sent you to Chicago to look after our little girl, and you go and let her fall in love with an older man."

In the summer of 1952, Marvella attended summer school at Indiana State, eventually transferring there. On one of their drives from Terre Haute to St. Louis so she could take the train back to Enid, Birch and Marvella decided that they wanted to be married. Marvella, born on February 14, 1933, was just nineteen, and Birch was twenty-four. They were married on August 24, 1952, and immediately moved to the Hollingsworth family farm to be with Grandfather Hollingsworth. Grandmother Kate had died the year before. Once married, Birch decided to stop pursuing his dream of being a professional baseball player. He was earning a good living on the farm, and with his grandfather's advancing age, he was increasingly in charge.

Just before the Bayh–Hern wedding, the couple fell victim to a shivaree, a prank popular at that time in rural Indiana. It involved separating the couple and transporting them to strange locations. A Purdue friend, Don Foltz, came to the farm to take Birch away, and Birch cooperated. After all, he had participated in this custom at the expense of others over the years. They blindfolded Birch, took off all his clothes, handcuffed his wrists behind his back, drove him around for quite a while, and carried the future bridegroom to a shed. He had no idea where he was.

Left alone in the shed, he wriggled around until he got his hands in front of him. Pulling off the blindfold, he saw that he was in a shed filled with feed grain. He struggled upward on the mound of grain until reaching the roof, which he pushed upward with his head. Looking outside, he got his bearings and could see the highway in the distance. Being naked, there was no way he could head to the highway, but soon he realized that Billie Marie's uncle Oren lived about two miles from where he was. He struggled out through an open window and dropped to the ground. Later, recalling the event, he said, "I could have castrated myself on a damn nail."

He walked the two miles, naked, handcuffed and barefoot. Arriving at Uncle Oren's house, he pounded on the door, calling out, "Uncle Oren, it's me, Birch."

The response came back: "Wait a minute, let me turn on the light."

"No, don't do that," said Birch. Uncle Oren came out and assessed the situation. Shivaree was known to Uncle Oren, so he took Birch to the barn and chiseled off the handcuffs. Given clothes to wear, Birch took off to find his betrothed. The female counterparts of Foltz and company had taken Marvella to a distant location, but eventually Birch found her, and they returned to the farm.

The next day, Don Foltz showed up at the shed, looking for Birch. Alarmed to find him missing, he went to the Hollingsworth farm and found Birch working in the field. "So, you got away," he commented and told Birch he needed the cuffs back, which had been borrowed from his policeman uncle. Birch gave him the pieces of the handcuffs, and the alarmed Foltz said, "We need to get them replaced." Birch said, "What do you mean, 'we'?" Don Foltz, a fellow Purdue grad from neighboring Vermillion County, would win election to the Indiana General Assembly on the same day as Birch.

Although Birch was close to his grandparents, he had the good fortune to have another surrogate mother, a woman he had always known as Auntie Katherine. Katherine Henley had been a sorority sister of Birch's mother and her closest friend. When he moved back to Terre Haute after his mother's death, he established a close, filial relationship with Katherine. Married to Henry Henley, a prominent Terre Haute florist, she was always someone Birch could turn to for advice. It was Auntie Katherine in whom Birch first confided that he had fallen in love with Marvella, and it was with Auntie Katherine that Marvella stayed when she paid her first visit to Terre Haute.

John Hollingsworth, Birch's grandfather, loved Marvella Bayh. He loved watching her come down the stairs in the morning and loved the breakfasts she made for him. In December 1952 he fell ill and went into the hospital, where he died the following month.

America in the 1950s was a different place than in the previous decade, which had seen both war and depression. Domestically, industry was booming, and the generation of former soldiers and their families was building the suburbs and assuming leadership in the country. Internationally, the Cold War was dominant, with the fear of communism constant. Even with the end of the Korean War, the nation remained almost on a war footing, always vigilant to the threat of communism. The fifties were rife with McCarthyism and the Red Scare; the Soviet Union developed the atomic bomb, Red China was on the rise, and eventually the Cuban revolution took place, with its leader Fidel Castro announcing that he too was a communist. It was the decade of General Dwight D. Eisenhower, who ascended to the presidency following the unpopular Truman presidency. The US economy prospered; even Europe was recovering, in part because of the Marshall Plan. By the end of the decade, America had become a country of fifty states, with the admissions of Alaska and Hawaii. The Soviet Union launched Sputnik, giving it the lead in the space race. Before the decade was done, the war in Vietnam had begun, though few Americans knew of its existence.

Indiana was very much a microcosm of the country. Between 1940 and 1950, its population grew almost 15 percent. Ten years later, in 1960, it was 19 percent larger, with over a million more people than in 1940. It was booming with suburbs sprouting up everywhere. While steel mills in the northwest corner of the state supplied industrial jobs, farmers throughout the state provided the food for much of the rest of the country. Birch was one of those farmers, but as he began his married life, he found himself thinking about politics.

Jake Maehling's brother Walter was a prominent businessman and politician in the Terre Haute area, serving as minority leader in the Indiana General Assembly. Facing reelection in 1954, Maehling had serious political difficulties, and Birch found himself wondering about the possibility of running for one of the three House seats serving Vigo County. Maehling served as part of a three-member district and suggested to Birch that he consider running as well. The other two seats were held by a tavern owner and a brewery worker, and Maehling figured it would be a good idea to have someone running who wasn't connected to the liquor business, especially if he were to stand a chance of being reelected himself. The fact that Birch was young and handsome, had an unblemished reputation, and was from a good family did not hurt. Late in 1953, Birch and Marvella decided to go for it. He would run.

In February 1954, while driving to Oklahoma to celebrate Marvella's birthday, the Bayhs were in a serious automobile accident. A car driven by an itinerant farmer and loaded down with a family of seven wandered across the median into their lane. Birch swerved to get out of the way, and the farmer turned his car in the same direction. They hit head on.

Marvella was bending forward at the time, pouring Birch a glass of lemonade from a jug. Her head hit the windshield, the dashboard broke her collarbone, and whiplash jerked her head leftward into the steering wheel. Birch was braking so hard that his foot went through the floorboard, and he gripped the steering wheel so tightly that he bent it in half. He only suffered bruises, but the near-comatose Marvella was rushed to Oklahoma City Hospital. She stayed there for two weeks. The farmer and his family escaped serious injury. It was a miracle no one died.

Marvella's parents moved into the hospital to care for her and encouraged Birch to return home, tend to the farm, and continue his campaign. Although Marvella recuperated and returned to Indiana, she would have episodes of double vision, as well as bouts of leg and back pain, for the rest of her life. This was an era before seat belts, and we can only speculate on the difference they might have made.

For Democrats in Indiana, winning was almost always less than a sure thing, except in the northwest corner of the state, Lake County, where the culture was similar to Chicago, the behemoth city to the north. In Vigo County, which included Terre Haute and Shirkieville, Democrats were often elected but not as automatically as in Lake County. There was a May primary in the three-member district where Birch sought a seat. The top three Democratic winners would compete in the fall against the top three Republicans.

Primary elections in that era were largely controlled by the party leaders. Birch met with the Vigo County Democratic chair, an old German named Lawrence V. "Dutch" Letzkus, who was described as having big ears and jowls. He told Birch that his rule was to support incumbent Democrats, adding, "If you win, two years from now I'll be supporting you." However, before leaving the meeting, Letzkus gave him a copy of the precinct committee list. County organizations were made up of precinct committeemen and vice committeemen, one a male and the other a female. There were 56 precincts in the county, so those 112 names were the focus of the campaign work for Birch and Marvella. The week before the May primary, the precinct committee people assembled at the county courthouse and the chair gave them their list of slated candidates. The Bayhs knew what they had to do—visit 112 precinct committee people as well as the courthouse to shake hands with each of the patronage employees.

Most of the precinct committee people were old enough to be his grandparents but had never been visited by a candidate who asked directly for their support. They turned out to be extremely friendly, inviting the Bayhs into their homes, and if they made a commitment, it was usually honored. One of the campaign issues was known as "local option," a position that each Indiana county should be allowed to determine its own liquor laws rather than complying with actions by the state legislature. Birch did not drink alcohol, but he knew that people who did drove out of the county, got liquored up, and then drove home. It didn't seem like a good idea to him, but his own habits led to a whispering campaign dubbing him "Dry Bayh." It threatened him in most of the county, except in Sugar Creek Township, where a prominent Catholic, Tom Curley, told Birch he would take care of the "Dry Bayh" problem.

Curley took Bayh to a saloon near one of Terre Haute's factory gates, a place many workers stopped before going home. Curley called for everyone's attention, introduced candidate Bayh, and announced that the "next drink is on Birch Bayh." Birch left much more popular than when he went in.

Birch also benefited from an act of political courage by Teresa Turner and Camilla McCarty, two black precinct committeewomen. Both women were patronage employees controlled by the county party and had reputations for delivering the vote in their precincts. Birch got to know them, and they liked him. When told they were not to support him on primary day, they did the opposite and delivered their precincts for Birch, clearly risking their jobs by doing so. They liked him, promised their support, and were good for their word.

The typesetter at the *Terre Haute Tribune* also played an important role in Birch's political career. As a member of a typesetters' union, part of the AFL-CIO, the typesetter wanted Birch to obtain the support of organized labor. As a farmer, Birch had not been exposed to labor issues and wasn't even familiar with the AFL-CIO. His typesetter friend introduced him to the county president of the AFL-CIO, who immediately asked the candidate how he felt about right-to-work laws, an issue that had never been on his radar screen. Right-to-work laws prevented union security agreements between companies and organized labor, agreements that would require employees of a company to be members of the union that the company has signed such an agreement with; they were anti-union laws.

Not fully understanding the issue, Birch wisely asked the labor leader to tell him about it. Instead of killing his relationship with labor by expressing his uninformed reaction to the name "right-to-work," Birch began a long and fruitful alliance with organized labor, locally, statewide, and nationally. He soon came to sympathize with the plight of the working man and woman. During this time, he also met the head of the United Auto Workers (UAW), Dallas Sells, and learned that Indiana was a larger producer of auto parts than any other state. Additionally, the automaker Studebaker had a plant in South Bend. Birch became a champion of organized labor throughout his political career. Of the six candidates seeking the three nominations, Birch ended up leading the ticket. Acknowledging that he had worked harder than everyone else, he also felt it didn't hurt to be named Birch Bayh; many people might have confused him with his father, who was well known and popular.

In November 1954, Birch was elected to the Indiana General Assembly. Sworn in at age twenty-six, he was the youngest member of the House. When State Representative Bayh was introduced to Governor George Craig at a reception, the governor told him, "Your father was the only man who ever threw me out of a basketball game."

Birch responded, "You probably deserved it, didn't you, Governor?"

When Birch arrived in the legislature, his annual salary was $1,200; he felt that his farm income was enough for them to live on, so the Bayhs set aside the money for future reelection campaigns. At the end of his first year in the state legislature, they became the proud parents of Birch Evans Bayh III, born on December 26, 1955, in Union Hospital in Terre Haute, the same birthplace as Birch and Aunt Mary Alice. The baby would be known as Evan.

One of the early pieces of legislation that Birch cosponsored and sought to pass was a constitutional amendment known as the "Home Rule Bill." The purpose of the bill was to allow municipal governments to determine their own relationships with their employees. All the states around Indiana operated in that manner, but Indiana firefighters and policemen did not support the amendment. They felt that home rule would subject them to the vagaries of local politics when the interests of public safety should make them immune from local political maneuvering. Their opposition gave Birch his first experience of a legislative defeat.

A defeat like this didn't hurt Birch's reputation because he didn't hold grudges. He fought amiably and hard. He learned important lessons for his future, lessons about compromise and that today's adversary may be tomorrow's ally and vice versa. He quickly grasped how the institution worked and began forming alliances among his colleagues. Also that year, he sponsored another constitutional amendment to lower the voting age in Indiana, which was also defeated. Regardless, before the end of that first session of the legislature, the Indiana press voted Birch as the "Outstanding Freshman."

Indiana's Democratic political organization was built upon the county chairs and vice chairs, who were elected by the precinct committeemen and women who ran in the primary elections. The precinct committee person has been described as "the backbone of the system."[11] Precinct committeemen and women were responsible for knowing the voters in their communities. They took their jobs seriously. The county leadership came from their ranks and depended on them. The precinct committeemen and women elected county chairs and vice chairs in a caucus the Saturday after the Tuesday primary. The county officials, in a similar caucus the following Tuesday, elected the chairs and vice chairs within each congressional district. Finally, the district chairs and vice chairs met at the Democratic State Committee the Saturday after their own election to elect the state chair and vice chair, who were normally chosen in advance by the sitting governor, if that person was a Democrat. But the party also helped determine who ran for office and was slated for nomination. Underlying all of this was the Indiana system of patronage, where public employees at all levels "donated" 2 percent of their pay to the party treasuries. Understanding and navigating this system was instrumental to advancing in Hoosier politics.

In the General Assembly, Birch formed a relationship with John Stacy, who was the Ninth District Democratic chairman. This relationship was one of a number that became crucial to Birch when he ran for minority leader in 1957, following the 1956 election.

Running for minority leader was not an easy or natural thing for Birch to do. The Democratic leader was Walter H. Maehling, brother of Birch's mentor Jake Maehling, with whom he had grown close. Birch often commuted with Rep. Rex Minnick, from neighboring Clay County. Known as "Red Button Rex" for pushing the red "no" button so often, consistently casting negative votes from the floor, Rex suggested that Birch run.

"How can I do that, Rex?" asked Birch, alluding to his close relationship with Maehling.

"If you'll consider it a possibility, leave the rest to me. The troops are restless out there. They want to get rid of Walter. Wouldn't it be best to get rid of him with someone he is close to?"

About the same time, Walter Maehling came to realize that his position was in jeopardy. He came around to the idea of Birch as minority leader. "How he got the idea that it ought to be me, I don't know," Birch remembered. There were only twenty-four Democratic members of the General Assembly that session, and its youngest member, at age twenty-eight, became their leader.

The 1956 election was a good year for Republicans, largely because of President Dwight Eisenhower's reelection campaign. Birch ran his campaign for leadership very much like he first ran for the legislature. Driving around the state during the period between Election Day and the start of the new session, he visited every Democratic member's home personally and asked for their support. When Birch was elected minority leader, the Speaker of the House was Republican George Diener. He and Birch worked well together, teaching him a valuable lesson about civility in politics. When Diener was defeated for reelection in 1958, he offered Birch advice on how to run the House.

Birch and his colleague from Vermillion County, Don Foltz, developed a useful and lifelong friendship. Also a Purdue grad, Foltz was smart and hardworking but had the unenviable tendency to "piss people off." Birch, who felt that Don was probably smarter than he was, remembered Don fighting against state bonuses for Korean War veterans, of which Foltz was one. He just didn't think it was a wise use of taxpayers' money. Birch knew that bonuses had been approved for veterans from both world wars, and he felt they should not treat the most recent veterans differently. As expected, Foltz's effort failed, but taking unpopular positions made him noticeable. Foltz had previously run against Birch for the position of minority leader, but when Birch became Speaker, Foltz served as majority leader.

Birch always felt that the legislature was a great learning experience for him. Just as farming taught him to love the land and livestock, serving in the General Assembly taught him how to get along with his colleagues and what made the legislative process viable. He realized he loved the process and felt he was "pretty decent" at it as well. One of the important lessons learned in the Indiana legislature, a lesson he took with him to the US Senate, was never to bring up a bill if you did not have the votes. However, there were times as a US senator when he believed it was right to bring up a measure even if it was not likely to become law.

As Democratic leader in the Indiana House of Representatives, Birch considered running for Congress as a natural progression up the political ladder. The first time he did so was in 1958, though running for the US Senate was clearly an ambition. It felt unlikely in 1958 because the mayor of Evansville, Vance Hartke, seemed to have the nomination sewn up. State political leaders encouraged Birch to run for Congress against the incumbent, Cecil Harden, a Republican woman who also served as the postmistress of her town. Birch had lots of local support from Democratic leaders, but he knew that if he won, he would have to run every other year and could lose two years later, which was worse than not running at all. Instead, he applied and was accepted to law school at Indiana University in Bloomington. Deep inside, he knew he was passionate about politics and not cut out to be a farmer for the rest of his life. Knowing the law better would help him in elective office, and lawyers always had more flexibility in their schedules to be able to campaign for office. Farming was every day with no letup. As much as he loved the farm and considered it a part of him, politics was replacing it as his first love, and he knew he couldn't do both. He paid the tuition by selling 120 acres of his farm. In that election, Congresswoman Harden was defeated by Fred Wampler, who failed to be reelected two years later—exactly Birch's concern.

Election year 1958 was a good one for Democrats. Vance Hartke, the Evansville mayor, was elected to become a US senator, the incumbent William Jenner retired, and Democrats in Congress picked up forty-eight seats in the House and thirteen in the Senate. Typically, in a midterm election, the party that won the presidency loses seats, but the recession and the advent of Sputnik made it a very good year for Democrats. With the Soviets launching the first space capsule, Americans were disturbed about falling behind the communists in the space race, and the specter of bombs falling from outer space became a typical American fear. Additionally, the Eisenhower administration's efforts to pass right-to-work laws galvanized the Democratic Party. In Indiana, the number of Democratic state representatives swelled to seventy-nine out of one hundred, a substantial increase from the twenty-four seats they had held. In January, they elected Birch Bayh, a thirty-year-old law student, to be Speaker of the House. In 1959, he was named the Indiana Jaycees "Outstanding Young Man."

Another important memory from Birch's years in the legislature involved ethics in politics. When discovering that one of his colleagues was soliciting bribes, he showed a mastery of the legislative process by moving every piece of legislation his colleague was responsible for to committees controlled by others; the colleague in question did not find out until it was too late.

It was in the General Assembly that Birch first came to understand how his actions might really make a difference in people's lives. He led an effort to redistrict school boundaries, changing the way state funds for education were allocated to make sure that rural students had the same opportunities as students living in wealthier, urban areas. Teacher salaries were increased, and a framework was created for secondary school consolidation. Commentators wrote that the education measures represented the "most massive reform of Indiana school law in history."[12] Under his leadership the General Assembly passed legislation on a wide variety of issues, including juvenile delinquency, patronage, pari-mutuel betting, secrecy of welfare records, price controls to protect small merchants from being undercut by discount stores, the establishment of intermediate corrections facilities that would focus on academic training, and flood control. Most of the bills they passed failed in the Republican-controlled Senate. His record of accomplishment as Speaker would have been notable for one who held that post for many years. The fact that Birch was in his early thirties makes his record even more notable.[13]

Two of his General Assembly colleagues, Andy Jacobs Jr. and Jack Bradshaw, became lifelong friends. Early on they supported Birch for the leadership position. He later credited them for making him Speaker of the House. Rep. Bob O'Bannon and his wife, Faith, were close friends of Marvella and Birch. O'Bannon's son, Frank, would be elected to the state legislature and later as governor of Indiana (in 1996).

There had been little exposure to African Americans in Birch's life. Outside of those he had been friendly with in the army, he knew very few members when he arrived at the legislature. His early friendship with the only two African Americans in the General Assembly, Jessie Dickinson and Jim Hunter, had meaning that has stayed with him through the years. While Birch was Speaker, he once saw Jim Hunter sitting in his seat on the floor of the General Assembly and realized that Hunter probably never had presided over the General Assembly. Birch motioned him to the podium and asked him to preside while he took a bathroom break. He quickly saw how much Hunter appreciated the gesture, and he repeated it later with others, including a senior Republican who not only appreciated the honor but also told Birch, "My own people have never asked me to do that."

It was a magical time for the Bayhs. Birch was Speaker and a law school student. While he was in law school, the family lived for three years in a two-room apartment. Jim and Gaynell Poff, lifelong friends of the Bayhs, ran the farm. Often while driving back and forth between Indianapolis and Bloomington, Birch spent the night sleeping in his car on the side of the road. It was among the happiest times in his life. The Fort Wayne *Journal Gazette* later printed an article pointing out that Birch had finished in the "top 10 percent of his Indiana University law school class in 1961, while serving as speaker of the Indiana house and carrying a full-time load at the State Legislature."[14] He graduated with distinction. Birch's only obstacle during this time was failing the bar exam, which he passed the following year. That fact would be used against him years later as he fought nominees to the Supreme Court and championed constitutional amendments. After his bar exam failure by a single point, he said, "I thought my life had come to an end. Marvella Bayh went to her grave" believing that the Republican bar examiners had failed him intentionally and illegitimately. He had been second in his class while also serving as Speaker, and he had felt he was prepared for the exam despite the other pressures he was experiencing.

In 1960, while John Kennedy was winning the presidency, though losing badly in Indiana, Democrat Matthew Welsh was elected governor. Because of the extensive patronage system, the governorship was tantamount to kingship—a hugely important role in the political parties. At that time governors had a one-term limit, and it was not until later in the decade that the limit was changed to two terms. Since holding the governor's seat was so important to the Democratic Party, many in the leadership wanted Birch to run for governor in 1964 rather than for the Senate in the midterm election of 1962. But Birch considered himself a legislative animal and wasn't interested in having responsibility for the patronage system. He could have a greater impact on more people by moving to a larger stage, and increasingly he thought about trying to become a US senator.

2

US Senator

For of those to whom much is given, much is required. And when at some future date the high court of history sits in judgment on each of us . . . our success or failure, in whatever office we hold, will be measured by the answers to four questions: First, were we truly men of courage? . . . Second, were we truly men of judgment? . . . Third, were we truly men of integrity? . . . Finally, were we truly men of dedication?

JOHN F. KENNEDY[1]

THE 1960 ELECTION WAS A GOOD ONE FOR JOHN KENNEDY AND Matthew Welsh, but Democrats lost twenty seats in the US House of Representatives and one in the Senate. The Democratic majority that in 1959 had elected Birch as Speaker of the Indiana House of Representatives shrank to only thirty-four seats.[2]

Birch again sought the nomination in a three-member district in Terre Haute and got the most votes. Birch won 12,417 votes; his friend and colleague Walter Maehling received 8,536, and Jack Neaderhiser received 7,832. Unfortunately, Maehling died shortly thereafter, a big loss for Birch, both personally and professionally. Politics in the Terre Haute area were not always easy, especially for Democrats. During the 1960 campaign, Kennedy detractors dropped leaflets from airplanes over the city warning the citizens that a vote for Kennedy would "open the floodgate for papal dominance of the federal government."[3] While much of the ticket was going down to defeat, Birch was easily reelected and once again became minority leader, as the Speaker's chair went to the Republicans.

In December 1961, when planning to seek the Democratic US Senate nomination to oppose the Republican incumbent Homer Capehart, Birch made a pilgrimage to Independence, Missouri, to meet former president Truman. Birch's father-in-law, Delbert Hern, was Democratic county chairman in Garfield County, Oklahoma, and knew the ex-president. He called Truman's secretary, Rose Conway, and made an appointment for the Bayhs to meet him. Marvella had already met Truman when she represented Oklahoma at Girls Nation in Washington in 1950. The American Legion Auxiliary sponsors a civic training program in most states each year known as Girls State. Female students excelling in government and civics compete in elections to Girls State. The representatives chosen in Girls State elect two of their own to represent each state in Girls Nation when it convenes in Washington, DC. Marvella had been one of the two Oklahoma representatives in the fourth year of Girls Nation's existence.[4] Birch recalled that his son Evan, not yet six years old, was dressed in his cowboy suit when they drove to Independence. Evan was strictly admonished to be quiet and told not to speak a word in the presence of the great man. They met in the study of the Truman Library.

Squirming in his seat, Evan attracted the attention of the former president, who realized the young boy had to go to the bathroom. Truman asked him if that was the case. Evan nodded, and Truman told him, "So do I." He took Evan's hand and walked him down the hallway to the bathroom. Later, he gave Evan a silver dollar.

At the meeting in his presidential library, Truman expressed an urgent desire to see Capehart defeated. He told a story about the time when he nationalized the steel industry. Capehart came to the White House for a private meeting and told the president he had to stop this program of nationalization. Truman asked why, and Capehart, who was a big investor in steel, told him, "Mr. President, you have to withdraw that order."

"Why is that?" asked Truman.

"Every day that order is in effect costs me $10,000," responded Capehart.

"Stand up," ordered the president.

"What's that?"

"Stand up, Senator. Now get your ass out of here, and don't ever come back."

After Birch won the 1962 nomination in May, he called Truman and asked him if he'd come to Indiana and tell that story. Truman felt it was a private conversation and that it wouldn't be proper to tell it in public. Nonetheless, near the end of the campaign, when Birch arrived late in Evansville for a major rally, Truman was already there, telling the large crowd of Democrats the exact same story.

Truman was not the only president who had no use for Homer Capehart.

John F. Kennedy had been a US senator from Massachusetts; when he served in Congress, the Indiana senators were William Jenner and Homer Capehart, two isolationists and far-right conservatives. Kennedy disliked both men intensely.

While serving in the Senate, Kennedy wrote the popular book *Profiles in Courage*. Many questioned whether he was the true author, with some believing that Ted Sorenson, a top White House aide to Kennedy, might have been the actual author. Sorenson once wrote, "Conservative and intellectually challenged Republican Senator Homer Capehart of Indiana, while debating Kennedy on another issue on the Senate floor, tried to score points and unsettle his Democratic colleague by inserting into his remarks the question, 'Who really wrote that book?' Kennedy deftly replied: 'Well, there's one thing certain. I'm confident that no one has prepared for you the remarks you are delivering now.'"[5]

The Moderate Voice, a blog describing the 1962 election, wrote about Capehart:

> Capehart, who had only a high school education, had become a successful coin-operated record player, jukebox, and popcorn machine maker for Packard in rural Indiana. His respect among GOP elders grew after he hosted a "Cornfield Conference" on his farm in 1938. But he was still a political novice when picked to run for the Senate in 1944. . . . He had the good fortune of appearing on the ballot as Thomas E. Dewey carried Indiana.
>
> Capehart won his second and third terms easily and by 1962, there seemed little reason to believe he wouldn't win a fourth.
>
> Capehart was not a hard-line conservative but rhetorically established a pugnacity. . . . The "Toledo Blade" called him "one of the hottest tempered of Republicans." He sparred with everyone from fellow members of Congress, sons of Presidents, and Chief Executives themselves. . . . He once told LBJ "I'm going to rub your nose in shit."[6]

Senators Capehart and Kennedy had squared off in the Senate Foreign Relations Committee over issues of defense spending. Capehart's enmity toward the Massachusetts senator was exacerbated by his distaste for JFK's brother Robert. Robert Kennedy had been counsel on the Senate committee investigating organized crime when he focused on rackets within the jukebox industry, something Capehart saw as a backhanded way of attacking him. Capehart had earned his fame and wealth as the father of the jukebox industry. At one point, Capehart described one of JFK's Senate speeches as "a good high school debate." After Kennedy became president and the Bay of Pigs incident took place, Capehart made Cuba and the removal of Castro as its leader his cause célèbre.[7]

The sixty-five-year-old Capehart had been elected to the Senate three times, and many thought he was unbeatable. Less obvious to most observers were troubles in the Indiana Republican Party. Before retiring in 1958, the far-right senator Jenner controlled the party machinery and feuded publicly with his colleague. Calling Capehart a "New Deal sonofabitch" because he supported President Eisenhower, Jenner passed control of the party to Lieutenant Governor Harold Handley.[8] Handley had been stained by scandals in state government, and he lost to Vance Hartke. With the election of Democrat Matt Welsh as governor in 1960, it became clear that with the internecine struggles among Indiana Republicans and demographic changes in the state, the fortune of the Republican Party was in trouble. Writing in the *Indianapolis Star*, political reporter Ed Ziegner said,

> The state Republican machine, so efficient and sleek in 1952, was unraveling at the seams through much of the 1950's. . . . The GOP percentage of the off-year election vote slid from 56.2 in 1946 to 53.9 in 1950 to 51.4 in 1954 and a disastrous 44.2 in 1958. The Republican primary vote declined and was surpassed by the Democrats in 1958, and the rural, conservative counties on which the GOP still relied so heavily counted for less and less in the total vote picture; by 1960, 25 percent of Indiana's people lived in just two urban counties; 40 percent in only five counties, and 60 percent in just fourteen out of the state's ninety-two counties. . . . The bedrock of the old GOP power base was gone.[9]

Additionally, there were 157,000 more voters in Indiana in 1962 than in 1956; many were young suburbanites who did not consider themselves Republican. Once Jenner left the Senate and Capehart tried to exercise leadership over the party, a luncheon was held for the Republican members of the congressional delegation. The Republicans were unaware that their public address system had not been turned off, and reporters outside the room could hear Capehart declare that the Hoosier GOP was "split right down the middle." He called on the state officials to support Eisenhower and other "modern Republicans." But the old guard wanted nothing to do with that form of modernism.[10]

Despite Democratic successes and the decline of the GOP, the Indiana Democratic Party did not seem poised for victory. It was revealing that no member of Congress was vacating his seat to take on Capehart. Congressman John Brademas from northern Indiana had been seriously eyeing the Senate seat but was unwilling to pull the trigger. Indianapolis mayor Charles H. Boswell, Marion mayor M. Jack Edwards, and Judge John S. Gonas of South Bend were all in the race. Governor Welsh, however, publicly stated that the nomination should be decided by the Democratic Party, not by him.[11]

The Senate seat held a great attraction for Birch Bayh. In 1961, he began traveling the state, and by mid-September he had visited with two-thirds of the county chairs. Birch urged them to contact the governor if he felt they were firmly in his corner.

Birch formally announced his candidacy for the Democratic nomination for the Senate on October 18, 1961, in the Ben Franklin Room of the Claypool Hotel in Indianapolis. His wife, Marvella, and son, Evan, were by his side. Senator Capehart had already announced his campaign for reelection. Lawrence V. "Dutch" Letzkus, chair of the 6th District Democratic Committee and of the Vigo County Democrats, became the head of the Bayh for Senate Committee. His staff was made up of a few volunteers, including Larry Conrad, a twenty-seven-year-old friend of Birch's. They had met at a Young Democrats Dinner in Muncie and later engaged in a bar exam study session in Indianapolis.[12] Conrad served as campaign coordinator. Bob Boxell, the thirty-one-year-old chairman of the Indiana Young Democrats, handled the advance work. Bob had been holding Kennedy dinners in his district to celebrate the new president and had met Birch at one of those dinners. Bob Hinshaw was a twenty-six-year-old 6th District Young Democrat chairman; starting in 1961 he became Birch's driver for eighteen months. Like Boxell and Conrad, Hinshaw agreed to work on the campaign for free for as long as he could afford to do so. From Hinshaw's standpoint, he was in a win–win situation. He would meet many people and have a great experience and lots of fun. In the beginning, winning was not as important to him as it would become. He laughed when recalling the first time he saw Birch Bayh. For the first time in his life, Birch was dressed in a white suit—clothes he would never wear again, while Capehart was often seen wearing a similar white suit. Soon after they began traveling together, he described Birch as a candidate who appealed to both men and women.[13] Earl Hawkins, another volunteer, also drove Birch, whom he called "Boss," around the state, but he got lost once too often and his driving was discontinued.

Like Birch, Homer E. Capehart was raised on a farm; similar to Birch's tomato-growing expertise, he had been a champion corn grower. But the similarities between the two candidates ended there. These two candidates represented a generational difference as well as a clear choice in political philosophy and approach to politics. What Birch didn't know was that Capehart had been suffering recent health problems, assumed to be heart-related, and his wife, Irma, had been having health issues as well. It was public knowledge that Capehart had experienced a heartbreaking personal loss just two years before, when his son Tom and his wife were killed in a plane crash, leaving four children behind.[14] Similarly, Birch would experience his own personal loss in the years leading up to his last Senate campaign.

When Birch announced his candidacy for the Senate nomination in October 1961, he had just begun practicing law. A day after he announced, he was called into the law office where he was working and was told by one of the senior partners, Howard Batman, that his services were no longer needed. It seems that the law firm was closely affiliated with Senator Capehart on a number of issues.

The nomination would take place at a state party convention in June 1962, and Boswell presented the biggest challenge, not only because he was mayor of the biggest city in the state and its capitol, but also because of the extensive patronage system under his control. It was like having a huge campaign staff that could fan out across the state and campaign as his surrogates. Ironically, Boswell was probably more conservative than Capehart. Governor Welsh's advisers wanted Boswell to be the nominee. They felt that no one could beat Capehart but that Boswell on the ballot would help the Marion County legislators seeking reelection. Welsh, though, was hearing from increasing numbers of Democratic leaders around Indiana that the idea of electing an Indianapolis mayor was anathema to many. The farther from Indianapolis one was, the more anti-Indianapolis people seemed to be.

Birch recalled that often while he was in a crowd, Boswell campaigners would try to swarm around him and occupy his attention, keeping him from talking to others in the area. On occasions like that, Hinshaw was known to "accidentally" step on one of their feet, turning their attention to him while Birch made his escape. Another Hinshaw task was to make use of the portable record player they carried in the car. While Birch was speaking, Hinshaw would set it up in the back of the room and play a recording of "America the Beautiful," turned down low at first and gradually made louder. As it reached its crescendo, Birch also reached his.

Birch had grown close to the leadership of organized labor while serving in the Indiana legislature, particularly with Dallas Sells, who served as president of both the state's AFL-CIO and UAW at different times. The AFL-CIO was the largest labor coalition in the state and the United Auto Workers (UAW) was among the largest unions in Indiana. When Sells endorsed Bayh for the Senate nomination, Boswell got into a protracted dispute with Sells. The dispute became important because it inflamed other leaders in other unions across the state, many of whom would be elected as delegates to the convention or would have a lot to say about those who were, or both.

Birch took a straightforward approach to the campaign for convention delegates. About 2,400 delegates were at stake, with many of them having served previously, so getting the list of previous delegates defined his task. He crisscrossed the state, meeting as many delegates as possible, one on one, many of whom had never been asked directly by a candidate for their support. Governor Welsh had advised Birch to ask the key question during those meetings. Don't let people get away with homilies and polite gestures, he told him; ask them directly for their vote. "People will tell you you're a great guy," said Welsh. "People will tell you they need young blood like you in the Senate and you'd make a good senator. But before you leave, you have to ask them, can I count on your vote? If the answer is yes, for most people you can count on that."

While the governor was giving advice to Birch, Welsh's lieutenants were working vigorously on behalf of Mayor Boswell. State Democratic chair Manfred Core and the governor's chief political operatives, Clint Green and Jack New, wanted Boswell on the ticket. They felt that Capehart would win; Boswell would still be mayor and in control of the Marion County patronage, something they knew to be valuable in electing other Democrats in the city and county.

Birch's mentors in the campaign were Bob Risk, head of the Indiana Civil Liberties Union, and Merle Miller, a prominent Indianapolis attorney. They brought in foreign-born industrialist Miklos "Mike" Sperling, who would become very important to the Bayh family. A Hungarian immigrant, Sperling had developed the first disposable syringe, going on to build a successful manufacturing business making machine parts. Late in 1961, the Bayhs participated in a forum for all the Senate candidates, and coincidentally Marvella was seated next to Mike Sperling. After each of the candidates had addressed the session, she heard Sperling exclaim, "I like that young Bayh boy." Marvella asked him if he wanted to meet Birch. That was the beginning of a long relationship. Mike and his wife, Gladys, would become hugely important to Birch. Mike was famous for reacting to financial questions by saying in his broken English, "It's only money."

In the spring of 1962, the Jefferson–Jackson dinner was fast approaching. The dinner was a famous conclave in Indiana where the most important Hoosier Democrats came together. Birch and Mike Sperling talked about the need to do something unique at the dinner. They decided on an attractive cameo of the First Lady, Jacqueline B. Kennedy, which cost two dollars each in bulk. Two thousand dollars seemed like an awful lot of money to spend on one thousand giveaways. "It's just money. Buy!" was the response from Sperling. As the cameos were passed out, they became the hot item at the dinner, generating huge excitement in the arena. When Birch rose to speak, he received the loudest and most positive reception he had yet experienced, something Governor Welsh couldn't avoid hearing.

With Boswell controlling Indianapolis and Marion County, the largest block of votes for Birch to seek was in Lake County. His friendship with Jim Hunter helped him with the black community, and his colleague Joe Klen, later mayor of Hammond, and county chair John Krupa became key supporters. Lake County politicians were traditionally hostile to the mayor of Indianapolis, whoever that might be, and Birch understood how to capitalize on that sentiment.

In January 1962, it was reported that a high-level Welsh aide was telling people that Welsh supported Boswell. In a meeting of the Democratic State Committee, Welsh was stunned by a rebuke he received from Fannie Mae Hummer of the 5th District. Responding to rumors about the governor endorsing Boswell, she spoke up loudly: "Governor, you're not going to do this to my boy Birch." By April, Welsh had done little to show support of Boswell, thought to be running substantially behind Birch. Boswell made the mistake of expressing his frustrations with the governor more openly than was prudent.[15] Capehart embarrassed Boswell during the campaign when both attended a Gridiron Dinner, an event during which politicians typically roast one another. Capehart asked the mayor to stand, and when he did, Capehart embarrassed him by giving him a tongue-in-cheek scolding.[16]

As the convention date grew closer, it was clear that Birch was the favorite, and the governor came around with his support. On May 10, Welsh endorsed Birch, talking about the preponderance of Democratic organizational people who were supporting him.[17] Once the governor's support was made known, the three Bayh staffers were put on the Democratic State Committee payroll at the rate of $500 per month. Soon others were added to the staff. Marcia Murphy was brought on to handle the speakers' bureau, and Harry Cain was put in charge of press relations. Larry Cummings, the twenty-one-year-old chair of the Young Collegiates for Bayh, was hired to drive Marvella to her speaking engagements. Both Birch and Marvella thought it made sense for them to travel separately, doubling the impact they would otherwise have.

At the convention, the Bayh campaign brought a donkey onto the floor, adding to the spectacle that had already been created by the Bayh Belles and a number of trumpeters and trombonists. Before the votes were actually cast, Birch's election as the Democratic nominee for the US Senate was a foregone conclusion. On June 22, 1962, the votes were counted; Birch had garnered more than 75 percent of the votes. Out of 2,578 votes, he had 1,982 delegates to Boswell's 459 and Edwards's 106, with 31 abstentions. Birch also received 90 percent of the votes from the Lake County delegation.[18]

Conrad and Boxell headed up the Bayh campaign team. Hinshaw stayed on as Birch's driver. Ollie Miller, a photographer who at age forty was considered old by others in the campaign, joined the team on the road. Marvella was a polished speaker and became enormously popular on the campaign trail. Birch was still doing some work for Tony Hulman, father of the Indianapolis 500. When he told Hulman that he was going to run for the Senate, Hulman reached in his desk drawer and threw him a set of keys. "You'll find this down in the basement," he said. "Might be helpful to you." Birch descended to the basement and found out that the keys were for a car, a Mercury with only 8,000 miles on it. They would add 70,000 more before Election Day.

Marvella and Birch were truly a team, and Birch long held that he never would have been so successful without her. Birch had a reputation for his even-handed temperament. But Marvella, though beautiful, smart, charismatic, and articulate, had a fiery temper. She was impatient with staff and often lost her temper with them. She would have described herself as a perfectionist. But campaigns are often messy, disorganized affairs, and she learned that in many ways over the next several years.

There were tensions among the party pros and the brash newcomers from the Bayh team. Nonetheless, a plan for the fall campaign was drawn up, and a full integration with the state Democratic Party took place. The budget for all Democrats, funded by the state party, was $500,000, the majority earmarked for the Bayh campaign. There were three sources of money: about $290,000 had been gathered from the 2 percent patronage money, another $125,000 from assessments on the county party organizations, and about $100,000 from individual donors.[19]

The biggest shot in the arm for the Bayh campaign came when Mary Lou Conrad, wife of the Bayh campaign coordinator Larry Conrad, came up with a campaign version of a popular song. "Hey, Look Me Over" was sung by Lucille Ball in the musical *Wildcat,* which opened on Broadway in 1960. The idea was first raised by Hinshaw, who heard the song on the radio and suggested it to Birch, coming up with the first two lines of the new song version. He and Birch came up with the next two and left it at that. Mary Lou sat down at the piano with the catchy tune and wrote the rest of the lyrics. Her version not only became widely popular in Indiana but also helped solve the problem of mispronunciations of the Bayh name. The song caught on big time and was often played as Birch walked into rallies across the state in 1962 and for many years to come.[20]

Hey look him over
He's my kind of guy
His first name is Birch
His last name is Bayh
Candidate for Senator
Of our Hoosier state
For Indiana he will do more than anyone has done before
So hey look him over
He's your kind of guy
Send him to Washington on Bayh you can rely
In November remember him at the polls
His name you can't pass by
Indiana's own Birch Bayh

Bob Long, an Indianapolis public relations executive, produced the song, which was recorded with the Chicago band the J's and Jamie. Jamie Silvia was the singer. The Broadway tune had been written by Cy Coleman. Long got through to Coleman about licensing the song and was told it was free for a political campaign, something hard to believe today. The impact of the song was impossible to predict, though it was later estimated to have played on the radio 6,000 times. Birch first recognized that it was having the intended effect while stopping at a Dairy Queen in Muncie. He and Hinshaw had arrived there in a car with a "Bayh for U.S. Senator" sign emblazoned across the door. Two boys rode up on their bicycles, and they could hear one ask the other if he knew who the name on the car door was. When the second boy couldn't answer, the first said, "That's the guy we're supposed to be looking over."[21] Later Birch would say that the song was exactly what they needed, because Capehart enjoyed an 85 percent name recognition factor, while his was about 23 percent.

Evan Bayh's earliest memories were of the house at 630 Jackson Boulevard and a big collie named Duffy, but his only memory of the Senate campaign was the song, which they often heard on the car radio.

He also remembers that, as if in a parody of the idyllic Midwestern life, the Bayhs' next-door neighbors in Terre Haute were the Waltons.[22]

The television ad that Long produced was a montage of Birch Bayh photos while the song played in the background. Similarly, radio ads featured the song, and as much as the campaign could afford, it filled the airwaves with "Hey, Look Him Over." When the American Commercial Television Festival was held in 1963, the ad and song won an award as the "best locally produced television commercial of 1962."[23]

Capehart, on the other hand, was running television ads that may have been hurting him more than helping. They showed a big black car driving up, its door opening to release a cloud of cigar smoke. Out came the balding, overweight senator with cigar in hand, underscoring the contrast in the two candidates' ages. In the ad Capehart waves to bystanders, but the camera scans the area, revealing no bystanders at all. Capehart, emulating the Bayh jingle, produced his own, which never caught on. Birch believed that Capehart might have won if he had simply stayed off the airwaves.

William B. Pickett, in his biography of Homer Capehart, described the contrast between the two candidates:

> Capehart was a caricature of an old-fashioned senator: overweight, bespectacled, cigar-smoking, sparse hairline above his large round face, and double chin. The senator's speeches seemed to reflect his appearance and played into the hands of his opponent. Making a fist with one hand and then the other during speeches, he would pound his right fist into his left, then open both hands and spread his arms. He would act coy: "I don't intend to, but I might get a little politics into this speech." He believed in "good common horse sense." Bayh made an excellent platform appearance. Before a speech he would move through the audience to shake hands. He opened his speeches with a flurry of jokes and anecdotes.

Capehart described his simple campaign plan: "If President Kennedy is popular by campaign time, I'll run on my own record in the Senate. If he is unpopular and his program bogged down, I'll run against the administration. The people of Indiana like me, because they always know where I stand."[24]

No published poll in 1962 ever had Birch Bayh ahead.

Two debates were agreed to, though Birch continually called for more. The first was at the Sigma Chi journalistic society on September

19, and an hourlong debate took place on WFBM-TV on November 4. When Birch charged that Capehart had revealed the details of a Foreign Relations Committee secret briefing, Capehart lost his temper and grabbed his younger opponent by the lapels. This was reminiscent of an incident many years earlier where Capehart did the same with Hubert Humphrey, inspiring concern that there might be fisticuffs. Birch's campaign against Capehart largely ignored the Cuba issue and focused on more local issues. He called for property tax relief, a judicious federal tax cut, updated depreciation allowances for business, expanded world trade, and reduced federal controls on farmers. He blasted Capehart's consistent opposition to increases in the minimum wage and charged him with failing to "propose one single piece of legislation for working men and women." Capehart's record on civil rights was criticized, with Birch pointing out that Capehart had supported anti–civil rights filibusters on four occasions and had voted six times against civil rights legislation. He also insisted that the "aging senator was behind the times."[25]

While seeking election to the Senate in 1962, Birch met John F. Kennedy, his second presidential introduction. Originally, Birch was determined not to ask JFK to visit Indiana during the campaign, since Kennedy had lost so badly in the state two years before. But Ray Berndt, the UAW head in South Bend, had leaned on Walter Reuther, the UAW's national president, to get Kennedy to come to Indiana, and Reuther persuaded the president to make the trip. Kennedy was scheduled to speak at an airport rally in Indianapolis on October 13, and Birch flew to Pittsburgh to board Air Force One and fly with the president to Indianapolis. He had previously met Kennedy in the White House Oval Office, where he stood in line with other Senate nominees, each to have a photo taken with the president at his desk, with the famous PT-109 coconut in the image. Before arriving in Indianapolis, Birch was ushered to the private section of the plane where the president rode. The president asked Birch what he should say in Indiana. Birch modestly said that he wouldn't want to tell the president what to say but that he would appreciate any kind words about his own campaign. He also mentioned the brewing controversy over the development of the Burns Harbor Port on Lake Michigan, opposed by those wanting to protect the Dunes National Lakeshore.

The president asked him how he should handle it, and Birch suggested he not touch it, which he didn't.

Governor Welsh was strongly in support of the Burns Harbor Port and, after the speech, was angry with Birch for getting the president to ignore it. "The president of the United States doesn't seek advice from me on what to say," Birch replied disingenuously.

Long before the missile crisis in October 1962, Capehart had been promoting the use of troops in Cuba to get rid of Castro. President Kennedy referred in his speech to "those self-appointed generals and admirals who want to send someone else's son to war."[26]

One week after the JFK visit, the campaign was interrupted because of the Cuban missile crisis. The Indiana stop was on Saturday, October 13. "Black Monday," the day the Cuban missile crisis became public, was October 22. On Sunday, October 21, the president was to make an appearance in Chicago, but it was announced that he was returning to Washington because he had a cold.

On that Sunday, there was a meeting in Indianapolis with the finance team that raised money for the Bayh candidacy. Chief among them were Merle Miller and Miklos Sperling; the latter had begun driving his Mercedes around town with a Bayh campaign sign on it. Birch stopped in during the meeting, where they committed to quickly raising $70,000 to buy television time. Birch would participate in live call-in shows, taking all questions phoned in by the viewing audience, usually right after the kids left for school or before they got home in the afternoon. One was scheduled for prime time.

On Black Monday, Birch concluded that Capehart's warmongering made him look prescient and justified. Capehart's campaign oozed "I told you so." Birch said, "I thought that was the end of the campaign." Bob Boxell said that until the Cuba crisis, "it never occurred to us that we could lose."[27] The Bayh campaign virtually came to a stop. Capehart suspended his campaign and retired to his farm, making his victory lap very public. It was said that he pulled his advertising off the air and that the Bayh campaign bought the time that was made available. Capehart did, however, participate in the previously agreed-upon debate on November 4.

President Kennedy apparently felt it was over as well, exclaiming in the White House to Kenneth O'Donnell that they had defeated Bayh and reelected Capehart. Hugh Sidey, *Time* magazine's longtime reporter covering the White House, wrote in his book about JFK, "Kennedy felt, with the worst about Cuba being confirmed, many of the Democratic candidates were apt to be defeated. There was no better example than the Senate race in Indiana, where the incumbent Republican, Homer Capehart, was crying for an invasion of Cuba. The young Democratic hopeful, Birch Bayh, had stayed with Kennedy, accused Capehart, with some effort, of warmongering. What would happen now that Capehart had been proved right? It seemed, at the White House at least, that Bayh was doomed."[28] But the resolution of the crisis changed everything.

The Bayh finance team reconvened and decided to hold fast. They purchased enough TV time for sixty-nine hours of call-in shows in the remaining days. The fact that Capehart was reported to be pulling back his TV money may have made this purchase easier. Birch always felt that the spectacle of the young candidate answering every question without hesitation or obfuscation had to be a major factor in the campaign's success. The entire campaign cost something around $465,000.

That October, the Cuban Missile Crisis was dominating Kennedy's every waking thought. It was eating away at JFK that congressional Republicans like Capehart, who had been demanding that the US government invade Cuba and remove Castro, would be the winners in this crisis. Sorenson wrote that the president asked sardonically, "Would you believe it? Homer Capehart is the Winston Churchill of our time!"[29]

Birch was in Michigan City when he was handed a note to call Ted Sorenson at the White House. "Birch, just wanted you to know," Sorenson told him, that the quarantine of Soviet ships had led to those ships turning around and going back twenty-five minutes earlier. The campaign jingle was quickly pulled from the air and replaced with a new radio ad saying, "Vote Democratic. Back the president for his stand in Cuba. Vote for Birch Bayh." Soon there were full-page newspaper ads showing JFK and Birch, ads saying "Support the President," who was now a national hero.

History has since taught us that Fidel Castro wanted a last-stand confrontation with the United States, whether it led to a nuclear war or not. Had we attacked Cuba, a great many lives would have been lost, and history has revealed that there were thousands of Russian troops in Cuba, not just missiles. Many of them would have died as well. In retrospect, the Kennedy solution proved to be the wise one. Its effect on the life and career of Birch Bayh is incalculable.

Out of more than 1.8 million Hoosier votes, Birch Bayh was elected United States senator by 10,944 votes, about one vote per precinct.

Birch and Marvella watched the election returns in an Indianapolis hotel with Miklos Sperling and other friends. Evan remembers that it had gotten very late and no winner was declared. The six-year-old Evan went to bed, and his grandfather slept on the floor outside his room so he was not awakened.[30] Hinshaw remembered watching the results at about 2:00 a.m. and seeing Sen. Homer Capehart arriving by himself at a television station to concede. It was a poignant and sad sight as the eighteen-year incumbent retired from public life . . . alone.[31] Bayh's ally and friend Lee Hamilton, who was later elected to Congress, described election night as "joyous beyond belief."[32]

The day after the election, Birch and Marvella scheduled time to be alone at their home in Terre Haute. Finding private time had been exceedingly difficult, and they were not to be disturbed. During lunch there was a knock at the door. When Marvella opened the door, Birch saw the chief of police on the porch. The police chief told Marvella that there was a message for Birch, and she remarked that she would put it at the bottom of the pile. "With all due respect, Mrs. Bayh," the chief responded, "I think you might want to put this one at the top of the pile." It was a note saying to call the president. When Birch called the number, expecting to ask for the president's secretary, he heard the familiar voice with the New England accent say, "Hi, Birch, you old miracle maker."

Hugh Sidey summed up the election this way: "Birch Bayh's win in Indiana could be attributed to just about any cause and every cause—Cuba, young candidate versus old candidate, emphasis on medicare. The boys back in Washington gave a sly smile. 'Indiana just happened to have the best-organized and best-financed campaign of any state in the union,' said one. Indiana's Governor Matt Welsh, holding the party reins, had sweated his party into top condition. A half-million dollar war fund didn't hurt."[33]

As much as he wanted to defeat Capehart and disagreed with his voting record, Birch never felt any animus toward him during the campaign and later described the campaign by saying, "There was no blood left on the floor." Some years later, Birch was appearing at a horse show in Western Indiana. It was raining outside, and everyone had been moved to drier quarters in a large building. He was handed a note that said someone wanted to see him, and he headed to a back room away from the stage. There he found former senator Homer Capehart. The two shook hands and embraced. Capehart introduced his granddaughter, Sally Mae Jones, saying, "I want her to meet you." Many years later, Birch looked back on the warmth of that exchange and the nature of human relationship, saying, "Man, that's what it's all about."

Two months later, he would be a United States senator. While traditionally off-year elections mean a loss of Senate seats by the party holding the presidency, this election resulted in a gain of three seats for the Democrats. It was the second-best showing in midterm elections by a political party in power in a hundred years. It was made possible, of course, by the fact that President Kennedy was enjoying a 74 percent approval rating late in 1962 after the Cuban Missile Crisis.[34] There would be a total of eight new Democratic senators in January. Joining Birch Bayh was Abraham Ribicoff of Connecticut, Daniel Inouye of Hawaii, Daniel Brewster of Maryland, Thomas McIntyre of New Hampshire, George McGovern of South Dakota, Howard Edmondson of Oklahoma, and Gaylord Nelson of Wisconsin.

Ira Shapiro, in his book *The Last Great Senate,* describes in detail the institution that Birch joined, one he describes as populated by giants, a Senate with a record of major accomplishment. He writes, "The twenty-one Democratic senators elected in 1958 and 1962 formed the heart and soul of the Great Senate."[35]

Before being sworn in, during the first week of December, Birch traveled to Washington to meet with the Senate Steering Committee, which decides on committee assignments, while Marvella began hunting for a home. He met with the Senate majority and minority leaders, Mike Mansfield of Montana and Everett Dirksen of Illinois, as well as the powerful

Richard Russell of Georgia, as part of his campaign to obtain the committee assignments he coveted, Public Works and Judiciary. Later, he and Marvella paid a visit to Vice President Lyndon B. Johnson in the office he maintained off the Senate floor. LBJ asked them, "What are you kids doing for dinner?" They looked at each other, realizing they had no plans, and the vice president said, "Let me call Bess [the maid] and have her heat up some leftovers." The vice president of the United States invited them to dine at his home, known as the Elms, that night.

The Elms was a palatial estate in northwest Washington that had been previously owned by Pearl Mesta, the doyenne of Washington socialites. A residence considerably larger than should have been afforded by LBJ on his government salary, it was largely financed by the wealth that his wife, Lady Bird, had generated as the owner of several television and radio stations, mostly in Texas. The vice president accompanied the Bayhs in his limousine; after dinner he took them house hunting. Birch remembered being advised by the vice president to buy a home because it might end up being all that they owned when the time came to leave office. At one point, Attorney General Robert Kennedy took them house hunting as well. They eventually bought a home across the Potomac River in McLean, Virginia, on Chesterbrook Road.

The day after meeting the vice president, Birch and Marvella lunched at Hickory Hill, the home of Robert Kennedy, his wife, Ethel, and a large number of kids and pets. The men played touch football after lunch. The New Frontier was in full bloom, with Washington populated by young officials, either elected or appointed. "Youth" and "vigor" were commonly used terms. The Bayhs represented that image to the Kennedys. To the Bayhs, they had fully arrived in a town of vast excitement and promise. They both hoped that they would measure up.

3

Assassination and Amendment: 1963

It is not the critic who counts; not the man who points out how the strong man stumbles, or where the doer of deeds could have done them better. The credit belongs to the man who is actually in the arena, whose face is marred by dust and sweat and blood; who strives valiantly; who errs . . . if he fails, at least fails while daring greatly, so that his place shall never be with those cold and timid souls who know neither victory nor defeat.

THEODORE ROOSEVELT[1]

THE SUCCESS OF THE BAYH CAMPAIGN IN INDIANA WAS THE biggest upset in the Senate contests that year. The new senators arriving in Washington in 1963 would prove to be a very distinguished group. The Moderate Voice blog posted this about the 1962 election:

> 1962 saw the end of several longtime Republican Senate careers. The results were surprising . . . and occurred in several states that were favorable to Republicans. But the defeats of Indiana's Homer Capehart and Wisconsin's Alexander Wylie, along with the Democrats' ability to peel off longtime Senate seats in South Dakota and New Hampshire not only sent major Capitol Hill institutions packing, but it gave Democrats and the Kennedy administration some unexpected crowing rights with their midterms.
>
> It also gave us a new generation of sorts. Birch Bayh may very well have been the most important Senator who wasn't part of leadership and Gaylord Nelson, as well as George McGovern and Tom McIntyre, also had more impact on today's world and everyday life than anyone realizes.[2]

Chris Sautter, a media consultant who grew up in Indiana, wrote, "Bayh ran the quintessential underdog retail politics campaign. Bayh had heavy labor backing and 'his fresh-faced appeal and non-stop courthouse campaign appearances in a Mercury' proved a welcome touch."[3]

Birch Bayh began his career as a United States senator on January 3, 1963. The second youngest senator elected in 1962, he was number 100 in seniority. The youngest at the time was Edward M. "Ted" Kennedy of Massachusetts, the president's brother. Elected in 1962 in a special election to fill his brother's unexpired term, Kennedy had been sworn in late that year.[4]

Birch served in the Senate alongside another Indiana Democrat, Vance Hartke, elected in 1958. He described his early relationship with Hartke as "cooperative." Before committee assignments were announced by the Senate Steering Committee, Lady Bird Johnson threw a reception at the Elms that Birch and Marvella attended. Vice President Johnson asked what committee assignments he wanted. Birch told him Public Works and Judiciary. Looking inside his suit pocket, where there was obviously some kind of list, LBJ replied, "I think you'll be satisfied." Both committee assignments happened.

Birch wanted to be on the Public Works Committee so he "could get some dams built in Indiana" and the Judiciary Committee because of his interests as a lawyer. Public works are tangible accomplishments that remain vivid examples of the effectiveness of the sponsors. While the most senior senators could be on four or five committees, freshmen senators in those days were appointed to only two. Later the Senate changed the rules so that each senator sat on two standing committees only. Richard Russell of Georgia, chair of the Steering Committee, was a lifelong bachelor and had been in the Senate since 1933. Russell was close to Vice President Johnson. Birch also grew close to Russell, one of the only colleagues he always addressed as "Senator," though they rarely saw eye to eye and Birch often opposed Russell's filibusters. Their friendliness was the result of Birch's tendency to get along with older men, something that would serve him well throughout his career.

Birch became part of the Senate's liberal bloc, led by Philip Hart of Michigan, who Birch considered "an outstanding gentleman." A fellow member was Ted Kennedy. The majority leader in the Senate was Michael "Mike" Mansfield of Montana. A man of few words, usually a yes or no, he was very good to Birch. Senators Bayh and Mansfield were members of the same college fraternity. Birch described him as remarkably different from LBJ. Johnson was known for his sharp elbows, while Mansfield was easier-going, making fewer demands on people. But when Mansfield needed something, he normally got his wish.

Birch felt that senior senators liked him because of his "go along to get along" attitude, which he developed as a member of the Indiana General Assembly. Senators Mansfield, James O. Eastland (D-MS), chairman of the Judiciary Committee, and Everett McKinley Dirksen (R-IL), Mansfield's counterpart as minority leader, while not always in agreement with Bayh, liked him and put him in a position to be effective. The Senate of the 1960s operated in a collegial manner, in stark contrast to the Senate in the twenty-first century.

That difference is best illustrated by a conversation Birch had with Senator Everett Dirksen early in 1963, Birch's first year in the Senate. Attorney General Robert Kennedy invited the members of the Senate Judiciary Committee for a cruise on the Potomac on the presidential yacht, *Sequoia*. It was an opportunity for the attorney general to get to know the senators better on a personal level. At one point during the cruise, Dirksen sat next to Birch, leaned over, put his arm around the shoulder of his junior colleague, and said, "Birch, you know what we need to do here? We need to start right now figuring out how we're going to get you reelected."

"What's that, Senator?" asked the startled Hoosier. "Any suggestions you have, I'd be glad to hear."

Dirksen repeated himself and went on to offer suggestions on how to use the franking privilege for sending newsletters and how to obtain federal projects to help constituents, as well as how to take advantage of the Senate recording studio. The leader of Republicans in the United States Senate was counseling a junior member of the opposition party on how to get himself reelected.

It was symptomatic of an era when comity was king, when getting along with each other in the Senate was valued highly, a real priority. A guiding philosophy seemed to be that today's adversary can be tomorrow's ally, a philosophy Birch had learned years before, and relationships between senators were considered important. The partisanship that has taken over the Congress in the first decades of the twenty-first century makes this exchange sound quite incredible.

When Birch arrived in Washington, the oldest senator was eighty-six-year-old Carl Hayden of Arizona, more than fifty years his elder. Elected to Congress during the presidency of William Howard Taft, Carl Hayden became the first Arizona congressman to hold the at-large congressional seat. Hayden had served in the US Congress for the entirety of Birch's life. He rarely spoke on the floor of the Senate, but his effectiveness was demonstrated in his committees, in his role as chairman of both the Senate Rules and Appropriations committees. Hayden retired in 1968 at age ninety-one and lived to be ninety-four.

Because of the Senate seniority system, Hayden became president pro tem of the Senate, next in line after the Speaker of the House to be president. After the assassination of President Kennedy, when the Twenty-Fifth Amendment was being considered to address issues of presidential disability and succession, Hayden's age was a factor. With a vacancy in the office of vice president, only the speaker stood in line between Hayden and LBJ, who had experienced a heart attack in his past. After Kennedy's assassination, Speaker John McCormack was seventy-two years old and Hayden fourteen years older. The effort to pass the Twenty-Fifth Amendment was the first great debate that Birch Bayh became immersed in; serving with personalities like Hayden helped frame the issues and colored the experience to a large degree.

Another story involving Dirksen was during the debate on the Twenty-Fifth Amendment. The House Judiciary Committee, headed by Emanuel Celler of New York, was promoting a version of the amendment that required Congress to decide on the process to replace a vacancy in the office of the vice president, a role Birch felt Congress already had. Birch was appointed chair of a conference committee to resolve the differences between the House and Senate bills, the first conference committee he would ever attend. He knew that it would help to have fellow conference committee member Dirksen on his side. When he went to Dirksen's office to discuss the issue, he waited in an office next to the senator's private suite. On a table next to his chair were stacks of postcards, perhaps six inches high, supporting the Prayer Amendment. That measure was an effort to overturn the Supreme Court's decision to outlaw government-sponsored prayer in the public schools, something that could only happen with a constitutional amendment. As chair of the Subcommittee on Constitutional Amendments, this tough issue was one of Birch's responsibilities, something he would have preferred not to be involved with. He knew the existence of those postcards was not an accident, and he brought up the prayer matter with Dirksen during their meeting, agreeing to hold hearings on the issue. That promise helped to cement Dirksen's support in the conference committee.

One reason senators like Birch were able to get along with the older senators and the Southern barons was the relationship among Senate wives. Marvella invited senators and their wives to dinners at the Bayh home, and Birch felt this was why he had a close personal relationship with John McClellan, a longtime senator from Arkansas and Senate Appropriations chair. This may have helped Birch withstand an uncomfortable political moment when McClellan faced a primary challenge from Congressman David Pryor in 1972. Birch had campaigned for Pryor years earlier, and they had developed a friendship. When the close primary results required a runoff election, Pryor and McClellan would face off one on one. After the primary, Birch called Pryor to congratulate him on his good showing. Unknown to Birch, Pryor took his call in the middle of a press conference. McClellan won the runoff election by a narrow margin, 52–48.

On the first day of the next Senate session, Birch got onto a subway car in the basement of the Senate, where he joined Senator McClellan for the ride to the Capitol. Birch greeted him by saying, "Congratulations, Mr. Chairman."

McClellan replied, "Well, you didn't get me, did you?" Birch was startled, and his senior colleague repeated himself.

Birch replied, "My father didn't raise a fool" and said that McClellan ought to know that he had enough sense not to get involved in a primary, especially with two of his friends running against each other. Acknowledging his friendship with David Pryor, he reassured McClellan about their own friendship.

Later that day, the Steering Committee met to fill a vacancy on the Appropriations Committee, an appointment that Birch sorely wanted. With McClellan's support, he became a member of the committee.

Another senator serving when Birch arrived and who had an outsized impact on his career was Thomas Dodd of Connecticut. Dodd chaired the Senate Subcommittee on Juvenile Delinquency, and Birch was a member. He asked Birch to his office and showed him a cardboard display of handguns, explaining why he was introducing a bill to outlaw the manufacture of inexpensive handguns known as "Saturday Night Specials." They were not good for target practice but were often used in crime. Birch agreed to cosponsor the bill, the opening salvo of a long battle with the National Rifle Association (NRA), which opposed all firearms regulation.

Birch felt that guns should be licensed, that concealment should not be allowed, and that these Saturday Night Specials should be outlawed. To the NRA, being sensible on gun legislation was akin to being "partially pregnant." It was all or nothing. After Dodd left the Senate, Birch chaired the same subcommittee and continued to champion the gun control effort, paying a political price for it. Periodically during his career, Birch participated in public shooting events, which showed off his shooting skills and temporarily muted the pro-gun vitriol. He remembered one such event in southern Indiana, where guns loaded with black powder were being used in a shooting competition. A barber from Shelbyville owned many of the guns on display and in the competition, and Birch was invited to shoot one. He did, and the owner looked at the target through his binoculars and said, "You should quit while you're ahead." Birch asked if he could see the target, and they walked out to it and saw that his bullet had neatly pierced the center of the bullseye. When Birch asked if he could have another shot, the barber again said, "You should quit while you're ahead." Eventually he gave in, and Birch took another shot, this time expanding the hole he had already created in the target, somewhat like Robin Hood splitting the arrow. "I'll be a son of a bitch," gasped the barber. The participants and spectators loved it.

The colleague who came to impress him the most, an impression he still had more than fifty years later, was Hubert Humphrey of Minnesota.

As majority whip, Humphrey closed the Senate each day, and often, as a freshman, Birch was assigned the role of presiding officer. He would watch Humphrey listen calmly to the debate, and regardless of the subject being discussed, Humphrey would join the colloquy, routinely demonstrating an impressive level of knowledge and expertise. As Birch described him, "He knew more about the issues than any other senator."

Mississippi was the home of two staunchly conservative Democratic senators and major powers in the institution, James O. Eastland and John Stennis. Eastland, its senior senator, had arrived in the Senate in 1941, when Birch was just thirteen years old. A committed segregationist, Eastland had been chairman of the Judiciary Committee since 1956 and held that post until the end of his career in 1978. He was almost never seen without a cigar (smoking was not permitted on the Senate floor, but as Eastland so often demonstrated, clearly it was allowed in committee sessions). Both Eastland's philosophical bent and his image were in stark contrast to Birch's. Yet Eastland became enormously helpful to the Indiana senator. When Senator Estes Kefauver died and the chairmanship of the Subcommittee on Constitutional Amendments opened up, Birch approached Eastland about the possibility of being appointed to the post. Eastland, who always referred to Birch as "boy," told him that he was going to let the subcommittee die a quiet death as a means to save money. A day later, he called Birch to tell him he had changed his mind and would let him have the post. That appointment would change Birch's career in major ways.

In fact, the Subcommittee on Constitutional Amendments was created to be the graveyard for those measures that senators periodically introduced to overturn unpopular Supreme Court decisions. These were generally matters that most senators preferred would never see the light of day.

Birch remembered one occasion when he was at his seat on the floor of the Senate when Eastland sat down next to him. The elder senator said, "Boy, you've got that abortion thing, you've got that prayer thing, you've got that busing thing. How do you survive all of that? Any one of those things could kill me at home."

On another occasion, Birch had introduced a measure to make the Eugene V. Debs homestead in Terre Haute a national monument, and later he introduced a resolution restoring Debs's citizenship, which was lost because he had been imprisoned. A famous socialist and candidate for president, Debs was among the most famous of Hoosiers and remained popular in Terre Haute. Republican senator William Scott of Virginia took on an aggressive role to prevent the pro-Debs resolutions from passing. His constant objections prompted Birch to speak to Eastland, and the next time Scott began to object in the Judiciary Committee, Eastland interjected, "Bill, sit your ass down. This is in the boy's hometown." Scott went silent.

Vance Hartke, Indiana's senior senator, was a flamboyant politician with a personality very different from Birch's. Birch and Hartke never were close, and there often seemed to be an unspoken competition between them. Bayh's popularity rankled Hartke, and he often accused him and his staff of activities designed to damage him. Hartke once danced on a table at an Indiana political event; he was known to stop at funeral homes to shake hands with the bereaved. One of his most unpopular and controversial moves was to refuse to be searched by airport security. Birch found himself in the difficult position of being asked by high school students if he would also refuse such a search and was forced to respond truthfully in the negative. One day he visited four high schools, and the question was the first one asked at each of the four schools. Despite their differences and their lack of affection for each other, they and their staffs often worked well together, particularly for government projects affecting Indiana.

The congressional ballgames were always enjoyable, and Birch would take part in one during his first year in the Senate. It was played at DC Stadium, later named RFK Stadium following the assassination of Robert Kennedy. Before the game, Birch met congressmen Gerald Ford and Donald Rumsfeld in the locker room. After that game, the *Washington Post* sports page on July 31 ran the headline "Lemon Homers" to describe a highlight of the Washington Senators game from the previous night but with two Bayh photos next to the headline. One was a picture of Birch in mid-swing, the other of him leaping in the air to make a catch at shortstop. Described as "a one-man show," Birch hit

5-for-5, including a home run and two doubles, driving in six runs. The Democrats won 11–0.[5]

Prior to Birch's first congressional baseball game in 1963, the Democrats had been coached by Harry Kingman, a former New York Yankees outfielder and civil rights activist. He wrote a letter describing his regret at not coaching in 1963, adding, "Prior to my return to California, I did discover some new candidates for the team such as Senator Birch Bayh of Indiana . . . the very attractive young senator from Indiana was the sensation of the contest. He fielded brilliantly at short-stop and hammered out half a dozen hits, including two homers."[6] Despite Kingman's inaccurate hitting summary, it was impressive.

Former congressman Lee Hamilton, a close friend of Birch's since college days, fondly remembered those games, in which he played outfield. Hamilton had been a star basketball player in Indiana during high school and college and was a good athlete. He recalled how the Republicans had won for many years in a row until Birch arrived. He described Birch as a great hitter and fielder but said that sometimes his throws to first were erratic. According to Hamilton, winning those games was difficult because it was so hard to find a pitcher who could throw strikes. When the Republicans were able to produce pitchers who had been major leaguers, such as Vinegar Bend Mizell of North Carolina and Jim Bunning of Kentucky, the latter in the Hall of Fame, the Democrats were compelled to change the rules.[7]

Birch's memories of those games include two incidents that meant a lot to him. At one game, he was hitting during batting practice, and the Washington Senators manager, the legendary Ted Williams, was watching. When Birch left the batting cage, Williams told him, "We ought to sign you up." He had another thrill during a game when Detroit Tigers star Al Kaline asked him for an autograph.

The Senate of the 1960s and 1970s was very different from the Senate of the first quarter of the twenty-first century. Senators quickly learned that it was better to listen and to defer to their elders rather than to start introducing legislation and pontificating just after arriving in the Senate. Birch's maiden speech wasn't until October 8 of his first year, expressing his views on pending legislation, H.R. 4955 on vocational education. The mentality that today's adversary was tomorrow's ally or that you had to "go along to get along" helped create an institution that got things done, one that made a huge impact on American life during those decades. It was an era before the twenty-four-hour news cycle and before the explosion in campaign finance that forced a senator to raise money virtually full time during the six-year term. And it was a time when the Senate helped change America, as it reacted to civil rights, women's rights, an unpopular war, assassinations of our leaders, the Watergate scandal, the environmental movement, the emergence of gay Americans, the increase in drug use, and divisive issues like the Panama Canal Treaty, busing, abortion, and gun control.

It was a time when a senator like John Kennedy could become a celebrity on television and use that celebrity to build a political following. Prior to the influence of television, a senator had to work hard in the institution and build a record of accomplishment in order to become well known. The role of television in politics was not yet understood. Birch was a member of the Kennedy generation and used TV to his advantage but also worked hard at gaining a record of accomplishment.[8] No one at the time would ever claim that he was as smooth and sophisticated as JFK. Whatever rough edges might exist from growing up on a farm in Indiana existed in full force in 1963. A possibly apocryphal story circulated early in those years that illustrates his lack of sophistication. The story alleges that when visiting his sister Mary Alice at a restaurant in New York City, he ordered Mogen David wine to accompany a Caesar salad with Thousand Island dressing.

The first year of the Bayh tenure meant an office for the first few months in room 1205 of the New Senate Office Building before settling for the next several years in the Old Senate Office Building, initially in room 304 and then for most of his career in room 363. The building is now the Russell Senate Office Building, named after Richard Russell of Georgia. The New Senate Office Building was later named after Everett Dirksen; years later when another office was built, it was named after Philip Hart. Birch's first administrative assistant was Bob Keefe of Huntington, Indiana, and his first secretary was Virginia Crume, who had worked for Governor Welsh. Room 304 can be seen in the 1962 film *Advise and Consent*, where it was the office of Sen. Brigham Anderson, played by Don Murray. The staff referred to Birch as B2, which was how he initialed staff memos after reading them.

Early in his Senate career, Birch's new colleagues encouraged him to pursue a relationship with Bobby Baker, secretary to the majority leader under Johnson and Mansfield. Birch recalled his administrative assistant, Keefe, advising him, "If there's one person you need to get to know in this town, it's Bobby Baker." He was known as the man who "knew where the bodies were buried." He held storied parties for senators at the Carroll Arms Hotel and was reputed to be an expert at providing liquor and women for those who were receptive. Baker was Birch's age with vastly more experience in the Senate. In May 1963, Birch took him to Indiana for the Indianapolis 500 race.[9] He remembered him as "a pleasant guy" but added that he only got close to him to have a better chance of getting things done in the Senate.

Later that year, Baker was embroiled in scandal from events under investigation that had largely taken place before Birch was elected, ending his Senate relationships. When asked about it, Birch replied, "I didn't owe Baker any favors, and I didn't get a television set." Baker offered Tom McIntyre, one of Birch's close friends in the Senate, money right after he arrived in the Senate to help pay off his campaign debt. McIntyre declined the offer after he was advised that taking the money would make him beholden to the oil interests. Birch had experienced virtually the same thing with Baker—an offer to pay the campaign debt by calling on his friends in the oil business—and he too had politely declined. By early 1964, Baker was the subject of hearings in the Senate Rules Committee on alleged financial improprieties and influence peddling as an officer of the Senate. Later he stood trial, prosecuted by the Justice Department of his mentor and friend Lyndon Johnson.[10]

Bobby Baker came to Washington to be a Senate page while early in his teens. He learned the ropes and established personal relationships with many of the most important senators. Part of that education included gaining knowledge of some senators' private peccadilloes and details of illegalities in which they participated. Whether it was protecting senators in their sexual affairs, in their drunken escapades, or in other moments of personal weakness, Baker personified the adage that information is power. That information brought him great power in the Senate. His close associations with senators Bob Kerr of Oklahoma and Lyndon Johnson made him a force to be reckoned with. When he fell from power, his fall was far more rapid than was his rise.

Baker wrote a memoir called *Wheeling and Dealing* that described in colorful fashion many of Birch's colleagues. He wrote about two men who had an oversized influence on Birch's career, James Eastland and Lyndon Johnson. "Senator Eastland's proximity to power," he wrote, "and his real friendships with the Kennedys and other national leaders—has always puzzled outsiders who know only his image as a fat, cigar-chomping, mushmouth who looks like—and often is credited with acting the part of—a political Neanderthal. . . . Yet, I found Jim Eastland to possess one of the quicker, more brilliant minds in the Senate. It's a tribute to his political genius that he's managed to satisfy the most reactionary element of his Mississippi constituency and, at the same time, remain a working power and influence among his Senate colleagues. No dumb man could do that."

He wrote about Lyndon Johnson, then majority leader and working with the Eisenhower administration to try to put together an effective committee to investigate the excesses of Joe McCarthy, leading to his censure. His characterizations of LBJ's comments about Sam Ervin (D-NC) and Karl Mundt (R-SD) are particularly memorable:

> "I've tried to convince Dick Russell and Walter George (then a senator from Georgia) to lead the fight from the Democratic side, but they're afraid of sticking their heads in the noose. They say McCarthy's still strong medicine down in Georgia and they've just got no stomach for it . . . Dick Russell don't want to go to the mat because he's afraid he'll get his hands dirty. You know who they're offering me?" he snorted in disgust. "Sam Ervin of North Carolina. The lightweight son-of-a-bitch. Goddamn windbag . . . But I guess I'll have to take him. He's one of the insiders and he's a Southerner and I've got to have one. . . . I also recommended they put Karl Mundt of South Dakota on the committee," Johnson said with a diabolical grin. "He's so goddamned dumb he won't know what's going on, and he's so ineffective he can't make anything happen, but it will look good to have a staunch pro-McCarthy man sitting in judgment of him." In this, too, LBJ got his way—and he proved right.[11]

Sen. John Williams of Delaware, often referred to as the "Conscience of the Senate," by the press was heading an investigation into the affairs of Bobby Baker that was growing increasingly close to producing major accusations against LBJ. The part of the investigation into the vice president's activities was quietly closed in November 1963 when Johnson was thrust into the presidency. Birch's recollection of Williams was that he was "as cold as ice" and that he was a "self-appointed Conscience of the Senate."

The hiring of Bob Keefe brought immediate experience to Birch's office, as he had worked for both Hartke and Congressman Roush (D-IN) previously. Keefe met Birch while he was working for Roush, and when Birch offered him the top staff job, he took it with enthusiasm. Keefe said he had his eyes on a presidential campaign for Birch from the moment he got there. It was not clear if Birch had ever given a presidential run any thought. Keefe's new boss was what he described as a "hot property," being invited to speak at Democratic events around the country immediately after taking his seat in the Senate. He typified the youth and vigor represented by the Kennedy administration and was becoming known as a dynamic public speaker.[12]

Larry Conrad and Bob Boxell had come to Washington with Birch. Conrad would be the first legislative assistant, and Boxell would have responsibility for Indiana political affairs. Soon, problems appeared in the office that needed to be rectified. Keefe was unprepared for the volume of mail received by the Bayh office. It was vastly different from his experiences in the Roush office or what any senator experienced prior to the election of JFK. In the early months of 1963, mail bags began to stack up in the office with no staff mechanism set up to deal with them. It came to a head in 1964 when Marvella began to hear about it. Ray Scheele, who wrote a biography of Larry Conrad, wrote, "Marvella discovered the problem when she was informed that 'somebody in Birch's office had declined an invitation to a party for us.' It was the first time she heard about the invitation. Upon investigating, she discovered that 'one secretary had about a thousand unanswered letters stashed away.' This incident resulted 'in the biggest' fight Marvella and Birch ever had and increased her resentment of staff members who did not run 'a tighter ship.' Birch turned his attention to office administration, made changes in the internal procedures and hired people who could better help run that 'tighter ship'."[13]

In July of that first year, *Life* magazine ran a feature on Marvella Bayh called "Imagine Me Here in D.C.!" In that story, she said, "Bobby Kennedy drove us around one evening to see the various neighborhoods. What we wanted was a house with a lot of trees around, in a neighborhood we could afford. Many people seem to have the idea that all senators are independently wealthy. Some of them are, of course, but it just wasn't the case with us. We live entirely on Birch's salary, $22,500 a year . . . that's not very much at all. But finally we found what we were looking for, right across the Potomac, in McLean, VA, and we're very happily settled."

When asked about the future, Marvella replied, "I have no ambition beyond the one that Birch has . . . That is for him to build up his seniority and be the best senator he possibly can so we can stay in Washington."[14]

Years later, when asked why *Life* magazine did a feature on Marvella during that first year in the Senate, Birch mused that there was lots of competition for new men on the scene in Washington but Marvella didn't have much competition.

Birch's first year in the Senate was one of great importance for the country and the world. One of the issues of little impact in the country or the world but significant to Birch was the closing of the Studebaker automobile plant in South Bend, Indiana. Keefe learned that the plant was due to close five days before Christmas, putting thousands of employees out of work. Birch was working as hard as he could to find another corporation that could benefit from assuming ownership of the existing facility while keeping employed the people whose jobs were in jeopardy. He worked diligently to save the plant and the jobs, having continuous talks with both Secretary of Defense Robert McNamara and Attorney General Robert Kennedy on the possibility of having the Defense Department utilize the facility. He worked closely with Representative Brademas to help the Studebaker workers and had reason to expect that the Kennedy administration would offer its assistance. Not only did Indiana have two Democratic senators, but also Birch's election over Capehart was important to JFK, and the Bayh victory was sweet. Birch and Marvella had been invited to events at the Kennedy White House and to a cruise on the *Sequoia* that Robert Kennedy organized.

On November 22, Birch was flying to Chicago to meet with the president of International Harvester about their possible interest in the Studebaker operation. The last conversation he had before heading to the airport was with Robert Kennedy. When the plane was landing at O'Hare Airport, the pilot announced the assassination of President Kennedy. Birch rushed from the plane to the nearest phone booth and called Marvella. She was in tears and confirmed that the president was dead. Ironically, Birch's trip had necessitated that he skip his assignment to preside over the Senate that day. His friend Ted Kennedy substituted for him and was sitting there when the news from Dallas arrived.

All of America was obsessed with the events of the next four days, ending with the burial of John F. Kennedy at Arlington National Cemetery. Those alive at the time will always know where they were when the assassination took place, and most were glued to their television sets for the next four days. The arrest of Lee Harvey Oswald; the casket lying in state in the Capitol; the assassination of Oswald; the service at St. Matthew's Cathedral and the salute by the President's son, three-year-old John F. Kennedy Jr.; the march to Arlington and the burial. The Bayhs attended the funeral service. Marvella wrote in her autobiography that an Indiana politician at the ceremony said to Birch, "You are the only one with the Kennedy-type charisma. You owe it to your country to reach for higher things." She also recounted extending an invitation to Bobby and Ted Kennedy and their wives to dinner on December 16. Bobby Kennedy declined, saying he couldn't accept invitations like that until at least early in the new year. Joan Kennedy, Ted's wife, called Marvella to tell her that ABC was broadcasting a film of Theodore White's *The Making of the President*, a book about the 1960 campaign, to be premiered for the family at Jackie Kennedy's house in Georgetown. Ted and Joan came for dinner and then departed with Marvella and Birch for Jackie's, where they watched the program together. With them were Franklin D. Roosevelt Jr., British ambassador David Ormsby-Gore, JFK aides Ted Sorenson and Ken O'Donnell, and others. They all cheered like it was a sporting event when the film showed JFK scoring points against Nixon on the stump and in debate. Marvella recalled the last time they had watched a movie with the Kennedys, "with John Kennedy lounging on his couch in the White House Theater, Jackie laughing at Birch." She added that it felt like a decade since they were first at Bobby's house and in JFK's Oval Office.[15]

Only days after the tragedy, leaders of the American Bar Association contacted the Bayh staff about the Bar Association's work on proposing an amendment to the constitution to deal with presidential disability and succession. Two major issues presented themselves to the country. One was that there was nothing in the Constitution addressing the possibility that a president could be disabled and unable to serve. It had happened a few times in American history. President Garfield lingered close to death for eleven weeks after being shot by an assassin. President Wilson had a stroke, and the only communications from him were by his wife, whom many considered to be the de facto president during that period. Had Kennedy lived with the injuries he received, he would have been unable to serve as president. The specter of a disabled president in the nuclear age was ominous.

The other major issue was another vacancy in the office of vice president, the sixteenth in American history. The country had been without a vice president for thirty-eight years in its history. There had been seven vice-presidential deaths, one resignation, and eight instances in which the vice president succeeded to the presidency, the most recent being Lyndon Johnson, who had previously survived a serious heart attack. Those eight presidential deaths included four assassinations: Lincoln, Garfield, McKinley, and now Kennedy. In December, Chairman Bayh of the Senate Subcommittee on Constitutional Amendments introduced a twenty-fifth amendment to the Constitution to deal with these two matters.

According to Keefe, they had been looking for an issue for the subcommittee to address even before the assassination. Keefe expanded on the steps that Birch took to get the chairmanship of the Subcommittee on Constitutional Amendments. They hadn't acted to get the subcommittee as quickly as they should have, but when its chief counsel, Fred Graham, told him he was being released because Eastland was dissolving the subcommittee, Birch decided to request a meeting with the Judiciary chairman. Birch went to see Eastland and was immediately offered a drink by Eastland's aide, Bill Taylor. It was clear to him that Taylor was a Bayh fan as well. In fact, Taylor at one point told Birch that if he ever decided to go national, "I'll take care of Mississippi for you." The meeting with Eastland about the vacancy in the subcommittee chairmanship—with the death of Senator Kefauver—went well into the night. When Birch returned to the office, Keefe saw that he was inebriated, the first and only time he ever witnessed it. Eastland was famous for serving scotch in his office, and the two men had imbibed for hours.[16]

The result was that Birch got the post, but not that evening. Eastland insisted that there were too many subcommittees within Judiciary and said that although he appreciated the fact that Birch was the only Democrat on the committee without a subcommittee chairmanship, he still needed to close it down. The morning after their long evening, Eastland called him and said, "Birch, I think you'll make a splendid chairman of the Subcommittee on Constitutional Amendments." The appointment was made official on September 30, 1963. But Eastland also told him that there was no money available for staff unless the subcommittee was pursuing an issue, a potential constitutional amendment. After the assassination, the issue presented itself.

Remembering the Eastland phone call, Birch asked, "When else could a guy like that do a favor for a guy like me?" The Eastland favor was not the only one given to Birch, but that first one resonated throughout his career.

Birch's first full year in the Senate was a time for learning, not just about the Senate but about representing Indiana as well. There is always a conflict in a job like this: family versus office versus travel to the state. It is a strange existence where one is serving multiple masters, more than just family versus work. Working in the Senate means long hours at the Capitol but also necessary time spent in the state, as well as trips to other states on behalf of colleagues.

Press releases from 1963 show a great deal of emphasis on Indiana matters, but national concerns were also appearing. Birch made statements on national fiscal policy, the testing and evaluation of the cancer drug Krebiozen, support for the equal time provision, passage of the civil rights bill, and a Paris trip as a US representative at the NATO conference.

Looking back on his first year in office, Birch remembered how there seemed to be a natural progression from the state legislature to the Senate. He learned in the legislature about the need for reservoirs and was part of the state's efforts to build the Monroe and Wabash Valley reservoirs. When he got to the Senate Public Works Committee, he was able to secure authorizations for the same reservoirs, beginning the federal government's participation in the projects. Once he moved to the Appropriations Committee, he could see to it that the funding for the projects was there. Similarly, he would later seek to lower the voting age to eighteen by constitutional amendment, an effort he tried in the legislature but failed by a single vote.

Birch's birthday on January 22 was the occasion for a fundraising birthday gala in Indianapolis, something that would be repeated many times throughout his career. Only a year earlier, he had not attained the celebrity status to be able to attract well-known talent to these events. At this first one, the celebrity guests were actress Janet Leigh and singer Vic Damone.

4

Crash and Constitution: 1964

The dogmas of the quiet past are inadequate to the stormy present.
As our case is new, so we must think anew and act anew.

ABRAHAM LINCOLN, AS QUOTED BY BIRCH BAYH, ADDRESSING THE DEMOCRATIC NATIONAL CONVENTION, AUGUST 25, 1964

ON JANUARY 25, 1964, THIRTY-SIX-YEAR-OLD BIRCH BAYH WAS part of a televised ceremony honoring the "10 Outstanding Young Men of the Nation," of which he was one. The designation was made by the US Junior Chamber of Commerce. It was likely only recognition of his being elected to the Senate at such a young age, since he had yet to record any legislative victories. Others on the list included Dr. Zbigniew K. Brzezinski, age thirty-five, a Columbia University professor that Birch would come to know well as President Carter's national security adviser; A. Leon Higginbotham Jr., also thirty-five, a member of the Federal Trade Commission who would go on to serve as a federal judge for many years; George Stevens Jr., age thirty-one, director of the US Information Agency who would go on to produce films and the Kennedy Center Honors; and James Whittaker, age thirty-four, the first American to climb Mount Everest.[1]

In 1964, Birch was largely devoted to the effort to pass the Twenty-Fifth Amendment. He became convinced of its importance and saw an opportunity to make a name for himself and establish the legitimacy of his subcommittee. It was a watershed year, one that would resonate for years to come. In January, President Johnson announced the War on

Poverty and the surgeon general stated that smoking cigarettes could be harmful to a person's health. One of the only two women in the Senate, Margaret Chase Smith of Maine, announced that she was running for president. In February, the poll tax was outlawed with the ratification of the Twenty-Fourth Amendment to the Constitution. In May, America witnessed the first burning of a draft card to protest the Vietnam War, and in June came news of the murders in Neshoba County, Mississippi, of James Chaney, Andrew Goodman, and Michael Schwerner, three young men who had traveled there to fight for civil rights.[2]

During the early months of 1964, Birch continued working to save the jobs at Studebaker, and he needed the new administration to commit itself to the cause. On a ski holiday in West Virginia, he received a call from Johnson City, Texas. It was a call from the president, who assured him that the administration was as committed as he was to saving the jobs in South Bend.

Johnson had another priority in 1964, though: getting reelected. Birch chaired Young Citizens for Johnson–Humphrey and toured much of the country with the Johnson daughters, Linda Bird, age twenty, and Lucy Baines, age seventeen. From that campaign experience, a lifetime friendship grew between Birch, Marvella, and the Johnson daughters.

Sometime after Hubert Humphrey joined the ticket with LBJ, the United Auto Workers held their convention in Atlantic City with Birch as the keynote speaker. They announced a surprise guest, and it was none other than the president of the United States. Johnson invited Birch to ride back to DC with him on Air Force One. On that ride, they discussed the Twenty-Fifth Amendment, and the president advised Birch that his effort to pass it would force a vote in the House to remove Speaker McCormack from the line of succession. If he would wait until after Johnson and Humphrey were elected later that year, the president could more easily move forward in support of the amendment without risking offense to the Speaker. Birch knew that his advice was sound.

On June 19, 1964, the Senate was deliberating the Civil Rights Act of 1964, a bill that would become one of LBJ's landmark accomplishments. As the Senate approached its final passage, following a filibuster lasting seventy-five days, Birch's schedule to leave Washington was in jeopardy. The Senate invoked cloture, a process to end the unlimited debate, or filibuster. Birch attributed the cloture vote's success to President Johnson calling Sen. Richard Russell to the White House to persuade him that the filibuster was an embarrassment to him and that the time had come to pass this legislation. As a result, cloture was invoked on June 10, and the vote on the Civil Rights Act finally took place on June 19. The president had been an important member of the Southern caucus in the Senate, of which Senator Russell was its leader. When the president took pro–civil rights positions in the first year of the presidency, he lost the affection he had enjoyed from others in that caucus. Russell, however, was not one who shared that disaffection. He remained a stalwart friend and ally of the new president.

Because of the lateness of the vote that day, Birch was forced to delay his departure to Springfield, Massachusetts, to be the keynote speaker at the Massachusetts Democratic State Convention. His colleague Ted Kennedy had been elected to the Senate in 1962 to fill his brother's unexpired term. The 1964 election was for a full six-year term, and Ted would be accepting the Democratic nomination. The original plan was for Birch and Ted to join other Kennedys heading to Springfield at 2:00 p.m. on the *Caroline*, the Kennedy family plane; among its passengers were Ted's wife, Joan, and others in the Kennedy clan. But the Senate was still debating; as the hours ticked away, another plane was arranged, an Aero Commander 680. Before departing for Massachusetts, Birch and Ted spoke to the convention on the telephone from the Senate cloakroom, telling those assembled about the vote to pass the Civil Rights Act and saying that they were on their way. The two senators joined Marvella in a car for the mad dash to the airport and flew north at around 7:00 p.m.

Also on the Aero Commander 680 were the pilot, Edwin Zimny, and Kennedy aide Edward Moss. Zimny, age forty-eight, was the owner and operator of Zimny's Flying Service of Lawrence, Massachusetts. Not only a commercial pilot, he was also a flight instructor and a mechanic. The twin-engine plane was owned by Daniel E. Hogan, who had lent it to Senator Kennedy for the flight. Kennedy sat on the co-pilot's side, facing the rear of the plane. Across from him were the Bayhs, facing front with Marvella on Birch's right. Moss sat in the co-pilot's seat. The weather bureau reported the visibility at less than two and a half miles with conditions considered "marginal." Birch remembered bouncing around in the clouds, then emerging to see the moon before bouncing around in the clouds again. Through the windshield he saw a black line stretching horizontally in the distance and assumed they were heading into a cluster of storm clouds. As the plane approached Barnes Municipal Airport in Westfield, Massachusetts, it made a sharp right turn. Kennedy later recalled feeling "that the plane had been hit by lightning. I saw black things outside my window." When the flight got rough, Birch thought they had entered the storm clouds, but they were not storm clouds. The plane had flown into an apple orchard and crashed about three miles away from the airport.[3]

When the pilot jerked the plane upward, Birch thought he was trying to get out of the clouds, but it was an orchard he was trying to escape. The plane stalled, hit the apple trees, and crashed to the ground. Had it not hit the trees, it would have crashed directly into a large boulder. Hitting the trees probably saved the lives of those who survived. Ted Kennedy had been leaning out of his seat, most likely untethered from his seatbelt, urging the pilot to get them to the convention. Birch thought it would have shown better judgment if the pilot had not flown into what he thought were storm clouds. With two United States senators on board, most likely the pilot felt compelled to continue the flight through the clouds.

When the plane crashed, Birch's seatbelt caused him to lose his breath, and he lost consciousness momentarily. When he came to, he heard Marvella screaming. The windows on this plane were designed to pop out automatically upon crashing, and that is what happened. Birch tried to calm his screaming wife and said, "Let's get out of here."

He maneuvered her out through the window over the plane's right wing, guiding her off the wing and away from the plane. Marvella had broken her tailbone and was in considerable pain. The seatbelt had torn the muscles in Birch's abdomen, and his right arm was numb and useless. He yelled for Ted, and there was no reply. The plane's beacon was still illuminated and slowly turning. Fog rested only about ten feet above them. The beacon light bouncing off the dark sky and fog created an eerie feeling that Birch never forgot. As Birch and Marvella were heading for help, they could smell gasoline, which convinced Birch that he needed to return to the plane to make sure that all survivors had gotten out. Walking around the plane, he could see into the front seats and concluded that both men were dead.

Zimny, the pilot, was killed, and Moss was critically injured and near death. Marvella was frantic and urging Birch away from the plane, thinking it might explode due to the gas fumes. "The more reason to try and get Ted out," he replied. As he got closer to the section of the plane where Kennedy was, he yelled again and heard a mumble in reply. A light in the cabin was still dimly lit, allowing him to see fingers moving and then a cufflinked arm reaching toward him. Birch got back inside the plane and pulled the 230-pound Kennedy, who was nearly unconscious, up under his right arm "like a sack of corn" and pushed him out through the window. Birch was amazed that he carried Ted as he did, because two days later he still couldn't lift his right arm. The adrenalin running through his veins gave him strength he wouldn't have after the emergency was over. Ted remembered Marvella repeatedly screaming, "We've got to get help!" while he was calling out, "I'm alive, Birch!" though all Birch heard was a mumble. Ted lay in the weeds where Birch set him down. They asked if there was anything they could do for him, and though he was in great pain, Kennedy asked them to get an annoying weed out of his face.

Birch and Marvella made their way to the road to flag down a passing car for help. Nine vehicles passed them before one stopped. Robert Schauer, a nearby resident who had heard the crash, arrived in his pickup truck and drove the Bayhs to his home, where they called for help. They took blankets and pillows from Schauer's home back to the crash site to try to make Ted more comfortable. Eventually, the police and an ambulance arrived. Kennedy was now unconscious and was put into the ambulance with Birch, Marvella, and Moss, whose death rattle could be heard and was never forgotten. Blood from his neck was spurting throughout the ambulance. Moss died the next day during surgery. Ted's account was that the ambulance arrived an hour and a half after Birch pulled him from the plane.[4] Birch recalled that the entire event took no more than half an hour.

Birch called Robert Kennedy from the hospital to tell him they had bad news; he and Ted had been in a plane crash. "We haven't lost him too, have we?" was the response.[5] Bayh staff member Bob Boxell was in the car when he heard the radio reports that Senator Kennedy had been in a plane crash and there were two unidentified bodies at the site. He assumed the two were Marvella and Birch.

Bob Keefe flew to Massachusetts to see Birch immediately after learning about the plane crash. President Johnson dispatched an official from the Federal Aviation Administration to report to him personally on Kennedy's condition. Keefe remembered how the Kennedy sisters, "very, very tough women," rebuffed the official's every attempt to see Kennedy.[6] Larry Conrad and Boxell flew up the next day and remembered seeing Robert Kennedy standing outside his brother's room. "I'll never forget his expression," recalled Boxell. "He looked lost. Desolate." They went immediately to the Bayhs' room, where Marvella was dazed and under heavy medication but murmured, "You wouldn't be here if Birch were not a US senator." Boxell replied quietly, "I'm here because my friends were injured in a plane crash." The next day when Boxell and Conrad returned to the hospital, Marvella gestured to them as if she needed to whisper something. When Boxell leaned over to listen to what she had to say, she smiled and kissed him on the cheek.[7]

Birch spent a few days in Northampton's Cooley Dickinson Hospital in a wheelchair, without any use of his right arm and suffering pain from his severely bruised hip and stomach. Marvella was treated for a damaged coccyx and fitted with a circular cushion to help with the pain, staying a few days longer. Ted suffered a broken back, cracked ribs, and a collapsed lung, which had been punctured by one of the tips of his cracked ribs. He didn't leave the hospital until mid-December, almost six months later. The attorney general visited each of the senators in the hospital. Before the Bayhs returned to Washington, their luggage was delivered to the Senate office. Staff member Patty Rees remembered how the suitcases reeked of gasoline fumes.[8]

Birch was able to return to the Senate in a week or so and was warmly greeted by his colleagues of all political persuasions. He was able to attend the signing ceremony for the Civil Rights Act at the White House on July 2 and was given one of the pens the president used to turn the act into the law of the land.

Civil rights were a major part of the backdrop of political issues dominating the American landscape when Birch arrived in the Senate.

During the fifties, the Eisenhower administration was dealing with the aftereffects of *Brown v. Board of Education*, a unanimous decision by the Supreme Court that declared the segregation of public schools to be unconstitutional. The demand for equal rights and an end to segregation in the public schools gave impetus to a growing movement among African Americans seeking their civil rights. The school desegregation crisis in Little Rock, Arkansas, resulted in Eisenhower sending the National Guard to Little Rock to enable black students to safely enter and attend a formerly segregated school.

The Kennedy administration preferred civil rights leaders to operate quietly and to trust the president to do the right things on their behalf with "all deliberate speed." Instead, the Southern Christian Leadership Conference (SCLC) and its leader, the Rev. Martin Luther King Jr., kept up the pressure on JFK through an aggressive and unprecedented series of sit-ins and other protests and with a march on Washington that captured the attention of the American public in a way never seen before. On May 4, 1963, the country was shocked by the television images of black children in Birmingham, Alabama marching for civil rights while firefighters sprayed them with firehoses, knocking them off their feet. On June 11, 1963, Americans saw a televised confrontation between deputy attorney general Nicholas Katzenbach and Alabama's governor, George Wallace. Wallace refused to obey a court order allowing the admission of two African American students to the University of Alabama in Birmingham. On a June evening in Birch's first year in the Senate, President Kennedy addressed the nation on the pressing issue of civil rights, describing it as a moral issue. He soon introduced the Civil Rights Act. Later that night, Mississippi NAACP head Medgar Evers was assassinated in the driveway of his home in Jackson. Neither Evers nor Kennedy lived to see the Civil Rights Act become law. The mantle of leadership on civil rights was passed to President Lyndon Johnson, a former member of the Senate's Southern bloc but now president of all the American people. It was Johnson's formidable legislative skill, combined with his ability to cajole, encourage, and intimidate senators, that helped him gain passage of two landmark civil rights measures, the Civil Rights Act of 1964 and the Voting Rights Act of 1965. These victories were not without political costs. They changed the demographics of the national

Democratic Party. Now that it was the party of civil rights, the Southern conservative Democratic South existed no more.

Clay Risen, in his book *The Bill of the Century,* a history of the Civil Rights Act, characterized the new law by saying, "An entire social system built on oppressing and excluding blacks had been outlawed with the stroke of a pen." He went on to describe the measure: "The Civil Rights Act of 1964 was the most important piece of legislation passed by Congress in the twentieth century. It reached deep into the social fabric of the nation to refashion structures of racial order and domination that had held for almost a century—and it worked. Along with banning segregation in public accommodations, it banned discrimination in the workplace—and not only on the basis of race, but sex, religion and national origin as well."[9]

The day that the transformative Civil Rights Act of 1964 passed was only days away from Nelson Mandela beginning a twenty-seven-year imprisonment on Robben Island near Cape Town, South Africa. In the same week, three civil rights workers were killed in Mississippi. Both events would have long-term effects on the relationship between blacks and whites in the United States and around the world, though in very different ways. Also that year, the Nobel Peace Prize was awarded to Rev. Martin Luther King Jr. Birch never had a chance to know King, though they met once.

Growing up, Birch did not know any African American people. There were no blacks in Fayette Township, Indiana. But he played baseball with black soldiers in Germany; in fact, he had coached the team, and he could not recall any racial incidents. His military company was segregated, but the baseball team was not.

The Civil Rights Act was not popular in Indiana, but many members of Congress took a stand to do what was right instead of doing what was politically expedient. Birch was one of those, and it became clear to him that not only did the Senate face a moral issue but also it was important that the American government live up to its credo. It was also an issue where members had to choose to stand with President Johnson or with the conservative Southerners. Birch's growing awareness of racial discrimination, whether de jure or de facto, became distinct when he found himself serving in the Senate. His evolution about and understanding of race relations would play a major role in his career.

In July 1964, the Republican National Convention met in San Francisco and nominated Arizona senator Barry Goldwater to oppose President Johnson. At the Democratic Convention in Atlantic City, LBJ kept everyone in suspense almost up to the first gavel, when he finally announced Hubert H. Humphrey as his vice-presidential running mate. Humphrey had gained a national reputation when competing with JFK in the 1960 Democratic primaries, and he had successfully managed the passage of the Civil Rights Act. Liberals applauded his nomination. What Birch did not know was that presidential aide Bill Moyers had written a memo to LBJ on June 29:

> We need to select a keynoter for the Convention as soon as possible, and I would like to suggest Birch Bayh.
>
> (1) He is young, and his youthfulness will project a spirited image of the Democratic Party that will help offset the more tired image of older (much older) men like McCormack and Truman, who will rightfully be in the spotlight at the Convention. His youthfulness will also be of use in identifying you with the generation that felt they had lost a true champion, "one of their own," when Kennedy was murdered. By beating the ancient Homer Capehart in an upset, he established himself as a symbol of the "new order" in American politics.
> (2) He would be an effective counter to the image of the GOP keynoter, the young and able Mark Hatfield.
> (3) He is not hurt by entanglements with either the doctrinaire liberal or doctrinaire conservative wing of the Democratic Party.
> (4) He comes from the Midwest, a convenient geographical balance to the West and Southwest of Lyndon Johnson and the big state influence of New York, Pennsylvania and New Jersey.
> (5) He is a talented speaker, a fine and articulate platform artist.
> (6) His escape from near death in the Kennedy plane crash a week ago brought him a dash of national recognition.
> (7) He is an avid Johnson supporter and would be most co-operative in developing the speech to be delivered at the convention.
>
> Bill Moyers[10]

Ultimately, Birch was not asked to keynote the convention, a task given to Rhode Island senator John Pastore. Birch was given a speaking slot at the convention and asked to chair the Young Citizens for Johnson–Humphrey. Birch's administrative assistant Bob Keefe had first broached the subject to the Kennedy staff in 1963; someone of Birch's age might chair a "Young Citizens" committee. Subsequently, Birch talked with Bobby Kennedy about it. Birch told Kennedy that he had been speaking at a number of colleges and his sense was that JFK did not have the voter support he needed. The attorney general's response was, "Okay, help us do something about it." After the assassination, the need for orchestrated outreach to younger voters seemed even more sensible than before, and Keefe pursued the idea with the Johnson staff. Birch began to travel extensively during the campaign with the two Johnson daughters. They held campaign sessions with people in several cities, including Los Angeles. Birch remembered the reaction at that event of staffer Bob Hinshaw, who couldn't take his eyes off the voluptuous actress Janet Leigh. The famous folksingers Peter, Paul and Mary campaigned for LBJ and spent an evening at the Bayh home. Birch remembered Mary Travers in high heels shooting baskets in their yard at one o'clock in the morning.

The 1964 election was the first presidential contest since Birch's election to the Senate. His focus in the campaign was more national and less on Indiana than perhaps any other time in his career, with the exception of when he was a candidate himself. Indiana had a primary in May, and though President Johnson had no serious challenge to his nomination, Gov. George Wallace competed there against Gov. Matt Welsh, a stand-in for the president. Wallace amassed 30 percent of the vote. This could be interpreted as indicating the unpopularity of LBJ's promotion of civil rights, a harbinger of potential difficulties ahead for Birch. Further evidence was that in the racially divided city of Gary, Wallace carried every one of its white precincts. Despite that, however, LBJ carried Indiana, a rare occurrence for a Democratic presidential candidate.[11]

Barry Goldwater was not a colleague with whom Birch had much of a relationship. They became closer when serving together on the Senate Intelligence Committee many years later. But in 1964, Goldwater was successfully painted by the Democrats as a wild-eyed warmonger and lost the presidential contest to LBJ by a margin of 61–39 percent of the vote. No candidate had ever amassed such a large percentage of the popular vote. Ironically, LBJ began the escalation of the Vietnam War during the very month that he celebrated one of the greatest election victories in history. Vietnam would haunt the Johnson presidency and prevent him from being viewed by history, in Birch's words, as "one of the greatest ever."

The report of attacks by North Vietnamese torpedo boats against an American destroyer, the USS *Maddox*, enabled LBJ to escalate in Vietnam through passage of the Gulf of Tonkin Resolution. The resolution provided a legal basis for retaliatory air strikes and became a blank check for him to prosecute the war without a formal declaration by Congress. Only two members of the Senate voted against it, Wayne Morse (D-OR) and Ernest Gruening (D-AK). The latter said that he objected to "sending our American boys into combat in a war in which we have no business, which is not our war, into which we have been misguidedly drawn, which is steadily being escalated."[12] Birch was among many in the Senate to support the resolution and later regret the blank check they gave LBJ. And he, like most of his colleagues, had not yet begun questioning the very reasons the country was fighting there.

The 1964 election saw the defeat of former Kennedy press secretary Pierre Salinger, who was running for the Senate in California against former actor George Murphy. Democrats elected to the Senate were Fred Harris (OK), Joseph Tydings (MD), Joseph Montoya (NM), and, in New York, Robert Kennedy. Walter "Fritz" Mondale, who became a close Bayh friend and ally, was appointed in Minnesota to fill the unexpired term of Hubert Humphrey, the newly elected vice president.

Added to Indiana's congressional delegation were Andy Jacobs Jr. and Lee Hamilton. Birch had served in the state legislature with Andy, and Hamilton had been a Bayh county coordinator in 1962. Prior to knowing Birch, Hamilton knew Birch's father when he was a referee of Indiana high school basketball. Birch and Lee met when both were ATO chapter presidents, Birch from Purdue and Lee from DePauw University. They were at a national ATO convention in Gettysburg, Pennsylvania, where they learned that the fraternity had a whites-only covenant. Birch led an unsuccessful effort to remove the section of the covenant prohibiting African Americans from joining.

Many years later, Birch would describe his feelings about Lee in effusive terms. He felt Lee combined all the best qualities one could ever hope for in a congressman: bright, savvy, personable, and enormously effective at his job on behalf of the people of Indiana and the country. Their personal relationship remained close throughout their lives. Birch looked to Lee Hamilton as one would a brother.

When Birch came to Congress, the members of the Indiana delegation, in addition to Senator Hartke, were Democrats Ray Madden (who had been there since FDR), John Brademas, J. Edward Roush, and Winfield Denton. The Republicans were Charles A. Halleck, E. Ross Adair, Richard L. Roudebush, William G. Bray, Ralph Harvey, Earl Wilson (who was replaced by Hamilton), and Donald C. Bruce, who was replaced by Jacobs. Brademas, a Rhodes Scholar, would become majority whip in the House and an important Bayh ally in the coming years. He won his seat on the third try, after losing in 1954 and 1956, finally winning in 1958, the first Greek American to serve in Congress.

The Bayh office issued several press releases in 1964 dealing with the proposed Twenty-Fifth Amendment but also a number reflecting activities in Indiana as well as on the national stage, showing his support for the civil rights bill, the farm bill, American aid to Indonesia, flood control, veterans' benefits for widows, social security, and Vietnam.

Birch announced a trip to Los Angeles in August for Young Citizens for Johnson-Humphrey and made a speech at the Democratic National Convention on August 25. He gave at least twenty-one speeches in at least a dozen trips to several cities in Indiana, also campaigning in fourteen states in support of the Democratic campaign for president.

For Birch Bayh, 1964 was memorable for the plane crash, for civil rights, and for the presidential election but also for the Twenty-Fifth Amendment. He characterized his various accomplishments by saying they were the result of people and circumstances—people like Everett Dirksen, who took him under his wing during his first months in the Senate; like LBJ, who made him an ally during that first evening in Washington and the subsequent house hunting; and like James O. Eastland, a man of vastly different demeanor and philosophy who nonetheless gave Birch the subcommittee chairmanship he coveted. And that chairmanship became the springboard for most of what he accomplished as a senator. The tragic Kennedy assassination, an event that gripped the country and changed American history, also fueled the argument that the constitution needed to address presidential succession and disability. The chairman of the Senate Subcommittee on Constitutional Amendments was the logical person to take on this issue.

Birch wrote *One Heartbeat Away,* a book about the passage of the Twenty-Fifth Amendment. In the foreword, Lyndon Johnson describes the accomplishment as follows: "He initiated and brought to fruition the first major alteration of Presidential and Vice-Presidential succession procedures since the ratification of the Constitution." The book's preface is by former president Eisenhower, who writes about the sixteen times there had been a vacancy in the office of vice president—taken together, a period of thirty-eight years. But he also mentions the illnesses he experienced as president and the measures taken to authorize Vice President Nixon to act in his stead. He describes those steps as stopgap measures, and constitutional scholars might argue that Nixon's actions at the time could have been subject to legal challenge. It was clear that a permanent solution dealing with succession issues as well as presidential disability was necessary.[13]

The issue had been the subject of hearings by the subcommittee in June 1963, five months before the assassination, chaired by Sen. Estes Kefauver (D-TN). There was, understandably, no sense of urgency to the issue. In August, Kefauver died of a heart attack, and the following month Birch was appointed chairman. Within three weeks of the assassination, Birch introduced Senate Joint Resolution 139, a proposed twenty-fifth amendment to the Constitution. Thus began a campaign that took more than two years to complete. Birch recalled that most people did not care about the issue. There had never been the deaths of a president and vice president during the same four-year term, and even though Woodrow Wilson was extremely disabled, people seemed unconcerned about that. It was hard to get people to care about the campaign for this amendment, as important as it was. The American Bar Association (ABA) was a key ally in the effort to pass this legislation. A major figure in the ABA at the time was Lewis Powell, a prominent Richmond attorney who would later become a justice of the U.S. Supreme Court. The ABA mobilized lawyers across the country to lobby their senators on the importance of passing the Twenty-Fifth Amendment. Support from the legal establishment proved invaluable. Hearings resumed in the subcommittee in January 1964, but it wasn't passed by the Judiciary Committee until August 1964. As Birch recounts in his book, it was an intensive effort that required all the legislative and diplomatic skills he could muster. What he learned about how to succeed in the Senate was immeasurable.[14]

When working on the Twenty-Fifth Amendment, Birch met with former president Eisenhower. In a brief visit before an event, Ike revealed that he knew little about the pending amendment. But the same was not true about Eisenhower's vice president, Richard Nixon. The most impressive performance of all those who testified before the subcommittee was that of former vice president Richard Nixon. Birch was dazzled by his testimony, wholly without notes, commanding a level of expertise that the freshman Democrat hardly expected. Vice President Nixon had to perform as president while Eisenhower was hospitalized after a heart attack and was painfully aware that no rules existed to deal with disability. There were three such hospitalizations during the two terms of the Eisenhower presidency. Said Birch, "Articulate and well organized, his statement was, in my opinion, the most effective of our entire series of hearings."[15]

In spite of Eisenhower's hospitalizations, his disabilities were only temporary, and no crisis occurred. When President Woodrow Wilson had a stroke, however, he was extremely disabled, and every message emanating from the White House may have been decided by the only person with constant access to him, his second wife, Edith Wilson. This state of affairs would be untenable in the nuclear age, when the nation's life could hang in the balance should the president be unable to act due to disability.

The Senate passed the amendment in September 1964, and the advice LBJ had given Birch earlier that year was on the mark. The House of Representatives would not pass the measure while the Speaker was still next in line of succession, something that would change after the inauguration in January. That election also saw the defeat of Sen. Kenneth Keating in New York, a sponsor of a competing version of the amendment. Robert Kennedy defeated Keating. After the election, in January 1965,

Congressman Emanuel Celler, chair of the House Judiciary Committee, produced a version of the amendment very similar to Birch's; it was reintroduced in the Senate as Senate Joint Resolution 1.

The Senate acted swiftly, passing the amendment out of the subcommittee on February 1 and three days later out of the full Judiciary Committee. On February 19, the Senate voted 72–0 in favor of S.J. Res. 1. The House passed its version of S.J. Res. 1 in April, and a conference committee was scheduled to work out the differences. Birch had never been on a conference committee, and now he would be co-chairing one with Representative Celler. By June the House passed the conference version, and the Senate followed suit the following week. But the campaign had just begun, and for the next two years, Birch's life would be consumed in lobbying state legislatures to obtain ratification from the necessary three-fourths of the states. The process to amend the Constitution is meant to be difficult. It requires a two-thirds vote in both houses of Congress before going to the states. It was finally ratified in February 1967 when Minnesota, Nevada, and North Dakota approved it within a twenty-four-hour period. A symbolic signing by the president took place in a White House ceremony on February 23. A first-term United States senator had engineered the passage of an amendment to the US Constitution. This was only the fifteenth time since the Bill of Rights was passed in 1789 that an amendment was added to the Constitution.

Birch was showered with praise from editorial writers and newsmen around the country. Kenneth Crawford of *Newsweek* wrote a column about the amendment titled "Blessings on Bayh."[16] Don Bacon of the *Jersey Journal* in Jersey City wrote, "The young senator's adept handling of this vexing constitutional deficiency—which some of his colleagues had labeled unsolvable—has propelled his prestige and respect on Capitol Hill sharply upward, far out of proportion to his age and Senate tenure."[17] The Kokomo, Indiana *Tribune* editorialized, "Ratification of the amendment assures its author, Sen. Bayh, of a place in U.S. history. . . . Indiana people share their young senator's pride in this needed and historic step."[18]

The amendment's importance soon became clear. Only seven years later, in 1974, it was invoked twice, playing a significant role in the Watergate scandal that brought down President Richard Nixon. When Spiro

Agnew resigned the office of vice president in disgrace, there was for the first time a process in place to fill the vacancy. Agnew had been charged with taking bribes and tax evasion, leading to a no-contest plea. Without the Twenty-Fifth Amendment, Speaker Carl Albert would have been next in line for the presidency, making it harder for Republicans to urge impeachment or resignation as so many did in August 1974. Speaker Albert was not popular outside the Democratic House caucus, but regardless, his succession would have put the presidency in the hands of the opposition party. The processes of the Twenty-Fifth Amendment led to the vice presidency of Gerald Ford. Republican leaders who went to Nixon to persuade him to resign would have found themselves in very different circumstances had Ford not been next in line. When Nixon resigned and Ford became president, he again invoked the Twenty-Fifth Amendment to fill the office of vice president with Nelson Rockefeller. Ford acceded to the presidency as the only person to sit in that office without ever being on the ticket in a presidential election. Rockefeller joined him as vice president; neither had been elected.

In 1981, President Ronald Reagan was shot in an attempted assassination by John Hinckley. As Reagan was under anesthesia at George Washington University hospital, Secretary of State Al Haig declared that he was "in charge, here in the White House"[19] because Vice President Bush was on an airplane and not immediately available to lead. It was soon made clear to the country that under the Twenty-Fifth Amendment, Secretary Haig was not actually in charge. A process existed, and the country would not doubt that a path existed to replace the president if he were disabled. The Twenty-Fifth Amendment has served the country well.

5

Civil Rights, Guns, and Vietnam: 1965–68

This is essentially a People's contest. On the side of the Union, it is a struggle for maintaining in the world, that form and substance of government whose leading object is to elevate the condition of men to lift artificial weights from all shoulders; to clear the paths of laudable pursuit for all; to afford all an unfettered start, and a fair chance in the race for life.

ABRAHAM LINCOLN[1]

BIRCH BAYH'S FIRST TWO YEARS IN THE SENATE POSITIONED him well on his way to authoring a constitutional amendment. During that first year, 1963, the Washington press corps named him one of the "Ten Most Promising Men in Congress." *Life* magazine ran a feature on Marvella. They were attracting attention, and Birch was learning the ropes in the Senate and much about the colleagues with whom he served.

On January 20, 1965, Lyndon Johnson was sworn in as president for a full term and began the year with a State of the Union address announcing the "Great Society," a plan to eradicate poverty and give all Americans equal opportunity for a better life. He also called for passage of the Twenty-Fifth Amendment to the Constitution, an important detail added to the address by speechwriter Bill Moyers. In February 1963, Malcolm X was assassinated, filling the front pages of our newspapers with stories on civil rights. In March was Bloody Sunday, a violent clash between Alabama state troopers and civil rights marchers in Selma. James Reeb, a white minister, was beaten to death by white supremacists in Selma. That same week, LBJ gave his "We Shall Overcome" speech, soon followed by the murder of another civil rights worker, Viola Liuzzo, by the Ku Klux Klan.

In the early years of his Senate tenure, Birch was too junior to be able to do much about civil rights. He felt he had to nurture his friendships with his Southern colleagues, treating them "gently and politely." He considered the race issue a "no-brainer," believing there was no justification for segregation or discrimination based on race. What was right and how we should treat minorities seemed obvious. He also knew that this differed from the views of many of his constituents but was absolutely consistent with the teachings of his father. On the other hand, he understood where some of his colleagues were coming from on the issue of race and why they felt they had to support segregationist views. Racial awareness was new to him and would quickly become far more important to him than it had been.

As chair of the Subcommittee on Constitutional Amendments, Birch was front and center on the major issues of the day. In the process of enacting the Twenty-Fifth Amendment, he came to understand issues that divided the country and forced legislators to examine their personal philosophies and confront their responsibilities as lawmakers. The year Birch was elected, the issue of prayer in the public schools was developing into a firestorm. It was fully ablaze in 1965.

The cause of the prayer controversy was two Supreme Court decisions, *Engel v. Vitale* in 1962 and *Abington School District v. Schempp* in 1963. Both rulings determined that government-sponsored prayer in public schools was unconstitutional. Much of the country damned the Court for "taking prayer out of the schools," and a movement took hold to pass a constitutional amendment that would overrule the Court's decisions. On a personal level, Birch felt it appropriate for his son to say a prayer before each meal or before a day in school. But when he read the decisions and realized they dealt with prayers sponsored by local governments, often prayers offensive to those of different religions or those without a religion, he sided with the Court. Indeed, the issue registered with him that day in the office of Senator Dirksen when he saw stacks of postcards demanding that school prayer be put back in the classroom. At that moment he made a commitment to hold hearings on the amendment.[2]

As the issue grew in importance—with significant emotionalism on both sides—Birch became introspective, not just on the issue of prayer but also on his role as an elected official. When elected initially he considered it his responsibility to represent his constituents to the best of his abilities and to cast votes as he thought Hoosiers would want him to. But he wrestled with this approach, finding himself on an evolutionary path to a different professional philosophy, realizing that his duty was to study issues and exercise his judgment. Most people were too busy making a living to study issues, and Birch felt it was his job to do so on their behalf. "People might act viscerally," he said, "to how they feel without knowing the facts." His job was to know the facts. But on the prayer amendment, he said, "I thought I had committed political suicide." He opposed the proposed constitutional amendment but agreed to give it a hearing, a dangerous strategy because while it relieved some of the political pressure, the measure easily could have passed.

Many years after leaving the Senate, Birch would paraphrase from memory the philosophy of Edmund Burke, which he felt summed up his own views: "A representative owes more than his blind allegiance to the uneducated will of his constituents. He owes them his judgment as well and he betrays rather than serves them if he provides otherwise."

Indeed, this would be a recurring theme throughout his eighteen-year career in the Senate. But the bottom line remained: "It's a privilege to be here, but there are limits on what I'll do to stay here." Prayer was not the only issue consuming the Congress during those years. Early in 1965 Birch had introduced another constitutional amendment, this time to abolish the Electoral College in order to have presidential elections decided by direct popular vote. The 1965–1966 congressional session passed Medicare and Medicaid, the Voting Rights Act, and the Freedom of Information Act. Debate about the war in Vietnam raged on, heightened by the escalation that began shortly after LBJ's election to a full term as president. Civil rights shared the national spotlight with the war. In 1966, the country learned of the shooting of James Meredith. He was made famous in 1962 when the Kennedy administration interceded to force the University of Mississippi to accept him as its first African American student. Also in 1966, President Johnson appointed to the Department of Housing and Urban Development the country's first black cabinet secretary, Robert Weaver. And Edward Brooke of Massachusetts won election to the US Senate, the first African American to accomplish that since Reconstruction.

The Voting Rights Act of 1965 was passed but not without a significant struggle between Johnson and the Kennedys. Sen. Edward Kennedy, with the support of his brother Sen. Robert Kennedy, wanted to add a prohibition of the poll tax, commonly used in the South to prevent poor black people from voting. Johnson opposed the poll tax provision, since he felt adding it to the bill would doom it to defeat. Birch sided with the Kennedys and talked with the president by phone about the matter. The president was fully aware of the fact that Birch was close to the Kennedys but allowed some of his feelings about them to show in this exchange:

BAYH: I'm motivated because of a philosophical conviction and because I'm staring down the gun barrel of a sizable percentage of Negro voters in two of our largest metropolitan areas. . . .

LBJ: . . . I told the Attorney General (Nicholas Katzenbach) I was going to say at a press conference that I was against the poll tax. . . . And I had instructed him to go as far as he could within his constitutional limits. . . . Now, he feels that this is going to mess up the bill. . . . I'd hate to have the public interpret that we were a bunch of kids fighting among ourselves. I'm willing to move as far as a human being can. . . . I think I'm doing more for those Negro groups than anybody has ever done for them.

BAYH: No question about that.

LBJ: . . . You ought to try to pull them into some unified action, or else we'll be like the Dominican Republic. We'll have a party that's split half and half, and really no leader. They pay no attention to their President. They don't follow their Attorney General. . . . Now this man's (Katzenbach) been Bobby's lawyer, he's been Teddy's lawyer, and he's been my lawyer. They asked me to keep him and I kept him, and I'm following what he says. . . . One of the Senators was told this morning, "Oh, you better watch out, there's going to be a Kennedy President here." . . . That's not good. . . . I don't believe we ought to have a brilliant young Senator that we want to make our future out of, and a brilliant young Attorney General and we can't get together. . . .

BAYH: You've done more than any other to see that these differences have been pulled together.

LBJ: . . . there's no reason why I can't get you and McCarthy (Sen. Eugene McCarthy) together because basically Teddy and Mondale (Sen. Walter Mondale) and those people believe the same things. And Katzenbach is the champion of you all . . . I took (the Kennedys') lawyer, because I foresaw . . . that I would be charged with not being quite strong enough on civil rights. And I initiated . . . the voting rights (bill) myself. Nobody else did. No Negro leaders. I did that at the ranch in November. And I came up with that bill. . . . You and Teddy and me—we're all going to be off asleep somewhere, speaking at some barbecue. This guy (Katzenbach) is going to have to be (administering this bill). And (there's) no reason why the Kennedys can't get along with Nick Katzenbach. . . . He lo-v-ves them, and he's loyal to them, and he's devoted to them. And I'm devoted to him, and he has no orders or instructions from me except "do what's right."

BAYH: The most unfortunate thing about this is that apparently there are people . . . trying to put some of us who feel very strongly about this whole issue in a very untenable position. . . .

LBJ: . . . I have no time to fight with any Democrat. And particularly you're the last one I want to. . . . You represent what I want in my party and the future of my party. You're going to be doing just what I've been doing all through the years as I'm laid aside. And I'm pretty well crippled up anyway now with all these problems I've got. But I do know that there're times when you don't like what your younger sister thinks or what your older brother says. But in the interest of your mother, you-all ought to get in a room and try to find some area (of agreement). You just ought to do it. Because all of you're for the Negroes.[3]

The Voting Rights Act did not end up including a provision prohibiting the poll tax, but it directed the attorney general to challenge its use in the courts. The Supreme Court would rule the poll tax unconstitutional the following year.

The addition of Jason Berman to the Bayh staff in 1965 would have a major impact on Birch's career over the next decade. Berman was a twenty-six-year-old graduate teaching assistant at the University of Pittsburgh when he found his way to the Bayh office. Seeking some real-world political experience, Berman volunteered in the campaign of Congressman William Moorhead of Pennsylvania. After Moorhead was reelected, he received a fellowship that allowed him to work in the congressman's office without pay. Instead, Moorhead suggested to him that there was an up-and-coming young senator from Indiana that he liked and knew was close to LBJ. Berman was intrigued. Moorhead set up a meeting with Berman, Birch, and Keefe. Berman was hired to write speeches, provide issues research, and help out in the Subcommittee on Constitutional Amendments. But soon his responsibilities included foreign policy, an area the Bayh staff had previously neglected; with the growing controversy surrounding the Vietnam War, having a knowledgeable foreign policy expert was important.[4]

On April 11, 1965, the Palm Sunday tornadoes ravaged six Midwestern states, hitting Indiana hard from north of Indianapolis to South Bend and as far east as Fort Wayne. Birch remembered flying over Elkhart, where some of the largest poultry production took place. He saw endless chicken coops with all the chickens gone. The devastation seemed to be everywhere. Birch persuaded President Johnson to come to Indiana to survey the damage. During a ride with Birch and Indiana governor Roger Branigin, LBJ leaned forward and, pointing at Birch, said to the governor, "You know, someday, this young man over here is going to be sitting in my office." This was quite a humbling statement for the thirty-seven-year-old senator to hear. In fact, LBJ had been on the other side of a similar conversation in the previous decade. In discussing a measure in the Senate that would restrict presidential power, President Eisenhower remarked to Senate Majority Leader Johnson that he ought to be careful on issues like the one being discussed. He might be president one day, and the matter might impact him then.[5]

Birch's close relationship with the president seemed to be as strong as ever. Late in the spring of 1965, President and Mrs. Johnson invited Birch and Marvella to spend a weekend at Camp David, the presidential retreat near Maryland's Catoctin Mountain. At Camp David, guests took long walks, enjoyed the fireplaces in their cabins, and had dinner with the president and his wife. Birch recalled seeing the president walking with Supreme Court Justice Arthur Goldberg, his arm draped over the justice's shoulder. As Birch approached them, he heard LBJ talking about how important the justice could be to world peace. It soon became clear that the president was pressuring Goldberg to resign his seat on the Supreme Court to become ambassador to the United Nations, which he did. LBJ appointed his friend Abe Fortas to fill the vacancy on the Court.

Birch also recalled the bowling alley at Camp David, where he bowled a score around 265. Presidential aide Jack Valenti admonished Birch, saying, "I thought you were smart enough to know you shouldn't bowl better than the president of the United States." In subsequent games, Birch's score dropped precipitously.

In October the *Chicago Sun-Times* published an article title "Indiana's Bayh Hits Peak on Capital Prestige Scale." If the Bayhs thought they had "arrived" in those early days of 1963, they had really arrived by now. The article opens with a story about a telephone call to the Bayh residence that was answered by nine-year-old Evan. It was an invitation from the president and the First Lady to join them for a cruise on the presidential yacht. Evan apologized, saying that they were not available as they were in Hyannis with the Kennedys. The article describes their close friendships with other prominent Washingtonians and says that the Johnsons gave them a thirteenth wedding anniversary party. Indeed, their social standing had a practical aspect, making Birch a more effective legislator and representative of his state. Also mentioned in the article was Birch's importance as a member of the Public Works Committee: the article states that in 1965 the state of Indiana received $98 million, an unprecedented level of funding. By comparison, Indiana received no public works funding in 1962 and only $6 million in 1960.[6]

That same month, an article appeared in the Lafayette *Journal & Courier* titled "Freshman Sen. Bayh 'Like a Son' to President Johnson." It compared LBJ's relationship with President Roosevelt when he first arrived in Washington with the one LBJ now had with Birch. The article described the independence Birch had shown by opposing the administration on occasion and said that this had not damaged his personal relationship with President Johnson.[7]

In November 1965, *Newsweek* ran an article entitled "Congress: A Crop of Bright Young Men" in which Birch was singled out for praise.

No one in the "all-work-and-little-play 89th Congress . . . can match the potential of Teddy and Bobby Kennedy . . . a good number of other talented newcomers captured the fancy of the Congressional leadership and, thanks to their performances this year, are well on their way to bridging the anonymity gap."[8]

The Vietnam War dominated the news in 1965, with troop levels rising to 125,000 by July and the number of young men drafted increasing from 17,000 per month to 35,000. In April, the Students for a Democratic Society (SDS) organized their first march on Washington to protest the war. By November, as tens of thousands of protesters picketed the White House and marched to the Washington Monument, troop levels in Southeast Asia had ballooned to 400,000. In June 1966 the United States began bombing Hanoi and Haiphong in North Vietnam.[9]

On the civil rights front, Congress passed the Voting Rights Act of 1965, but the unrest continued that summer with riots in the Los Angeles community known as Watts. In the summer of 1966, Martin Luther King took his civil rights movement north to Chicago. Chicago-area whites' virulent reactions to King were worse than anything he had seen in the South.

With the backdrop of the Vietnam War and civil rights unrest in the United States, Birch was still focused on passing the Twenty-Fifth Amendment to the Constitution. In response to a Supreme Court ruling, he held hearings on reapportionment of congressional seats and introduced his amendment for direct election of the president. He responded to the events in Selma, Alabama, and supported LBJ's efforts to fight juvenile crime. In June 1965, he supported a measure to control gun purchases, an issue that dogged him throughout his career. On the issue of gun control, he said that he "had no idea what kind of hornets' nest" he was getting into. He urged the passage of disaster relief legislation after the Palm Sunday tornadoes and opposed an administration farm bill, saying it was "too piecemeal."

While most of his travels were to Indiana, where he spoke about the Twenty-Fifth Amendment and promoted public works projects, he also toured Latin America. His Senate press operation was becoming more professional and aggressive. The archives show 28 press releases in 1963, ballooning to 115 in 1964 and 120 in 1965, demonstrating that the Bayh operation felt it had something to announce to the public more than twice weekly in both years.

In January 1966, Birch launched a project that was dear to his heart. He had always felt there were too many awards and too much recognition for athletes in high schools and almost nothing for those who might excel in civics and government. He sponsored the Indiana High School Government Leadership Conference and continued it for many years. Each high school in Indiana was invited to send student leaders in civics or government, along with their teachers, to attend an all-day conference in an Indianapolis high school. The students attended four classroom sessions, and the speakers were famous names from the Washington world of government, politics, and the media. The first four speakers were Patricia R. Harris, the nation's first African American woman ambassador; G. Mennen Williams, former governor of Michigan and then assistant secretary of state; John T. McNaughton, assistant secretary of defense; and James Symington, executive director of the President's Committee on Juvenile Delinquency and Youth Crimes. Symington, the son of Sen. Stuart Symington, would serve many years in Congress and would become Birch's close friend.

Later speakers would include members of the news media, including newscasters Walter Cronkite, Dan Rather, Daniel Schorr, and Frank Reynolds; and political figures such as Andrew Young, Benjamin Hooks, Attorney General Griffin Bell, and Supreme Court Justice Potter Stewart.

Also in 1966, Birch introduced his third proposal to amend the Constitution, this time to increase the term length of members of the House of Representatives from its current two years to four years. He proposed the idea in January, and hearings took place in July, Birch encountering heavy opposition within the subcommittee from Sen. Sam Ervin of North Carolina. Ervin argued that the founders wanted the House to be closer to the people and that running every two years accomplished that. Birch argued that we were now living in a complex age and members of Congress were required to master a variety of complicated issues as well as serve an ever-increasing number of constituents. They could be far more effective with a four-year term than by being a member one year and a candidate the next, term after term. If a longer term decreased effectiveness, what did that say about the Senate with its six-year term? When senior members of the House opposed it, he let the matter drop.

Birch was evolving: the senator from Indiana was more and more a United States senator with interests and involvement that took him far beyond the borders of his state. He was now promoting multiple constitutional amendments and receiving honorary degrees at colleges. He spoke out on environmental issues such as air and water pollution, touted federal assistance to buttress local air service, and spoke about disaster relief, crime, and the involvement of young people in public service. As his role in national affairs grew, he also became a more effective advocate for the people of Indiana.

In June 1966, Sen. John Pastore of Rhode Island travelled to Indianapolis to speak to the Democratic Precinct Workers Convention held at the Coliseum on the State Fairgrounds. Pastore brought the Democratic partisans to their feet when he pounded on the podium and shouted, "You're not going to have Birch Bayh to yourselves for a very long time . . . the United States of America and the people of this great country have a claim on that kind of talent and we mean to use it."[10]

The country was experiencing a period of growing unrest over the Vietnam War and the fight for civil rights. Throughout that period, President Johnson was moving his Great Society agenda forward, unaware of how it would be undermined by the war. Birch continued to learn the ropes in the Senate and move up the ladder of seniority, now eighty-third at the beginning of the 90th Congress in January 1967.

Another activist movement, the women's movement, accelerated in 1966 with the creation of the National Organization of Women (NOW). Women represented a majority of the American population, while African Americans made up no more than 11 percent of the country. But during the 1950s and 1960s, the legacy of slavery and lynching of African Americans created a head of steam for the civil rights movement, which overshadowed the women's movement. Equal rights for women, however, was not a new issue.

In the 1775 Thomas Paine essay "An Occasional Letter on Female Sex," Paine states, "If we take a survey of ages and of countries, we shall find that women, almost—without exception—at all times and in all places, adored and oppressed." Later in the essay, he writes, "Man with regard to them, in all climates, and in all ages, has been either an insensible husband or an oppressor..."[11] Abigail Adams wrote to her husband John Adams while he was in Philadelphia debating a declaration of independence: "In the new code of laws which I suppose it will be necessary for you to make, I desire you would remember the ladies and be more generous and favorable to them than your ancestors. Do not put such unlimited power into the hands of the husbands. Remember, all men would be tyrants if they could. If particular care and attention is not paid to the ladies, we are determined to foment a rebellion, and will not hold ourselves bound by any laws in which we have no voice or representation."[12]

But the Founding Fathers didn't address the concerns of women, despite the important role women played in colonial America and in settling the West. The law and the government remained virtually silent on the issues of equality among the sexes.

Even Abraham Lincoln early in his career supported giving women the right to vote. While a member of the Illinois General Assembly in the 1830s, he promoted the idea of extending the franchise to women.[13]

The early feminists were focused on legal disabilities and suffrage. Feminist leaders fought for a woman's right to vote, led by Elizabeth Cady Stanton, Lucretia Mott, and Susan B. Anthony, culminating in the passage of the Nineteenth Amendment to the Constitution in 1920.[14] When America went to war after the bombing of Pearl Harbor in 1941, women took the jobs in American factories previously reserved for men, changing forever the perception of a woman's role in society.

The year 1963 saw milestone events in the growing movement for women's rights. Betty Friedan's book *The Feminine Mystique* questioned the popular assumption that a woman's place was in the home, arguing that women's potential could not be fully realized without equal treatment in the workplace. The Kennedy administration's Commission on the Status of Women released its report on gender inequality, revealing a substantial level of discrimination against women in virtually all aspects of American life. The Equal Pay Act of 1963 tried to end wage disparity based on sex. In 1965, the Supreme Court ruled in *Griswold v. Connecticut* that Connecticut's law outlawing contraceptives violated a right to marital privacy, a loud statement supporting the right of women to determine for themselves whether or not to bear children. And in that year, Friedan became the first president of the National Organization of Women, known as NOW.

As chair of the Subcommittee on Constitutional Amendments, Birch was a legitimate target for lobbying by the leaders of the women's movement. No law could have greater impact on the role of women in America than an amendment to the Constitution requiring that women be treated equally under the law. This presented no conflict for him. He had been raised by strong women, Grandmother Kate Hollingsworth and Auntie Katherine Henley, and was married to a strong, independent woman, Marvella Bayh. His natural response was to listen to these activists, a posture that would yield substantial results in the next decade and a half.

In the fall of 1966, after Everett Dirksen's entreaties, Birch held hearings on the prayer amendment. Dirksen went to the floor of the Senate to urge his colleagues to pass his amendment and overturn the Supreme Court decisions. Calling down the wrath of God, he decried the "social engineers" who opposed his measure. "I think of the children," he whispered audibly, "the millions whose souls need the spiritual rehearsal of prayer." His voice getting louder, he pleaded, "Mr. President, the soul needs practice, too. It needs rehearsal."[15]

Jay Berman remembered a valuable lesson he learned while working for Birch during the hearings. As Dirksen droned on from the witness table, Berman kept passing notes to Birch that contained questions he thought needed to be asked. All were ignored by Chairman Bayh. When the hearings concluded and he asked Birch why he hadn't asked those questions, Birch told him, "Never ask a question unless you know the answer."

The Senate took on new members with the 1966 election, men who would play important roles in Birch's life and career. Charles Percy (R-IL) defeated Sen.Paul Douglas; also elected were Mark Hatfield (R-OR), Howard Baker (R-TN), Walter Mondale (D-MN), and Ernest Hollings (D-SC). Both Mondale and Hollings were nicknamed "Fritz." And an event with future significance took place that year with the election of Ronald Reagan, the actor, as governor of California.

On February 10, 1967, the Twenty-Fifth Amendment was ratified. Birch was never worried about its ultimate ratification. He recalled that the American Bar Association was "a major force" for getting it into the Constitution.

Later that year, LBJ appointed Thurgood Marshall to the Supreme Court, its first African American justice. Just before that, the Court had ruled (*Loving v. Virginia*) that laws seeking to prohibit interracial marriage were unconstitutional. Across the country there were more race riots and many more antiwar demonstrations. Toward the end of November, as the world of politics began to look toward the next presidential election, Sen. Eugene McCarthy of Minnesota announced his candidacy for president, challenging President Johnson on the issue of Vietnam.

On January 7, 1967, President Johnson asked Birch and Marvella to meet with him at the White House. When they arrived, the president told them that he wanted to give Marvella a job. "We need to give the Democratic National Committee a new image. We need somebody on the committee who is Christian, young, sexy, and can talk to the Luci Johnsons of this country." Turning to Birch, LBJ said, "I want your wife to be vice chairman of the Democratic National Committee."

The Bayhs promised to discuss the matter. There was concern about the 1968 election and the unpopularity of the war. Being that close to the president could turn out badly. It became clear to Marvella that Birch and his staff did not want her to take the job. She sorely wanted it, perhaps as much as she had wanted anything before. But she decided to be a good soldier, a helpful wife and political partner, and eventually acceded to their judgment. Six days after the offer was made, the Bayhs went back to the White House to decline it.[16] Later, Birch would say that it was the worst mistake he had ever made. Stifling her ambition in the interests of his was a decision he would come to regret.

Staff members saw an additional, nonpolitical aspect of the decision. Johnson had a reputation with the ladies, and both Bayhs were concerned about that. While Birch could not recall that as a concern at the time, former staffer Allan Rachles insisted that there had been unwanted advances by LBJ and that both Bayhs were uneasy about creating a situation where Marvella and the president might be alone more often together.[17]

The year began with Birch announcing the second annual High School Government Leadership Conference. The speakers were Congresswoman Patsy Mink of Hawaii, Supreme Court Justice Potter Stewart, Peace Corps director Jack Vaughn, and Fred Flott, a Foreign Service Officer from the US embassy in Saigon. Birch also spoke out on direct election, lowering the voting age, crime control, toll increases on the St. Lawrence Seaway, and increases in education, highway, and disaster relief funding. He also promised to fight the renomination of Rutherford M. Poats to a post in the Agency for International Development (AID), an unusual stance of public opposition to an action by President Johnson.

Poats began his government career in 1961 with AID as program director and special assistant of the Far East Bureau. He was deputy assistant administrator (1963–64) and assistant administrator for Far East (1964–67). President Johnson's intention was to renominate him in 1967, this time as deputy administrator, and the *Washington Post* editorialized in favor of the nomination. Hoosier businessmen who manufactured corrugated steel had approached Birch and convinced him that the Korean-made corrugated steel being used in Vietnam was of poor quality and deteriorated quickly in the tropical heat and humidity of Southeast Asia, lasting only a few years. The businessmen contended that their brand of corrugated steel would last twelve to thirteen years. When Birch referred those questions in a letter to Poats, he never received a satisfactory response, so he invited Poats to his office. When asked about the Korean corrugated steel, Poats replied, "That's not what we're doing, Senator." Birch decided to investigate further and came to the conclusion that Poats had lied to his face. He announced his opposition to the nomination.

Opposing Poats put Birch at odds with his Democratic colleague from South Carolina, Fritz Hollings. Hollings, Poats's sponsor and friend, expressed his extreme unhappiness with Birch for opposing the nomination. Berman remembered a tongue-lashing he and Birch received from Hollings on the Poats matter that "almost made me cry."

The morning of the *Post* editorial, which singled out Birch for criticism, he was asked to go to the White House to meet with the president. On arrival, he was ushered into the anteroom adjoining the Oval Office.

There sat the president with Press Secretary George Christian and the *Washington Post* editor who had written that morning's piece. Birch felt as though he had been brought to the woodshed for a whipping. President Johnson asked the editor what he thought of a Hoosier hayseed that knew more than the president of the United States.

Birch interjected, "Mr. President, I thought you were more discriminating about what you read in the newspaper."

If it was a whipping, it was a good-natured one, and soon Birch was alone in the room with the president, who changed the subject to ask Birch if he planned to run for reelection in 1968.

"I expect to, Mr. President."

"Good," replied LBJ. "Then get Marvin (Watson) a list of the ten things you need most." Then, showing a side of himself not often seen by the public, the president said, "Let me suggest that you take some shots at me. Might help you in Indiana, and it won't affect our relationship at all."

Birch returned to the office and huddled with his staff, eventually developing a list of ten things the president might be able to do for Indiana, all in the area of federal projects or appointments. Birch recalled that they included a railroad crossing in southeast Indiana and developing a port facility in Clark County; all ten were approved.

Birch always felt a great deal of affection for Johnson, whom he thought would have been judged by history to be one of the greatest presidents if it weren't for Vietnam. An important rule of governance is that leaders must choose between guns and butter. Johnson wanted to have both and was forced to learn that he couldn't.

In March 1967, Birch held hearings on the need to balance the rights of the accused while protecting law-abiding citizens. The Supreme Court's 1966 *Miranda* decision required a defendant to be advised of her or his constitutional right to remain silent, a decision that reverberated negatively among law enforcement officials. Birch spoke about how critical it was to avoid situations where a person is accused unjustly and forced to submit to conditions that result in coerced confessions. At the same time, it was hard to accept the release from jail of hardened criminals because of technical violations concerning the right to counsel and against self-incrimination. In 1967 the crime rate was far too high, and people were calling for tougher treatment of criminals. Birch tried to demonstrate a quiet, judicious demeanor in dealing with the emotional issues at stake. He called for a reasonable response to issues raised by *Miranda*. While he stated his belief that the defendant should receive legal advice, he thought it was carrying things too far to require an attorney's presence at all questioning. "I know it's inconvenient to sit in the police station and answer questions," he said. "It's also inconvenient to sit on a jury . . . to register and vote . . . to pay taxes. If you are a suspect in a police case, interrogation is an inconvenience that is also the price of citizenship."[18]

A tragic fire in 1967 caused the first fatalities in the space program. Astronauts Gus Grissom, Edward White, and Roger Chaffee were killed when fire broke out in their Apollo spacecraft during a launch pad test. Gus Grissom, the second American in space, was from Mitchell, Indiana. He had been mentioned as a possible Republican opponent to Birch in 1968. They had been classmates at Purdue but got to know one another only after Grissom became famous. Birch knew that someone as famous as Grissom would be hard to beat. Years after the tragedy of Grissom's death, that story was mentioned to former astronaut, Sen. John Glenn of Ohio. Glenn was incredulous. He was amazed by the suggestion, saying that Gus Grissom was a fun-loving flyboy who had absolutely no interest in politics; he never would have considered a career in politics.[19]

Civil rights and the Vietnam War dominated the news in 1967. Birch traveled to Los Angeles for a dinner at the Century Plaza Hotel, where earlier in the day thousands of Vietnam War protesters clashed with LA police. The Bayh press releases that fall referenced the "Summer of Violence," largely reacting to unrest on the antiwar and civil rights fronts.

There were also issues related only to Hoosiers: the problem of alewife, a fish dying in Lake Michigan and washing up on the shores of northern Indiana, as well as disaster relief legislation introduced in the aftermath of the Palm Sunday tornadoes. The disaster relief legislation became law and resulted, years later, in the creation of the Federal Emergency Management Agency (FEMA).

Mayoral elections take place in Indiana during odd-numbered years, and in the Lake County city of Gary, Richard Gordon Hatcher was elected the first African American mayor in Indiana. Carl Stokes, also African American, was elected mayor of Cleveland. They became the first two African Americans elected mayor of any major American city.

Against the intense objections of the Lake County Democratic leadership, Birch endorsed Hatcher, establishing him as the first white politician in Indiana to do so.

For Gary, a segregated city, the election was a major development. Only a year before, an open housing ordinance was passed allowing blacks to live in Gary wherever they wanted. Previously blacks could only live in midtown; they were permitted to travel the city freely during the day but were required to return to midtown by sundown.[20]

Hatcher became a political powerhouse, both in Lake County where he was mayor and among national civil rights leaders. He and Birch worked together closely on a number of matters throughout the remainder of their service together.

At the end of 1967, there were nearly 475,000 American troops in Vietnam and 80 lives had been lost across the United States from student unrest and antiwar and civil rights riots. On December 20, Birch announced his forthcoming trip to Vietnam.[21] Before the end of 1967, President Johnson would be challenged by Sen. Eugene McCarthy for the Democratic presidential nomination, a challenge emanating from McCarthy's opposition to the Vietnam War.

Democrats supporting President Johnson in his far-reaching domestic initiatives found themselves in a conundrum over the Vietnam War. American involvement in the small country in Southeast Asia began during the Eisenhower administration after the forces of Ho Chi Minh successfully ousted the French. France had been in control of Vietnam for most of its colonial history, and its departure created a vacuum that Ho Chi Minh hoped to fill. When Ho declared himself a communist, he began receiving aid from the Soviet Union and China. In the same era, the United States was in the throes of a cold war with the Soviet Union. Vietnam was an area of the world where the forces of democracy could oppose communism without turning up the temperature of the cold war. Although this may have been viewed as a conflict between two ideologies, capitalism and communism, more accurately it was a civil war with Ho trying to unite Vietnam. Many American leaders hoped that this proxy war could be sustained without growing into an American war. Two schools of thought dominated conversation in Washington, DC. One school of thought recalled the appeasement of the Nazis, a policy enunciated by British prime minister Neville Chamberlain that virtually guaranteed the world's leap into World War II. No American leader could afford to be known as an appeaser. Conceding Vietnam to the communists would be perceived as an act of appeasement. The other school of thought was that Vietnam becoming communist was only the beginning. Other nations in the area would soon follow like dominoes. This "domino theory," as it was known, required the United States to hold the line wherever communism threatened.

Another domino theory developed in the halls of Congress. During the Kennedy presidency, Senator Mansfield openly questioned the US role in Vietnam. Soon after, other members of Congress followed suit. The American "dominoes," as congressmen and congresswomen were called, would fall more surely than other nations would.

Though JFK referred to Vietnam as a quagmire, LBJ wasn't persuaded that Kennedy really wanted out. History has recorded that Kennedy gradually turned away from his own policy in Southeast Asia but felt he couldn't act on it until after his reelection. Johnson also was not persuaded that he had to sacrifice his domestic policies in order to pursue successes in Vietnam. Instead of choosing between guns (military) and butter (domestic), he was convinced that he could have both. As the failure of LBJ's policies in Vietnam was becoming clear, his administration sought to keep the American people in the dark. The deceptions and outright lies emanating from Washington led to an erosion of support from the American people and, soon after, from Congress. The draft was instituted in order to generate the manpower thought necessary to prosecute the war. The draft grew increasingly unpopular, fomenting anger and disruption on college campuses. Students protested, clashing with the police. Massive antiwar marches on Washington and other demonstrations ensued. On college campuses, there were protests against corporations like Dow Chemical, to keep them from recruiting students on campus. Dow was the producer of napalm, a flammable liquid used in the bombing of Vietnam that after detonation spread a flaming gel that caused severe burns when it adhered to human skin. The protests against Dow resulted in confrontations with police and many severe injuries of students and police alike. Dow also developed an even deadlier substance used in Vietnam: the defoliant Agent Orange. Its long-term deleterious effects on the health of those who fought in Vietnam have continued through the decades since the war. In October 1967, a march on the Pentagon in Washington attracted over 50,000 people, far more than administration officials anticipated. Surveying the aftermath of the demonstration, LBJ muttered, "In the last two months, we've almost lost the war in the court of public opinion. These demonstrators are trying to show that we need somebody else to take over this country."[22]

In order to better understand the war, Birch traveled to Vietnam on January 6, 1968. He wanted to glean firsthand knowledge of the war that he could communicate to his constituents. While visiting Vietnam, he recalled climbing up a very high tower in order to talk with a young soldier from Fort Wayne. He found that when he traveled into the countryside, away from officials trying to control the flow of information, he better learned about the hopelessness of the effort being made by the United States in Southeast Asia.

In one instance, Birch asked to talk with the pilots and officers involved in the helicopter actions taking place in the country. He rode on a helicopter to survey the jungle where much of the war was taking place. He recalled experiencing the abject fear of an enemy attack.

Jay Berman flew in advance to Vietnam to meet with the American station chiefs there and devise a schedule to help inform Birch on the facts of the conflict. Senators Bayh and Ted Kennedy flew first to Hong Kong and then Vietnam, where they separated to individually assess the ravages of war. Berman recalled the unparalleled experience of flying on an F-14 to the aircraft carrier *Kitty Hawk*, where they spent the night. They later learned that nuclear weapons were aboard the *Kitty Hawk*. Regardless of whom they met in Saigon, they were given the official line. But in the countryside, the opposite was true. Military and intelligence officers made comments like "Can't win," "Gotta get out," and "It's a mess." The out-country experience was the accurate revelation of war for both senators.

Birch felt instinctively that he must make up his mind about Vietnam. He knew the war was a bad policy that he would have to oppose while delicately disentangling himself from the Johnson administration.

While staff members were concerned about presidential retribution, they were also surprised when none came. Birch had no recollection of ever speaking with the president about his opposition to the war, and he felt that LBJ wouldn't have been that surprised, given the fact that so many other Democrats were heading in the same direction. The president also consistently showed a willingness to maintain friendships regardless of political differences.

After leaving Southeast Asia on January 22, Birch stopped in California and appeared on *The Joey Bishop Show*, a late-night talk show then in competition with *The Tonight Show*. The Tet Offensive began shortly after that in Vietnam. The officers who briefed him in Vietnam had predicted that North Vietnamese forces and the Viet Cong would launch the campaign, and the officers were pleased that they would be bringing themselves out into the open. "We'll decimate them," Birch was told, and that was what happened. Initially, the military considered the Tet Offensive an American success, in that it got the guerillas out of the jungles so they could be more readily killed. However, the number of American casualties soon became the story, consolidating and accelerating the antiwar movement at home. This was America's first television war. The searing images of American boys in body bags had a devastating impact on American public opinion. It had become clear to Birch that the administration had been lying and that the war was out of the military's control. He knew the US role in Vietnam had to end.

Decades later, it is difficult to fully appreciate the gravity of the war and the challenge for elected officials like Birch in going against the president and publicly opposing the war. Never in American history had there been so much opposition to a war. Opposition was often portrayed as a failure to support American soldiers, never a popular position to take. Much of the country shared an attitude of "my country, right or wrong," leading to a level of allegiance many American leaders took for granted. Since Birch had a close relationship with LBJ, opposing the president was even more difficult. Once he concluded that the war was a mistake, he had no choice but to declare his opposition.

With the impending election, Birch turned his attention to the campaign. In January, the Bayh office announced birthday galas that would be held for him in Indiana, headlined by the singer Bobby Darin and comedian Jimmy Durante. Birch later described Durante as "a great human being." Birch had a very different view of another celebrity, Bill Cosby. He remembered when Cosby appeared in Indiana and Birch was there with his father, a big Cosby fan. When Birch asked the comic to meet his father, Cosby answered, "No, thank you," and walked away. He never felt good about Cosby after that event.

6

1968

Lincoln had said that he was ready to die, but that he "desired to live," to do something meaningful that would "redound to the interest of his fellow man."

JOSHUA WOLF SHENK, *LINCOLN'S MELANCHOLY*[1]

BIRCH TURNED FORTY ON JANUARY 22, 1968, AND THE NEXT day, Majority Leader Mansfield read a tribute to him on the floor of the Senate. Mansfield's lengthy remarks included biographical details and a list of Bayh accomplishments and were reproduced and distributed throughout Indiana.

Nineteen sixty-eight was no ordinary election year. The war in Vietnam and the civil rights movement dominated American life. Separately and together they created a series of cataclysmic events that made memories of that year seem dizzying, both at the time and in retrospect. It was an extraordinary year, a year of change and challenge, not just in the United States but also around the world. For those who lived through it, it felt as if the world were coming apart at the seams. There were fights for freedom and for equality and marches against the war and in its support. And there was a historical presidential campaign with a severely weakened incumbent.

In Europe, the stirrings of freedom manifested themselves in Czechoslovakia, where in January 1968 Alexander Dubček was chosen as the leader of the Communist Party, an event that caused severe repercussions later that year. The Tet Offensive had begun, and in the waters off

North Korea, the USS Pueblo was seized. North Korea claimed that the ship had violated its territorial waters while spying and began a crisis for LBJ to resolve.

For Birch, an eye-opening trip to Vietnam was followed by another Indiana High School Government Leadership Conference and a renewed focus on Indiana issues. The speakers for this third conference were the deputy commandant of the US Marine Corps, Major General Lewis Walt; the first woman ambassador to the United Nations, Marietta Tree; secretary of the Department of Health, Education and Welfare, John Gardner; and FBI director J. Edgar Hoover's administrative assistant, Fred Stuckenboker. Birch also called for the speeding up of FHA loans and an end to the freeze on public works funds; he joined the effort to fight the alewife problem affecting Lake Michigan. The actions begun by Birch to combat the alewife issue resulted in a brand of trout being introduced into the St. Lawrence Seaway, where the lamprey eels were killing the alewives, whose bodies were washing up on the shores of the Great Lakes. The trout attacked the eels, and the problem went away, an example of how government could use science to solve a social problem very simply.[2]

In February, a photo of a gruesome execution of a Viet Cong officer made headlines around the world. The wrenching photo fostered an increase of antiwar sentiment in the United States. Civil rights protests created disturbances in several colleges across the country. Birch made several public statements about Vietnam in support of General Westmoreland, who headed up the American forces there, and underscored the responsibility of the United States to "stop the Communist aggression." He also called on the government to get the USS *Pueblo* back.

The New Hampshire primary on March 12 shocked American politicians, as President Johnson barely defeated the antiwar candidate, Sen. Eugene McCarthy. This highlighted the deep divisions over the war in Vietnam, both in the country and in the Democratic Party. Speculation grew about the possibility of Senator Robert Kennedy entering the race for the Democratic nomination. On March 16, four days after the primary, he did just that, announcing his candidacy in the same Senate caucus room where his brother had announced his candidacy eight years before. On the same day as Robert Kennedy's announcement, American troops killed scores of civilians at My Lai, Vietnam, a story that would not be made public until late in 1969.

American concerns about the war in Vietnam fueled the campaign of McCarthy for president. Actor Paul Newman and other celebrities boosted McCarthy's national profile, marking the entrance of celebrities into American politics. Among the first celebrities to speak out publicly against the war in Indianapolis in 1966 was the actor Robert Vaughn, star of television's *The Man from U.N.C.L.E.*,[3] at a dinner in support of President Johnson's likely campaign for reelection. Vaughn switched to McCarthy and was joined by a long list of celebrities.

Tony Podesta was a college student and a McCarthy supporter who was assigned to Paul Newman's detail. Years later, Podesta would become a prominent DC power broker. Podesta said, "Until that point, McCarthy was some sort of quack not too many people knew about, but as soon as Paul Newman came to speak for him, he immediately became a national figure."[4]

In many respects, McCarthy was an atypical politician. An erudite, cultured man who sometimes seemed ill at ease while glad-handing with voters, he showed a quick wit and was known to write poetry. Jay Berman remembered an evening on the Senate floor when a vote was taking place for final passage of an appropriations bill. McCarthy arrived in a tuxedo, having interrupted his evening plans. When he was asked sarcastically about his attire, he responded, "I always dress up for final passage."

The McCarthy candidacy provided a structure for millions of young people in the country who were energized by their opposition to the war. These were the same people Birch had tried to mobilize for Johnson–Humphrey four years earlier, but for McCarthy, getting them to sign on to his campaign took little effort.

March 1968 saw demonstrations against US involvement in Vietnam in London and Paris, the latter bringing France to the brink of revolution. Back at home in Washington, DC, Howard University students staged a five-day sit-in, shutting down the university in protest over its ROTC program and the Vietnam War. This form of student activism included demands by the civil rights movement for a more Afrocentric university curriculum. While African Americans were ratcheting up the pressure on civil rights, their elected officials were only slowly coming around in support, with leading politicians coming to understand that simple justice was at stake.

At a meeting in Indianapolis, Birch's remarks were interrupted by a large man who was hard to ignore. A thirty-one-year-old six-foot-three albino African American with a shock of white hair and a booming voice, his name was Gordon Alexander, and he questioned Birch's commitment to civil rights. After the meeting, Birch asked to speak to him. Birch instinctively felt that Alexander was the kind of man who could help him on his own journey to better understand what was happening in the black community—someone he had been looking for. Birch asked him if he was interested in coming to Washington to work for him, to which Alexander responded that he might not be reelected. Birch asked if he would help organize minority communities for him and then come to DC after the 1968 election, once he was safely in office for another six years. Alexander was sympathetic to the idea and years later described his early relationship with Birch by saying he was the "first color-blind white person I ever met, the first white man I ever knew who got it." Birch felt that his heightened sensitivity to the cause of civil rights and against the scourge of racism was due to his relationship with Gordon Alexander and what he taught him. Birch was keenly aware, Alexander remembered, that he wasn't sufficiently sensitized on issues of race. Until that moment in time, the hiring practices in the Senate office had not demonstrated any race consciousness at all, which naturally changed after he joined the staff.[5]

On March 31 came the next major bombshell in American politics when President Johnson announced he would not seek reelection. The Democratic Party was in turmoil, and Richard Nixon was establishing himself as the Republican front runner, opposed by governors George Romney of Michigan, Nelson Rockefeller of New York, John Volpe of Massachusetts, and Ronald Reagan of California. Nixon entered eleven primaries and lost only to Rockefeller in Massachusetts. California was conceded to Reagan.

Johnson's withdrawal was a shock, but another body blow was felt four days later with the assassination in Memphis, Tennessee of

Martin Luther King Jr., the nation's undisputed leader in the battle for civil rights. With the death of MLK, spontaneous riots erupted in many of the largest American cities across the country, lasting for several days.

Birch's remarks on the death of King echoed the horror and sadness felt by most Americans at this tragic event. "The hearts of Americans are heavy with sorrow; our heads hang in shame. The brutal and tragic assassination of Dr. Martin Luther King will place him among the most revered martyrs in American history. That this man of nonviolence should meet a violent death will remain an indelible smear on the face of our nation."

President Johnson made every effort to devote his attention to the tasks before him, signing the Civil Rights Act of 1968 in April and spending most of his waking hours seeking an end to the war. In April there were weeklong antiwar student protests at Columbia University that ended up with students taking over administration buildings and shutting down the university. Birch made statements in March and April in support of those fighting communist repression in Poland and Czechoslovakia and to schedule hearings on a proposal to lower the voting age. In between Indiana trips, he attended the funeral of Martin Luther King Jr.

There was a Democratic primary in Indiana on May 7. Birch's nomination for another term in the Senate was never in doubt, but the presidential primary attracted a great deal of national attention and was a challenge for the junior senator.

The backdrop for this election year would present a challenge to any Democratic incumbent. Average Americans supported the war because that was what you did; America—right or wrong—my country. Yet increasingly, Americans grew uneasy watching the steady diet of nightly news covering the carnage of war, body bags and all. They were unsettled by student protests, and public opinion determined that American kids were spoiled. With regard to African Americans, many believed that those uppity Negroes needed to be put in their place. As for women, the focus was on bra burnings and rebelling against being housewives. The America they had grown up in was under assault, and many Americans found it difficult to endure the daily onslaught of events. The culture was changing, and there wasn't anything you could do about it. Indiana is the heartland of America, and these views were widely shared across the state.

The Indiana primary would be remembered for the battle between McCarthy and Kennedy but also because of the celebrities swarming the state in support of both candidates. McCarthy had the support of Paul Newman, who said in Indiana, "I am not a public speaker. I am not a politician. I'm not here because I'm an actor. I'm here because I've got six kids. I don't want it written on my gravestone, 'He was not part of his times.'" Robert Kennedy also competed in the celebrity sweepstakes as dozens of actors, entertainers, and athletes flocked to his side.[6] The stakes were high. Both candidates felt that winning in Indiana was critical. McCarthy needed to demonstrate that he could beat Kennedy. Kennedy said, "Indiana is the ball game. This is my West Virginia," referring to the most critical primary state for his brother's campaign eight years earlier. West Virginia had become the test of whether a Protestant state would elect a Catholic candidate. Indiana was RFK's test to show that he could win in a conservative state.[7]

Many politicians steer clear of endorsing a candidate in their party's presidential primary. Birch considered endorsement a no-win situation, a policy he maintained throughout most of his career. Endorsing a candidate usually resulted in little appreciation from the endorsed candidate and lots of enmity from the unendorsed. The Indiana Democratic primary saw two of Birch's Senate colleagues competing against each other, McCarthy and Kennedy, one of whom he had a lot of history with, plus Indiana governor Roger Branigin as a stand-in for the president. Birch remembered Branigin as being "very smart, but not nearly as smart as he thought he was." Vice President Humphrey, who was also close to Birch, entered the race too late to compete in the primaries, though many viewed a vote for Branigin as a vote for Humphrey as Johnson's vice president. Bayh supporters across the state were dividing up among the various campaigns. Tilting toward one candidate or another would likely do damage to Birch's own chances for reelection. In most situations like this, all candidates are ultimately unhappy with a popular incumbent that they felt could be helpful to them. *Time* magazine printed a quotation from McCarthy: "Birch Bayh could find a way to hide in a field of stubble."[8] Since he felt he had been bending over backward to avoid taking sides, even though the Bayh campaign through Bob Keefe surreptitiously passed information along to Kennedy throughout the campaign, Birch was extremely unhappy with the remark. When he called McCarthy to ask about it, the Minnesota senator denied ever saying it and offered to issue a press release to publicly deny it. Birch thought that might only make matters worse, so he dropped the matter. Vance Hartke was openly supporting Kennedy, who ended up winning the primary. Kennedy would get 42 percent of the vote, with Governor Branigin receiving 30.7 percent and McCarthy 27.4 percent.

Kennedy's victory was in part fueled by an act of courage that he demonstrated upon hearing of the death of Martin Luther King. After campaigning in Muncie, Indiana, RFK was preparing to fly to Indianapolis when news of King's death reached him. He chose to speak spontaneously to a crowd of two thousand people—mostly African Americans—at an inner-city park in Indianapolis. To the horror of those assembled that night, Kennedy somberly announced the news of the assassination. His remarks calmed the crowd and saved Indianapolis from the racial unrest that resulted from news of MLK's death and spread quickly in major cities throughout America.

> In this difficult day, in this difficult time for the United States, it is perhaps well to ask what kind of nation we are and what direction we want to move in. For those of you who are black—considering the evidence there evidently is that there were white people who were responsible—you can be filled with bitterness, with hatred, and a desire for revenge. We can move in that direction as a country, in great polarization—black people amongst black, white people amongst white, filled with hatred toward one another.
>
> Or we can make an effort, as Martin Luther King did, to understand and to comprehend, and to replace that violence, that stain of bloodshed that has spread across our land, with an effort to understand with compassion and love.
>
> For those of you who are black and are tempted to be filled with hatred and distrust at the injustice of such an act, against all white people, I can only say that I feel in my own heart the same kind of feeling. I had a member of my family killed, but he was killed by a white man. But we have to make an effort in the United States, we have to make an effort to understand, to go beyond these rather difficult times.
>
> My favorite poet was Aeschylus. He wrote: "in our sleep, pain which cannot forget falls drop by drop upon the heart until, in our own despair, against our will, comes wisdom through the awful grace of God."[9]

Diane Meyer from Nappanee, Indiana, who joined Birch's staff after the campaign, was present at that speech and mesmerized by the event. Three days after the speech, Ted Kennedy visited her campus at Butler University to meet with the small Young Democrats chapter, seven people in all. Shortly after that meeting, the twenty-two-year-old Meyer became a volunteer and traveled with the Kennedy campaign as an assistant to the traveling press secretary Dick Drayne until it ended. She gravitated to Birch on the recommendation of Drayne, becoming a key member of the Bayh staff and, during certain periods, a member of Larry Conrad's staff.[10]

As the Vietnam War continued to dominate the news at home, it also enraged people in other countries, with one million marching in the streets of Paris.

The presidential campaign was in full swing with a Kennedy victory in Nebraska on May 14. In a surprise turn in the election, McCarthy defeated RFK two weeks later in Oregon, the first election defeat ever suffered by a Kennedy. Robert Kennedy would again face McCarthy in the California primary on June 4.

Prior to the Indiana Democratic State Convention in June, Birch was scheduled to formally announce his candidacy for reelection. All eyes were on the California primary, which Robert Kennedy won. Late that night at the Ambassador Hotel in Los Angeles, in a shock to the nation, RFK was assassinated by a Palestinian named Sirhan Sirhan.

Much of the country for the next several days was consumed by events surrounding the latest Kennedy tragedy: the funeral services in New York City, the final train trip he would take from New York to Washington, his interment at Arlington Cemetery next to his slain brother.

Few paid much attention to Birch's postponed announcement of his candidacy for reelection. It was impossible to accurately express his feelings about this assassination, one that he felt on a very personal level. "Like most Hoosiers and I think like most Americans, I find it difficult to believe that something like this happened in the United States of America." He continued,

> Another light has gone out of our lives.
>
> I remember the overwhelming feeling of loss that my wife and I shared with the nation in 1963 when John Kennedy was taken from us.

> I remember our profound relief when, after we had gone down together in a plane in Massachusetts in 1964, we learned in our own hospital beds that Edward Kennedy would recover.
>
> I suppose we believed then that the tragedy with which the Kennedy family had lived for a generation had ended, and that only the blessings they had reaped through the unstinting public service would remain.
>
> And now, this.

Birch won nomination to a second term on June 21, and Bob Rock, a friend from his state legislative days, would be the nominee for governor. The Republicans nominated a thirty-six-year-old state legislator, William Ruckelshaus, for the Senate.

Ruckelshaus was two years older than Birch had been when he became a senator, yet he was described at the time as the youngest GOP contender to ever seek a US Senate seat. His campaign would focus on three areas where he believed LBJ had failed the country, with the help of his ally Birch Bayh. The first he described as the "war–peace problem," something he saw as a challenge facing the country since World War II and that had become a crisis with Vietnam. The second issue was urbanization, a development causing the rise of racism, crime, and pollution, another failure of the Democrats in charge. The third was seen as an economic problem, with the United States failing to protect the "stability and integrity of our dollar."[11]

Bob Keefe took a leave of absence to run the Bayh campaign. He knew that Ruckelshaus would be formidable but felt they had organized the best possible campaign. He had the best candidate with an effective and charismatic wife, an experienced staff, and a solid record to run on. Keefe remembered that the total cost of the 1968 campaign did not reach $1 million, a small sum for a Senate race today. Birch remembered Ruckelshaus as being smart but "not a warm and cuddly type." He felt the campaign was aggressive but said it never got nasty or personal. He traveled to Indiana two or three weekends a month until kicking off his full-time campaign activity.

Jay Berman was the "issues guy" in the campaign and handled all the position papers. The most important aspect of Berman's job during the campaign was getting to know Birch's donors, especially the prominent Jewish donors and fundraisers. Chief among these were Herb Simon and Morrie Katz. Simon was the youngest of three brothers who had been enormously successful in developing shopping centers across the country. Katz had a company that manufactured bags. Both became close and important friends to Birch and Berman. Other key donors and supporters included Sam Smulyan and Mike and Gladys Sperling. Jay's memory of the campaign mirrored that of Keefe in that "Birch Bayh had a personality and Ruckelshaus didn't."

The slogans for the campaign were "The Promising Young Senator Kept His Promise" and "Senator Bayh for Senator," with ads produced by adman Don Nathanson. The first was meant to underscore a substantial record of accomplishment in a single term.

The year of turmoil continued with the arrest on June 8 of James Earl Ray, believed to be the assassin of Martin Luther King Jr. Also, Chief Justice Earl Warren announced his plan to retire. To replace him, President Johnson nominated his close friend and former adviser Associate Justice Abe Fortas. Birch and Marvella pursued reelection with a vengeance, traveling across the state as aggressively as they had done in the first campaign six years earlier. The couple had earned a reputation as hard workers who never let up. Marvella was particularly effective meeting with women and participating in ladies' teas organized by Luella Cotton, a prominent Democratic activist. The sold-out gatherings were billed as an opportunity for Hoosier women to meet Marvella Bayh, and they were remembered as events that showed off Marvella's poise and her abilities as a public speaker and political campaigner.

While July was consumed by campaign travel and speeches, the book Birch wrote about the passage of the Twenty-Fifth Amendment, *One Heartbeat Away,* was published. He appeared on *The Today Show* on July 18 to publicize the book. He sprinkled his Hoosier-oriented public statements with press comments on national issues—the farm bill, the space program, public works projects—and restated his support for the appointment of Justice Fortas. On the last day of the month, Senator Thomas McIntyre, speaking on the floor of the Senate, endorsed Birch for vice president.

This had been a subject of rumor for some time, and it put Birch in an awkward position because he was running for reelection to the Senate at that time. The turmoil within the party also made it appear unlikely that the Democrats could win in November, meaning that Birch would have to buck the Republican tide, always strong in Indiana even when there wasn't an unpopular Democratic administration in Washington. But he couldn't deny being flattered by the attention, and a number of his aides and friends were silently moving among those who would be delegates to the convention to promote the possible candidacy. On August 12, he made a statement that he was "startled and pleased" when Vice President Humphrey mentioned him as a vice presidential prospect. Birch used the occasion of the upcoming convention to call for greater involvement of young people at the convention, and he planned to testify in the platform committee in support of his direct election amendment. He was slated to introduce the convention's keynote speaker, his colleague from Hawaii, Sen. Daniel Inouye, on August 26.

That same month, the Republican National Convention nominated Richard Nixon for president. Nixon chose for his running mate a little-known governor from Maryland, Spiro Agnew. But before the nation's attention turned to the coming Democratic convention, events abroad forced their way back into the headlines. The Prague Spring of political liberalization ended as Russian-sponsored Warsaw Pact troops invaded Czechoslovakia in the biggest military operation in Europe since World War II.

Days before the convention, while campaigning near North Vernon, Indiana, Birch was heading to the next campaign stop after dark and saw in the distance a barn on fire. He had his driver take the car to the scene, and he realized that the owner of the farm was unaware of the fire. A chicken coop not far from the barn was on fire and dangerously close to a pickup truck, not to mention the house. Birch went to the truck, saw that the key was in it, and put it in gear so they could roll it away and out of danger. Then he ran to the farmer's door and knocked continuously; someone yelled out the window for him to go away. When they kept knocking, the man of the house opened the door, clearly just out of bed, nearly naked, and unhappy about being awakened. "What do you want?" he asked angrily. Birch told him what was going on; the man scrambled to put on some pants and ran to grab a twenty-foot hose. They used it to fill a few pails, and Birch and the farmer created a bucket brigade to take the water from the house's faucet to throw onto the fire. Birch's aide was dispatched to drive toward town, hoping to find a fire department. The firefighters eventually arrived and helped extinguish the fire. The barn was mostly saved. The farmer, who recognized Birch while passing the buckets, hesitated, asking, "Hey, aren't you. . . ?"

"Yes," Birch replied. "Keep passing the buckets."

The Democratic National Convention met in Chicago from August 26 to 29, nominating Vice President Hubert H. Humphrey for president with Senator Edmund Muskie of Maine as his running mate.[12] There had been stirrings for a Bayh candidacy for VP, but they were ineffective. When Birch found out that materials promoting his candidacy were being distributed in the hall, he immediately ordered that they be collected and thrown away. Promoting oneself for vice president was the best way to ensure that the presidential candidate will not select you, he thought. Some former staff members describe the event differently, saying that Birch was totally behind the effort and helped decide to have materials about him delivered to every delegate's hotel room. He may have gotten cold feet as the decision date grew closer. As Birch remembered it, he didn't think there was a law in Indiana that prevented him from being on the national ticket while also on the ballot for senator. He laughed when thinking how weird it would have been to win the vice presidency and lose the Senate seat on the same day. In 1960, Lyndon Johnson had a law passed in Texas that would allow him to run on the ticket for both vice president and senator.

Birch also felt that Humphrey had assured him that he was his first choice for a running mate. As much as he admired Humphrey, he later felt that Hubert always had a problem saying no and thought he probably gave the same assurances to others rather than telling anyone he was not his choice for the ticket.

The real story of the tumultuous convention was not what took place inside the convention hall but instead what happened on the streets of Chicago. In what was later judged a "police riot," Chicago mayor Richard Daley's police forces clashed violently with thousands of antiwar protestors, creating a spectacle for all the world to see. This black eye for the Democratic Party and for the country played out to the chant "The whole world is watching," which it was.

Ironically, it was in Chicago's Grant Park only four years before that LBJ had announced his Great Society. The riots in Grant Park in August 1968, in reaction to Vietnam War policy, effectively pronounced the end of the Great Society and the unraveling of Johnson's domestic program.[13]

Birch was one of a number of prominent Democrats invited to the NBC studio above the convention floor to be interviewed by well-known anchors David Brinkley and Chet Huntley. *The Huntley–Brinkley Report* was the flagship news program on NBC for fourteen years. Huntley threw a fundraising event for Birch in New York that fall, something unheard of today. Allan Rachles vividly remembered the remarks made by Huntley to his wealthy friends in a restaurant near Central Park: "Birch Bayh is a man whose head is in the clouds, but his feet are planted firmly on the ground." That kind of commingling of journalists and politicians is unthinkable in today's political world.

The Bayh campaign emphasized his accomplishments. A summary it produced begins with the Twenty-Fifth Amendment and the drive for Electoral College reform. The remainder is a litany of accomplishments rare for any senator who had yet to complete his first term. But despite his record of accomplishment and as much as Birch wanted the campaign to focus on his record, the major issues of 1968 were national in scope—the unrest, civil rights, gun control, the war.

Gun control has always been an emotional issue, one that stirs intense feelings on both sides, but the issue has been dominated by its most effective lobby, the pro-gun National Rifle Association (NRA). The NRA built a national mailing list and cadre of activists that made it a difficult issue for gun control supporters across the country. Indiana was one of those areas, a state with substantial hunting and gun ownership. Birch's support of a bill to prohibit the mail-order sale of handguns and to mandate registration and licensing of concealed weapons put him squarely in the crosshairs of the NRA, which opposed all gun control legislation. But the assassinations of 1968 provided an opportunity for gun control advocates, and during the campaign many Indiana newspapers ran full-page ads sponsored by a group called Indiana Emergency Committee for Gun Control. The ads called for stronger gun control laws and included coupons to fill out, cut out, and send to Senator Bayh. Thousands were received in the Senate office.

But the single biggest issue was the war. On August 15, Birch went to the floor of the Senate to draw his own line in the sand on the issue of Vietnam. After brief remarks about the history of the conflict and the manner in which blame and responsibility were widely shared, he called for a number of things to happen. As before, he emphasized the importance of ending the deep-seated corruption in the leadership of South Vietnam and reiterated the need for that government to assume greater responsibility for the war. He also called for a halt in bombing and for accelerated negotiations. The crux of his message can be summed up by Birch's comment, "The South Vietnamese must do a great deal more, or the United States will be forced to do a great deal less." He concluded by stating that if they were incapable of assuming control over their own future, the United States should no longer shoulder that responsibility.

On September 6, feminists protested at the Miss America Pageant in Atlantic City, calling the pageant exploitative of women. This was one of the first large demonstrations by what would become known as the women's liberation movement. On September 24, *60 Minutes* debuted on CBS.[14] And on September 28, Evan Bayh was elected vice president of his seventh grade class, a successful Bayh VP effort.

The fall campaign included a debate with Ruckelshaus, someone who seemed to agree with Birch on most issues. He supported two of the constitutional amendments Birch was promoting: direct election of the president and giving eighteen-year-olds the right to vote. Ruckelshaus's strategy was to tie Birch closely to LBJ, to the unrest in the country, and to the war. The country was in turmoil, and the Democrats were in charge. Running for office in 1968 was a daunting task for incumbent Democrats and even more difficult in GOP-leaning states like Indiana.

In the debate, Ruckelshaus charged Birch with targeting different gun control messages to different Hoosier audiences depending on where they stood on the issue. His charge was disputed by a number of Indiana newspapers. They pointed out that there were in fact different letters from the Bayh office but only the emphasis, not the substance, differed. The Bayh campaign reproduced those newspaper articles and did its best to distribute them to constituents who cared about gun control.

The debates were broadcast on the radio. Birch wasn't willing to debate him on television. It's not unusual for incumbents who are running ahead to avoid TV encounters, especially when the competition is attractive, smart, and articulate, as Ruckelshaus was. But Ruckelshaus persisted in publicly calling for a debate. Congressman Andy Jacobs Jr. offered to stand in for Birch. Amazingly, Ruckelshaus assented, and a televised debate was held between Ruckelshaus and Jacobs. Andy Jacobs was known for his quick wit and rapier-like humor. Ruckelshaus had his hands full. When Jacobs criticized one of Ruckelshaus's positions, he denied that he felt that way and was certain about it because "I know what I'm doing." Jacobs responded by wondering if Ruckelshaus remembered the football game when Ruckelshaus broke Jacobs's ankle. "You didn't know what you were doing then, and you don't know what you are doing now," he said—a line that would be repeated many times that fall.

Senator Muskie campaigned in Indiana for Birch, as did Vice President Humphrey. Joan Kennedy made a highly publicized campaign visit to the state. Birch's schedule throughout that hectic month included a surprise drop-in at an Indiana State University ceremony where Birch Bayh Sr. was receiving an award.

Senator Bayh received the endorsements of the *Evansville Press*, the *Evansville Courier*, the *Louisville Courier-Journal*, and other newspapers in Gary and Fort Wayne. Baseball legend Jackie Robinson traveled to Indiana to endorse him, and they rode together in an open convertible in an Indianapolis parade. The campaign produced a thirty-minute film called *The Promise of the Land*, which aired on TV. Birch was extremely proud of this film. He and Marvella traveled relentlessly, just as they had in 1962, separately covering twice as much ground as they could together. Late in the campaign, Humphrey separated himself from the president by calling for a bombing halt in Vietnam, and Birch publicly supported him in that decision. As Humphrey gained momentum and the race tightened nationally, polls in Indiana showed Bayh leading Ruckelshaus. Birch raised the volume of the campaign with attacks on Ruckelshaus's legislative career. Remaining on the offensive, Birch touted his own record, calling for passage of his constitutional proposals, claiming credit for federal programs in the state, and emphasizing his opposition to the war.

On October 22, Congress passed the Gun Control Act of 1968, an unusually controversial bill to be considered in the middle of a hotly contested presidential campaign. Ruckelshaus did all he could to pound

Birch on the issue of guns. As the campaign developed, the two Indiana candidates differed little on the war, with Birch calling for a halt to bombings and Ruckelshaus expressing a concern about possible military buildup by North Vietnam during such a pause, though he didn't oppose it directly. On the final day of the month, President Johnson provided a boost to the Humphrey campaign as well as to Democrats like Birch by announcing an end to the bombing of North Vietnam.

Birch charged Ruckelshaus with absenteeism in the Indiana House, singling out missed votes on benefits for widows, a pay raise for police, flood control legislation, and a law that would make LSD illegal. Religion reared its ugly head, with reporters speculating that Ruckelshaus's Catholicism might hurt him in certain parts of Indiana.

As Election Day grew closer, the Bayh campaign purchased television time for a twelve-hour telethon. Voters could call in from anywhere in the state, and Birch fielded all questions for the entire time, something that is almost impossible to imagine in the twenty-first century. Hosting the telethon was Sid Collins, "The Voice of the 500," the Indianapolis 500 race announcer for many years. Collins asked if he would be paid. He accepted a $1,000 fee, but after the event he asked Allan Rachles to agree that Indianapolis Speedway owner Tony Hulman would never know he took money for the job. Hulman would have wanted him to do it for free to help his friend Birch.

Hulman was always good to Birch and came through for him once again late in the campaign, when emergency money was needed for TV and radio. Rachles was dispatched to seek his help, and when Hulman asked him how much they needed, the response was $10–$15,000. "Can you take cash?" Hulman asked.

"Yes," responded Rachles, accurately in a far different era of fundraising for federal campaigns. Hulman took him to a room behind his office where there were large bookcases stacked with cash—not bound, crisp stacks of cash but crumpled bills stained with coffee, mustard, and ketchup, requiring several large bags to be filled before reaching the amount needed.[15] Birch was fully aware of the cash room, as he had worked for Hulman, driving a scooter around the track on race day to collect cash from the concession stands. How the currency looked was no surprise. In fact, Birch recalled one occasion when he drove a battered old car to the bank with a quarter of a million dollars in cash in its trunk after a good day at the races.

Gordon Alexander fondly recalled a day just before the election when he was with Birch and they encountered a man wearing both George Wallace and Birch Bayh campaign buttons. Wallace was the conservative Alabama governor running as an independent. Birch asked the man to explain how he could be for both. Pointing at the Wallace button first and then at Birch's, he said, "This guy says what he means. This guy means what he says."[16]

During the final week before the election, copies of a two-page handwritten letter from Marvella were mailed to all precinct committeemen and women urging them to work even harder for her husband. Monday, November 4, was election eve, and it culminated in a gala honoring Marvella and Birch at Loughner's South Side Cafeteria in Indianapolis.

Birch campaigned throughout Election Day, November 5, casting his vote at the New Goshen Firehouse near Shirkieville. Election night was spent in Indianapolis. Marvella, Evan, Marvella's father, Delbert Hern, Miklos Sperling, and an assortment of friends and staff joined Birch in his hotel suite. Once the polls closed and the television coverage began, it was soon clear that Birch had bucked the tide in Indiana, defeating Bill Ruckelshaus with 1,060,456 votes to 988,571, a margin of 51.7 percent to 48.2 percent. One of the congratulatory phone calls he received was from Ted Kennedy. Nixon carried Indiana by a larger margin than almost every other state, making the Bayh election noteworthy. Indiana voting machines made straight-ticket voting easy, but thousands of voters split their tickets to support Birch. Bob Rock, the Democratic nominee for governor, lost to Republican Edgar Whitcomb with a margin of 53–47. The presidential election was not settled until very late that night, with Nixon beating Humphrey by a little more than 500,000 votes out of more than 70 million cast. His popular-vote victory represented 43.4 percent to Humphrey's 42.7 percent. George Wallace mounted one of the most successful third-party candidacies ever, accumulating almost 10 million votes and 13.5 percent of the electorate; he received almost a quarter million votes in Indiana, just less than 12 percent of the total.

Allan Rachles was with Birch on election night when Rachles picked up the phone and was asked to hold for the president of the United States.

He thought it was a joke, and when Johnson came on the phone asking to speak with Birch, Rachles hung up on him. Eventually, the congratulatory call was made successfully.

A few years after the election, Ruckelshaus told reporters how his campaign bought the rights to "Hey, Look Me Over," the song from Birch's 1962 campaign. He said, "Literally everybody was whistling or singing it. It was the most effective song I've ever heard." Larry Conrad commented, "Sixty-eight wasn't a singing year." The Bayh campaign had no intention of rolling it out again anyway.[17]

Like with Homer Capehart in 1962, Birch never felt any animosity toward Bill Ruckelshaus in 1968. He recalled that he liked Ruckelshaus and considered him a very credible candidate.

Nationwide, Republicans enjoyed a net gain of five seats in the Senate, with a sixth added in December when Democratic senator Bob Bartlett of Alaska died and was replaced by Republican Ted Stevens. At that point, the Democrats controlled fifty-six seats. Of the new senators, Bob Dole (R-KS) deserves additional comment; his name and Birch's would be forever joined by the success of the Bayh–Dole Act.

It was a year that many were glad to see go. With riots in the streets and the burning of many American cities, with assassinations in America and a youth rebellion abroad, the turmoil of 1968 would never be forgotten. The Russians crushed rebellion, but dissent was growing in the United States. The opposition to the Vietnam War that marked the beginning of the year had not weakened by year's end.

Joe Kernan of South Bend, Indiana, recalled an amusing story from that time involving constituent service and Vietnam. Applying for a commission to the navy to become a navy flyer, he contacted the Bayh office for help in expediting his background check. Many years later, Kernan was Indiana's governor and talked about how Birch had intervened to help him get into the navy as desired. When talking about his service in the war, he added facetiously, "Birch Bayh is responsible for getting me shot down over Vietnam."

Ironically, while public opposition to the Vietnam War grew steadily, the only two senators to vote against LBJ's Tonkin Gulf Resolution, Wayne Morse and Ernest Gruening, were defeated. Barry Goldwater was returning to the Senate after leaving to run for president in 1964, and the Democrats had elected future presidential aspirants in Harold Hughes (IA), Mike Gravel (AK), and Alan Cranston (CA). Two future vice-presidential aspirants would emerge with the elections of Tom Eagleton (D-MO) and Richard Schweiker (R-PA). Dole would run for vice president as well as for president. And Birch's "class of 1962" lost its first member with the defeat of Dan Brewster (D-MD). Lyndon Baines Johnson would return home to the LBJ Ranch in Texas.

Three weeks after Election Day, Birch appeared on *Meet the Press*. Amid the discussion about the tumultuous year coming to an end, his main topic on the program was his direct election amendment, a topic with currency due to the results of the recent election. Reform of the present system, he said, would be his "first order of business."

He was prepared to begin a new term in the Senate, one in which he felt far more comfortable than six years earlier, with a portfolio of items he wanted to accomplish. Congress remained in Democratic hands, but the presidency did not. Change was still in the air as the turbulent year drew to a close. Much of that change would greatly impact Birch's life and career. In November, for instance, Yale University announced that it would begin admitting women. A new phenomenon appeared on the national scene when four men hijacked a Pan Am flight from New York's JFK Airport, demanding to be taken to Havana, Cuba. And on Christmas Eve, the moon mission captured the nation's attention as Apollo 8 entered its lunar orbit. The three astronauts became the first people to see the far side of the moon as well as our own planet from far above. The nation watched as they read from the bible.[18]

All photos of former presidents were taken by White House photographers and are considered to be in the public domain.

Other photos without attribution are from the author's personal collection or were provided by Katherine and Birch Bayh.

Photograph used for 1980 campaign poster. *Courtesy of Tim Minor.*

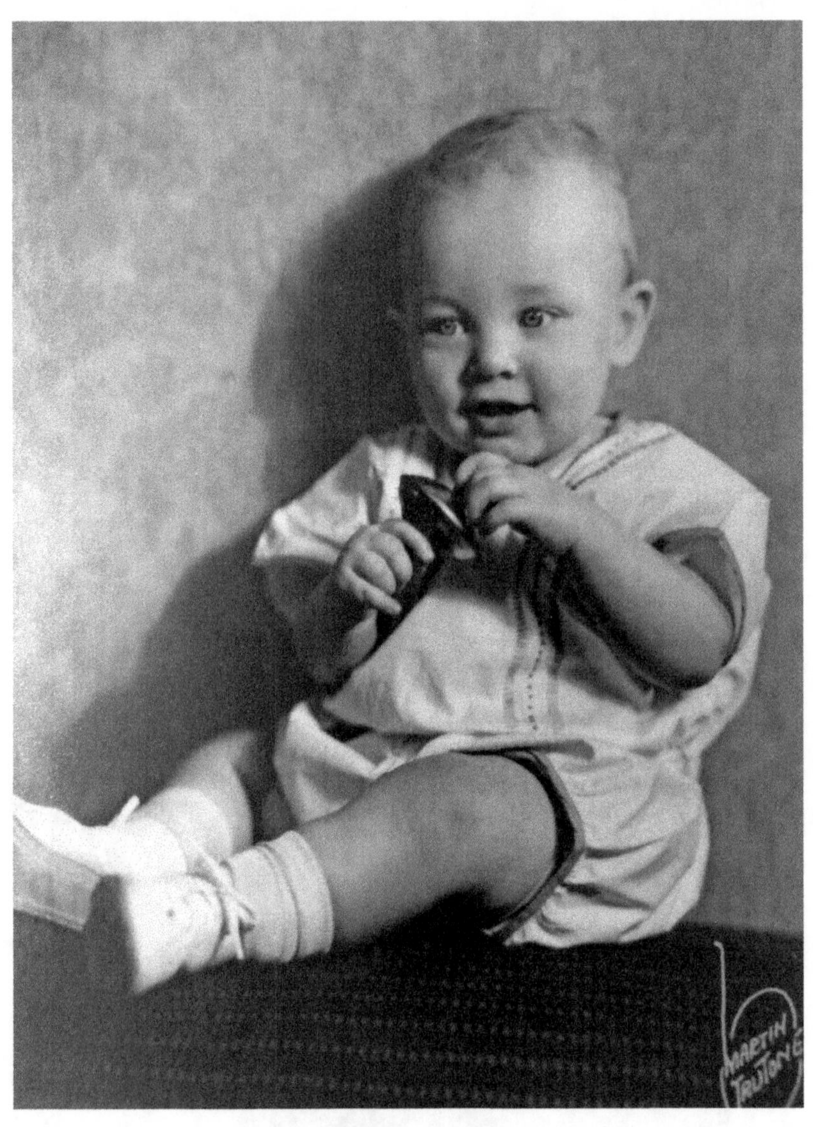

Birch Bayh as a baby.

Facing: Birch Bayh with his mother, Leah Hollingsworth Bayh.

Above: Bayh farmhouse in Shirkieville, Indiana. *Courtesy of Susie Marvel.*

Facing: Left to right: Birch, grandmother Kate Hollingsworth, father Birch Bayh Sr., sister Mary Alice Bayh.

Above: As a young farmer.

Left: PFC Bayh, 1946.

Marriage to Marvella Hern, August 24, 1952.

Birch, Evan, and Marvella at the 1962 Indiana Democratic State Convention.

Facing: Poster from 1962 campaign for the US Senate.

BIRCH BAYH

FOR U.S. SENATOR

JOURNAL COURIER
BAYH UPSETS SEN. CAPEHART
GOP Tightens County Hold
Plurality of Four Years Ago Eclipsed
Official County Vote

Above: Marvella and Birch with President John F. Kennedy.

Facing top: President John F. Kennedy campaigning for Birch Bayh, October 13, 1962.

Facing bottom: 1962 Senate victory, Herb Smith, photographer, Delphi, Indiana. *Courtesy of Jay Myerson.*

Above: Plane crash with Ted Kennedy in which two people died, June 19, 1964. *Getty Images.*

Facing top: With former president Dwight D. Eisenhower during the debate over the Twenty-Fifth Amendment.

Facing bottom: The signing of the Twenty-Fifth Amendment, 1967.

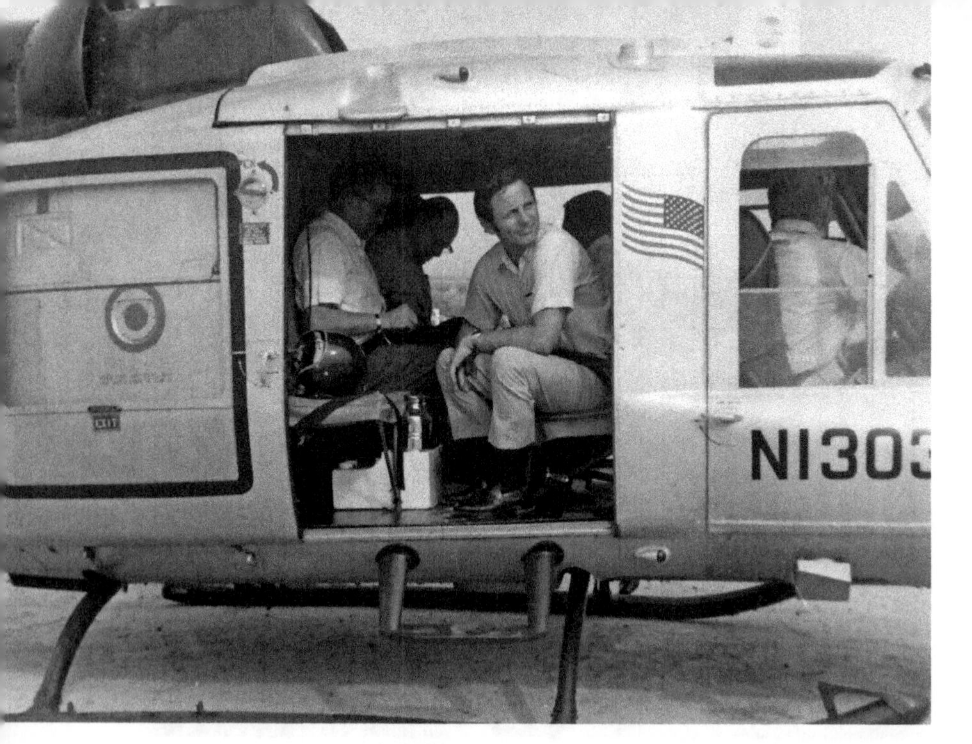

Birch Bayh's visit to Vietnam, January 1968.

With President Lyndon B. Johnson in the Oval Office.

Birch Bayh with his father, Birch E. Bayh Sr., in the Senate office, 1968.

Birch Bayh addressing the 1968 Democratic National Convention, August 1968.
Courtesy of the Library of Congress.

7

Haynsworth: 1969–70

New social conditions bring new ideas of social responsibility.

FRANKLIN DELANO ROOSEVELT[1]

AMERICANS CLOSED OUT 1968 WITH A SENSE OF RELIEF. SO much had happened during the year that made it memorable, but many would have preferred to forget it. With a presidential campaign that had the incumbent drop out, two assassinations, endless antiwar and civil rights riots, cities on fire, and a Democratic convention described as a "police riot," the year was more than a little unsettling to most Americans. The year ended with the election of Richard Nixon, who, after losing the campaign for California governor in 1962, had told us, "I leave you gentlemen now . . . just think how much you're going to be missing. You won't have Nixon to kick around anymore, because, gentlemen, this is my last press conference."[2] A candidate described in 1968 as "The New Nixon" was back and now in charge.

The presidential election results reinforced Birch's argument in favor of direct popular election of the president and abolishing the Electoral College. Nixon defeated Humphrey by one of the narrowest margins in history, only 43.2 percent to 42.7 percent, with Wallace receiving 13.5 percent of the vote. Yet Nixon's total in the Electoral College was 56 percent to Humphrey's 35.5 percent and Wallace's 8.5 percent, a landslide by any definition. A margin of less than half of one percent of the popular vote translated to an electoral vote margin of more than 20 percent. It was conceivable that Humphrey could have lost in the Electoral College while winning the popular vote, something that has happened during other elections in American history, most recently Clinton versus Trump in 2016. A shift of 42,000 votes in three states might have resulted in no candidate winning an electoral vote majority, which would result in the House of Representatives choosing the president. Surely that is not what the country expects in an election. The Electoral College, it has been argued, is a relic of colonial times providing a guarantee that the aristocracy would have primacy over the uneducated electorate. Theodore H. White, who chronicled several presidential elections in books beginning with *Making of the President 1960*, described the Electoral College as "an anachronistic survival of a primitive past—as useless as a row of nipples on a boar hog."[3]

As established by the Founding Fathers, the system for electing the president was flawed in ways that became obvious early on. George Washington established the tradition of a two-term president. When he stepped down, John Adams and Thomas Jefferson, political rivals at the time, garnered most of the electoral votes, becoming president and vice president in 1796. A system with the second-place candidate becoming vice president was immediately seen as flawed and was rectified by the Twelfth Amendment, proposed in 1803. The 1800 election, a battle between Jefferson and President Adams, ended with a result worse than four years earlier and something never predicted. Jefferson's running mate was Aaron Burr, but Jefferson and Burr were tied in the Electoral College; the election was then decided by the House of Representatives. The contest was not between Adams and Jefferson this time but had Jefferson pitted against his erstwhile running mate Burr. Adams had come in third. The intrigue was substantial, but eventually Jefferson was elected president and Burr, who ran against him in the House, was elected vice president.

Both elections caused results unintended by the framers. The presidential election process was soon changed when the Twelfth Amendment was added to the Constitution. It replaced the original procedure of the Electoral College with a requirement that the electors cast separate ballots for president and vice president, preventing them from competing with each other.

The Electoral College is made up of electors, each state's quantity determined by the combined number of its senators and representatives. It was originally designed so those Americans involved in public affairs could make this critical choice for the country. What Americans today consider the right to vote was not actually established by the Constitution and didn't exist in the early years of the republic. No right to vote was ever written into the Constitution, and the first elections were conducted largely among adult white males who owned property. Each state determined who could vote, and most only permitted white adult males to vote if they owned property or had taxable accounts. There were eleven states whose electors unanimously elected George Washington president in 1789 with only a smattering of votes from the white adult male landowning public being cast for the electors at the time.

By the 1828 election of Andrew Jackson, the twenty-eight states composing the United States had largely dropped the requirement of land ownership. Four years earlier, Jackson had won the popular vote but was denied the presidency when John Quincy Adams was elected by the Electoral College. The fact that Jackson was the people's choice became an obsession for him, and he proposed a constitutional amendment to remedy what he viewed as an undemocratic electoral process. He called for the abolition of the Electoral College, saying, "To the people belongs their right of electing their Chief Magistrate." He declared, "The majority is to govern," and he called for Americans to "amend our system" so that the "fair expression of the will of the majority" would decide who became president.[4]

As the country grew from twenty-eight states to fifty, the number of people who were eligible to vote also increased substantially. A constitutional amendment was passed to enfranchise black American men, both former slaves and freedmen, in 1870. While African Americans were given the right to vote, several Southern states charged a poll tax to prevent them from voting. It took another constitutional amendment in 1964 to bar the use of the poll tax. The franchise virtually doubled in 1920 when women were given the right to vote. Birch helped the franchise grow once again by steering to passage the Twenty-Sixth Amendment to the Constitution, lowering the voting age to eighteen (a story to be told later). In other words, since those first elections, the trend in American political history has been to expand democracy, to broaden the franchise, and to give more Americans the right to vote in presidential elections.

For the growing American electorate, the right to vote for president was implicit in the right to vote. Throughout history, however, there were a number of instances when a candidate with the most votes lost the election in the Electoral College.

Because of the Electoral College, as a state becomes solidly Republican or Democratic, voters in those states are effectively removed from the presidential election. Campaigning in those states becomes a waste of time and resources, meaning millions of Americans are excluded from the process of electing a president.

As the 91st Congress convened, Birch's seniority had risen to number sixty-eight. Pushing for enactment of the direct election constitutional amendment was a high priority for him in 1969, and a great deal of time and effort was devoted to it. He felt that any analysis of history showed that the Electoral College was an anachronism and that America in the twentieth century would be served far better with a direct popular vote, just as it decides every other election. He chaired the first hearings on the matter in January, and he traveled to several states throughout the year to drum up support for the issue, visiting colleges, women's groups, and Democratic events.

The year began with the Nixon inauguration on January 20, which Birch did not attend. Much of that period was consumed by the Senate hearings to confirm Nixon's cabinet.

In February, Birch held hearings on the proposed constitutional amendment to lower the voting age from twenty-one to eighteen. In several congressional sessions, Sen. Jennings Randolph of West Virginia had proposed legislation to accomplish that goal. Since Birch had introduced the same measure in the Indiana legislature several years earlier, losing by a single vote, it was an issue dear to his heart, and Randolph could support his introduction of the issue. The chairmanship of the Subcommittee on Constitutional Amendments presented Birch with the best opportunity to pass the bill since Randolph had first sponsored it. The Vietnam issue also buttressed his argument, with so many young men going to war who couldn't vote on the policies that were sending them there. That same month, Birch also introduced a bill to give congressional representation to the District of Columbia.

The Indiana High School Government Leadership Conference was held once again. This time the speakers were Virginia Mac Brown, chair of the Interstate Commerce Commission; VISTA director Padraic Kennedy; Deputy Assistant Secretary of State for Public Affairs Robert McCloskey; and Peace Corps director Joseph Blatchford.

While the Nixon era was just getting under way, Birch went about his business working on a variety of areas before the Senate simultaneously. In addition to the proposed constitutional amendments, he also spoke out on a bill to benefit policemen and firemen and announced his opposition to deployment of the anti-ballistic missile system (ABM) and his support for the ratification of the Nuclear Non-Proliferation Treaty. His interest in disaster relief led to legislation extending the benefits to disaster victims. His interest in agriculture led him to speak out on the federal response to fight hog cholera. And in May, he introduced a bill to fight the spread of pornography.

Unrest in the country caused by the continued US involvement in Vietnam persisted. In April, students seized Harvard University's administration building, leading to many arrests and injuries. Nixon had campaigned as the candidate with a "secret plan" to end the war, which has still never been revealed. However, he did take actions to reduce the American troop commitment in Vietnam before eventually prosecuting the war more aggressively. In June, he flew to Midway Island to meet with the president of South Vietnam, Nguyen Van Thieu. Afterward, the president announced that the United States would withdraw 25,000 troops by September. The troop withdrawal began in July.

On July 25, the president announced a policy that would become known as the Nixon Doctrine. It declared that the United States ultimately expected South Vietnam to take responsibility for its own defense; this later became known as the "Vietnamization" of the war. A few days later, he made a surprise visit to South Vietnam to meet with the political and military leadership there. Nixon's new national security adviser, Henry Kissinger, began secret negotiations to end the war.[5]

The citizen activism that fueled the civil rights and antiwar movements seemed to allow Americans to get involved in ways they never had. The post-WWII generation was showing signs of social consciousness and a willingness to get active in ways the country had rarely seen. Another social movement that got its initial impetus at that time concerned the environment. On June 22, the Cuyahoga River essentially caught on fire. The fact that this critical but terribly polluted river near Cleveland was burning awakened the public and spurred activism to fight water pollution. The senators heading up those efforts were Gaylord Nelson (D-WI) and Henry "Scoop" Jackson (D-WA), who wrote legislation in 1969 that became the National Environmental Policy Act, the first of several major environmental laws eventually passed, paving the way for national policy to protect all aspects of our environment. Among those measures was the creation of the Environmental Protection Agency (EPA) and the Council on Environmental Quality. To implement the national policy, detailed environmental impact statements would be required for all major federal actions significantly affecting the environment. The first EPA administrator was the man Birch defeated in 1968, William Ruckelshaus. Birch testified on his behalf during his confirmation hearing.

The previous year, Chief Justice Earl Warren had announced his plan to retire. To replace him, President Johnson decided to nominate Associate Justice Abe Fortas. But LBJ's withdrawal from the campaign made him a lame duck, leading many conservative senators to express their extreme dissatisfaction with the actions of the Warren Court. Fortas was a staunch liberal, and his nomination presented a wealth of opportunity for many senators. Chairman Eastland of the Senate Judiciary Committee told President Johnson that he had "never seen so much feeling against a man as against Fortas."[6] Fortas became the first nominee for chief justice to appear in person before the Senate, and he was greeted with substantial hostility. Revelations that he had accepted $15,000 in speaking fees from American University's law school doomed his nomination. Not only did his payments exceed fees paid by the university to any other speaker by a factor of seven, but also it turned out that the money had come from private corporate entities and not directly from the university. The potential for conflict of interest and this apparent lack of ethical sensitivity fueled the opposition, leading to a filibuster. When another Fortas scandal came to light in 1969, he resigned his seat on the Supreme Court.[7]

The Nixon administration was well aware of the controversies and rumors surrounding Fortas. The new attorney general, John N. Mitchell, informed Chief Justice Warren of all that the Justice Department knew about Fortas's troubles and suggested that impeachment proceedings were a possibility. As a result, Warren helped persuade Fortas to resign. As John P. Frank writes in his book about Judge Clement Haynsworth's nomination following Fortas's resignation, "It is the only time in history a justice has resigned under pressure from the press and the executive branch of the government." Frank goes on to suggest that the "politicization of the Fortas nomination by the Republicans undoubtedly contributed to a willingness by the Democrats to politicize the Haynsworth nomination a little later."[8]

Nixon nominated Warren Burger of Minnesota to be chief justice. His nomination sailed through the Senate, and he was sworn in on June 23. To replace Fortas, he nominated appeals court judge Clement Haynsworth of South Carolina on August 21.

That summer, Birch remained busy, opposing efforts to reduce the size of Indiana Dunes Park and supporting efforts to require financial disclosure by federal judges, a reaction to the Fortas matter. He also participated in the Commission on Delegate Selection and Party Structure in Jackson, Mississippi, a body the Democratic Party created in response to the chaos in the 1968 nomination process. It was headed up by South Dakota senator George McGovern.

In the 1968 campaign, Nixon employed a "Southern strategy" to capture former states of the Confederacy that had always been solidly Democratic. President Johnson's civil rights legislation had created the environment for that kind of strategy to work. But implicit in it was a level of racism used to capture and solidify white support for the Republicans. Kevin Phillips, a Nixon strategist, was interviewed by the *New York Times* in 1970 and discussed the strategy:

> From now on, the Republicans are never going to get more than 10 to 20 percent of the Negro vote and they don't need any more than that . . . but Republicans would be shortsighted if they weakened enforcement of the Voting Rights Act.

> The more Negroes who register as Democrats in the South, the sooner the Negrophobe whites will quit the Democrats and become Republicans. That's where the votes are. Without that prodding from the blacks, the whites will backslide into their old comfortable arrangement with the local Democrats.[9]

Part of that strategy has been described as a way to fend off a threat from the right by Gov. Ronald Reagan, a potential presidential candidate in 1972. Another part of that strategy led the new president to conclude that his next Supreme Court nomination had to be from the South—thus the Haynsworth nomination. Because of Haynsworth's nomination, the next three months would substantially change the course of Birch Bayh's life.

The summer of 1969 was eventful in a variety of ways. On July 16, Apollo 11 lifted off on a mission to become the first manned moon landing. While the country waited with bated breath for the four days it would take until its landing on the lunar surface, an event at home dominated the headlines.

On July 18, Sen. Ted Kennedy left a party in Chappaquiddick, Massachusetts, that had been arranged to thank staff members of his brother Bobby's presidential campaign of the previous year. But he didn't leave alone. One of Bobby Kennedy's former aides, Mary Jo Kopechne, was with him, and they were in an auto accident that would end Kopechne's life and forever change Ted Kennedy's political fortunes. They drove off the side of a bridge, their car submerged in the murky waters late at night. Kennedy escaped from the car and made his way back to Edgartown, where he got to his hotel and went to sleep. Mary Jo Kopechne, still in the car, drowned. Kennedy's bizarre behavior and his culpability in her death became a national scandal. Birch felt badly for his friend and colleague, knowing that even if a friend did something wrong, it was okay to feel badly for him. He worried about Ted and assumed his behavior was the result of the trauma of the accident, which reminded Birch of the auto accident he and Marvella had been in years earlier and how disoriented he had felt at the time.

The day after the Kennedy news broke, the country turned its attention back to outer space. All Americans old enough to comprehend what was happening know where they were when Neil Armstrong stepped onto the moon, the first time a human had ever done so. Armstrong was a Purdue University graduate and became Birch's friend. There is no question that on July 20, he was the most popular man in the world. It has been estimated that over half a billion people worldwide watched—the largest audience for a live broadcast at the time—as he took his historic first steps on the moon at 10:56 pm Eastern time.[10]

Birch's life was consumed by the Haynsworth nomination. When allies from organized labor came to him to discuss rulings Haynsworth had made that hurt workers in South Carolina, a closer look at Haynsworth's career was warranted. This was not the kind of thing Birch wanted to pursue. He felt that confronting the popular Nixon administration was foolhardy. At first he only agreed to ask questions of Haynsworth that were important to labor leaders. But looking at the lineup of Democrats on the Judiciary Committee, it was clear that if Haynsworth needed to be opposed, it would be up to Birch to lead the effort. Chairman Eastland was supportive of the conservative Southern jurist, as were the next two in line by seniority, John McClellan and Sam Ervin. Tom Dodd followed in seniority, but he was under an ethical cloud himself and in no condition to step up in opposition. Phil Hart was a senator of sterling reputation and would be an ideal leader of the opposition, but he was up for reelection in 1970 and did not want to risk controversy in an election year, particularly as he had been a strong supporter of Fortas. Ted Kennedy was under a serious cloud because of Chappaquiddick. That left Birch next in seniority, and he felt that if opposition was necessary, it had to be his to lead. He felt he had no choice. "I couldn't punt on it," he said.

Opposing Haynsworth might be a doomed effort from the outset. The last time the Senate had rejected a Supreme Court nominee was decades ago, in 1930. Judge John Parker, chief judge on the Fourth Circuit, the same court Haynsworth sat on, was defeated largely due to two issues. The first was based on comments he had made when a candidate for elective office, arguing that America shouldn't allow black people to vote. The second was his ruling that validated agreements that prevented those who signed from joining a union, known as "yellow dog" contracts. The coalition that successfully opposed Parker was almost exactly the same as the one that would assemble against Haynsworth, though the specific issues were different. Parker and Haynsworth were very close personally, with Haynsworth expressing extreme admiration and devotion for Parker, who was very much his mentor as well as his friend.[11]

Clement F. Haynsworth Jr. was a fifty-six-year-old federal judge from Greenville, South Carolina, serving on the Fourth Circuit Court of Appeals. As a conservative Southerner, he engendered a certain level of opposition because of rulings on civil rights and on matters of importance to organized labor. But it was unlikely that the level of that opposition was sufficient to keep him from being confirmed. Neither partisanship nor liberal versus conservative conflict would be enough to turn away the Supreme Court nominee of a newly elected president. Ultimately, his greatest source of problems came from a lack of sensitivity on ethical issues and an unwillingness to treat those issues with candor. Ethical sensitivity alone might have been insufficient to defeat him, except that the nomination was to fill the vacancy resulting from the resignation of Justice Abe Fortas, with ethical issues front and center. Some would argue that the ethical issues provided ample cover for those senators who wanted to vote against the nominee based on philosophy but otherwise wouldn't be able to justify such a position.

John P. Frank wrote a book called *Clement Haynsworth, the Senate, and the Supreme Court,* which spells out in great detail the high drama that took place from August 1969 until the Senate vote in November. Birch assumed the mantle of leadership and put together a coalition of labor and civil rights leaders that helped build the successful case against Haynsworth. He and his allies were helped by a series of revelations that showed the judge to lack the necessary concerns about creating an appearance of impropriety while on the bench:

- Haynsworth joined a ruling favoring a textile company while owning stock in a vending company, Carolina Vend-a-Matic, which did considerable business with the textile firm. In his confirmation hearings before the Senate Judiciary Committee, he admitted that he had remained a director of Carolina Vend-a-Matic through 1963, even though he had told the Senate a few months earlier, "Of course when I went on the bench [in 1957]

I resigned from all business associations I had, directorships and things of that sort."[12] It was later discovered that he had purchased stock in another company, Brunswick Corp., just after he was involved in a decision affecting the company but before that decision was announced.

- Organized labor was energized over a case concerning collective bargaining. On a case ultimately overturned by the Supreme Court, Haynsworth ruled on a question of when authorization cards signed by union members could be used to give collective bargaining rights to a union where, due to unfair labor practices, it was impractical to hold an election. His ruling was that the cards could not be used, an unpopular view with which other federal courts had also disagreed.[13]
- The National Labor Relations Board (NLRB) found that Darlington Mills had been "closed because of anti-union animus." It ruled that "the Deering Milliken Company was liable for Darlington's unfair labor practices and accordingly ordered back pay for discharged Darlington workers and other remedies." When the Textile Workers Union won its campaign to represent its workers at the bargaining table, a decision was reached to close the company. Haynsworth voted with the 3–2 majority on the Fourth Circuit Court of Appeals to set aside the NLRB determination, stating that the company could close its operations regardless of any motives it might have had concerning organized labor.[14]
- Haynsworth affirmed a decision by officials in Prince Edward County, Virginia, to close the public schools rather than integrate them. He also ruled to uphold the "constitutionality of granting tuition to white students seeking to enroll in schools"[15] that were not integrated. Joseph L. Rauh, legal counsel for the Leadership Conference on Civil Rights, described Haynsworth as "a sort of laundered segregationist."[16]
- An injured seaman filed a claim against Grace Lines for $30,000, and the court awarded him only $50. Haynsworth, judge on that court, owned stock in the parent company of Grace Lines.[17]

The *Washington Post* summed up a portion of the debate as follows:

> In the labor-management field, there was no question in terms of his background, his professional and personal associations, Haynsworth had come to the Court of Appeals as a "management man."
>
> He had large stock holdings, valued last month at more than $1 million. He was a director of several corporations and owned a one-seventh interest in Carolina Vend-a-Matic, whose sales rose from $20,000 a year in 1950 to more than $3 million in 1963 when he sold out his holdings for $437,000.
>
> His law clients had included such firms as J. P. Stevens Co., a textile manufacturer that had fought unionization and that was described as "the classic violater of federal labor laws, which has in effect . . . told Congress by its behavior . . . [that] it does not recognize federal labor law."[18]

The labor and civil rights coalition, composed of 125 groups representing the interests of religion, labor, welfare, and civil rights, virtually camped out in the Bayh Senate office during most of the three months of the battle against the Haynsworth confirmation.[19]

White House press secretary Ron Ziegler responded to the various ethics charges by saying that White House knew all these facts and did not expect them to be "interpreted" the way they were. He said there had been a "full review" of Haynsworth's background and that after the charges surfaced there was a further review, according to wire service reports by the UPI. "The President did not learn anything additional of major significance in Judge Haynsworth's background," Ziegler went on to say.

The Nixon White House put on a full-court press and had donors and local politicians around the country pressuring Republican senators. It turned out to be a ham-handed effort, likely hurting the cause rather than helping it.

On September 7, an event would have a major impact on the Haynsworth nomination and would affect all senators and untold events to come. Senate Minority Leader Everett Dirksen died after surgery at Walter Reed Hospital. The post was not filled until late September, well after the Haynsworth hearings had begun. Hugh Scott of Pennsylvania became the new Republican leader, with Dirksen's seat in the Judiciary Committee taken by Michigan senator Robert Griffin.

The passing of Dirksen signaled a change in the Senate. Birch never forgot the time early in his career when Dirksen discussed the importance of helping Birch get reelected, a level of cooperation across the political aisle that would be unheard of in the Senate of the twenty-first century. A reporter described him as one whose "face looks like he slept in it." Dirksen had a mellifluous voice that drew attention to his colorful phrases. Often when he rose to speak on the Senate floor, a murmur would circulate in the press galleries: "Ev's up!" He maintained a close friendship with both presidents Kennedy and Johnson. He remarked that when he was in the hospital during Johnson's administration and absent, three Republican bills were narrowly defeated. "To my bedridden amazement," he stated in his uniquely colorful manner, "my pajama-ruffled consternation, yes, my pill-laden astonishment, I learned they were victims of that new White House telephonic half-Nelson known as the Texas twist." But he also showed a remarkable ability not only to compromise but also to change his mind in a very public way. His support was indispensable on issues like the Civil Rights Act, the Voting Rights Act, the war in Vietnam, and the nuclear test ban treaty. Once, when asked how he could support a bill he had aggressively attacked so recently, he replied, "On the night Victor Hugo died, he wrote in his diary: 'Stronger than all the armies is an idea whose time has come.'"[20] He was remembered by many for his sarcastic remark about the Senate's big spending: "A billion here, a billion there, pretty soon we're talking about real money."[21] His like would not be seen again, and his departure from the Senate would have a marked impact on events to come.

With hints of defections by Republicans like Jacob Javits of New York and Assistant Minority Leader Griffin, who announced his support with "great want of enthusiasm," defeating the nomination seemed increasingly possible. Birch, however, predicted the success of the nomination as late as September 24, the same day that a poll listed as many as twenty-five senators in opposition.[22] The following day, Chairman Eastland expressed pessimism about the nomination's chances to Senator Mathias of Maryland.[23] At the end of the hearings, Birch had still not made up his mind to vote against Haynsworth, much less take on the mantle of leadership. He finally did so on October 3, including in his remarks the portion of the Code of Conduct for United States Judges that calls on a judge to avoid even the appearance of impropriety.

The drama heightened when two Republican leaders in the Senate, Griffin and Margaret Chase Smith of Maine, announced their opposition.

Many years later, Birch described his role in the Haynsworth matter as "hard duty." He talked about questioning the judge while his wife, children, and maybe even his parents sat behind him. "I didn't want to be against the guy," he remembered, but when giving him every opportunity to tell the committee that, in hindsight, he should have recused himself, the judge instead kept repeating, "I haven't done anything wrong." Had he simply admitted that if he had to do it over again he might have acted differently, he could have been home free.

As leader of the opposition, Birch had allies in the labor unions and the civil rights movement. The AFL-CIO dedicated forty full-time lobbyists to the cause, and full-throated efforts across the country were mounted by the NAACP, the Urban League, and the Leadership Conference on Civil Rights. After receiving a report on Haynsworth's labor decisions, the AFL-CIO president, George Meany, made an early commitment to go all out to defeat the nominee. He aligned with NAACP president Roy Wilkins and the counsel for the Leadership Council on Civil Rights, Joseph L. Rauh Jr.[24]

Birch developed a five-point set of claims, called a bill of particulars, that all centered on the judge's ethical problems. It claimed that (1) on at least four occasions, Judge Haynsworth sat on "cases involving corporations in which he had a financial interest;"[25] (2) the Code of Conduct for United States Judges required that a judge avoid investments in those kinds of enterprises that are liable to be involved in litigation; (3) in *Vend-a-Matic* and five other cases involving its customers, Haynsworth's participation showed a lack of sensitivity to ethical considerations and demonstrated an appearance of impropriety; (4) as a trustee of Vend-a-Matic, Haynsworth failed to report on the retirement funds he shared in, a violation that Birch conceded was most likely an oversight; (5) and the judge showed a lack of candor. Birch added his recommendation that nominees for the nation's highest court should show greater sensitivity to potential ethical problems.[26]

In the US Senate, there is a tradition that presidential nominees appearing before a committee for confirmation should be introduced or accompanied by their respective state senators. Most often, the nominations are initiated by a recommendation of one or both of those senators, but at the least they are nominees that both senators can support. In the case of Haynsworth, South Carolina senators Fritz Hollings and Strom Thurmond were enthusiastic supporters. Hollings had become friendly with future attorney general Mitchell when he was governor of South Carolina; Hollings was someone Mitchell wanted in his law firm when he finished his term. With the death of Sen. Olin Johnson, Hollings chose a Senate seat over returning to the practice of law. In May, Hollings met with President Nixon to recommend Haynsworth for the vacancy created by the Fortas resignation. Nixon was reminded at that meeting that he had met Haynsworth years earlier. Having a Democrat like Hollings promote the nomination was the seal of approval needed by the White House, while Thurmond's support would be downplayed, as he was viewed by the public and his Senate colleagues as the most racist member of the Senate.[27]

Birch's opposition affected his relationship with Hollings in a major way, something that probably never fully healed. The relationship between the two personable and gregarious senators had been very friendly at the start but had already been damaged during the nomination of Rutherford Poats, another Hollings-sponsored nominee. Birch remembered riding on the plane with Hollings to the Dirksen funeral, and it was obvious that there were problems between them.

On October 9, Hollings challenged Birch to a televised debate, referring to Birch's nine-page bill of particulars issued in opposition to the Haynsworth nomination. The next day, after Birch refused the debate, Hollings sent him a telegram:

> I have your refusal to debate. As you know we have received invitations to appear on various universally respected national television programs Sunday. I am anxious to accept these invitations because I believe that an answer to your charges against Judge Haynsworth is the only way we can protect the reputation of the United States Senate and of the Court. I have been begging for weeks that the debate be limited to the floor of the Senate. However, it was you who took the debate from Senatorial processes to the headlines and television and now there is no alternative because both public opinion and Senate opinion is formulating. Not too

long ago, the hit and run tactic of questionable charge and inuendo [*sic*] became known as "McCarthyism." Let's not have in an age of "tell it like it is" a revival of "McCarthyism." Some have suggested that we allow the Republicans to fight this out. However, I am sure that you agree with me that we love our country more and honor the change [*sic*] of serving in the United States Senate more than any political party consideration. I urge you to reconsider in the name of fair play and in the interest of the good name of the Senate.

Fritz Hollings

That evening, Birch had a reply drafted, typed, signed, and hand-delivered to Hollings's office.

Dear Fritz:

I have just received your telegram relative to our debating the Haynsworth matter on national television. In re-reading it a second and third time, I feel very much as I did in response to your first inquiry.

Fritz, so many sources have been responsible for various statements and allegations throughout this entire distasteful affair that I am returning your wire without disclosing its contents in the hopes that someone other than yourself was the author.

I have too much respect for you and for the service you have performed and will continue to perform for your state in the Senate to let others goad us into this type of personal acrimony.

Best regards,

Birch Bayh

Nothing more about the matter was said publicly by either party, but Birch knew that the experience affected his relationship with Hollings for the rest of his career. Hollings later publicly blamed the White House for the defeat of Haynsworth, complaining that the administration was late in putting its strategy for confirmation into play and ultimately played too rough.

Tom Connaughton, a lawyer from South Bend, Indiana, joined the Bayh office in May 1969. His first experiences were covering Judiciary hearings on the extension of the Voting Rights Act. He found the

Judiciary Committee a very smoky place because of Eastland's ever-present cigars and Phil Hart's cigarettes. His full immersion came with the Haynsworth nomination.

During the Haynsworth hearings, Connaughton and others on Bayh's staff were digging into the judge's past to find out as much as possible about what stocks he had and where he had sat on cases. Connaughton recalled Judge Haynsworth speaking with a stutter, making it difficult for him to clearly answer questions from the committee members. As the questioning from Bayh and his colleagues became increasingly sharp, the tension in the committee grew, and Haynsworth's presentation became awkward.

Connaughton believed that Nixon officials should have said, "Look, these were relatively small cases and deals compared with the Supreme Court decisions." It would have added perspective to the instances where he held stock that created the appearance of a conflict. Or had Haynsworth responded, "I overlooked these, and I should have been more careful, and I can assure you that I'll be more careful in the future," that would have been the end of it. But they and he kept insisting that he had not done anything wrong. And the more the staff dug, the more cases they found that were troubling. Connaughton later said, "I have no reason to think that Judge Haynsworth ever decided something because he had those shares of stock. I think he was an honorable man. But at any rate, it went on and then Birch decided—and that was pretty traumatic—would he really oppose him or not, and he decided yes, he would oppose him."[28]

Before the matter was debated on the floor, Birch fulfilled a previous commitment to participate in the Interparliamentary Union in the Soviet Union. He traveled there with senators B. Everett Jordan (D-NC), John Sparkman (D-AL), Ralph Yarborough (D-TX), John Monagan (D-CT) and Lee Metcalf (D-MT). At the Interparliamentary Union meeting, the six senators found themselves sitting across a conference table from six officials from the Kremlin. The trip was historic, as these talks were the first ever held between US members of Congress and leaders of the Supreme Soviet.

Birch returned to Washington feeling very anxious from being away. But he also knew that the nomination made in late August was in serious trouble by early October. When *Newsweek*'s congressional correspondent Sam Shaffer approached him on October 7 and said, "I'm convinced you have it won. Do you intend to win it?" Birch answered in the affirmative. The next issue of the magazine predicted the rejection of the nominee.[29]

By October 9, reported head counts showed at least forty senators opposed to Haynsworth. That same day, the *Washington Post* urged withdrawal of the nomination. On October 20, President Nixon stated that he would not withdraw the nomination even if Haynsworth asked him to.[30]

The Senate debate was intense for everyone close to it, with the galleries packed almost daily. Just before the vote on November 21, two of the Senate's most influential Republicans, John J. Williams of Delaware and John Sherman Cooper of Kentucky, announced their opposition to Haynsworth. Their reason was the Brunswick case. Sen. Charles M. Mathias of Maryland soon followed suit for the same reason. There is no question that there would not have been so many Republican senators voting against Haynsworth had it not been for the ethical issues. Seventeen Republicans ended up voting no, including the new leader, Hugh Scott. Haynsworth's confirmation failed by a vote of 55–45.

The big loser in this battle was President Nixon, and there were many stalwart and distinguished supporters of Judge Haynsworth who were more than chagrined by the result. Future Supreme Court justice Lewis Powell, a Southern nominee who successfully found his way to the court, wrote about the Haynsworth confirmation fight in the foreword of John Frank's book. "The defeat of this eminently qualified jurist was 'purely political' and reflected adversely on the Senate rather than on Clement Haynsworth."[31]

The winners in this battle were clearly the leaders of organized labor and of the civil rights movement, Meany, Biemiller, Rauh, and the leader in the Senate, Birch Bayh. Labor and civil rights movement efforts made it a local issue in many states, turning on the pressure from home just as the Nixon administration had done among Republican donors and party officials. John Frank issued this description: "By his victory, Bayh, a young second-term senator at the time, established himself nationally as a major tactician. From the standpoint of one wishing to defeat a major nomination, his tactics were flawless. First and foremost, he raised the doubts, and then he kept them alive."[32]

The Senate operated differently in that era. After Haynsworth's confirmation was defeated. Birch was leaving the Senate floor when Marlow Cook, who was a leading supporter of Haynsworth, came over and put his arm around him, saying, "Let's find something we can work on tomorrow or next week together." And they did.[33]

Birch once described the Haynsworth matter as "the most distasteful thing I have ever had to do in the Senate," but there is no question that his role thrust him more clearly into the national spotlight. He might have thought he had experienced fame with the passage of the Twenty-Fifth Amendment or because of the plane crash, but this was different. Articles sprung up everywhere; he was in demand on television news shows across the country and invited to college campuses as never before. Over six feet tall with dazzling blue eyes, forty-one years old with a full head of dark hair without any gray, an athlete's physique, and an "aw, shucks" demeanor, he was being noticed not just for his image and appearance but also for his effectiveness, his leadership, and his ability to articulately address the issues and state his case. Reflecting on that period, he recounted that the Haynsworth defeat made him a national senator, "though that wasn't my intention."

Yet there is always the other side of the coin. Birch felt instinctively (and the mail to his office confirmed) that his role in the Haynsworth matter was unpopular at home. Indiana had, moreover, given Nixon one of the largest margins of any state only a year earlier. And there was a late-night phone call at home with the unidentified caller saying, "Too bad you didn't die in that plane crash with Ted Kennedy."

Jay Berman, in discussing the Haynsworth matter, said that Birch had learned during the Twenty-Fifth Amendment experience that he could rely on the weight and integrity of the American Bar Association (ABA). After labor leaders raised issues about the judge, Birch heard candid opinions from ABA leaders that helped inform his own thinking on the matter. Berman said, "We won because of Birch Bayh's leadership efforts."

While the Haynsworth matter was unfolding, another event would impact Birch's career as well as the history of the United States. On

September 5, a court charged Lieutenant William Calley with the 1968 murder of 109 Vietnamese civilians in what became known as the My Lai massacre. The country was split apart on the issue, with great sympathy for Calley by many and supreme disgust and anger demonstrated by others. It typified the divisions in the country caused by the Vietnam War, divisions that played themselves out in different ways during a very troubled era in American history. Birch was conflicted over the initial reports, recoiling from the killings but wanting to show support for America's soldiers.

On October 15, hundreds of thousands of people participated in demonstrations across the United States, the largest in Washington, DC. This mass protest was called the Moratorium to End the War in Vietnam. President Nixon responded to this unrest by going on television to ask what he referred to as the "silent majority" to support his efforts in Vietnam. Vice President Agnew created a role for himself as the president's attack dog by characterizing Nixon's critics in a series of alliterative insults. Referring to "an effete corps of impudent snobs" and "nattering nabobs of negativism," Agnew did little to bring America together, which was Nixon's promise before he entered the White House.[34]

On September 18, the House of Representatives passed Birch's proposed constitutional amendment on direct election of the president. Birch publicly praised President Nixon for his support of the amendment.

The Calley matter took a turn with the publication of investigative journalist Seymour Hersh's articles detailing the massacre at My Lai. In late November, the Cleveland *Plain Dealer* published explicit photographs of dead villagers. Meanwhile, antiwar protests continued, with almost half a million protesters demonstrating in Washington on November 15. The war struck closer to home when the draft lottery took place on December 1, the first since World War II.

As the Calley matter grew in intensity, Birch took a lot of heat by calling for Calley's prosecution. He felt that far too many people were willing to dismiss the massacre as just more killing of "gooks" and that it was important for the country to stand by its values and to require a higher standard of our men in uniform. The negative mail from Indiana picked up once again.

The antiwar protests on November 15 were memorable because a number of demonstrators in DC that day came to the Bayh office to discuss the war with him, as they did with other senators. It was a very cold day, and it had begun to rain when Birch asked the young people—seventeen in all—if they had a place to stay. They looked at each other uncertainly. Birch offered them the office as a place to sleep if they needed it, and his staff had to scramble to make sure the Capitol police would allow all-night access to them and any others who specifically said they were coming to the Bayh office. The Bayh office was the only one in the Capitol open to demonstrators that night.

The year 1970 began with hearings on comprehensive disaster relief legislation. Birch served as chair of the Special Subcommittee on Disaster Relief of the Public Works Committee and kicked off the hearings in Biloxi, Mississippi. He went from there to Palm Beach to give a speech on Electoral College reform. A week later, he found himself in Oklahoma City, speaking out on judicial reform, the Haynsworth nomination, and how to make our system work better in "turbulent times." In January, he took the disaster relief hearings on the road again, this time to Roanoke, Virginia, followed by hearings in Washington on lowering the voting age to eighteen.

Those early days in 1970 were also consumed by Bayh staff meetings to plan for a possible presidential campaign in 1972. If Birch decided to do it, it was important to do the necessary spade work to see if it was even remotely possible. Could he raise the money? Were there politicians who would want to sign on in key states? Did he have the fire in his belly? There were no answers to these questions in those first weeks of the year.

On January 19, President Nixon announced his intention to nominate G. Harrold Carswell for the vacancy still remaining on the Supreme Court. Fifty-year-old Carswell had been a district court judge in Florida when confirmed for the Fifth Circuit Court of Appeals only the previous year. Birch knew nothing about Carswell and dearly hoped that this nomination would be easy to support. Southern senators quickly announced their support for Carswell, but soon criticism arose about his high reversal rate as a district court judge—an astonishing 58 percent.

Larry Conrad left his post as chief counsel of the Subcommittee on Constitutional Amendments to return to Indiana to pursue his own political career. In 1970, he was elected Indiana secretary of state, and instantly he became a major political force. On February 3, Paul J. ("PJ") Mode replaced Conrad on the subcommittee. Mode immediately had to turn his attention to the Carswell nomination while also preparing for hearings on the voting age amendment and later on the proposed Equal Rights Amendment. The Carswell hearings had begun in January, and they continued into the first days of February.

A few days later, Birch was in Los Angeles speaking at the California Democratic Conference, this time on the environment and pollution. His national activities and profile were on the rise. This speech was followed by one on Electoral College reform before the California State Legislature. Press releases that month demonstrate his broad range of interests; despite Birch's desire to enter any number of battles in support of those interests, the growing concerns about Carswell kept intruding. He spoke out on issues of ethics and judicial reform, of unrest at the Indiana State Reformatory and the need to provide food aid to the victims of war in Eastern Nigeria, soon to be known as Biafra. He expressed concerns about the administration's policies toward Israel, on the need to establish a White House office of consumer affairs and his continued efforts to gain support for the eighteen-year-old vote, Electoral College reform, and the equal rights amendments.

8

Carswell: 1970

I have always felt that a politician is to be judged by the animosities he excites among his opponents.

WINSTON CHURCHILL[1]

BIRCH KNEW THAT OPPOSING THE NIXON ADMINISTRATION again might jeopardize his legislative priorities. Plus in order to wage a successful battle against Carswell's confirmation, he would be posing a challenge to the moderate Senate Republicans, making life difficult for them by asking them to oppose the president again.

His instinct was that Carswell's nomination was looking considerably worse than Haynsworth's, but he wanted to avoid another battle if he could, and he believed such a battle was doomed to defeat. He told Richard Harris, author of a book on the Carswell nomination called *Decision*, "When a bad thing is before the Senate and it has the support of the President, any effort to defeat it has to be immense to succeed. At the time, there seemed no chance that an effort of that magnitude could be pulled off—even though the Carswell nomination was clearly bad—because the senators' mood was 'God, don't put us through that again!' Also, there were other things for me to consider. One was that I had spent eight years here trying to build an image of myself as someone who isn't divisive, who isn't vindictive, who can get along with all factions. If I took on Carswell after having taken on Haynsworth, that could all vanish, because a lot of people would figure I was just out for blood."[2]

During the Haynsworth debate, Birch attended a meeting in the office of Sen. J. William Fulbright, who was concerned about the ethical issues surrounding Haynsworth. Fulbright, truly one of the Senate's giants, made a comment that would resonate over the years. He said he felt he knew Richard Nixon and said, "If you beat this guy, the next guy is going to be worse."[3]

Birch was reluctant to take a leadership role in the Carswell hearings as he had with Haynsworth, despite the fact that it was increasingly obvious to everyone how bad Carswell was. He wanted Joe Tydings to lead the battle, and for a while it seemed like Tydings would.[4]

On February 25, Aaron Henry, head of the NAACP in Mississippi, testified before the Subcommittee on Constitutional Rights about renewing the Voting Rights Act of 1965. Birch was presiding as chair that day. Henry didn't talk about the attempts on his own life, the bombing of his home, or the threats he continually faced. He talked about the deaths of civil rights workers, black and white, and how effective the Voting Rights Act had been. His stoicism and quiet bravery provided resounding evidence that there were good reasons to reauthorize the act. Attempts by the Nixon administration to gut the effects of the act provided solid evidence of the continued Southern strategy, which had resulted in the nomination of Haynsworth and now Carswell. After that, as Birch muttered to an aide, "How can you listen to these stories and then let Carswell go on the Court?"[5]

One of the first news items that soured many senators on Carswell came from reports about Carswell's campaign for the Georgia legislature in 1948. In a speech he had said, "I am a Southerner by ancestry, birth, training, inclination, belief and practice. I believe that segregation of the races is proper and the only practical and correct way of life in our states. I have always so believed, and I shall always so act. I shall be the last to submit to any attempt on the part of anyone to break down and to weaken this firmly established policy of our people. . . . I yield to no man as a fellow-candidate, or as a fellow-citizen, in the firm, vigorous belief in the principles of white supremacy, and I shall always be so governed."

Senators would be among the first to forgive political missteps and statements made at a much younger age that they would love to take back. The question was whether Carswell still had those views; if he claimed he did not, did his record support such a change in viewpoint?[6]

A leading Republican who ultimately voted against Carswell characterized the White House as rubbing "the Senate's nose in the mess it had made of the Haynsworth nomination. I learned that the Justice Department had rated Carswell way down below Haynsworth and a couple of other candidates. That made it clear that the choice of Carswell was vengeance—to make us sorry we hadn't accepted Haynsworth—and implement the Southern strategy. The Attorney General obviously believed that we had no stomach for another fight after Haynsworth, and that we would accept any dog, so he took this opportunity to show his disdain for the Senate. He and a lot of the other fellows downtown seem to feel that they, and they alone, constitute the government of the United States."[7]

Birch always suspected that after Haynsworth, Nixon had ordered Attorney General Mitchell to find another Southerner to nominate, one who owned no stock. Nonetheless he said, "The last thing I needed was another fight." He was concerned that Carswell would be confirmed simply because Haynsworth was not—and for that, he knew he would hold himself responsible.

Another piece of evidence circulated that confirmed to many the suspicion that Carswell remained a racist, though he continually disavowed that campaign statement made thirty-two years earlier. As a private lawyer, he had drawn up "incorporation papers for a white-only fraternity . . . of which he was a principal subscriber and charter member. Then, in 1956, while he was serving as a United States Attorney in Florida, he helped incorporate a Tallahassee golf course that was being transferred, on a ninety-year lease at a dollar a year, from a public, city-owned facility, which had been built with $35,000 of federal money, to a private club—a move that was clearly made to circumvent a Supreme Court decision handed down about six months before prohibiting segregation in municipal recreation facilities."[8]

Congressman Charlie Rangel and Clarence Mitchell, along with two or three other very prominent African American leaders, came to the

Bayh office for a meeting. They felt very strongly that Carswell was a racist. They appealed by saying, "Birch, you've got to do this. We've got nobody else to turn to. You've got to help us with this." That had a major impact on his decision to lead the opposition. Ten to fifteen Yale law students also came to see Birch, wanting to help. They started researching all of Carswell's cases that had been overturned, and they first uncovered his high reversal rate. Those students provided the manpower to research the cases that the Bayh staff could not do by themselves. Birch often told that story to students over the years to illustrate how they can make a difference.[9]

Birch was invited to Washington's Statler Hilton hotel to speak to the Leadership Conference on Civil Rights, subbing for Senator Tydings, who had the flu. Like Aaron Henry, members of the Leadership Conference were in town to urge extension of the Voting Rights Act. It soon became abundantly clear that the dominant issue at the convention was the Carswell nomination. Birch began his prepared remarks on the Voting Rights Act but soon set them aside, turning to Carswell. He "launched into a free-swinging attack on Carswell's nomination. The audience was with Bayh from the start, and as he warmed up they stayed with him. In conclusion, he shouted that Carswell not only could be defeated but *had* to be defeated, and the audience rose to its feet and stamped, cheered, whistled, and applauded for several minutes."[10]

That night, Birch had trouble sleeping. Clarence Mitchell, of the Leadership Conference on Civil Rights, had given Birch his personal assessment of Carswell's speech when a candidate for the legislature. He said it was his experience that if a person believes what Carswell did, though he was only twenty-eight years old, that person rarely changes his mind. Yet Birch was dogged by the conclusion that they couldn't win such a battle. At 2:00 a.m. he found himself reaching a conclusion. With all the unrest in the country—riots, assassinations, and violent confrontations with police—he had always told Evan that we didn't solve our problems in that manner, that we worked within the system. But if his son's face had been black, how could he tell him to work within the system when sitting atop that system—for life—was G. Harrold Carswell? He made a personal commitment to take on the nomination once again, assuming that they probably wouldn't muster more than twenty-five votes. But he had to try.

The next morning at the office, Birch met with his staff to discuss the confirmation fight. Once they were done, he "got up from his desk, smiled, and said, "O.K., let's crank it up."[11]

When he reassumed the mantle of leadership, it was alongside many of the members of the anti-Haynsworth coalition. Because Carswell seemed so much worse, the intensity was greater. A key addition was Marian Wright Edelman, a young lawyer heading up the Washington Research Project, a DC-based civil rights group. She already knew a great deal about Carswell, having opposed his elevation to the Fifth Circuit the previous summer, and was frustrated by the lack of support from other civil rights leaders in that effort. Now she found the necessary support. Her research on Carswell was critical, as was her ability to motivate others within the movement.

The next several weeks were dominated by the Carswell nomination, a battle that raised Birch's stature nationally but also took place during a time of severe personal crises. The details of the Senate debate are described expertly by Richard Harris in *Decision*.

Birch was not the first senator to oppose Carswell; Democrats William Proxmire (D-WI) and Fritz Mondale were early opponents, as was Republican Charles Goodell of New York. George Meany of the AFL-CIO announced his opposition, as did John Gardner, former Health, Education, and Welfare secretary and head of the Urban Coalition.[12] One of the key Republican opponents was Edward Brooke of Massachusetts. Before Brooke, no African American senator had been elected since Reconstruction; he was the eighteenth senator to announce opposition, only the second Republican to do so. Brooke had been an important Bayh ally in the Haynsworth debate, contributing greatly to the task of rounding up seventeen Republicans to vote no. Brooke delayed a decision until he was convinced that there was ample evidence to warrant opposition, and he wanted to be sure that at least one third of the Senate was with him, enough to show blacks and young people crying out for a no vote that they were being heard. He also knew that such a stance, particularly after Haynsworth, would make any influence he had at the White House evaporate.

Brooke felt that Carswell's 1948 speech reflected deeply held views. Carswell may only have been twenty-eight, but by that age Brooke had already spent five years in the military in World War II. He considered himself to be a fully grown adult by then. Looking for a change in Carswell's views, Brooke couldn't find it. He found the opposite. When Birch asked for the floor, Brooke yielded. The Indiana senator talked about the problem of convincing the disenfranchised and discontented that there was hope for them within the system. He feared the violence that would erupt if people were not convinced. "The thing that concerns me is how are these people going to look at the system if they know that a man who unfortunately has this background is sitting at the very top of it?"

Brooke nodded and added, "I do not think this nation can afford G. Harrold Carswell on the Supreme Court."[13]

Carswell's white supremacy speech, his high reversal rate, and his role in the whites-only covenant were bad enough, but things got worse for him when he was found to have lied at his confirmation hearing. He denied taking part in the incorporation of a golf course to protect its whites-only status, but it was soon revealed that he had discussed the matter with ABA officials the night before.[14] Claims of support from prominent lawyers and judges were discovered to be untrue, digging him a deeper hole. Then it was learned that he gave a speech only two months before his nomination was disclosed, one riddled with racism. He began the speech with an anecdote. "I was out in the Far East a little while ago, and I ran into a dark-skinned fella," he reported. "I asked him if he was from Indo-China and he said, 'Naw, suh, I'se from outdo 'Gawgee.'"[15]

The Nixon administration demonstrated a remarkable inability to learn from the previous nomination battle. Its ham-handed efforts were repeated as before, often characterized by overkill that repulsed potential supporters.

The month of March found Birch and his allies marshaling their forces in hopes of defeating the confirmation. An early April vote was likely to follow the Easter recess, which began March 27. During the recess, Birch was scheduled to travel with other senators to the Interparliamentary Union meetings in Monaco.

As the opposition intensified in the Senate and in the country, one of Carswell's chief supporters shot himself in the foot, severely wounding the nomination itself. Nebraska senator Roman Hruska, responding to questions just off the Senate floor, said, "Even if he were mediocre, there are a lot of mediocre judges and people and lawyers. They are entitled to a little representation, aren't they, and a little chance? We can't have all Brandeises and Frankfurters and Cardozos and stuff like that there." Berman later described the Hruska comment as the "kiss of death."[16]

One of the eight senators attending the Interparliamentary Conference in Monaco with Birch was Minority Leader Hugh Scott. Scott remained in Monaco until April 4, with the recommittal vote scheduled for April 6. There was an effort for recommittal of the nomination, sending it back to the Judiciary Committee as a way of killing it indirectly. Some senators wanted to avoid a final vote, and others thought a vote to recommit was cowardly. Birch, accompanied by Marvella on the trip, was too worried about the matter, and he returned to DC without her on April 1. During a refueling stop, he was given a message that Marvella's father had died. Unimaginably, her father had killed her stepmother and himself. Marvella didn't get word until after Birch left Monaco, and quickly she made arrangements to return to Washington.[17] She was given the sad news by Sen. John Sparkman, the chair of the delegation and father of Marvella's close friend Jan Shepard.

Marvella's father, Delbert Hern, had first shown evidence of mental illness shortly after Birch's nomination for the Senate in 1962. While Marvella was visiting in Oklahoma, he overdosed on pills and ended up hospitalized, largely for depression. After the election, she went home, only to see her father under the influence of alcohol for the first time in her life. When Birch was sworn in the following January, Marvella's parents were there for the celebration, along with Birch's father and sister. The only thing that marred the night was the drunken behavior of Delbert in front of everyone. That was when Marvella's mother, Bernett, revealed that he had begun drinking shortly after Marvella left home, was drinking when he tried to commit suicide, and had a severe problem. These issues, along with the physical effects of the 1954 automobile accident and their increasingly stressful life, resulted in Marvella developing serious sleep issues. By this time, she was taking tranquilizers at night, estrogen shots and thyroid medication during the day.

In April 1964, Bernett died, and Marvella heard from relatives about her parents' troubled marriage. Her father had fallen in love with a younger woman and wanted to marry her. There were suspicions of domestic abuse. A week after Bernett's funeral, Delbert left Enid, Oklahoma, to be with his lover, whom he married five weeks later.[18]

These events were incredibly hard on his only child, Marvella. He continued drinking to excess. Of course, no one knows what goes on inside a marriage, but the result of their last day was that Hern shot and killed his wife and himself. Marvella and Birch traveled to Oklahoma, spending most of the weekend of April 4–5 there, returning to DC on Sunday night for the recommittal vote on Monday, April 6. When they arrived at Marvella's home in Enid, they were greeted by prominent fresh bloodstains on the living room carpet.[19]

Once Birch had returned to Washington, a vote to recommit was imminent. Hugh Scott told Mike Mansfield that Carswell would win on recommittal and lose on the floor, causing Mansfield to go to Birch and suggest he call for a vote on confirmation right after the recommittal vote. Hruska was "so flabbergasted by the Bayh motion—a clear sign that Scott hadn't mentioned it to him—that he jumped up and shouted, 'The Nebraska from Senator objects!'"[20]

The vote on recommittal failed 52–44. Knowing which senators were supporting recommittal and would oppose Carswell, as well as those who intended to vote no on both, allowed Birch and his allies to begin to feel optimistic. He predicted to his staff that the confirmation would fail by a vote of 51–45.

The afternoon before the final vote, Birch flew to Houston to speak to the Texas League of Women Voters. Returning early the next morning, he had to go to a hospital and give his father a shave; he had suffered an angina attack while Birch was in Texas. Given the events involving Marvella's father and now his own, his emotions were frayed beyond description. But he knew that he had to focus during the time remaining and deal with not only his own strategy but also the possible strategies of the pro-Carswell forces.

Prior to the vote, Ed Brooke pointed out to Birch that Margaret Chase Smith had entered the chamber and suggested that he huddle with her for some last-minute lobbying. Birch disagreed and told him to leave her alone. "I was not close to her at all," Birch remembered, but he felt in his gut that she would be fine, that she would do what she thought was right, and that lobbying her could be counterproductive. Brooke did approach Smith to tell her that the White House was telling others that she had agreed to vote for Carswell. He assured her that he was not going to try to persuade her, but he thought she should be aware of what was being said. Unknown to Brooke, she went to a phone and called chief White House lobbyist Bryce Harlow to ask him if the story was true. Harlow tried to obfuscate; she slammed down the phone and returned to her seat on the floor.

Birch's final speech before the vote was described in an article in *Women's Wear Daily*. "In ringing tones he told his colleagues: 'Today we have the opportunity to tell our children and their children that the advise and consent responsibility given to us by our Founding Fathers nearly two centuries ago still has meaning today.' That was a direct slap at President Nixon's letter demanding that his nominee be rubber-stamped forthwith."[21]

Vice President Agnew entered the chamber and took his seat as presiding officer, calling the Senate to order. He announced that the matter before the Senate was to advise and consent to the nomination to the Supreme Court of Judge G. Harrold Carswell. With the galleries filled to capacity and about as many staff on the floor as could be jammed along the back walls, the scene was intensely dramatic. The corridors outside the chamber were as crowded as they had ever been, with far more people than could fit into the galleries trying to wedge their way close to the action. Most of the senators were on the floor as the buzzers for the quorum call echoed throughout the Senate buildings.

The one o'clock vote commenced, and the first four senators cast votes for the nominee. Because the roll was called alphabetically, Birch's was the fifth vote cast, and he was the first to vote no. At the end of the roll call, only 90 senators had voted and the nomination was losing, 46–44. Four senators were necessarily absent. When the remaining six senators made their way to the podium to cast their votes, five of them voted no. Birch always gave the credit for that fifty-first vote to Margaret Chase Smith, the Republican from Maine.

The galleries erupted in cheers, and the vice president gaveled to clear the galleries. But it didn't matter, as everyone was leaving for celebrations in the hall outside the chamber.

Thirteen of forty-one GOP senators voted no. The two whose votes made the largest difference were Cook of Kentucky and Smith of Maine. For Cook and Smith, it was the falsehoods in Carswell's testimony about the golf club that determined their vote.[22]

John Frank wrote, "Carswell must be viewed historically as a bad legal joke. Like Fortas, he helped to dig his own grave with his mouth. Want of candor to a senatorial committee is rarely rewarded, and Carswell was almost disgustingly wanting in candor. But he was so demonstrably incompetent that he might well have lost no matter how much truth he told; and I believe that he, unlike Haynsworth, really was a racist. Political triumph in Carswell's case goes partly to the senators, particularly to Senator Bayh, but also very markedly to Joseph Rauh and Marian Wright Edelman, the civil rights leaders who organized the country."[23]

Once again the Nixon staff did an awful job managing the nomination and lobbying the Senate. As with the Haynsworth nomination, they overreached and damaged their own case. Senators were not pleased to be lobbied by Chief Justice Warren Burger, which Richard Harris insisted took place but Burger later denied. It would be a clear violation of judicial ethics. Bob Dole of Kansas fought hard for Carswell; when the vote was over, Dole described the White House aides as "those idiots downtown."[24]

The day of the vote, April 8, Marvella asked me to write down my recollections of the events leading up to the vote. What I wrote provided me with details that I had long since forgotten:

> I have been asked by Mrs. Bayh to give my account of the events leading up to the defeat of Judge Carswell's nomination as one who has spent a good deal of time throughout the last few days before the vote with Senator Bayh.
>
> The weekend of April 4th and 5th, while the Bayhs were in Oklahoma, it was clear that the Monday (6th) vote for re-committal to Judiciary would fail. I picked up the Bayhs at National Airport on Sunday evening. The next morning when I picked up the senator he said, "Well, Bob, we're going to lose this one."
>
> I attended the Monday re-committal vote and left disappointed but not surprised. After the vote, the senator was due at the Sheraton Park Hotel for a Convention of the United Mine Workers. We sped there with columnist Mary McGrory in the back seat interviewing the senator.

The Convention was similar to a National Party Convention. It was loud and boisterous, one of the most tremendous responses I've ever seen the senator receive. His speech was extemporaneous. The hall was quiet during the speech. McGrory agreed with me that the crowd readily grasped hold of the "mediocrity" argument. They cheered the senator on and after much pushing and shoving, the three of us left in good moods.

Conversation continued at McGrory's apartment over a gin and tonic. Our talk centered on the whole drawn-out battle and how senators like Hatfield and Cooper had decided as they did. I tried to convince McGrory that Birch had begun rolling the anti-Carswell ball and not her favorite—Ed Brooke. At 5:40 PM we rushed off for WTOP-TV with a promise by Mary that she would light a candle till the vote on Wednesday was over.

The senator did a spot on the Martin Agronsky show and proceeded upstairs to the offices of Tom Braden and Frank Mankiewicz. After about 20 minutes there, we were all convinced our chances were good for the showdown.

Tuesday evening we took off for Dulles Airport in rush-hour traffic. I asked the senator how things looked and he assured me it looked good. He said Percy would be with us and that he'd heard rumblings from Mrs. Smith and Cook that sounded good. His best comment was, "I had the audacity to tell Keefe coming down the stairs that we'd get 50 or 51 votes." I was elated.

We got to Dulles after a real race, three minutes early for the flight. The senator was in a good mood and we laughed about the fact that I'd meet him twelve hours later, to Houston and back.

Wednesday morning at 6:00 AM I met him at Dulles. The *Washington Post* predicted a 48–46 edge for Carswell opponents with Cook and Smith as the keys to a tie. Not wanting a role for Agnew, the senator said there was one discrepancy in the *Post* report. He said if we got either Cook and/or Smith, Prouty would come around. This would make it 51–45 if we got all 3. And it was.

By 6:30 we were eating breakfast at a Waffle Shop just off Pennsylvania Avenue [522 10th Street NW].[25] Coming out, he mused, "Senator Bayh and his right hand man, Bob Blaemire, were seen rounding up anti-Carswell votes at 6:30 AM at the Waffle Shop on Pennsylvania Avenue." Surprisingly enough we were in good moods despite the hour. We got to the office and both slept till 9:00 AM. Then we found out that Percy announced for us. We headed out to see how Col. Bayh was doing and then back to the office as early as we could.

> After the victory, Mrs. Bayh and I pushed our way to the Old Supreme Court Chambers where the senator was being interviewed. Once in the hall, we were mobbed. Mary McGrory came up to say, "How about that candle, Birch?"
>
> As we prepared to go to the airport again, our staff mobbed "The Boss" on the Senate steps and he was thoroughly pleased. In the car he said, "This is a great day for the Senate." His excitement was somewhat dampened by the sickly condition of his father but when he left for the gate, he was jubilant and happy.

This second defeat of a Supreme Court appointment made Nixon's staff irate. They mounted efforts to make as many of the disloyal senators as possible pay dearly for the Carswell vote. In Texas, Ralph Yarborough was defeated in the Democratic primary. In the fall, Albert Gore Sr. was defeated for reelection in Tennessee, as was Joseph Tydings in Maryland. In Nevada, a challenge to the reelection of Howard Cannon failed narrowly. In Florida, Carswell resigned as judge in order to run for the US Senate. Newspapers ran a photo of Carswell holding up a bumper sticker saying, "Bye, Bye, Bayh—Heah Come De Judge." He was defeated 2–1 in the primary.

Nixon's view softened with time. In his autobiography, he said, "Looking back I have no quarrel with some of those senators who voted against Carswell because of their belief that he lacked the superior intellectual and judicial qualities to be a Supreme Court Justice. But I still believe that many of the senators who voted against him used the issue of his competence as camouflage for their real reason, which was their disapproval of his constitutional philosophy." He acknowledged that when researching his past, he and his staff had missed Carswell's 1948 statement extolling segregation.[26] Nixon could not have known about another aspect of Carswell's life that was revealed some years later. In 1976, Carswell was arrested after soliciting a homosexual act from an undercover policeman in a men's bathroom at a Tallahassee shopping mall. Three years after that, he invited a man to his hotel room and was subsequently beaten up. No one in the Bayh operation had any inkling about this side of Judge Carswell, and any hints about homosexuality at the time surely would have scuttled the nomination. Even rumors of homosexuality in 1970 would have prevented him from being chosen.[27]

Jay Berman commented, "I honestly believe that no other senator would have come up with the tactical way to beat this guy." Birch had created public furor over the nomination that created pressure on his colleagues. And again, the ABA was helpful. But Berman's summation of the affair was that "Birch Bayh's approach to Carswell was genius" and that he hasn't received enough credit for the manner in which he got it done.

There was no question that Birch raised his national profile with the Carswell victory as well as earning respect among Washington's politicos. Few had thought it possible to turn back a Nixon Supreme Court nomination a second time. Birch was clearly in charge of the strategy, coordinating the activities of staff and marshaling the resources of the allied groups. Because of his leadership in the Haynsworth effort, Carswell opponents naturally gravitated to him, and he took advantage of their energies in an expert manner. This was unprecedented. When had the Senate had two successive victories against a newly elected president in this manner, both engineered by the same person?

In a period of less than 5 months, only 138 days, the Senate had twice denied the president a Supreme Court nomination, with both efforts engineered by the same member of the Senate. It is hard to exaggerate the difficulty of this—and its rarity.

The commentary across the nation speculated about Birch's future and the chances that he would end up on the presidential ticket in 1972. Chet Huntley, one of NBC's television anchors, editorialized about him:

> It was obvious this week that the Democratic Party and the Carswell struggle in the Senate has produced another potential and serious candidate for national office. If there were individual winners and losers in the Carswell affair, the junior Senator from Indiana, Birch Bayh, was the former. Senator Bayh's opposition to the Carswell nomination to the Supreme Court was observed to grow and develop over the weeks. . . . Ultimately, the Indiana senator says he had the choice to make . . . [to] look the other way and let the confirmation take its course or pursue his conscience, even though it might be unpopular in sometimes conservative Indiana. As the controversy developed, the Senator reminded us of one of the heroic lines of our childhood: "The boy stood on the burning deck, whence all but he had fled, etc." Early in the contest, Senator Bayh was by no means on a crowded deck. This farmer-lawyer from Indiana, by the strange and unpredictable workings of political argument, could one day muse that he was nominated to higher office by a little-known federal Circuit Judge

> from Florida. The list of Democratic possibilities . . . now includes the singular name Birch Bayh.[28]

Eric Severeid, a noted national commentator on *CBS News*, editorialized about the Carswell matter and referred to Birch as a "Midwest John Kennedy." He wrote that Nixon had intentionally

> lifted up from the disorganized ranks of the Democrats a man who may very well turn out to be running against him in '72, either as a Presidential or Vice Presidential candidate. The 42-year-old Senator Birch Bayh of Indiana has now become an undisputed national figure. A broad path has opened up before him by that unarrangeable combination of personal quality and good fortune that seems to attend the fate of the chosen very few in our political history. He looks more and more like a Midwest John Kennedy, but with a personal background oriented much more closely not only to the country's heartland, but to the life of most ordinary citizens. His record of accomplishment as a Senator already outshines the Kennedy record in the Senate. . . . His origins are straight out of Horatio Alger and the mythology of the All American Boy. Farmer, Lawyer, class president, star athlete, Army Veteran, leading State Legislator and, in '62 the boyish David who felled the local Goliath—Senator Capehart. Handsome, strong, married to a natural born political wife with a record as a formidable campaigner, the image is almost too good to be true. If it holds a serious flaw, that hasn't shown up yet.[29]

That evening, Evan Bayh, age fourteen, answered the phone at home and was told, "You tell your father that if he is going to be the next John F. Kennedy, I'm going to be the next Lee Harvey Oswald."[30]

There were also troubling letters from Indiana, reminding Birch that opposing Nixon in a state where he was very popular would have its political costs. He began to realize that taking a stand would always make enemies as well as friends and that he had to be willing to accept the criticism along with the praise. But Evan's experience with the phone call made it hard on a very personal level.

A slew of mostly positive articles profiling Birch continued to be published:

- *Christian Science Monitor*: "Birch Bayh, Candidate?"
- *Women's Wear Daily:* "Bayh—Could Be Household Word in White House Race"
- *The Washington Star:* "Bayh Emerges Only Victor in Bitter Court Fights"[31]

In a magazine piece, investigative reporter Robert Sherrill characterized Birch as "cautious, cagey, constantly sliding around questions—a tough but modest professional who spent his formative political years learning how to maneuver a few progressive statutes through a moss-backed legislature by cozying up to all sides and saying nothing for the record until he had strained it through several layers of friendly ears." Birch was also invited to appear on the *Dick Cavett Show,* a popular nightly television interview program. Birch's life seemed to proceed on two tracks, one with the normal routines of Senate legislative life and the other contemplating a possible presidential race.

Among the normal routines was an announcement of another Indiana High School Government Leadership Conference. The speakers would be Curtis Tarr, Selective Service director; A. Leon Higgenbotham, judge and vice chair of the Eisenhower Commission on the Causes and Prevention of Violence; Harry McPherson, special counsel to President Johnson; and Jean Worth Matthews, public information officer at the Department of the Interior. Two speakers had to be found at the last minute because President Nixon intervened when the initial group was announced, forcing the cancellations of Deputy Assistant for Consumer Affairs Elizabeth Hanford (later to be known as Elizabeth Dole, following her marriage to Sen. Bob Dole) and astronaut Michael Collins. This was revenge for Haynsworth and Carswell.

While pressing forward with the proposed constitutional amendments, Birch spent considerable time on Indiana issues. He issued statements on disaster relief, farm subsidies, and measures to deal with violent crime. He again spoke out against the Vietnam War; antiwar fervor was growing in intensity as the American role expanded.

The national celebrity generated by Haynsworth and Carswell caused speculation about the next presidential race to continue. With the midterms still months away, questions circulated about which Democrat would dominate the discussion in early 1971. Ed Muskie, the vice presidential candidate two years before, was an early leader in polls, though he was still second to his running mate, Humphrey. A Harris poll showed a surprising new entry to the speculation, Republican mayor of New York John Lindsay. His intention to switch parties was a poorly kept secret, and he seemed to be filling the vacuum left by Ted Kennedy when his potential candidacy also drowned at Chappaquiddick. Humphrey, Muskie, and Lindsay were all registering in double digits, followed in single digits by Bayh, George McGovern, and Fred Harris.

April 22 was the first celebration of Earth Day, with events scheduled to focus Americans on the environment. Birch was invited to speak at Georgetown University, and I drove him there from Capitol Hill. As we left the office, the car began spewing thick white smoke from the tailpipe, hardly an example of clean auto emissions that one should be touting on Earth Day. We were both appalled and unhappy about the smoke, though in retrospect it seemed oddly funny. I dropped him off a few blocks from campus so he would not be seen climbing out of an environmentally unfriendly car.

But the war in Vietnam still dominated the news, with growing unrest in the country. On April 29, the United States invaded Cambodia to try to clean out sanctuaries used by the North Vietnamese and the Viet Cong. Immediately, large antiwar protests occurred across the country, as people feared a widening of the conflict, this time into another country. Birch joined those calling on the president to stop the escalation of the war.

On May 4, a day of demonstrations across the country, particularly on many American college campuses, Ohio National Guardsmen confronted student protesters at Kent State University and opened fire. Four students were killed and nine wounded. Outrage in the country, especially among those already opposed to the war, was intense and loud. Five days later, demonstrations against the war brought 100,000 people to Washington, DC.

On Capitol Hill there were swarms of protesters in the halls and offices. Many were college students. The Bayh staff determined that a place and time should be designated for those in town to be able to hear from senators who shared their concerns. A session was scheduled in the New Senate Office Building at which Birch, Gene McCarthy, and others addressed the throng of people crowding into the hearing room. It received wide coverage on television and in the newspapers and magazines like *Time* and *Newsweek*. Birch addressed the crowd by saying, "I understand your frustrations. I understand your rage. When year after year this nation seems incapable of arresting the deterioration of its cities, incapable of cleaning up its streets and its streams, incapable of stopping racial discrimination, incapable of halting a war that has cost us 40,000 American lives—the urge to explode in violent rage as a result of persistent and apparently hopeless frustration is understandable. But I urge and implore you who have come to Washington to express your dissent—and others throughout the country—to refrain from violence, to resist the temptation to become radicalized by recent events. America has and America will again respond to moral pressure."[32]

The protests continued, particularly on campuses. Many of them turned ugly. After two days of demonstrations at Jackson State University in Jackson, Mississippi, demonstrations that often turned violent, law enforcement officers opened fire at the protesters. Only eight days after Kent State, two more students were killed and twelve injured. The national reaction to these killings didn't compare with those at Kent State, causing many to wonder whether the black faces of the dead in Jackson were the reasons why. Aaron Henry, a leader of Mississippi's NAACP, called Gordon Alexander and told him they needed a federal presence there. Gordon made the case to Birch, who invited Walter Mondale to go with him. Birch got to know Mondale very well and genuinely liked him. The only two senators traveling to Jackson in the aftermath of the shooting were Mondale and Bayh, along with future secretary of the army Clifford Alexander. They toured the scene and jointly wrote to Attorney General Mitchell, asking that a grand jury be convened to investigate the incident. Back in Washington, they conducted their own probe, which revealed that ambulances for the wounded students weren't summoned until after the police finished picking up their spent shells. Jackson authorities argued that the police had not been involved.[33]

Gordon Alexander traveled with Birch to Mississippi. He later said, "Birch Bayh was a white man who would listen." Alexander felt that once you made your argument, Birch decided what to do based on its merits, not on race or politics. He added that Birch went to places and challenged customs more than other liberals. At Jackson State, said Alexander, "he could have been assassinated."

In between the incidents at Kent State and Jackson State, Birch reconvened hearings on his proposed Equal Rights Amendment. The opening testimony was by Rep. Shirley Chisholm of New York. Chisholm was the first African American woman elected to Congress and became the first woman to ever have her name placed in nomination by a major American political party, at the Democratic National Convention in 1972. She would serve seven terms in the House. Her Equal Rights Amendment testimony was memorable because she pointed out that she had surely suffered discrimination as a black person, but it didn't compare to the amount of discrimination she experienced as a woman.

During those months, Birch's life was consumed by activity. The news media continued to speculate about his future. His national travel was on the rise as he increasingly moved toward a decision on whether or not to be a candidate.

A nagging issue in the state distracted Birch from his travels. If he were to be a candidate for president, he would need the support of the Indiana Democratic Party. While state chairman Gordon St. Angelo supported him publicly, behind the scenes the opposite was true. Party leaders in other states passed messages to Birch saying that he was being undermined by St. Angelo, a powerful and shrewd party leader. St. Angelo aspired to be DNC chair and was campaigning for it around the country while denigrating Birch's chances as a presidential candidate. The 1968 nominee for governor, Bob Rock, urged Birch to join him in ousting St. Angelo at the state party reorganization meeting on May 16, an event to take place after the May 5 primary.

Bob Mooney, a political reporter for the *Indianapolis Star*, known to be a St. Angelo ally, wrote an article that seemed to underscore the problem facing Birch. After mentioning the favorable comments by Eric Severeid, Mooney went on to say, "There is much resentment in the Hoosier state over his role in the Haynsworth and Carswell controversies. Some Bayh detractors continue to point to his flunking of the Indiana bar exam in 1960, although he ultimately passed. They also pointed out he had relatively little or no experience in a courtroom. . . . Bayh's critics say he pinned 'racist' charges on Carswell because of a 1948 statement of belief in white supremacy, a 22-year-old incident. At the same time, they say Bayh that same year joined the Purdue University chapter of Alpha Tau Omega, a fraternity which in its charter restricted membership to 'white Christian males.' This restriction was rescinded by the fraternity in 1962, 11 years after Bayh was graduated from Purdue." The article included a statement by Birch's press secretary, Bill Wise, who said that when Birch was regional officer of the ATO, he led a floor fight at the fraternity's national convention to eliminate the racial clause. His region supported the move unanimously, though it was not passed by the convention. Birch had no doubt that St. Angelo had fed these comments to Mooney.

It was clear to Birch that St. Angelo didn't really want a Democrat to be governor. Such an event would displace him as statewide Democratic leader, and the 1968 campaign was proof of that. St. Angelo's behavior in 1970 made the problem more personal, but Birch knew enough about state party politics to know that he couldn't do it himself. Larry Conrad was seeking the office of secretary of state and had no desire to get involved in a battle over chairmanship of the state party. This was the first fissure in their longtime relationship. There would be others. Bob Boxell, who left the Bayh staff after the 1968 campaign but remained close, expressed similar reservations about taking on St. Angelo.

Birch contacted Vance Hartke to solicit his support for removing St. Angelo, an important step because Hartke was facing a tough reelection battle. Hartke promised his support, and Birch had his staff run a campaign to replace St. Angelo after the primary with a Bayh ally, Ken Cragen. There would be twenty-two votes for state chair, one from each district chair and vice chair of the eleven congressional districts. When the vote came, Cragen garnered only one vote of support, from Bill Trisler, a county chair who would figure prominently in Birch's relationship with the state party eight years later. St. Angelo's votes were largely the result of his support from Hartke. I picked up Birch at the airport that evening and had never seen him so angry, mainly over the fact that Hartke had lied to him. Most of the party leaders were his friends, and he had been humiliated by his own party, a burden he would have to endure if he continued to pursue national office.

At the end of May, Birch rode in the pace car around the track at the Indianapolis 500 and was booed roundly. Even though he heard cheers of "Give it to 'em, Birch" as he passed the less expensive seats, those in the higher-priced seats were clearly incensed with him. He heard the crescendo of boos as his name was announced, increasing in volume as he passed by the center of the grandstand. It was a reminder that Nixon was extremely popular in the state and that his actions in the Senate had struck a negative chord at home. Arriving in DC after the race, he said, "You haven't lived until you've been booed for a third of a mile."

It is startling to review all that took place during the nine weeks ending with the Indianapolis 500. From a historical perspective, certainly enough happened to deserve note; from Birch's personal emotional standpoint, it seemed almost overwhelming. Marvella received news of her father's murder-suicide on April 1. On April 7, Colonel Bayh entered the hospital. The following day was the Carswell vote. April 22 was the first Earth Day, and Nixon's incursion into Cambodia took place on April 30. The Kent State shootings followed only four days later, and the ERA hearings took place the following day. A massive anti–Vietnam War demonstration in Washington happened on May 9. The Jackson State shootings took place on May 14, and two days later was the Indiana state party's reorganization, handing Birch a bitter defeat. And then he was booed at the Indianapolis 500.

As Nixon's presidency moved well into its second year, his secret plan to end the war was still a secret; in fact, the war was widening. With that expansion came tremendous growth in the antiwar movement as well. Congress was not immune to this sentiment and in June repealed the Gulf of Tonkin Resolution of 1964, which had provided LBJ the blank check he needed to prosecute the war in Vietnam. The Senate also introduced the Cooper–Church Amendment (John Sherman Cooper and Frank Church) barring the president from involving the military in Cambodia after July 1. Birch supported the amendment.

His travels around the country continued, where he talked about electoral reform, economic decline under Nixon and his failed policies in Vietnam and Cambodia, the political participation of African Americans, and the need for national Democratic leadership. Civil rights leader Hosea Williams endorsed him for president, and the American Veterans Committee presented him with the Eleanor Roosevelt Citizenship Award. He spoke about the need to override Nixon's veto of the Hill–Burton Hospital Construction Program, about proposed legislation requiring airbags in all automobiles, and about the need for broader minority participation in the construction industry. And he was working on a proposal to reform the military justice system, all the while raising money and gathering support for a possible presidential run.

About this time, NBC announced a new role for its popular program *The Bold Ones* during the 1970–71 season: Hal Holbrook would play a United States senator. The actor told reporters that he was an admirer of Birch Bayh. "He is a brilliant man. I see him as the prototype of the senator I'll be playing next season." Holbrook spent time in Birch's Senate office, preparing for his TV role.

Also in June, Birch played in another congressional baseball game at the former DC Stadium, renamed RFK Stadium after Bobby Kennedy, the late senator. Birch's photo was prominently displayed in the *Washington Post*, which also published a photo of Rep. Barry Goldwater Jr. (R-AZ) in a pickoff play at second. More notable was the description of a triple Birch hit to the 410-foot sign in right center in the third inning, which followed an earlier run-scoring triple in the first.[34]

Problems facing the nation were highlighted when it was discovered that a family was living in a cardboard box literally in the shadow of the Capitol, just a few blocks away. It was a family of three, including a ninety-year-old woman suffering from untreated serious burns. Birch visited the site, which was shocking and sad, even if it hadn't been so symbolic. The Bayh staff helped provide food to the family, alerted DC human resources officials, and directed press attention to the matter, both to help find better shelter for the family and to make people aware that this was taking place. In the *Washington Post* article about the cardboard home, Birch was quoted as saying, "I thought it was terrible. I had never seen anything like it except in Lima, Peru. It does not represent what America is to most of us, but this kind of condition still exists here."[35]

The women's movement ratcheted up a notch with the Women's Strike for Equality on August 26, a New York City march celebrating the fiftieth anniversary of women's right to vote. The rally attracted thousands of women from several American cities with calls for equality in the workplace as well as for political rights for women.[36]

In September were more antiwar rallies, highlighted by the appearances of political activist and Vietnam veteran John Kerry and stars from Hollywood, including Jane Fonda and Donald Sutherland. There were also efforts in Paris by the Nixon administration to negotiate an end to the war. On October 7, the North Vietnamese delegation rejected the Nixon's cease fire proposal as "a maneuver to deceive world opinion." Two days later, the Khmer Republic was proclaimed in Cambodia, beginning a civil war in a country that was receiving increasing attention from Americans. In an effort to seek an end to the war, President Nixon announced a planned withdrawal of 40,000 more US troops before Christmas.[37]

The Bayh operation was spending substantial time planning for a possible presidential campaign. While he was performing so aggressively as a legislator as well as an Indiana politician, it was obvious to all how difficult the effort to mount a presidential campaign was. National politics created a constant tug of war with the rest of his agenda. Birch was unwilling to relax his efforts to enact constitutional amendments, and even while traveling around the country, he still returned to Indiana as much as possible. In August, he introduced legislation to reform the codes of military justice. His Military Justice Act of 1970 provided changes to the current set of codes. This effort resulted from Birch's concern that soldiers accused of crimes often take the blame for the orders of their superiors because of the control of the system by commanding officers. A day after introducing that bill, he introduced the Judicial Disqualification Act of 1970 and the Omnibus Disclosure Act for Public Officials, the two representing prodigious efforts to address needed reforms in our systems of civilian and military legal affairs. He was later awarded the 1970 Legislative Award of Merit by the American Trial Lawyers Association.

In mid-October, Birch condemned Nixon's veto of a campaign spending bill. On the other hand, he showed his support for the president when he spoke out against the stoning of a Nixon motorcade in California at the end of the month.

The debate on Birch's amendment for direct popular election hit a major snag in September. Having already passed the House, it was filibustered in the Senate by Southern senators led by Sam Ervin. A cloture vote ending Senate debate required a two-thirds vote, the same requirement as for the amendment itself. Yet there were several senators who would never support cloture, regardless of the legislation being blocked by the filibuster. On September 23, Birch announced that he would hold up all other Senate business, including committee meetings and other floor deliberations, making those who were filibustering pay the price for their dilatory tactics. The *New York Times* reported on the dispute in the Senate, quoting senators who felt Birch was grandstanding and only seeking publicity for his possible presidential race. It also addressed the views of his allies in the effort and printed a complimentary assessment of his role in the Senate and his career. One quote in the article was of former postmaster general and JFK campaign manager Lawrence O'Brien, who said that Birch was "a political realist with a full understanding of the art of the possible."[38] The *Washington Post* editorialized that Birch's move represented "A Welcome Showdown in the Senate." When the Senate tried to cut off debate with another cloture vote a few days later, Birch's effort fell five votes short. He could read the writing on the wall, and he knew that the measure would not be passed in 1970.[39]

Once the cloture effort failed, Birch stepped up his activities around the country. By Election Day 1970, he had visited forty states in a flurry of speaking, traveling, meeting people, and trying to determine whether or not a presidential contest made sense. The *Washington Post* printed a major article about him during this period that began by pointing out the nature of the task ahead. "The stewardess had been alerted that a VIP would be boarding at National Airport, and she approached the group of three men to ask breathlessly, 'Which one of you is Senator Blight?'"[40] He was certainly not yet a household word.

The election on November 3 produced results typical for the first midterm after a presidential election: The opposition party normally gains seats. Congressional Democrats gained twelve seats in the House but lost three in the Senate. Among the notable results in the races for governor were the reelection of Ronald Reagan in California and the election of Jimmy Carter, a peanut farmer from Georgia.

Democrats who were reelected included Indiana's Hartke, who squeaked by with 4,383 votes, or a margin of two-tenths of a percent. His opponent, Rep. Richard Roudebush, contested the election, which would not be resolved by the federal courts until 1972. Because the Senate has the right to seat whomever it wants, Hartke was assumed the winner and maintained his seat. Most noteworthy in Indiana was the election of Larry Conrad to be secretary of state, a traditional stepping-stone to the governorship. Conrad's victory was pure delight to Birch and his staff.

The war in Vietnam was a major issue in the November election, and it dominated the news. On November 5 the US military reported the fifth consecutive week with fewer than 50 American soldiers killed, the lowest number in five years, but 431 soldiers were wounded. The next week, there were no combat fatalities at all. Nixon reacted to that welcome news by asking Congress to provide more than $150 million to the Cambodian government to prevent its overthrow by the Khmer Rouge and its allies in North Vietnam.[41]

Following the election, *U.S. News & World Report* ran a three-page article speculating on the state of the race to replace Nixon in 1972. Still listed among the Democratic front-runners were Muskie, Humphrey, and Kennedy. The "dark horses" it listed were McGovern, Bayh, Hughes, and Scoop Jackson of Washington; the article also recognized Mayor Lindsay, should he decide to formally switch parties. While Birch was not willing to say that he was actually running, his interest was obvious; he had visited more states than any of the others during October. In December, he opened an office for the exploratory committee, preparatory to a presidential run. Muskie and McGovern had already set up theirs.[42]

In November and December, in addition to extensive travel, Birch was preoccupied with many issues: suggesting modifications to the Equal Rights Amendment to facilitate its movement through Congress, urging Congress to adopt anti-inflationary measures, urging the Senate to override the presidential veto on campaign spending, challenging the president to clarify his position on anti-poverty programs, calling for approval of the Consumer Protection Act, opposing the supersonic transport plane (SST), proposing child care legislation, criticizing the treatment of Jews in the Soviet Union, and proposing an amendment to the Family Assistance Act. On December 18, the Senate approved the Bayh Disaster Relief Act of 1970.

Also in November, the trial of Lieutenant William Calley for the My Lai massacre began and French president Charles de Gaulle died. Two days before Christmas, construction crews in New York City completed the north tower of the World Trade Center. Standing 1,368 feet high, it was taller than any other building in the world.[43]

9

Campaign and Cancer: 1971

> Franklin Roosevelt . . . said that "This generation of Americans has a rendezvous with destiny." I think they met that rendezvous. I am asking this generation of Americans in 1960 to do the same, to do in its time what those generations before us did, to maintain freedom and serve as an example and a bright light to the world around us. That is our opportunity, and I think that is our destiny.
>
> JOHN F. KENNEDY[1]

FOR BIRCH BAYH, 1970 WAS A YEAR OF SENATE VICTORIES, legislative accomplishments, and the beginning of a presidential campaign. But it was also a year of personal tragedy with the death of Marvella's father and stepmother and the illness of Birch's father, Colonel Bayh. And Birch had spent far less time with Evan than he ever would have imagined. The year 1971 was also marked by legislative accomplishment and a nascent presidential campaign of great promise, and it too was a year in which father and son spent far too little time together. And it would be another year of personal tragedy.

Members of Congress have an ongoing need to balance their working lives in Washington with their family lives in the home state or districts. A senator often needs to maintain homes in both places. The farther from the Capitol they have to travel, the more difficult it is to strike an easy balance. A few of them augment those difficulties by adding a campaign for higher office to an already demanding schedule. So began Birch's life in 1971. He represented Indiana in the Senate, spending most of his time in Washington but often traveling to his home state. His family lived in

DC, where they maintained a home, adding new responsibilities and activities to their lives that further demanded Birch's time. Marvella was increasingly in demand as a public speaker, while Evan, age fifteen, was an active student at St. Albans, a private Episcopal school near their home. At first they lived in McLean, Virginia, and then they moved to a house at 2919 Garfield Street in northwest Washington. Birch's elderly father was ill in a nursing home, requiring more regular visits.

Evan aspired to be an athlete at St. Albans, and Birch wanted to attend as many of his games as possible, whatever the sport was. Travel was a constant challenge but adding a presidential campaign meant increased time away, putting additional stress on his family life. Scheduling him became harder for the staff, and the staff often seemed to compete with his wife Marvella, who cultivated an active social life for the senator and herself. Hard feelings surfaced regularly and were inevitable. Birch valued his relationship with Marvella and Evan and tried mightily to satisfy his professional ambitions in a way that could meld with family responsibilities. But no solution seemed to work. He felt a need to protect his family as well as his staff; both wanted the best for him but with different ways of achieving that. When Birch was able to attend Evan's basketball games, he and Marvella often sat in the bleachers with Barbara and George H. W. Bush, the parents of Evan's classmate Marvin and future First Lady and US president. Bush was then a Texas congressman.

Despite the conflicts inherent in his life, Birch did his best to make all parties happy. But the fact that he could only be in one place at a time guaranteed that the inherent challenges would not go away. In her autobiography, Marvella wrote of the marital strains brought on by the presidential campaign.

Birch was clearly affected by the buzz being generated about him around the country. He surveyed the Democratic field and considered himself as good, if not better, than the others, and he was motivated by an urgent desire to remove Richard Nixon from the presidency. He also harbored self-doubt. Birch's natural humility, coupled with a dislike for the typical puffed-up politician who thought he was better than everyone else, gave him pause. His motivation came largely from the conclusion that Nixon was beatable at that point in time and that if someone could beat him, he would rather it be Birch Bayh than anyone else.

A commitment to make a difference for the country and world was a driving force, while he constantly fought off the temptations brought on by the intoxicating levels of support sprouting up around the country. He realized he had to keep his ego in check and reminded those around him that he still put on his pants one leg at a time.

The first candidate to declare his presidential candidacy that year was Sen. George McGovern. On January 18, he delivered letters to his Democratic Senate colleagues with his speech enclosed. McGovern's office, previously occupied by Sen. John Kennedy, was directly across the hall from the Bayh office, which had previously been occupied by Nixon. In McGovern's reception room was a plaque noting the office history; no such plaque was in the Bayh office. McGovern was a friend of Birch's but not the candidate who most concerned him. That was Muskie, who had distinguished himself as Humphrey's running mate in 1968 and was often described as "Lincoln-esque." Muskie was also a close personal friend of Birch's. Even though McGovern announced first, Muskie's campaign was more seriously organized. McGovern's campaign was being managed by Denver attorney Gary Hart.

The year 1971 began with the introduction of a new legislative effort to completely overhaul the nation's approach to juvenile delinquency. As the new chair of the Subcommittee to Investigate Juvenile Delinquency, Birch was energized to change a criminal justice system that was failing youthful offenders. He introduced the Juvenile Justice and Delinquency Prevention Act to try to enact necessary changes. One story that resonated with him became public during congressional hearings. A young girl named Debbie was incarcerated in an adult facility because she repeatedly ran away from home, where her stepfather was continually abusing her. This made no sense to Birch, and he was committed to changing a system that allowed that to happen.

He promoted his disaster relief legislation and his proposed direct election and voting age amendments. As an increasingly national figure, he made his voice heard. He attacked Nixon's effort to reduce business taxes, called on the Soviet Union to help reverse the deterioration of Soviet–American relations, responded to the president's State of the Union address, urged Congress to control inflation if the president would not, and asked the Civil Service Commission to investigate reports that political clearance was being required for career civil service positions. He criticized the president for suspending the Davis–Bacon Act as it pertained to wage contracts on federally assisted construction projects. He also spoke out on the right of citizens to see and correct information in governmental data banks, and he urged Congress to upgrade unemployment insurance, increase payments toward the burial costs of veterans, oppose the extension of the draft, and reform the military justice system. Because of departures from the Senate after the 1970 election, he was sixtieth in seniority.

Birch's chairmanship on the juvenile delinquency subcommittee melded with his desire to become a prominent national politician. Marvella urged Birch to speak out about the challenges of finding adequate child care so women could work outside the home. In February 1971 at an event in Philadelphia, she spoke about day care for children and the specific problem she encountered seeking good child care for her son, Evan, so she could attend college on a part-time basis. For Marvella, the issue of child care was inseparable from women's rights. If women were to work alongside men on an equal basis, our society needed to provide excellent day care for children.[2]

As reported in a February 22 article in the *Houston Post*, Birch took it up a notch, stating that most of the problems in the country were the result of "inadequate attention to children. . . . We have too many kids being crippled mentally and emotionally because of too little attention at an early age."[3] He introduced the Universal Child Care and Development Act to provide day care facilities for all children. He spoke about estimates of twelve million working mothers in America, half of whom had children under six. The act would create child care centers funded by the government and run by parents. Child care was only one of the issues for the new chairman. The problem of juvenile delinquency was a larger focus of the subcommittee's agenda. Birch used this new platform to urge the nation to turn its attention to statistics: "More than 25 percent of all arrests are of persons under 18 . . . more than 50 percent of all those arrested and 75 percent of those arrested for serious crimes are under 25." He went on to describe how 70 percent of individuals released from prison end up being rearrested but said that the number was higher for juveniles who end up being repeat offenders after spending time in juvenile correctional institutions. "Can you imagine how deplorable a correctional institution must be if our expert criminologists are suggesting that young offenders would be better off on the streets?" he asked. His intention was to focus on two areas of the problem: how to keep that first brush with the law from ballooning into a future of continued lawbreaking and how to correct problems within correctional institutions.[4]

As a presidential aspirant, he traveled to New York City to urge local Democrats to work for unity within the party and took part in the Democratic Party's efforts to change its rules on apportionment. Birch had encouraged guidelines to satisfy those who wanted quotas for minority participation at the next convention, even though quotas had been previously rejected. He said, "There should be some reasonable relationship between the representation of delegates and the representation of the minority group in the population of the state in question." This passed, and soon others sought the same protections for women and youth. A vote supported that position as well. At the 1972 convention, state parties would have to seat a delegation that reflected the state's percentages of these groups, opening up challenges on the Credentials Committee to delegations that did not.[5] While the concept of quotas did not affect the process of winning the nomination in 1972, it became one of those images that seriously damaged the Democratic campaign against Nixon in the fall.

On the Indiana front, Birch once again announced the guest speakers at his Indiana High School Leadership Conference. The headliner was the man he had defeated in 1968 and the new director of the Environmental Protection Agency, William Ruckelshaus. Joining him were FCC commissioner Nicholas Johnson, Deputy Assistant Secretary of State on African Affairs W. Beverly Carter Jr., and Erma Angevine, executive director of the Consumer Federation of America.

In the fall of 1970, a Bayh presidential headquarters was opened at 1225 19th Street NW in Washington, and the staff grew rapidly, with many people moving from the Senate office to the campaign. Materials from the Bayh Committee included reprints of the Eric Severeid commentary after the Carswell vote and included a list of Birch's positions on key issues. On the issue of peace, he wanted to prohibit funds for involvement in Cambodia, terminate the Tonkin Gulf Resolution, and limit deployment of anti-ballistic missiles; he supported troop withdrawals from Vietnam and opposed military aid to the junta in Greece. On civil rights, he opposed a measure allowing governors to veto legal services to the poor; he also wanted to add $400 million to the War on Poverty, extend the Voting Rights Act, and deny states "freedom of choice" plans for their schools, seen as a disguise for the right to segregate schools. Also distributed were his views on fighting crime, on inflation, and on unemployment and some describing his environmental efforts and his support for education funding. One flyer listed his opposition to Haynsworth and Carswell and his support for the ERA and an amendment lowering the voting age to eighteen years. His campaign materials characterized him as "Birch Bayh—The Man Who Defeated Nixon Twice!"

He described a president who promised to "bring us together' but was sending Spiro Agnew around the country to cause divisions instead, "a sort of sinister Johnny Appleseed sowing hatred, prejudice and fear wherever he passed. This was the same President who orchestrated the Southern strategy, a carefully calculated effort to reopen old wounds and to pit one region of America against another."[6]

In February, Birch renewed his efforts to pass a constitutional amendment that would lower the voting age from twenty-one to eighteen. Having already steered to passage the Twenty-Fifth Amendment, he could become the first person since James Madison to have the principal responsibility for passing more than one amendment. It was an issue that Birch had sponsored in the Indiana legislature. Sen. Jennings Randolph (D-WV) had introduced the amendment eleven times in previous sessions of Congress, the first time in 1942. Birch co-sponsored the bill with Randolph early in 1971. It had wide support in the country, largely because of the number of young people fighting the war in Vietnam, who were old enough to be drafted into military service but too young to have voted for those who sent them there. On March 2, Birch achieved the first milestone: passing the amendment in his subcommittee; two days later it passed in the Judiciary Committee. From there, he became the floor manager for the legislation in the Senate. It passed unanimously in the Senate on March 10 and in the House on March 23.

Birch combined his campaign travels with visits to states where ratification was under consideration. On July 1, the amendment was ratified;

the process had taken only three months, the quickest ratification in history. It was estimated that over eleven million people age eighteen to twenty-one would be able to vote for president in 1972. It is likely that Richard Nixon would not have been elected president had the amendment been in place four years earlier.

Birch was first exposed to advocates of an equal rights amendment during the hearings on the voting age amendment, which were disrupted by women loudly advocating for equal rights. Birch almost had the Capitol police remove them. As he left the room for a Senate vote, he asked what the ruckus was about; after being told, he sent a message to the protesters, saying that if they would quiet down, he would agree to see them to discuss their concerns.

The matter of Lt. William Calley burst onto the front pages again on March 31, when a six-member military officer jury convicted Calley of the premeditated murder of twenty-two Vietnamese civilians, sentencing him to life imprisonment. The next day, President Nixon ordered him released from prison and placed into house arrest; the president declared that he would personally review the case. This was inconsistent with normal military procedure and seen by many as obvious pandering to those supporting Calley. Standard military justice procedure allowed for review by two military courts, followed by the secretary of the army and finally the president.[7] Birch was among the first to speak up forcefully in opposition to the president's actions, and he received substantial coverage in the nation's press as a result.

In an article in the *Washington Post* on April 8, Birch is quoted pointing out that there were eighty other servicemen facing convictions for serious crimes while serving in the armed forces and none of them had received this presidential intervention. All of those crimes were less severe than the twenty-two murders of which Calley was convicted. "By his premature actions, the President has made a truly impartial, equitable review of the Lt. Calley case impossible. . . . Reluctantly, I have concluded that the President is determined to play politics with the Calley decision and the entire My Lai tragedy." Birch held a press conference on the matter in which he read from a five-page statement that also pointed out the political risks he was taking. The Bayh office had received 1,029 letters on the matter, only 3 of which supported Birch's position.[8] A United Press International (UPI) story highlighted the role Birch assumed in the Calley debate and added, "If the first impulse of the American public suggests that the murder of captive women, children and the aged by Americans should not be punished, I hope we will reach a more thoughtful conclusion in the course of careful reflections." It went on to say, "Bayh is wagering the public will reach that 'more thoughtful conclusion.' When public opinion shifts, as many predict it will, Bayh's forthright stand will be taken correctly as an act of political courage, rather than a gamble."[9]

Newsweek ran a full-page article about the Bayh campaign in April, describing its attention to detail as well as its aggressiveness. The campaign staff had risen to seventy people, with organizational efforts being mounted in twenty-three states. *Newsweek* said he had visited eighteen states thus far in 1971 and that Miami mayor David Kennedy was throwing his support to Birch—a blow to Muskie, who had assertively sought the mayor's endorsement. Amid the flattering descriptions of Birch's effort were less favorable descriptions of him by a California legislator and a leading Texas Democrat. The former was quoted as saying, "Nominating Birch would be like nominating one of us. We like him but we don't see much chance for him." The latter said, "No matter what he's doing, he always comes off like a college cheerleader." The article speculated that Birch might really be after the vice-presidential nod in 1972, putting himself in a good position to seek the top slot in 1976. "Other Democratic chieftains, however, find themselves yielding gracefully to Bayh's charm and to his broadly liberal, anti-Vietnam, pro-labor record. Again and again, the senator's name comes up as a good possible second choice by Democrats. . . . This leads to discussion of Bayh as the ideal Vice Presidential candidate, but Bayh resists it. 'There's a distinction between being everyone's No. 2 choice and everyone's choice to *be* No. 2,' he told *Newsweek*'s Richard Stout. 'One does not do what I'm doing if all he has in mind is the Vice Presidency.'"[10]

In May, *Washington Post* political reporter David Broder wrote a front-page piece about Birch entitled, "Hustle, Detail, Money: Bayh Picking Up Steam." It described Bayh's staff as well organized and detail-oriented. Broder concluded the article with a quote from Birch: "If we do as well in the next six months as we have in the last, I'll be well pleased."[11]

By the time the Broder article appeared, Birch had been in twenty states during 1971, many more than once.

Although the Bayh campaign and Senate offices were separate operations, they were not as distinct as they should have been, particularly after the Watergate revelations a few years later. Bob Keefe moved out of the Senate office to work full time in the campaign office. Jim Nicholson, the campaign manager, was a former Indianapolis lawyer and a member of the Federal Trade Commission. Jay Berman and P. J. Mode headed up legislative activities and campaign policy. Bill Wise, a former *Time* magazine correspondent and former Middle East bureau chief for *Life* magazine, ran the press operation. Wise's press department sent out radio feeds with Birch's comments on every issue coming before the Senate. Howard Paster, who had been on the staff of Rep. Lester Wolff (D-NY), was one of the new hires to bolster the press department.

Berman also remained involved with fundraising. He recalled the 1971 effort with fond memories of working with Milton Gilbert and Nat Kalikow, two beloved Bayh fundraisers, among many others. He said that they never wanted anything from Birch; they simply had a personal affinity for him. Because of them, Berman came to know others in the Democratic fundraising world that would be important to Birch and to himself for years to come. Among those were Arthur Krim of United Artists, Steve Ross of Warner Communications, and Lew Wasserman of both Universal Studios and MCA.

Although the Bayh effort was generating little support in national polls, Birch understood that the more he picked up steam and national exposure, particularly through the free press, the more he would garner support. The staff concentrated as much on likely and potential delegates as on former delegates, anticipating the changes due to happen as younger activists were increasingly involved in the process. As the organization continued to grow and the fundraising was proving successful, the press notices were increasingly notable.

The energy and momentum seemed to be on target. By mid-March, Birch had visited forty-four states in the previous year as well as Israel, the latter during a Middle East fact-finding trip in February, where he met with Prime Minister Golda Meir and a number of other Israeli leaders. He even made a pilgrimage to Johnson City, Texas, to discuss his campaign prospects with former president Johnson. While the Bayh Committee trumpeted Birch's campaign activities, it also highlighted his legislative actions, such as the recently introduced Universal Child Care and Child Development Act of 1971. During these first months of 1971, Birch criticized President Nixon on a variety of fronts: reduced business taxes, his budget, inflation, unemployment, Laos, the draft, the SST, the army spying on citizens who opposed his policies, and Nixon's intervention in the Calley matter. Each issue seemed to strengthen his resolve to become the Democratic nominee in 1972.

Birch's opposition to Nixon was deeply felt and fueled his ambition to run, as it did for the others in the race. Nixon's policies were bad enough, but his assault on civil liberties made his presidency appear particularly dangerous. During this period, Birch began using a quotation he often repeated in the ensuing decade. It was from Pastor Martin Niemoeller, a German anti-Nazi theologian during the rise of Hitler: "In Germany they came for the Communists, and I didn't speak up because I wasn't a Communist. Then they came for the Jews, and I didn't speak up because I wasn't a Jew. Then they came for the trade unionists, and I didn't speak up because I wasn't a trade unionist. Then they came for the Catholics, and I didn't speak up because I was a Protestant. Then they came for me, and by that time no one was left to speak up."[12]

In April, the Supreme Court created a future hot-potato issue when it announced its unanimous decision in Swann v. Charlotte-Mecklenburg Board of Education. The ruling allowed the busing of students to be ordered as a legitimate means to achieve racial desegregation. The war continued to roil the country as hundreds of thousands of people in Washington, DC, and San Francisco marched in protest. A Harris poll showed 60 percent of Americans in opposition to the Vietnam War. The May Day protests were next, with demonstrators in Washington trying to disrupt government business. Several thousand protesters were arrested.

The busing issue grew in intensity, and as chair of the constitutional amendments subcommittee, Birch found himself in the crosshairs. Opponents of forced busing wanted an amendment to permanently remedy what they saw as an egregious way to fight segregation. Birch did not agree with that solution, but politics required that he tread lightly. It was emotional on both sides. While Birch wrestled with busing proposals before his subcommittee, Chairman Eastland called him and told him to pass a busing amendment out of the subcommittee without recommendation. In other words, send it up to Chairman Jim's Judiciary Committee and let him deal with it. Eastland assured him that the amendment "would not see the light of day," adding that he was doing this for Birch "because I like you."

By mid-year, Birch's frenetic lifestyle was taking its toll. There was no letup in the Senate schedule while he was aggressively traveling the country to raise money and secure support for the presidential campaign. The price he was paying was not noticeable to the Senate staff. While it was clear that he was away from the Capitol more often, when in town he always walked through the office, stopping at most desks to chat with the staffers present, and treating each staff person or volunteer like he was deeply interested in what he or she had to say. It was one of the things that bound staffers to him. He paid a higher price on the home front.

Birch described his political success as the result of his partnership with Marvella. They ran as a team, whether it was for the General Assembly or the US Senate. By this ninth year in Washington, the strains on that partnership were becoming apparent. When Birch first arrived in the Senate, Marvella hired a social secretary to help handle the volume of mail she received, to help manage her travels to Indiana for Birch, and to assist in maintaining their social lives in the Capitol. They agreed on the importance of getting to know Senate colleagues and their wives, and they hosted regular dinners at their home to foster relationships and support Birch's career. They also attended dinners at the homes of other senators.

Marvella and her husband were opposites in many ways. Far more intense and less likely to see the humor in a situation, Marvella was a perfectionist. She cared about money, and Birch did not. He encouraged dissent from those around him; she spurned it. The trappings of office also seemed to mean more to her than to him. Lee Hamilton recalled a day when Marvella, Andy Jacobs, and he flew from DC to Indiana, Marvella sitting in the middle seat between them. As she talked on about how "the senator wants to do this and the senator wants to do that," Jacobs joked with her, "Marvella, senator who?"[13] She did not see the humor in it.

As the years went by, the push and pull on Birch's time intensified. What developed as a competition between the Senate staff and Marvella grew into a feud. Given her perfectionism and the level of professionalism that Marvella required, hard feelings and enmity among staffers often resulted. As she increasingly saw herself in competition with the staff, the lines were drawn more distinctly. If Birch defended an accused staffer, tempers grew hot and the words became harsh. She had persistently locked horns with administrative assistant Keefe and the senator's personal secretary, Fran Voorde. Both had moved over to the campaign, making it even more untrustworthy to her.

Exacerbating all of this was her life as the mother of an adolescent son, one who was pushing the boundaries against a woman who was very set in her ways and, like many mothers of teenage boys, perhaps not as tolerant as she might be.

Additionally, Marvella worried about her health. It began when she felt a pain in her right breast in February. She consulted doctors who found nothing to be concerned about. In the same month, the health of Colonel Bayh, Birch's father, was noticeably deteriorating, adding to Birch and Marvella's many anxieties.

Birch's desire to protect the partnership began to waver. He discussed few details of the presidential planning with her and cast aside her concerns about incurring debt they couldn't afford if he failed. She became obsessed with possible debt and in the process pushed Birch farther away. He no longer wanted to share any details of the planning for his presidential race, which normally would have involved her. They both behaved in a way that intensified her feuds with the staff and turned their few evenings home together into misery. Her autobiography reveals that June 1971 was when she first believed their marriage was in trouble. Birch recalled reading this in her book and feeling ashamed that she felt that way but also because he was unaware of how deeply she was feeling.

On May 18, Senate Democrats scored another victory when Congress voted to end funding of the supersonic transport (SST) aircraft. The SST was a recent invention that captured the fascination of the public as the first commercial aircraft to break the sound barrier. It caused sonic booms that could be heard for miles, but its exhaust was reputed to threaten the ozone layer. While Nixon and much of the business community supported its development, public concern about its environmental effects was widespread. Environmentalists applauded the action by Congress. Birch's opposition had been made very clear to anyone watching the early presidential nomination contest.

The first week of June, Birch voted for the Hatfield Amendment to end the draft and several days later gave a speech saying that the only way for the war to end was for Congress to set a date for it to end. No date would ever be set by the Nixon White House. Birch's call for Congress to terminate the war was overshadowed by another event that sent ominous reverberations throughout the country. On June 13, the *New York Times* published the first of a series of articles that would come to be known as the *Pentagon Papers*. Derived from classified documents, the articles explained how the Pentagon had put together a report on the history of the Vietnam conflict, and they revealed the deception and outright lies committed by the Johnson administration, not only to the public but to Congress as well. Some within the Nixon White House argued that publication should be permitted because it embarrassed the previous two administrations, not the current one. But Nixon's national security adviser Kissinger felt that leaks of that nature were dangerous and that it would be a terrible precedent to allow them to be published without a fight. The report was leaked by Daniel Ellsberg, later prosecuted for his role in leaking the documents. The dispute centered, however, on Nixon's attempt to halt publication by the *New York Times*. The prior restraint issue was argued before the Supreme Court, which decided in favor of the newspaper. History has revealed that the Nixon administration efforts to investigate Daniel Ellsberg were among the illegal activities that later would be known as the Watergate scandal. But in mid-1971, there were few news items more prominent than the Pentagon Papers matter. During this same period, the Senate passed Birch's Juvenile Delinquency Prevention and Control Act.

Birch traveled to California to hold hearings on disaster relief. He was chairing a special subcommittee of the Public Works Committee, and California had just experienced earthquake-related destruction. Allan Rachles, who traveled with him on that trip, retained an indelible memory from the experience that reinforced his feelings about Birch as a tough leader. Senator Bob Dole was part of the entourage, as was freshman senator John Tunney of California, who was invited as a senatorial courtesy. Dole was also serving as chair of the Republican National Committee at the time. Every time Tunney opened his mouth, Dole interjected a nasty comment or simply put him down. After tolerating more than he probably should have, Birch put a stop to it, telling Dole in no uncertain terms that his discourtesy and lack of grace to a fellow colleague would stop right then, or the matter would be referred to the Senate for review once they returned. Dole remained quiet.[14]

Also in June, Birch stepped more deeply into the fight for women's rights by introducing legislation extending the provisions of the 1964 and 1968 Civil Rights Acts to cover sex discrimination. Throughout the summer Birch made statements on pending legislation as well as a wide variety of national issues. He applauded the president's planned trip to the People's Republic of China. Given Nixon's strident anti-communist credentials, Nixon watchers were startled by the China visit but saw it as a positive step for world peace.

On August 11, Birch ventured into territory that would exact a heavy price on his political future when he introduced a measure to amend the Gun Control Act of 1968 by prohibiting the sale of the inexpensive handguns known as "Saturday night specials." While it was easy to argue that this type of weapon was not used in hunting and was the weapon most often used in the commission of crimes, gun control was a hot-button issue for those against it. The National Rifle Association (NRA) was noted for generating heat against any legislator who supported gun restrictions of any kind, and its Indiana membership was not something a politician could ignore. But Birch felt strongly about the issue and took heat from Hoosiers who vehemently opposed it. He compared the issue to being "partially pregnant." You were either for gun control or not; there were no halfway reasonable positions in between. As much as he might consider some people to be "gun nuts," they represented a force to be reckoned with.

On August 26, Birch Evans Bayh Sr. died at age seventy-seven. It was the culmination of months of tragedy in the Bayh family, made less tolerable given the strains in Birch and Marvella's marriage. They both adored Colonel Bayh and took his passing hard. Colonel Bayh's body was flown to Terre Haute for the funeral service and burial.

As a father of two preteens who had lost their mother, Colonel Bayh consistently provided examples of how to live well with common sense and compassion. He was a role model as an educator, a soldier, and a strong father. As director of physical education in Washington, DC, he set an example of how one conducts oneself in athletic pursuits. His influence on Birch was strong. His service in the military provided an example for his son and one that Birch sought to emulate. While the military did not become his chosen field, public service did. When Birch's mother passed away, Colonel Bayh took his son and daughter back to Terre Haute to live with their loving and devoted grandparents, who had a warm and positive impact on the development of Birch and Mary Alice. Birch never would have gravitated to agriculture if it weren't for his grandfather. His grandmother was a model of the strong woman for him. And when Birch began in politics, he knew how beneficial his father's sterling reputation was. Two days after Colonel Bayh's funeral, Birch and Marvella went to Paris together. As a member of the Interparliamentary Union, Birch was encouraged to bring his spouse. Each trip was a good experience, and this one was no different, becoming a welcome elixir for their marriage. Marvella wrote that the trip to Paris did a "world of good" for both of them. This particular trip was important because of a meeting with the North Vietnamese delegation to the peace talks. Birch made sure that Marvella was included in the meeting, but back at their hotel, tempers flared once again. They argued about the staff and about the campaign debt, which then stood at $200,000, of which $60,000 was considered theirs personally. She told him that she would refuse to campaign at all until the debt was paid, adding, "I wish so much that you'd be satisfied with being a great senator."

Paris can be magical, and it worked its magic on this occasion. They sat at sidewalk cafes unique to Paris, holding hands and talking. She relented and assured him that she would travel for the campaign to Wisconsin and Oregon as he wanted her to, while he would be traveling in California, Wisconsin, and New Hampshire. They discussed the planned announcement schedule, starting in Washington and proceeding to Tallahassee, Milwaukee, Lincoln, and Los Angeles the same day, each city in a key primary state.

Back home, Birch resumed his travels for the presidential campaign and remained aggressive in his efforts to enact legislation. On September 10, he opened hearings on his "Saturday night special" bill; in a statement at this hearing, he lambasted the administration for its inaction on the issue of violent street crime. He called again for withdrawal from Vietnam and for the immediate resumption of deliveries of Phantom jets to Israel. He introduced the Omnibus Correctional Reform Act of 1971 to change the direction of the nation's correctional system, and he announced hearings (scheduled for October) on a constitutional amendment to limit future presidents to a single six-year term.

The matter of Supreme Court nominations raised its head again with two resignations occurring almost simultaneously. Justice Hugo Black admitted himself to the hospital in ill health on August 28, and his retirement was announced on September 17. He suffered a stroke shortly thereafter and died ten days after retiring. Also suffering from deteriorating health, Justice John M. Harlan retired on September 23. Birch desperately wanted the next nominees to be acceptable and had no stomach for another battle. He had other priorities, and it would be interpreted as presidential politics for him to oppose another nominee, given his recent activities. He called for the president to appoint a woman to the Supreme Court. The frenetic travel continued while his campaign staff prepared the announcement of his presidential candidacy in November.

On September 23, Marvella once again felt discomfort in her right breast, which was swollen and periodically caused a "zip of pain, like a small electric shock." She made an appointment with her physician, Dr. Sanford Hawken, to discuss it and to have a mammogram after returning from a campaign trip to New Jersey. Her follow-up appointment was a week later.

Dr. Hawken said that the X-rays were clear, but he wanted to schedule a biopsy. That morning, Birch called Marvella from New York, excitedly telling her that he had the pledges he needed to erase the campaign debt. When he called back later that afternoon from Ohio, she told him her news.[15] That was a sobering conclusion to a rather humorous day, as Gov. Jack Gilligan of Ohio had hosted Birch in the governor's mansion, serving baloney sandwiches.[16] Birch interrupted his plans and returned home to spend the night with Marvella, leaving again the next morning. The doctor had told her that there was an 80–85 percent chance she would be fine.

The following night, Marvella performed in a show at the Women's National Democratic Club with Birch in the audience. She wore a strapless flapper dress, low cut with spaghetti straps and no bra, and that night she found herself wondering whether she would ever be able to dress like that again. Birch left for New York and Florida the next morning, and Marvella was driven to Columbia Hospital for Women, the same place where Birch's mother had died almost exactly thirty-one years earlier. She was focused on the doctor's 80–85 percent chance that everything would be okay. She spent that afternoon preparing for the biopsy the next morning. Birch returned from Florida that evening, and they ate steaks together in her hospital room. She assured him that if everything went as they hoped, she would do whatever she could to help the campaign.

As scary as life was for Marvella at that moment, it was exciting for Birch politically. The New York trip concluded with his fundraisers handing him checks totaling $500,000. Flying to Florida, he arrived on the tarmac in Miami where Mayor Kennedy and the mayor from Tampa, Dick Greco, met them. Birch handed them the checks to be used for the Florida campaign, got back on the jet, and returned to Washington.

Gail Alexander, a staff member who was close to Marvella, remembered that Marvella had insisted that she was only in the hospital to find out if she had breast cancer; she would decide later about surgery. She expressed her fears about having a mastectomy, but when she had expressed those fears to Birch, he told her that he had lived his whole life without breasts, which made her laugh.[17]

The next day, awakening from the biopsy, she found Dr. Hawken and Birch at her side and was told that she had breast cancer and that they needed to remove her breast. She sobbed at the news, terrified of the word "cancer," and found herself worrying about dying and about Evan's future. On October 8, she had a modified radical mastectomy, in which Dr. Hawken removed the right breast, portions of her chest muscle, and the lymph glands in the armpit.

The afternoon after the surgery, I sat with Marvella in her room, trying to crack jokes when she grew panicky about Birch's whereabouts.

She wanted him with her but mainly wanted to be sure that the doctors were not communicating information to Birch that she would not be told. I assured her that he was undoubtedly in a phone booth somewhere and went to find him. When I did, he was sitting in a room and crying, with doctors present. Wiping his eyes, he made it clear that she was not to know that he was talking with the doctors. I suspected bad news had been shared with him, but he did not tell me any details. Many years later, I recounted that experience to him, and he said that the doctors had told him that the cancer was aggressive and terminal. When he asked what they meant by that, they told him, "She'll have one good year."

He returned to the room, smiling and appearing happy, setting her at ease. At one point she said, "Look at me. I only have one breast."

He replied, "That's one more than I have."

Birch's travel schedule was canceled for the next four days. He met with his campaign staff in a hospital sunroom to discuss the meaning of her surgery and how it would impact their plans. Deep down, everyone knew what was going to happen next. On the night of October 11, Birch called me at home to ask if I'd pick him up early for some staff meetings at the Senate office. The next morning, we left his house for Capitol Hill, and he read me a statement he had written to announce the end of his presidential campaign. I was saddened but not surprised. The surprise came when he revealed that he had two meetings scheduled in the office, one with Bob Keefe and the other with Fran Voorde. Both were going to be told that they would not be brought back to the Senate staff from the campaign. Each meeting was private. Keefe was there when we arrived, and Voorde arrived soon after. When Keefe left from one end of Birch's suite, Birch came out the other and invited Voorde in. Both left in tears.

The announcement speech ending the campaign took place in the ornate Senate Caucus Room, a site where other presidential campaigns began, including those of John and Robert Kennedy. It was a room that would become well known to most of the country when the Watergate hearings took place there.

There was an outpouring of concern for Marvella, expressed both to her and to Birch. Immediately after the news broke, Sen. John Stennis of Mississippi was the first of Birch's colleagues to stop in at the office to express concern about Marvella. It served as a reminder of how the relationships between colleagues were important, even those with differing political positions.

That afternoon, Birch held a press conference, one jammed with reporters from every major news organization. His remarks were as follows: "During the last several months, I have seriously considered becoming a candidate for the presidency. . . . I have made this effort because of my concern for the problems that confront our country and each of us as individuals." He went on to talk about Vietnam and about the economy and added,

> We had been encouraged by citizens all over the country declaring their support. . . . Whenever I have had an important decision to make during the seventeen years I have had the good fortune to serve in public life, my wife, Marvella, has always been there. But Marvella is not here today.
>
> She is not here because she underwent critical surgery for a malignancy. We have every reason to believe the operation was a success. However, her complete recovery may require a lengthy period of recuperation. During this time, I want to be at her side—not in Miami, Milwaukee or Los Angeles. . . . Therefore, I am not a candidate for the presidency.

Birch recalled that this was one of the only times he had done something of importance in his career without consulting Marvella. When he told her that he had scheduled a press conference to end his campaign, she replied, "Birch, are you sure you want to do this?"

"The rightest decision I ever made," he told her. He knew how important it was for him to be there when she checked out of the hospital. Remaining a candidate was an obstacle to be removed.

Marvella Bayh was the first public figure to experience and share her breast cancer publicly. Other, more famous women would share this experience in coming years, but her illness shined a light on a problem for women that had only been discussed behind closed doors. The Bayhs did not have the option of keeping her cancer private.

Marvella's health and the demise of the Bayh presidential campaign were major news across the country. Not long after the surgery, Allan Rachles got into a cab near the Madison Hotel on Fifteenth Street in DC. To his surprise, there was another rider in the taxi, and of all people, it was Judge Clement Haynsworth. Allan introduced himself and told the judge where he worked. Haynsworth replied that he had just written Birch a letter expressing concern over his wife's health.[18]

Ironically, the House of Representatives passed the Equal Rights Amendment on the same day that Birch abandoned the presidential campaign. He could now spend more time and energy urging the same step in the Senate. While Marvella's recovery was his top priority, he also turned his attention to the pending Supreme Court nominations, urging the president to avoid politicizing the process the way he had with Haynsworth and Carswell. Additionally, the staff of the constitutional amendments subcommittee was focusing on a new, potentially explosive issue.

The nation's constitution was written in the late 1780s during a convention called to revise the Articles of Confederation. Few would argue that the nation hadn't benefited by the wholly new document that resulted, but the point remained that a constitutional convention, should one be called again, could pretty much do what it wanted. Since the Constitution was ratified, there had not been another convention, but the procedure written into the Constitution for adding amendments included the option that a convention could be called for that purpose. Birch saw this as a ticking time bomb. No rules had ever been passed by Congress determining the process, should such a convention be called. Article V of the Constitution spells out the two procedures for passing amendments. The only process ever used was for Congress to pass proposed amendments by a vote of two-thirds in the House and Senate with ratification by three-fourths of the states. The other spells out how amendments are to be proposed—"[Congress,] on the Application of the Legislatures of two-thirds of the several States, shall call a Convention for proposing Amendments"—also requiring three-fourths of the states to ratify them once proposed.

Serious constitutional scholars were suggesting that this was an invitation for chaos and potentially very dangerous for the republic. The kinds of questions raised were as follows:

- Did the wording of the amendments requested by the states need to be identical?
- Who decided the persons representing each state, and how should they be chosen?

- Who would chair the convention, and for what duration?
- Did state petitions calling for a convention ever expire? For how long did they remain valid?
- How could the convention be limited to address only the subject for which it was called?
- How would the procedures for voting in the convention be determined and by whom?

The more the convention issue was debated internally, the greater the concerns about its potential implications. Clearly, more questions could be raised than those listed above, and many were concerned about how the process could be manipulated by whatever political movement happened to be stronger at a given time in history. It made sense to try to promulgate rules, to attempt to force the nation's leaders to consider all sides of the question. The reality to this day is that nothing of substance has ever passed the Congress to address this potential crisis.

Birch's travel ceased, and he turned his attention to taking care of Marvella and his legislative priorities in the Senate. He was concerned about the ERA, his proposal for a six-year term for presidents, the issue of the constitutional convention, Supreme Court nominations, "Saturday night special" legislation, public works proposals, and other matters.

On October 21, Nixon announced two Supreme Court nominations, Lewis Franklin Powell Jr. and William H. Rehnquist. On October 31, Birch appointed Mathea Falco chief counsel of the Subcommittee to Investigate Juvenile Delinquency, the first time a woman had held such a senior position in the Senate.

Once again, there would be no presidential nomination of a woman to serve on the Supreme Court. Nixon wrote in his autobiography, "Our search for a woman nominee was serious and intense, and accompanied, I might add, by Pat's cogent and determined lobbying on every available occasion. But we found that in general the women judges and lawyers qualified to be nominated for the Supreme Court were too liberal to meet the strict constructionist criterion I had established."[19]

The Supreme Court nominations again took center stage in Birch's Senate life. During the passage of the Twenty-Fifth Amendment, Birch came to know Lewis Powell, a man for whom he had great respect and affection. Rehnquist raised greater concerns, with rumors flying around the Capitol about his far-right views and anti–civil rights positions, both as a clerk for Justice Robert Jackson and later as assistant attorney general in the Justice Department. He had written a memorandum to Justice Jackson opposing school desegregation during the *Brown v. Board of Education* deliberations. Rehnquist argued that he was tasked with making an argument supporting Jackson's views, not his own, but senators were suspicious in light of the anti–civil rights actions of the Nixon administration. Birch called on Attorney General Mitchell to release Rehnquist from the ABA Model Code of Professional Responsibility so he would be free to answer questions about public policy and his judicial philosophy. Mitchell denied the request, and Rehnquist successfully obfuscated in his responses to questions aimed at pinning him down on matters pertaining to civil rights. Birch saw ample evidence to conclude that Rehnquist was opposed to using the law to promote racial equality, something he supported. The Haynsworth and Carswell victories, however, made it impossible for the next controversial appointment to be stopped. It is extremely ironic that the most intelligent conservative of them all, Rehnquist, ended up on the Court, something unlikely if the earlier nominations hadn't been defeated.

On November 11, Birch announced his support for Powell as Supreme Court justice, which prevailed in an 89–1 vote. His Subcommittee on Constitutional Amendments successfully passed the Equal Rights Amendment on November 22, sending it to the full committee. The following day, Birch announced his opposition to the Rehnquist nomination, though he did not try to assume a leadership role as he had done twice in the previous two years. He was pleased to be able to support Nixon's nomination of Earl Butz of Indiana for secretary of agriculture. Butz, also a Purdue alumnus as well as a former teacher of his, was confirmed in December. Butz was not popular with Indiana farmers but Birch voted for his confirmation because he felt a president should be allowed to have the cabinet members he wanted unless there are compelling reasons to oppose them. He soon regretted that vote. Butz turned out to be a cheerleader for corporate farmers, and he worked hard to push back reforms from the New Deal. He would make many political and personal gaffes in the coming years.

Gun control dominated Birch's work from September to mid-November 1971. When he took over the juvenile delinquency subcommittee, he succeeded Thomas Dodd, an early advocate of gun control who in 1961 introduced legislation to control the mail-order purchase of firearms. After the assassination of President Kennedy, who had introduced similar legislation as a senator in 1958, Dodd's effort was bolstered by the fact that JFK had been killed by a mail-order rifle. Nonetheless, he was unsuccessful until the 1968 assassinations of Martin Luther King Jr. and Robert Kennedy. Then Congress acted to pass the Gun Control Act of 1968. The act required those engaged in the manufacture or importation of firearms to be licensed, required records to be kept of all sales, prohibited most mail-order sales, required manufacturers to have their names stamped on the firearms they made, banned importation of surplus military arms or firearms not suitable for sporting purposes, and prohibited the sale of firearms to felons and certain other categories of persons.[20]

Birch supported the legislation and quickly learned what a difficult issue it would be in Indiana. The problem was that anti–gun control voters were obsessed by the issue and would vote on that issue alone. Pro–gun control voters tended to cast their votes on a variety of issues. Most other emotional issues, including abortion, had passionate advocates on both sides, but the gun control issue was a passion only among those who differed with Birch on the issue.

Birch convened hearings on amendments to the Gun Control Act of 1968. Law enforcement officials urged him to introduce an amendment to prohibit the sale of "Saturday night special" handguns. Police watched with alarm the growing presence of these inexpensive handguns on the streets of American cities. Birch opened the hearings by saying,

> This morning we begin hearings on the misuse of handguns which are usually called Saturday night specials—those small caliber, relatively inexpensive pistols and revolvers that have no sporting use, but which are widely used in crimes of violence, in murders, in robberies, and in aggravated assaults. . . . Because of the prohibition in the 1968 act, we witnessed the ingenuity of gun manufacturers and former importers, who began to make the basic frame of the gun here, import parts for that frame, and then assemble the finished product in the United States. In addition, other entrepreneurs began to manufacture these very guns in their entirety in this country. Thus, both of these operations clearly circumvented the intent of Congress necessitating these hearings, and the legislation that I propose today.

He went on to cite statistics: "In 1970, 65% of this Nation's murders were committed with a gun. Thus, over 10,000 of our citizens were murdered by gunmen last year . . . 35% of all gun murders . . . were victims of gunmen armed with Saturday night specials."[21] He added many more statistics before concluding his speech.

In November, Miklos "Mike" Sperling died. Sperling had been an early supporter and was probably more responsible for Birch making it to the Senate than any other person involved in giving or raising campaign money. Birch always felt that without Mike Sperling and Tony Hulman, he never would have been elected. Losing him was like losing a member of the family.

On December 2, Birch received a statement from twenty Harvard Law School faculty members who opposed the Rehnquist nomination. In the Senate debate on the Rehnquist nomination, he challenged Rehnquist's account of the Jackson memorandum; years later, this would seem prescient. According to a Jackson biography, the memo was not requested by Jackson to buttress his own views but instead was an argument by Rehnquist to convince the justice to vote against *Brown*. On December 10, Birch released a list of 271 law professors who opposed Rehnquist's confirmation. He included a letter from a participant in the 1966 National Conference of Commissioners on Uniform State Laws stating that Rehnquist had opposed provisions of a model Anti-Discrimination Act. Nonetheless, the Senate confirmed Rehnquist 68–26.

That same day Birch blasted the president for his veto of a child care bill, calling the act "heartless." He finished the year by announcing hearings on barbiturate use, testifying before an agriculture subcommittee against a predator poisoning program, and proposing the reintroduction of the two-dollar bill, but with Thomas Jefferson's face replaced by that of Susan B. Anthony.

The year had been a difficult one for the Bayh family. Birch's father had died in August, his wife had developed cancer and had a mastectomy in October, and Mike Sperling had died in November. There had been ongoing strains in their marriage over the campaign and his absences from home, the latter of which led him to miss far more of Evan's life than he would have thought possible.

10

Title IX: 1972

It is better to be making the news than taking it; to be an actor rather than a critic. Politics are almost as exciting as war, and quite as dangerous. In war you can only be killed once, but in politics many times.

WINSTON CHURCHILL[1]

IN 1972, DEMOCRATS WERE PREOCCUPIED WITH CHOOSING A candidate to defeat Richard Nixon. Birch was officially on the sidelines making Marvella's recovery from breast cancer his priority. He maintained a busy schedule in the Senate, continuing to move forward on several fronts. Because he had divided his energies with the exploration of a presidential candidacy, he was focused less on Indiana and needed to repair some fences back home. While some think a senator's constituents would be proud of the presidential buzz swirling about one of their own, it is not always the case. Many thought he had gotten too big for his britches, wondering how he could be worrying about their problems when he was spending so much time worrying about his own ambitions.

It was a year when the greatest scandal in American history was spawned and the year of one of Birch's greatest legislative accomplishments. It was a year buffeted about by international events and a continuation of the Vietnam War. It was a year of the Olympics, which included a tragedy that will ever mark memories of those 1972 games. And it was a year of great electoral achievement by Richard Nixon, but the seeds of his demise were planted and beginning to bear fruit.

The year began with President Nixon announcing the beginning of the space shuttle program, and it ended with the tragic death of a popular athlete doing charitable work. The Old Senate Office Building was renamed the Richard Russell Senate Office Building in honor of the Georgia Democratic senator who had died the previous year. And Shirley Chisholm of New York announced her candidacy for president, the first African American congresswoman ever to do so.

On February 2 was the final draft lottery, a sign of changes in the country. The Vietnam War seemed to be winding down, and the country was preparing for an all-volunteer military. In fact, the last candidates for the draft were never called to active duty.

Birch's highest priority in early 1972 was the Equal Rights Amendment (ERA), written and introduced by Alice Paul almost half a century earlier. It had gained political momentum, and there was a feeling that it was ripe for passage. Birch felt now was the time. But the effort to pass it should not have been necessary. Alice Paul, then eighty-seven, had been instrumental in the campaign to give women the right to vote in 1920, leading to passage of the Nineteenth Amendment, and had organized the National Woman's Party (NWP).[2] She wisely devised a strategy that linked the ERA with the interests of business. Business leaders had been reeling because of laws and court decisions that forced changes in the ways women were treated in the workplace. An ERA would level the playing field and could help the scores of women working in textile mills—most important to her friend Congressman Howard Smith of Virginia. Smith had been the House's version of the Senate's Southern committee chairman and stood in the way of most progressive legislation, particularly in the area of civil rights. But he agreed with Paul on this issue and added one critical word to Title VII of the Civil Rights Act, written to prohibit employers from discriminating based on "race, color, religion, or national origin."[3] The word Smith added was "sex." Smith wouldn't mind if his one-word addition scuttled the entire bill but felt that as long as black people were going to get anti-discrimination guarantees, let's add them for women as well. An arch segregationist and chairman of the House Rules Committee from 1955 to 1966, Smith was a major obstacle in the effort to pass the Civil Rights Act of 1964. But he was a surprising advocate of the Equal Rights Amendment, which had been introduced in every congressional session since 1923.[4] He forged an alliance with Paul and the NWP and supported the proposed amendment.[5]

As Clay Risen states in his book about the history of the Civil Rights Act, *The Bill of the Century*, "It remains a great irony of the civil rights story that one of the men most responsible for holding back the advancement of American blacks was also the man responsible for the single biggest advance in women's rights since the Nineteenth Amendment granted them suffrage."[6]

But after the bill passed, that portion of Title VII was largely ignored. On the floor of Congress, snide and disparaging jokes about women were made regularly, and even James Roosevelt, FDR's son, said that his mother would have opposed the ERA. His brother, Franklin Jr., had been appointed chair of the Equal Employment Opportunity Commission in 1965; at his first press conference, when asked, "What about sex?" he responded, "Don't get me started. I'm all for it."[7]

Melding women's rights with civil rights hadn't worked well for women, and Congresswoman Martha Griffiths, another former member of the NWP, became the leading voice for the ERA. There could be no mistaking the difficulty of passing a constitutional amendment, but its enforcement couldn't be avoided like Title VII had been. Betty Friedan, president of NOW, became a close Griffiths ally, and both found a friendly ally in the senator from Indiana, who had already shown an ability to shepherd an amendment through to passage not once but twice.[8]

Birch's year began with subcommittee hearings to establish temporary shelters for runaway youth and to establish an Institute for Continuing Studies of Juvenile Justice, culminating in the subcommittee's passage of the Runaway Youth Act in January. Then he got involved investigating allegations of abuse in procurement policies at Camp Atterbury Job Corps Center near Edinburgh, Indiana. Another Indiana High School Leadership Conference was scheduled. This time the speakers were Paul Perito, acting deputy director of the Special Action Office for Drug Abuse Prevention; Stephen Johnson, the State Department's country officer for North Vietnam and part of the Vietnam Working Group; Constance Newman, director of VISTA (Volunteers in Service to America); and Marianne Means, a King Features syndicated newspaper columnist.

In February, the Senate was considering renewal of the Higher Education Act of 1965. Birch devised a strategy to insert language prohibiting gender discrimination in that bill. It was a way of doubling down on the issue of women's rights, a backup in case the ERA failed. The education issue resonated with him because he remembered when Marvella had applied to the University of Virginia (UVA) and received a letter saying, in effect, "Women need not apply." Marvella and Birch thought that was wrong, especially for institutions accepting taxpayers' money. The University of Virginia was males-only until a lawsuit in the 1960s but had become completely "co-ed" by 1970. Given its heritage, UVA should have been a model of egalitarianism; in fact, it was integrated before the *Brown* decision. Its founders included Thomas Jefferson, James Madison, James Monroe, and John Marshall.[9] The university was hardly a place where one would expect sex discrimination, but its exclusionary policies inadvertently led to one of the most important legislative measures affecting the role of women in American society.

Now Birch saw an opportunity to do something about it. He announced this intention on February 10 and introduced it on the floor eighteen days later. It would be forever known as Title IX. In his remarks, he said,

> We are all familiar with the stereotype of women as pretty things who go to college to find a husband, go on to graduate school because they want a more interesting husband, and finally marry, have children, and never work again. The desire of many schools not to waste a "man's place" on a woman stems from such stereotyped notions. But the facts absolutely contradict these myths about the "weaker sex" and it is time to change our operating assumptions . . . While the impact of this amendment would be far-reaching, it is not a panacea. It is, however, an important first step in the effort to provide for the women of America something that is rightfully theirs—an equal chance to attend the schools of their choice, to develop the skills they want, and to apply those skills with the knowledge that they will have a fair chance to secure the jobs of their choice with equal pay for equal work.

Title IX comprises a total of thirty-seven words: "No person in the United States shall, on the basis of sex, be excluded from participation in, be denied the benefits of, or be subjected to discrimination under any education program or activity receiving Federal financial assistance."[10] Undeniably it is a sweeping concept, but it received little public attention at the time.

Birch had been taught that boys and girls should be treated the same. He recalled a breakfast when living in the DC area before high school. His father, then working in the DC public school system, told his sister and him that he was going to testify before Congress that day. He was going to tell them they needed to fund girls' physical education. "They need strong bodies too."

Rep. Patsy Mink of Hawaii introduced Title IX in the House, and it became law on June 23, 1972. Lee Hamilton remembered when Birch was promoting Title IX and was resolved to do what he could to help it pass in the House. Because of her substantial abilities, Lee was glad to see Patsy Mink shepherding it there.[11] Rep. Edith Green of Oregon had tried to do the same thing in previous years but had been unsuccessful. When President Nixon signed the Higher Education Act, he spoke mostly about another focus of the bill, desegregated busing, and never mentioned Title IX, which was enacted with his signature. In the early years after Title IX's birth, few people had even heard of it. Most people today are aware of it only as it applies to women's sports. It may have created organized women's sports, but the law addresses ten key areas. These are "access to higher education, career education, education for pregnant and parenting students, employment, learning environment, math and science, sexual harassment, standardized testing and technology."[12]

Birch met in his office with the presidents of Indiana, Notre Dame, and Purdue universities. They argued that Title IX would ruin men's sports if they were forced to treat women equally in that aspect of their academic lives. These academic leaders personified higher education in Indiana and Birch respected them greatly, but he told them that he simply didn't agree.

In the years since Title IX was passed, there have been efforts to challenge it or amend it, and it has even come under review by the Supreme Court. It has survived to this day as a living, breathing law.

Among Birch Bayh's accomplishments, it is the one for which he has been awarded the most honors over the years. He felt it had a greater impact on more individuals than any other legislation he touched.

Many years later, Birch was invited to the Supreme Court as a friend of the court. Title IX was being challenged, and the court often reviews cases by looking at the framers' intent. The framers are usually long since deceased, so his attendance was unusual in that respect. The challenge was unsuccessful, and the law remained in effect. When Title IX was celebrated at the White House on its fortieth anniversary, tennis great Billie Jean King was present, as was Birch. King loudly proclaimed Birch to be her hero, and she spoke eloquently to those assembled about the law's positive impact on the lives of millions of women. Indeed, its impact on women in all aspects of life has been profound.

While Title IX's original intent may not have been about sports, its impact on sports has energized overall social acceptance of the law. It is now customary for universities throughout the country to set up Title IX compliance offices to ensure that women are protected from discrimination in college life and that the university upholds the standards compelled by the law. A 2003 poll showed that 68 percent of the public supported Title IX. Its impact on female college enrollment has been substantial: 57 percent of college enrollments in 2003 were women. In 1971, 294,015 women participated in athletics at high schools and colleges. That number rose to 2,806,998 by 2001, a tenfold increase in thirty years. By 2011–12, that number was 3,173,549. American women athletes outnumbered men in the 2012 Olympic Games. Mark Emmert, president of the NCAA, said, "You can make a pretty strong argument that Title IX has had as big an impact on the landscape of the United States as anything."

Bernice Resnick Sandler, referred to as "the Godmother of Title IX" for her advocacy efforts that began long before its passage, said, "People have said that Title IX is the most important law passed for women and girls in the history of this country since 1920 when women obtained the right to vote."[13]

Now there is even a women's sportswear company named Title Nine.

Looking back on Title IX, Birch said, "Successes in my life have often been the result of failures." With the failure to pass the ERA, Title IX was born, and it has had an enormous impact on American society, particularly for millions of American women and girls.

In the meantime, the Judiciary Committee agreed to schedule a February 29 vote on the ERA. The amendment passed and was sent to the Senate floor, where debate began on March 17. The committee was also holding hearings to investigate allegations that Attorney General

Designate Richard Kleindienst had dropped an antitrust suit against International Telephone and Telegraph (ITT) in return for a $400,000 contribution to the Republican Party. Kleindienst had been named to replace Attorney General Mitchell, who was slated to run the president's reelection campaign. In the last two weeks of February, Kleindienst was nominated and the nomination was unanimously voted out of committee. But at Kleindienst's request, hearings were reopened because of the charges against him, and it would be months before he was confirmed. Hearings resumed in late April, and the nomination was voted out of committee on April 27, going to the full Senate once again.

The hearings were civilized affairs with little revealed about the nominee that would impede his confirmation. Birch led the questioning, which reflected his concerns about Kleindienst's commitment to protect citizens from unwarranted surveillance. He told reporters that Kleindienst would likely be approved and that he probably would vote for him, despite their many differences. Unlike a Supreme Court nomination, since the attorney general's term is limited, he felt the president should get the person with whom he felt most comfortable. But issues raised by the press prompted committee members to continue asking questions. The matter was still unresolved on June 2 when Birch called for the Senate to recommit the nomination to the committee for further study. But the Senate disagreed, and Kleindienst was confirmed.

Birch had not always felt as anti-Richard Nixon as he became. Clearly Birch had not supported him in 1960 when he ran against JFK, but the experience of Nixon testifying on the Twenty-Fifth Amendment had made him an admirer. He was mightily impressed with the former vice president and appreciated the support Nixon had given to the effort to pass the amendment. But as Nixon's presidency unfolded, Birch grew into an avid opponent. Considering Haynsworth, Carswell, civil rights, Vietnam, and the ways Nixon and Agnew were fostering divisions in the country, Birch became committed to seeing Nixon removed from the White House. That had been the driving force behind the Bayh candidacy for president the previous year and those negative feelings fueled his attitude toward Kleindienst.

At the same time, Nixon had been legitimately elected and was very popular in Indiana, and his right to a cabinet of his choice was an important one to preserve for all presidents. But as Kleindienst's misdeeds within the administration became more apparent, Birch's support wavered. He joined senators Kennedy and Tunney on the Judiciary Committee to oppose the nominee and to file a minority report for the Senate to consider.

The report challenged the Senate to "decide whether or not it is going to be a party to a whitewash. There is much that is wrong in the evidence received so far. There is much evidence not yet received. There is no justification for the Senate's failing to obtain it." It went on, "The absence of documentary and testimonial evidence to which the Senate is entitled precludes complete conclusions." The report concluded that a large contribution from ITT was received at the same time it was seeking to settle an anti-trust suit and that the White House staff was fully aware of the contribution and the suit. At the very least, it represented an appearance of impropriety. Despite that, Kleindienst played a determinative role in events prior to settling the ITT cases and "attempted to withhold from the public and the committee the full facts about his extensive participation."

As the Nixon scandals were unveiled, additional facts about Kleindienst surfaced: he had urged the nomination of both Haynsworth and Carswell; he had authorized the arrest of antiwar demonstrators without charges; he had accused House Majority Leader Hale Boggs of insanity when Boggs accused the administration of tapping his telephones. None of this would have been remembered had it not been for the Watergate scandal. Despite these facts and the views of Birch and his allies, Kleindienst was confirmed on June 8.[14]

Regardless of activities in the Senate, from women's rights to the Vietnam War to the Kleindienst nomination, most people were paying attention to the president, who was dominating the national news. He made an unprecedented trip in February to China to meet with Chairman Mao Zedong. His very public trip mesmerized Americans, who largely ignored the North Vietnamese negotiators' walkout of the Paris peace talks in protest of the ongoing American air raids. Some of the spotlight shifted back to Capitol Hill on March 22 when the Equal Rights Amendment finally passed Congress to begin the ratification process in the fifty states. Birch was thrilled. And as a symbol of the changing role of women in America, in the spring of 1972, the Boston Marathon allowed women to participate for the very first time.

Mathea Falco was the staff person responsible for managing the Equal Rights Amendment. After the Senate vote, she wrote about the various "crippling amendments" some senators tried to attach to the ERA, which Birch deftly turned away. Six were introduced and rejected, with the most humorous being what the staff referred to as the "potty amendment." Senator Ervin spoke often of his concern that separate bathrooms would no longer be required and certain sexual practices might no longer be prohibited. The bottom line, he said, was that women needed protections from men, and the ERA would remove those protections. Birch kept returning to the simple concept in the amendment that equal rights under the law cannot be denied on the basis of sex. The ERA had been proposed in nearly every session of Congress since 1923, but only now, with different leadership, did it pass both houses of Congress. Falco also noted that the debate itself seemed to be changing the country's view toward equal rights, with several legislatures considering legislation to ensure equal rights outside any consideration of a constitutional amendment. "Whatever the ultimate fate of the ERA, its impact has been enormous."[15]

After the ERA passed the Judiciary Committee, Senator Eastland was asked in a press conference how he had allowed such a thing to happen. He replied, "Birch Bayh played by the rules and beat me fair and square."[16] Now that it had passed both houses, the battle was on to achieve ratification by the necessary three-fourths of the states.

For the Republican presidential nomination, Nixon faced no serious opposition, though Rep. Pete McCloskey of California was doing his best to change that. For Democrats, McGovern was gaining momentum. Birch felt that his friendship with Muskie and his concern about the potential weakness of a McGovern nomination required him to make an endorsement, something he had never done before in the nominating process, at home or nationally. In private conversations Muskie hinted that Birch would make a good running mate should he prevail over McGovern. On March 16, Birch formally endorsed the senator from Maine. His timing, however, was questionable. One of McGovern's top campaign staffers and the former press secretary to Robert Kennedy was

Frank Mankiewicz, who referred to the Bayh endorsement as "the first time I've ever heard of a rat jumping on a sinking ship."[17]

Before the New Hampshire primary in February, Muskie faced what he later referred to as a "watershed incident." Reacting to the *Manchester Union Leader*'s editorial attacks on him and his wife, he got emotional in her defense and cried in front of the assembled reporters. The publisher, William Loeb, a far-right, take-no-prisoners type of newspaperman, had written the editorials and was the target of Muskie's wrath. Muskie won in New Hampshire, but his campaign began to unravel after the incident as McGovern surged ahead in the polls. A month later, Muskie withdrew.[18] Later, it was revealed that a letter Loeb published as evidence supporting his attacks on Muskie was a hoax perpetrated by the Nixon White House. Nixon strategists felt Muskie was the only credible contender in 1972 against the president and took great joy in helping to bring him down.

North Vietnamese soldiers crossed Vietnam's demilitarized zone as part of their Easter offensive; two weeks later, the Hue offensive by the North resulted in the resumption of American bombing of Hanoi and Haiphong; a bit later the harbor at Haiphong was mined. It seemed that the winding-down of the war was stalled. The presidential race was rocked on May 15 when Arthur Bremer shot and wounded Governor George Wallace at a shopping center in Laurel, Maryland. An independent candidate for president in 1968, Wallace had competed in the Democratic primaries in early 1972. The gunshots paralyzed Wallace for life. Bremer would be convicted and sentenced to prison for sixty-three years for the attempted assassination.

Birch's work in late spring and into the summer included measures to fight barbiturate abuse, control guns, reform foreign service grievance procedures, ratify the ERA, increase highway funding, and oppose the Kleindienst nomination. There was even legislation for a constitutional amendment establishing a single nationwide president primary, with hearings scheduled for June. He addressed the Indiana Disabled American Veterans Convention and introduced legislation with Sen. Frank Church to increase aid to Israel. And Marvella had started traveling again, working with Indiana congressman Bill Bray's wife to help out at Girls State.

In May, Nixon followed up his China trip with one to Moscow, signing an anti-ballistic missile (ABM) treaty and an arms limitation treaty, SALT I, with Soviet leader Leonid Brezhnev. Nixon's foreign policy credentials were burnished to a high sheen after these two visits, and his standing in the polls rose exponentially.

Also in May was the Indiana primary. The election would choose among candidates for president but also determine who the delegates would be at the state Democratic convention in the latter half of June. The governor's race confronted Birch with a difficult decision. Competing for the nomination were former governor Matt Welsh and longtime Bayh ally and friend, Secretary of State Larry Conrad. Birch retreated to his long-standing posture of neutrality in primaries, though the Muskie endorsement weakened that stance. Welsh had been very good to Birch throughout his career. They liked each other immensely, and it was Welsh's endorsement that had guaranteed the Bayh nomination for US Senate in 1962. But Conrad had managed that campaign, and they had a personal bond unmatched by those of other political allies. Birch urged Larry to run for lieutenant governor, putting him in line for governor four years later whether the 1972 election was successful or not. Conrad would have none of it, and Birch decided to stay out of it and avoid making a choice.[19] Welsh defeated Conrad at the convention but would lose in the fall. Conrad remained secretary of state and likely felt fortunate at the time that he had not been the nominee. He would have to turn his attention to a reelection campaign, however, if he were to be the gubernatorial nominee in four years.

On June 17 Birch was elected to serve on the Senate Appropriations Committee, something almost every senator desires. His role there could raise his profile on any issue, since all legislation needed funding, and it provided a path for bringing more federal money to Indiana. Among his highest priorities was to fund projects in Indiana that he had begun working on in the state legislature many years before. While serving there, he was a member of the Wabash Valley Association, a group of legislators that promoted reservoir projects to prevent flooding of the Wabash River upstream. He sponsored legislation to authorize the reservoir, to prevent flooding, to protect farmers' cropland, and to provide recreation, leading to the construction of new houses and thereby creating jobs. Birch always considered it a wise way to expend federal money, and he was proud of the legislation. When asked to support the Monroe Reservoir, Birch joined the two reservoirs together as a package, and they were authorized. But an authorization by a state legislature does not get reservoirs built; they require federal authorization and funding. Birch was able to secure the former as a member of Public Works in the Senate, now he was in a position to obtain funding.

As important as this was to him, an event of far greater importance occurred the same day: a break-in at the Democratic National Committee (DNC) headquarters in the Watergate office and apartment complex in Washington, DC. Five men were arrested. At that moment, no one foresaw the break-in's far-ranging effects on American politics and history. A third event also took place that day that would impact the actions of the Nixon administration: the formation of a new government in Chile by its president Salvador Allende, an avowed socialist.

While the country enjoyed its summer and the Democrats prepared for their convention, we now know that the White House spent that time heavily engaged in a cover-up of the various misdeeds that have since become known as the Watergate scandal. Because of the White House taping system that would come to light the following year, we learned that on June 23 there was a discussion at the White House involving the president and his chief of staff, H. R. Haldeman, concerning how the CIA might be used to obstruct the FBI's Watergate investigation. This was one of a number of conversations that would prove problematic for the president. Nonetheless, he seemed headed for a tremendous electoral victory in the fall, which made the Watergate break-in all the more ill-conceived. But without the taping system, we never would have learned the extent of the crimes.

The Democrats met in convention in Miami Beach from July 10 to 14 and nominated George McGovern for president. This was a result of his victories in the primaries following the collapse of the Muskie campaign. Birch attended the convention. McGovern's choice for vice president was Thomas Eagleton, a senator from Missouri.

Almost immediately, reports surfaced that Eagleton had undergone electric shock treatments years earlier to deal with depression and emotional issues. In this day and age, that hardly seems a sin, but it was big news in 1972, and the press wrote aggressively about this "shocking" development, with all sorts of hints being laid that mental illness was really the issue. McGovern's rocky campaign, which already seemed doomed to defeat, spiraled even farther away from any possibility of success. On August 1, despite McGovern's pledge that he was "one thousand percent" behind him, Eagleton withdrew his candidacy for vice president.

Lost in the news emanating from the convention was the vote in the Senate Judiciary Committee in favor of Birch's Runaway Youth Act, sent to the Senate floor on July 26.

In the second half of August, the Republicans held their convention in Miami Beach and renominated the team of Nixon and Agnew. It was a love fest, with Nixon celebrating the fact that voters under age twenty-one could now participate in electoral politics and would be voting for president for the first time. Watching the convention proceedings, it was hard to imagine what would happen to the Nixon presidency in the next two years.

Shortly after the Republican convention, the Summer Olympics began in Munich, scheduled to conclude on September 11. The world once again reeled from an incident of international terrorism, this time at those Olympics. A Palestinian organization known as Black September raided the area in Munich where the Israeli athletes were housed and killed eleven of them. Other Israelis were taken hostage; during a failed rescue attempt, the Israelis died, as did a policeman and five of the Black September guerillas.

Because McGovern's campaign started from behind and shot itself in the foot with the Eagleton fiasco, it lost precious time getting Americans to pay attention to his message while the world was riveted to the television, first by the Olympics, and then to watch the unfolding of the Munich massacre. Just a month before the election, McGovern named, and the Democratic National Committee approved, R. Sargent Shriver to replace Eagleton as the Democratic vice presidential nominee. The brother-in-law of John F. Kennedy, Shriver had been the first director of the Peace Corps and was most recently the US ambassador to France. But Shriver's appointment was overshadowed by what appeared to be a major breakthrough at the Paris peace talks. Henry Kissinger had been negotiating with North Vietnam's Le Duc Tho. After Kissinger went to South Vietnam from Paris, on October 26 he stated, "Peace is at hand."

The presidential election took place November 7 with the lowest turnout since 1948, and Richard Nixon was overwhelmingly reelected, 60.7 percent to 37.5 percent, with McGovern winning electoral votes only in Massachusetts and Washington, DC. In Indiana, Nixon's margin was literally two to one, a distinct warning to Birch about his own political health, to be tested in a reelection campaign two years hence. However, Nixon's coattails were nonexistent, and Democrats picked up two seats in the Senate. The Republicans did add twelve seats to their minority in the House of Representatives. One of those congressional losses was a painful one for Birch and his staff: Andy Jacobs lost to Bill Hudnut in Indianapolis.

Decades after McGovern's crushing defeat, his campaign manager Gary Hart was asked if he ever, for even a moment, thought they were going to win that race. He began his response by saying that every day you have to go to the headquarters being optimistic, and then he hesitated before answering, "No."[20]

Nineteen senators were reelected: seven Democrats and twelve Republicans. The Senate would now be composed of fifty-six Democrats, forty-two Republicans, a Conservative, and an Independent. Several of the new senators would play important roles in the remaining years of Birch's career.

Looking back, one wonders what could have been had Marvella not been touched by cancer and had lightning struck and Birch won the nomination. Defeating Nixon in the environment of 1972 would have been difficult, to say the least. The trips to China and the Soviet Union and the perceived winding down of the Vietnam War were large issues that would have placed any Democrat in an uphill battle. It's easy to argue that Birch might have been more successful than McGovern, though even that can never be said with certainty. But it was clear to him that Marvella's health had to come first, and he was perfectly satisfied with being the best senator he could, as she suggested.

Birch remained busy that fall, with lots of time devoted to hearings on barbiturate abuse but also hearings on highway funding, disaster relief, increasing support for the Indiana Dunes National Lakeshore, farm subsidies, military procurement, support for Israel, hog cholera, and family farm inheritance taxes. He continued his efforts to ban the sale of "Saturday night specials." He also denounced the treatment of Soviet Jews, continued to raise questions about Attorney General Richard Kleindienst and his role in the ITT contribution, and even got involved in an issue with Ted Kennedy in his role as chair of the Health Subcommittee. Birch requested that Kennedy investigate reports of an American public health experiment from the 1930s examining the effects of syphilis on black men. It was revealed that there had been no informed consent by the men being studied, and those who had syphilis were not treated for it and had subjected themselves to the study with the promise of free medical care. The Tuskegee Institute study had gone on for forty years, and the revelations eventually led to federal legislation and new standards in medical experimentation.

It was an eventful year for America and the world. The final months of 1972 were no different. In October, a small airplane vanished in Alaska with House Majority Leader Hale Boggs of Louisiana and three other men aboard. It has never been found. The first female FBI agents were hired that month, seemingly in homage to Title IX. Hijackers of a Lufthansa flight threatened to blow it up if West Germany refused to release three of the terrorists that took part in the Munich massacre. When the West Germans accepted the demands, they were roundly condemned by Israel. In December was the last manned mission to the Moon, Apollo 17, piloted by Harrison Schmitt, Ronald Evans, and Eugene Cernan. Their lunar landing would make Cernan the last man to walk on its surface to date.

On the Vietnam front, while the American army was turning over military bases to South Vietnam, White House Press Secretary Ron Ziegler announced that reporting on troop withdrawals would no longer be part of his briefings, since the number of troops had been reduced to 27,000. Nonetheless, on Christmas Day, the bombing of North Vietnam was resumed.

A plane crash in December was big news and became part of the Watergate scandal. In Chicago, the crash of United Airlines Flight 553 killed forty-three passengers plus two people on the ground. Among the dead passengers was the wife of Watergate conspirator Howard Hunt; in her purse was found over $10,000 in cash.

On the international front, a visit to Hanoi was in the news, during which singer Joan Baez and others delivered Christmas mail to American prisoners of war who were also endangered by the Christmas bombing. The United States broke off diplomatic contact with Sweden after its prime minister, Olof Palme, compared America to the Nazis because of the bombings of North Vietnam. Also that month, on December 18, Neilla Hunter, the wife of newly elected senator Joseph Biden of Delaware, was killed in an automobile accident. The accident also claimed the life of Biden's one-year-old daughter Naomi, leaving his two sons Beau and Hunter in the hospital fighting for their lives. Biden was not yet thirty years old when elected but would come of age before being sworn in. He was understandably devastated and virtually lived at the hospital until his sons' conditions improved and they could be released. Birch called Biden right after the accident and remembered going to see him to discuss, face to face, Biden's consideration of resigning from the Senate. He remembered it as a serious "Dutch uncle" talk that likely figured in Biden's ultimate decision to remain in his recently won Senate seat. The conversation was the beginning of a warm friendship between Bayh and Biden. Biden would remain in the Senate for the next thirty-six years, leaving after being elected Barack Obama's vice president. Tragically his eldest son, Beau, would die of brain cancer at the age of forty-six.

On December 26 in Kansas City, eighty-eight-year-old former president Harry S. Truman passed away. And on New Year's Eve, baseball great Roberto Clemente was killed in an airplane crash while trying to deliver aid to the victims of an earthquake in Nicaragua.[21]

11

Watergate: 1973

Keep your eyes on the stars, but remember to keep
your feet on the ground.

THEODORE ROOSEVELT[1]

RICHARD NIXON BEGAN HIS SECOND TERM AS PRESIDENT IN 1973, but the year is remembered as the year of Watergate. Before year's end, Watergate had become a household word and Nixon's presidency was hanging in the balance.

Birch Bayh had risen to number forty-eight in Senate seniority and was entering the last two years of his second term. Planning for Birch's 1974 reelection would soon begin. In his role as senator, there was a lot of unfinished business. When the Senate convened on January 4, a day when all new and reelected members of both houses spent the day celebrating with open houses and office parties, Birch introduced two proposals for amending the Constitution. The first was for doing away with the Electoral College, changing the presidential system to one of direct popular vote. The second proposed amendment was to lower the age for serving in the House and Senate. A week later, the trial of the Watergate burglars began, only a stone's throw from the Capitol.

While this book is the story of Birch Bayh's political career and not a diary of the Watergate scandal, it's important to realize how much Watergate dominated life in Washington and the nation in 1973.

Certainly, other political and legislative stories took place during the year, but a proper perspective cannot ignore the overwhelming role the scandal played. The trial of the burglars began on January 11 and exploded in late March 1973. To set the scene, the Vietnam War continued to be controversial, with the Christmas bombing indelibly imprinting itself in the minds of Americans. The peace process, however, was creating optimism, and the president announced the suspension of offensive actions by the United States in North Vietnam on January 14. It appeared that Richard Nixon might be remembered as the president who actually concluded the Vietnam War.

On January 22, Birch celebrated his forty-fifth birthday. It was a bittersweet day, as it was also the day of former president Lyndon Johnson's death at age sixty-four. One of the last meetings Birch had with the former president was during the Carswell debate. Johnson had arranged for Birch to meet him at DC's National Airport. They shared a limousine, which was taking LBJ to the Madison Hotel in downtown Washington, where he would stay. They discussed Birch's plans for a presidential race. Birch took me with him to meet the former president, and I remember that when I reached out my hand to shake his, he grabbed my thumb instead. I asked about that later, and Birch replied that LBJ had grabbed his thumb as well, in order to keep either of us from grabbing his hand. Johnson had contracted a form of malaria in World War II that periodically made his hands dry out, crack, and bleed. He was protecting himself in a way totally unexpected by me.

On a far different note, on Birch's birthday and the day of Johnson's death, the Supreme Court decided a case that would reverberate through the American political system for years to come and have an enormous impact on Birch's career. It would also have an enormous impact on the lives of many women in our country. The *Roe v. Wade* decision gave women the right to choose to have abortions during the first three months of pregnancy. On each subsequent anniversary of *Roe*, anti-abortion activists delivered roses to all Senate offices as a way of telling the senators to honor life by prohibiting abortion. It made for a strange birthday present for Birch each year. His own thinking on the issue had evolved, and he agreed with the decision but feared that the issue's heightened profile could be politically dangerous to those who shared his view.

The following day, President Nixon announced that an accord had been reached in Paris between the United States and North Vietnam, signed by Henry Kissinger and bringing, in Nixon's words, "peace with honor in Vietnam and Southeast Asia." North Vietnam released the first American prisoners of war on February 11.[2]

Also that January, Sen. John Stennis of Mississippi was mugged and shot outside of his Washington, DC home. Birch went to visit Stennis in the hospital and later mused that while they never voted the same way he liked the man and valued their friendship. Stennis would serve in the Senate for more than 41 years and was one of the southern barons who dominated the Senate committee system in those times. Most of those Southern committee chairmen were staunch segregationists, some virulently racist, others more quietly so. Stennis was among the latter. When Birch chaired the Appropriations subcommittee on the District of Columbia, Stennis once approached him about a swimming pool in the District that was going to be built with federal funds. The neighbors, Stennis claimed, were concerned about the noise such a venue would generate in their community. The fact that it would be racially mixed was unmentioned. Birch understood Stennis's real concerns, but he also had to respect his role in the Senate. As a result, the pool was constructed but as an indoor facility, mitigating any noise that might come from it.

Besides chairing the DC subcommittee on the Appropriations Committee, Birch was appointed to three other subcommittees and was fully engaged in Senate business. He held a hearing on the use of methadone in Louisville, Kentucky, on February 13 and another in Indianapolis the next day. In mid-March, he introduced the Methaqualone Control Act of 1973, an attempt to stem the use of the popular Quaalude. He offered legislation to prevent the diversion to illegal markets of legally manufactured barbiturates. Officials from two drug companies refused to appear at the hearings on methaqualone but were promptly subpoenaed by Chairman Bayh.

Next, he and Senator Kennedy introduced a constitutional amendment that would provide full congressional representation to Washington, DC. This was an issue that was hard to feel optimistic about. Arguing that the voters of the District should have full representation was not difficult, but because of DC's overwhelming Democratic voting, it was hard to envision Republicans allowing a permanent Democratic seat in the House of Representatives to be created. And as far as the possibility of two DC Democratic senators was concerned, that was even less likely.

Another growing issue concerned the discovery of oil in Alaska, and Birch found himself in a position familiar to many others in his party. The benefits to the country from increased domestic oil production were obvious, but the delivery of that oil to the lower forty-eight states could present severe environmental risks. This ended up being a long-term issue that required an enormous amount of time from Birch and the Congress at large. He and Senator Mondale introduced legislation to create a Canadian pipeline in order to expedite the use of Alaskan oil and natural gas.

There was another Indiana High School Government Leadership Conference, this one headlined by Dan Rather, who was then the White House correspondent for *CBS News* and would later become its anchor. Joining him was William H. Sullivan, the deputy assistant secretary of state for East Asian and Pacific affairs. Later, Sullivan would serve as the final ambassador to Iran, leaving there shortly before the Iranians took fifty-two Americans hostage in late 1979. The other speakers were Will Erwin, assistant secretary of agriculture for rural development, and Barbara Watson, administrator at the State Department's Bureau of Security and Consular Affairs.

During a flight to Indiana, Birch talked about the changes that had taken place in his life. He had become more prominent a figure in American politics than he had ever expected. One gets into politics, he mused, and does everything possible to make friends, to make people like you, to avoid making enemies. But he had come to the realization that during your life and career, if you stayed true to your values and held strong to your personal ideals, you were going to make enemies. He found himself in the big league, where he had carved out a place for himself as a distinctly liberal Democratic senator in the year of Watergate. That "liberal" label felt unusual to him as well.[3]

By mid-March, Watergate had become a national obsession, in striking contrast to the way the public ignored the story throughout

1972. Americans took note of the announcement on January 30 that two of President Nixon's former aides were found guilty of conspiracy, burglary, and wiretapping. While G. Gordon Liddy and James W. McCord Jr. awaited sentencing, five other men plead guilty. On March 20, McCord sent a letter to presiding judge John Sirica, claiming that he and his fellow defendants had been pressured to "plead guilty and remain silent" about the case. He implied that higher-ups in the Justice Department were involved, and speculation centered on former attorney general John Mitchell as the "overall boss" of the operation.[4] After the verdict but before the McCord letter on February 5, Senator Kennedy introduced Senate Resolution 60, a bill to create a Select Committee on Presidential Campaign Activities to investigate the alleged events surrounding the 1972 presidential election. The resolution would grant the committee the power to investigate the Watergate break-in and any "cover-up of criminal activity, as well as all other illegal, improper, or unethical conduct occurring during the Presidential campaign of 1972, including political espionage and campaign finance practices."[5] It was unanimously approved by the Senate two days later. Traditionally, Kennedy would preside over an inquiry resulting from his own resolution, but Majority Leader Mansfield thought that with Kennedy as chair, Republicans and the press would characterize the committee as unduly partisan. As a result, Mansfield asked Sen. Sam Ervin of North Carolina to serve as chair.[6]

Ervin was not viewed as partisan, and while he portrayed himself as a simple country lawyer, he held a Harvard law degree and was considered the Senate's constitutional law expert. Not known by many was the fact that he had actually graduated from Harvard backward. Having met his future wife before going to Harvard, he was concerned that she wouldn't wait three years for him to obtain his degree, so he sought and was granted the right to take the third year first. After completing that year, she told him she was willing to wait another year; he went back to Cambridge and took the second year. Finally, she agreed to wait once again, and he took the first year's classes and then got his degree.[7] As chairman of the Constitutional Rights Subcommittee of the Judiciary Committee, Ervin had developed expertise on issues related to surveillance and wiretapping, matters that would seem pertinent to the

Watergate scandal. Ervin, age seventy-six, was not one of the Senate liberals, nor did he have presidential ambitions. This made him a more acceptable chairman to handle a situation as complex and controversial as the Watergate break-in.

For those living in Washington, it seemed like the morning newspaper was exciting every day. *Washington Post* reporters Bob Woodward and Carl Bernstein were reporting tirelessly and regularly about the unfolding scandal, anticipating many of the events that would happen later. The ongoing drip-by-drip unraveling of the cover-up by the Nixon White House began to have dramatic effect. As Woodward and Bernstein would recount forty years later, the cover-up was necessary because of the extent of the crimes committed by the Nixon people. At this point in 1973, most of those crimes were not yet revealed, but one of the major bombshells came on the last day of April, with President Nixon announcing the departure of four major players in his administration. Chief of Staff H. R. Haldeman, domestic policy adviser John Ehrlichman, and Attorney General Richard Kleindienst had all resigned; White House counsel John Dean had been fired. The Watergate hearings, led by Sam Ervin, began on May 17, and much of the country was obsessed by the almost daily revelations that resulted from them.

Elizabeth Drew wrote *Washington Journal: The Events of 1973–1974*, a gripping account of the Watergate years with a perspective that has held up well. The book is a virtual diary, providing a detailed account that enlightens us to this day.

> There are those who believe that the cover-up has proved far more damaging to the President than the discovery of what was being covered up would have. . . . But the problem for the White House, as we keep learning, was that there was so much to cover up. The break-in at the Watergate combined the elements of covert operations, espionage, secret funds, hidden contributions and aggression against political opponents (real and perceived) which also characterized other activities carried out under the Administration. Perhaps the President was trapped all along. The indictment charges that the seven men and others "known and unknown" conspired "to obstruct justice . . . to make false statements to a government agency . . . to make false declarations. . . . It accuses them of seeking the release of one or more of the men imprisoned for the Watergate break-in; of destroying records; of "covertly" raising cash and paying it to the Watergate defendants; of making and causing to be made "offers of leniency, executive clemency and other benefits" to Hunt, Liddy, McCord, and Magruder; and

> of attempting to obtain financial assistance from the C.I.A. for the Watergate defendants. This indictment includes thirteen counts against the various defendants and lists forty-five "overt acts" to describe the conspiracy.

Later in March, as the scandal continued to dominate the country's news, Drew wrote,

> Richard Cohen writes in the *Washington Post* that as of now charges have been brought against twenty-eight persons formerly associated with the White House or the President's reelection campaign. "In addition," he writes, "ten corporations or their officers, or both, have been accused of making illegal contributions," eight of them to the Nixon campaign. And six people are now in jail for Watergate-related crimes: three of the seven original co-conspirators for the break-in; two who were sentenced for "dirty tricks" in the campaign.

Drew's account continues a week later:

> Someone who went through those White House years has pointed out to me that when the President began to run into trouble—after the Senate defeated the nominations of Clement Haynsworth and G. Harrold Carswell to the Supreme Court; after the war was extended into Cambodia and there were protests on the campuses—he drew closer to him those who spoke to his own suspicious nature: Haldeman and Mitchell. It was in 1970, after the student protests, that the Huston plan was drawn up, and one year later that the plumbers unit was established."[8]

In the midst of the Watergate scandal, few noticed Birch's introduction of legislation to overhaul the federal parole system.

For Nixon haters and many others, the allegations surrounding Watergate seemed to confirm long-held suspicions. Well known were comments former president Truman had made about Nixon: "All the time I've been in politics there's only two people I hate, and he's one. He not only doesn't give a damn about the people; he doesn't know how to tell the truth. I don't think the son of a bitch knows the difference between telling the truth and lying." He added, even more graphically, "Nixon is a shifty-eyed, goddamn liar, and people know it."[9]

In 1952, Nixon gave his Checkers speech, which will go down in history as one of the most maudlin and phony speeches ever given. The *Washington Post's* Herblock cartoons always made him appear a shady character. Nixon had been dubbed "Tricky Dick" early in his career, but the crimes that were becoming known went far beyond dirty tricks.

Responsible members of Congress had to take the charges seriously. Were federal agencies used to punish Nixon's enemies? Had the White House actually authorized break-ins and fire bombings? Were the CIA and FBI being used as part of a massive conspiracy to cover up those crimes? These were serious questions, and any reading of the events of 1973 must concentrate on Watergate.

On May 18, Attorney General Designate Elliot Richardson selected a special prosecutor for Watergate: former solicitor general Archibald Cox. Richardson was filling the vacancy created by the resignation of Attorney General Kleindienst. Nixon's people had reason to be concerned about Cox. He was competent, he would be spending all his time and resources investigating the president, and he was known to be a Kennedy ally. On June 3, the *Washington Post* reported that former White House counsel John Dean had told Cox's investigators that he and the president had discussed the Watergate cover-up on at least thirty-five occasions. Ten days later, the investigators found an internal White House memo that described plans to burglarize the office of Daniel Ellsberg's psychiatrist; Ellsberg was once again a defendant in the Pentagon Papers case, as reported by the *Washington Post*.

Early that summer, Birch pursued his own legislative agenda, in spite of the difficulty in getting the news media to pay attention to these efforts. On June 5, the Senate adopted a Bayh amendment to guarantee farmers adequate fuel for the next two years. Three days later, the Senate passed the Narcotic Addict Treatment Act and the Runaway Youth Act, both written by Birch. The following week, he announced hearings in his subcommittee on drug usage by athletes. This latter issue presents an almost eerie prescience, given the revelations about athletes and drugs in the coming decades.

But the bigger news that summer was the televised Watergate hearings taking place in the Caucus Room down the hall from the Bayh office and chaired by Sam Ervin. Virtually every suite in every Senate office had televisions turned on throughout the day. Staffers often went to the Caucus Room to watch the hearings but were riveted to their televisions when they could not. The Bayh office was no exception. Americans around the country were known to have purchased TV sets so they could watch the hearings during business hours.

John Dean, former White House counsel, was preparing to provide testimony before the committee and told Sen. Barry Goldwater about the many lies in the various Nixon statements made in his defense. Goldwater replied, "Hell, I'm not surprised. That goddam Nixon has been lying all of his life."[10]

On June 25, Dean testified before the Watergate Committee, and it was memorable. His encyclopedic memory allowed him to recount detailed conversations with the president and other White House officials, citing times and dates in fascinating detail. One of the most memorable parts of his testimony was the recounting of a conversation with the president and telling him there was "a cancer on the presidency." After Dean's testimony, on July 13, former presidential appointments secretary Alexander Butterfield revealed a bombshell: there was a taping system in the Oval Office. If Nixon were part of the cover-up, the tapes would reveal it. The nation's attention shifted to the efforts to obtain those tapes. Reportedly Nixon had the taping system disconnected a few days later.

The president was trying mightily to shift the country's attention away from the drama taking place on Capitol Hill by beginning talks on June 16 with Soviet leader Leonid Brezhnev. The following week, Brezhnev went on live television to address the American public, the first time a Soviet leader had ever done so.

The Watergate committee and the special prosecutor demanded that the president surrender the presidential tape recordings, and on July 23, Nixon issued his refusal. The drama continued on live television, with millions of people obsessed by the almost daily revelations. The hearings became a national event, broadcast live each day on commercial television. Initially, all three television networks broadcast the hearings, later changing to coverage on a rotating basis. Public television replayed them in the evening. More than 300 hours of Watergate testimony were broadcast overall, and it was estimated that 85 percent of American households watched at least some portion of the testimony. It was also a constant event on many National Public Radio stations, with gavel-to-gavel coverage, giving NPR a prominence it had not previously enjoyed.[11]

The hearings turned Senator Ervin and his co-chair, Howard Baker, into media stars. Senator Baker was a well-liked Republican from

Tennessee and a close Bayh friend. Ervin was celebrated for being folksy as well as serious. Baker projected a nonpartisan image and first asked a question asked repeatedly throughout the hearings: "What did the President know, and when did he know it?"[12] Another personality introduced to the public was that of minority counsel Fred Thompson, who would become much better known in later years as an actor, senator, and candidate for president.

John Dean's testimony and Alexander Butterfield's revelation about the White House taping system were among the unforgettable moments from the Watergate hearings. One of the revelations emanating from the hearings concerned the ways the Nixon administration was going after its "enemies." In fact, Dean's testimony ended up revealing an actual "enemies list" that had been compiled by White House counsel Charles Colson in 1971 and presented to Dean. The plan was to "screw" political enemies by auditing taxes and manipulating federal contracts, grants, and legal matters. Dean wrote a memo saying, "This memorandum addresses the matter of how we can maximize the fact of our incumbency in dealing with persons known to be active in their opposition to our Administration; stated a bit more bluntly—how we can use the available federal machinery to screw our political enemies."[13]

To the delight of the Bayh staff, Birch was one of the enemies. The *Washington Post* front page listed the first twenty names, almost completely without any members of Congress. Next came a "Master List of Political Opponents" with ten senators listed, and the first name was Birch Bayh.

Of course, there were other events happening in the country, even if they did take a back seat to the national drama unfolding daily. On July 26, the Subcommittee on Constitutional Amendments announced that it would hold hearings on seven proposed amendments dealing with school prayer, an issue that just would not go away. During the August recess, the Bayh office released a summary of his legislative activities in 1973, including congressional approval of his initiatives in the area of agriculture, crime control, drug abuse, education, the environment, and veterans' affairs. There were also highlights of a number of efforts with executive branch officials to solve problems in Indiana, from arranging for the emergency shipment of one million gallons of propane gas to farmers to relieve a shortage, to making sure jobs were restored to the Crane Naval Depot. A variety of successes were a result of "inside baseball," shorthand for knowing how to maneuver the bureaucracy to make things happen. He made successful efforts to fight presidential impoundment of appropriated funds, to increase the benefits for the widows of firemen and police who lost their lives in the line of duty, to protect the recreational use of black powder by muzzle-loaders in the state, and to provide increased funding to the Library of Congress to improve its programs for the blind and handicapped.

Birch supported the Case–Church Amendment prohibiting military operations in Laos, Cambodia, and Vietnam without congressional approval; the amendment passed in June. By the middle of August, the American bombing of Cambodia ended. On September 11, there was a military coup in Chile. Its elected president, socialist Salvador Allende, died in what was termed a suicide. For the next decade and a half, Chile would be ruled by an American-backed military junta headed by General Augusto Pinochet.

On a less serious note but with deep cultural significance, on September 20 Americans experienced what became known as "The Battle of the Sexes." Former tennis pro Bobby Riggs challenged female tennis great Billie Jean King to a match, Riggs declaring that the best female player couldn't beat a professional male player. The match was televised live from the Houston Astrodome, and King defeated Riggs in three straight sets. More than 30,000 people watched the match inside the Astrodome, and as many as 90 million people watched on television in dozens of countries around the world, a welcome respite from Watergate that underscored the changing role of women in American life.

On September 22, Henry Kissinger, Nixon's national security adviser, became the new secretary of state. On October 6, the fourth and largest of the Arab–Israeli conflicts, the Yom Kippur War, broke out; it lasted three weeks. The Middle East remained in the news when the Organization of Petroleum Exporting Countries (OPEC) cartel began an oil embargo that threatened the availability of oil to all Americans. But as challenging as these foreign problems might have been to President Nixon, his domestic problems overshadowed them all.

On October 10, Vice President Spiro Agnew resigned. He had been under federal investigation for taking bribes in office, going back to his days as county executive of Baltimore County and as Maryland's governor. Charged with income tax evasion, he entered a plea of no contest and was fined $10,000 and given three years' probation. More importantly, he resigned as vice president.

Amid the continued efforts of the Watergate Committee and the resignation of his vice president, Nixon could no longer tolerate the investigation by special prosecutor Archibald Cox. On October 20, Nixon ordered Attorney General Richardson to fire Cox. Unwilling to do so, Richardson resigned his office. Also refusing to carry out the president's order was his deputy attorney general, former EPA administrator and Senate candidate defeated by Birch in 1968, William Ruckelshaus. He was abruptly discharged. Next in line at the Justice Department was Solicitor General Robert Bork, who fired Cox as ordered, with the office of the special prosecutor shut down and all files sealed. The event, known as the "Saturday Night Massacre," raised immediate calls for Nixon's impeachment for obstruction of justice.

As these events unfolded, Birch was on a trip abroad with Sen. Tom Eagleton and others. When it became clear he needed to get home to play his part in dealing with the growing controversy, he wondered if he would be met at the airport by the military and secreted off to some location to be kept away from the matter. It was the only time he ever found himself seriously thinking that a coup was possible.

That night, I got a call asking me to go to George Washington University library and check out every book I could find on the impeachment trial of President Andrew Johnson, which had occurred 100 years earlier. I checked out the books and brought them all to the Senate office. Birch was already there with several staffers plus a member of Cox's staff, who talked about the FBI arriving at the office, how they had to leave and could take nothing with them. It was an incredibly sober moment that Sunday afternoon and was likely replicated by other Democratic senators and their staffs.

The outcry and condemnations didn't come from Democratic senators only. Network anchors, editorial writers and political leaders of both parties and from all corners of the country objected. Nixon eventually retreated a bit, reestablishing the office of special prosecutor and allowing Robert Bork, now acting attorney general, to appoint a replacement for Cox. On November 1, Bork chose Leon Jaworski, a prominent Texas lawyer, as the new Watergate special prosecutor. Jaworski had been a friend of LBJ's but was also a Nixon supporter, and his appointment allowed the intensity of the crisis to ease a bit.

In reporting on the developments of the Saturday Night Massacre, Elizabeth Drew noted that Nixon had been trying to prevent Cox from demanding certain papers based on claims of national security. A compromise was being sought that might satisfy the special prosecutor and Attorney General Richardson. Drew wrote, "Senators Kennedy, Birch Bayh, Philip Hart, and John Tunney, all Democrats, have issued a joint statement saying that the compromise may be acceptable to the Ervin committee but that it may 'cripple the role' of the Special Prosecutor, and that the President's instructions to Cox are an 'unjustified challenge' to his authority. The ostensible purpose of the statement is to bring pressure on the President to proceed with caution in the matter of firing Cox—to warn him that such an action would bring trouble."[14]

After the Saturday Night Massacre, Sen. William Saxbe of Ohio was nominated to replace Richardson as attorney general, and Leon Jaworski was named to replace Cox as special prosecutor. Elizabeth Drew talked with Birch: "Senator Birch Bayh . . . says that Saxbe and Jaworski are 'O.K'—probably as good nominees as there could be—but that Richardson and Ruckelshaus and Cox were good too, and 'they were canned.' Therefore, he will continue to press for legislation establishing a court-appointed Special Prosecutor."[15]

A week later, Congress acted in a manner that clearly underscored the diminution of Nixon's power over the affairs of his office. Nixon had vetoed the War Powers Resolution passed by Congress, a measure to limit a president's ability to wage war without congressional approval. Congress overrode the veto.[16] The president would seek other opportunities to display his leadership and importance in public affairs, though increasingly it was looking like he would not survive Watergate. On November 17, hoping to shore up support of his base on a campaign-like swing in Orlando, Florida, Nixon was addressing an assemblage of

Associated Press editors and describing the allegations against him when he declared, "I am not a crook." No president has ever felt the need to assert innocence by making a similar statement.

The new special prosecutor applied pressure to obtain copies of the White House tapes for specific dates, many of which were revealed by the televised testimony of John Dean before the Watergate Committee. But the tension in the crisis escalated once again when, on November 21, the president's lawyer, J. Fred Buzhardt, revealed to Judge Sirica that a Nixon–Haldeman conversation on June 20, 1972, had been erased; there was an eighteen-and-a-half-minute gap in the tape recordings that couldn't be explained.[17] White House chief of staff Alexander Haig suggested that "some sinister force" caused the erasure.

In the meantime, due to the passage of the Twenty-Fifth Amendment, a process existed for the first time to fill a vice presidential vacancy. The Watergate scandal underscored how important filling that vacancy was. Without a vice president, the next in line would be Speaker of the House Carl Albert. Albert, an Oklahoma congressman, was rumored to have a serious drinking problem—but more importantly, he was a Democrat. In other words, those seeking Nixon's resignation or impeachment needed Republican support, and that support would be a lot harder to obtain if their actions also meant a change of party in the White House. With the Twenty-Fifth Amendment, at this time only six years old, the president was required to name a new vice president who would have to be confirmed by both houses of Congress. Nixon chose House minority leader and Michigan congressman Gerald R. Ford. The Senate confirmed Ford on November 27, and the House did so on December 6, both by overwhelming majorities. Ford was sworn in later that day.

Watergate-style dirty tricks were also taking place in Indiana near the end of 1973, though there was no suggestion that Nixon's people were involved. A few days before Thanksgiving, a story burst onto the front pages of Indianapolis newspapers to expose the existence of a "master plan" allegedly written by Larry Conrad and his team, a document that laid out a plan to secure the governorship for Conrad. What was damaging about it was the incredibly negative manner in which virtually every prominent Democrat was described. One of the few left unscathed by these characterizations was Democratic state chair Gordon St. Angelo, widely believed to be the instigator of the "master plan." He denied writing the document but admitted to its distribution, and it quickly became clear to Conrad and his staff that it was written by no one close to Conrad and was designed only to hurt his career. In that, it was successful. Others close to St. Angelo eventually came forward to admit to participating in the creation of the "master plan," but the facts didn't seem to matter. Its existence became an ongoing burden to Conrad and remained an irritant throughout his career.

Events surrounding the "master plan" turned sinister. Death threats were received by Conrad and his deputy, Stuart Grauel, husband of Bayh staffer Diane Meyer at the time. They suspected that their phones might be bugged, and they hired an attorney from outside Indiana to investigate. Bud Fensterwald later represented James McCord of Watergate fame and Martin Luther King assassin James Earl Ray, but he was also deeply involved in investigations of possible conspiracies connected to JFK's murder. He seemed to navigate in the netherworld of the surreptitious quite comfortably. Fensterwald uncovered listening devices in both homes, according to Meyer, and was attacked in his Indianapolis hotel room by unknown assailants. Eventually, lawyers Bob Wagner and Ed Lewis approached St. Angelo with their own research into his private life, and the newspaper articles stopped.[18]

Birch perceived the political threat posed by the "master plan" right away. Immediately after it hit the front pages, he flew to Chicago with staff member David Bochnowski to meet with Conrad and his staff. Nothing was resolved that night, because so little was known about it. After they met, Birch and Bochnowski went across town to the AFL-CIO convention. When the Indiana AFL president, Willis Zagrovich, rose to speak, he droned on endlessly about the "master plan" and how it was going to wreck the Indiana Democratic Party. Very few in the audience had a clue what he was talking about. Zagrovich was a close ally of St. Angelo's and probably knew more than most about the document at that time. Years earlier, Birch had visited Ringling Bros. and Barnum & Bailey Circus at its winter quarters in Peru, Indiana; now, rising to speak after Zagrovich, he remembered the advice he was given by a ringmaster there. He began his remarks by saying that he had learned a long time ago never to follow an animal act. He never mentioned the "master plan."

When they departed, Birch asked Bochnowski if he wanted to have some fun. Despite their early flight the next morning and the late hour already upon them, David was excited to learn what Birch had in mind. They got in a taxi, and Birch asked the driver if he could find a place where they might get some ice cream.

12

Bayh versus Lugar: 1974

Government is more than the sum of all the interests; it is the paramount interest, the public interest. It must be the efficient, effective agent of a responsible citizenry, not the shelter of the incompetent and the corrupt.

ADLAI E. STEVENSON[1]

AS 1974 BEGAN, RICHARD NIXON WAS ONCE AGAIN IN THE forefront of American life. He had dominated much of US political culture since 1948, when he led the investigation of State Department official Alger Hiss as a possible spy. In 1952 he was chosen as the vice presidential candidate by Eisenhower and was memorable for the Checkers speech in which he defended himself against allegations of financial impropriety. He temporarily served as acting president when Eisenhower was hospitalized, and he was also famous for his kitchen debate with Soviet leader Nikita Khrushchev in Moscow in 1959. The Republican nominee for president in 1960, he lost to John Kennedy by a razor-thin margin, losing again as a candidate for governor of California two years later. His concession speech after that election was notable for the announcement that he was through with politics. As a candidate for the presidency in 1968, he once again occupied the headlines, winning the office against Hubert Humphrey, again by a very narrow margin. Now he was entering his sixth year as president and was dominating the news in ways he would have preferred to avoid. Love him or hate him, you couldn't deny that he was a giant presence in our lives and had become an American obsession, especially as 1974 began.

For Birch, it was a campaign year. There were still aftereffects from his presidential effort, but the Watergate scandal seemed to portend good things for Democrats running that year. He continued his aggressive legislative schedule while ramping up his Indiana travels; full-time campaigning wouldn't begin until June. But his priorities clearly had shifted, and his life was becoming increasingly Hoosier-centric.

In February, Birch received a letter from HEW secretary Caspar Weinberger saying that Pap smears to detect uterine cancer were not covered by Medicare because they were "preventative." Birch had lost his mother to uterine cancer at age twelve, and he knew that the disease might have been detected had the test been available; he thought Weinberger's policy was outrageous. As a result, he introduced legislation to ensure that Medicare covered Pap smears. Because the OPEC embargo still affected the country, causing long gas lines as well as job loss in occupations related to energy, Birch introduced a measure to extend unemployment benefits to workers affected by the crisis. It passed on February 8. The oil embargo ended March 18, after five months that impacted both the mood and the economy of the country. Long lines at the gas pump undercut a sense of security and well-being in light of the fact that a group of foreign potentates could take an action short of war that would have such a pervasive effect on our society.

The Bayh office announced the participants for the ninth High School Government Leadership Conference. This time the speakers made news because the headliner was the "most trusted man in America," CBS anchor Walter Cronkite. Also appearing were Dixie Lee Ray, chair of the Atomic Energy Commission; FCC commissioner Ben Hooks; and Rodger Davies, deputy assistant secretary of state for Near Eastern and South Asian affairs.

Also in February, the House Judiciary Committee began its investigation into whether there were grounds for the impeachment of President Nixon. The drama was building. There had not been a serious impeachment process in more than a century. And this one was serious.

Abortion was a major issue before the country after the *Roe v. Wade* decision. Both sides, those for and those against access to abortion, were becoming increasingly organized, with the women's movement working hard to ensure the right of a woman to choose whether or not to have an abortion. The pro-life movement was led by religious leaders who argued that the government needed to protect life from the moment of conception. The issue remained heated, with Congress drawn into the middle. The *Roe* decision gave a woman the right to have an abortion during a pregnancy's first trimester. Only a constitutional amendment could overrule a Supreme Court decision, putting the matter firmly in Birch's lap. His subcommittee gave the matter a public airing, opening hearings on the various proposed amendments on March 4. These hearings went on while the House Judiciary Committee was holding its hearings on impeachment.

Birch struggled with the abortion issue, finding the practice personally abhorrent but coming down firmly on the side of a woman's right to choose. Abortion and guns were the two most controversial issues he would face, neither of which ever went away.

These were not the only hot-button issues facing Birch. Forced busing of schoolchildren to achieve full racial integration became a heated issue, with those who supported court decisions on busing finding themselves in the crosshairs of millions of average Americans. He decided to face the issue head on, announcing that he would rent the Indiana Convention Center in Indianapolis for a day to speak to the pros and cons of busing and to entertain questions on the matter for as long as necessary from anyone who wanted to show up. One of the FBI agents providing security for the event remarked that he had never seen so many "fruitcakes" from their watch list all in one place at the same time. Two men with guns were escorted out. Starting early in the day, Birch took questions from each person who approached the microphone. When it got ugly, he calmly reminded everyone that emotions were high on all sides, but they needed to remain civil to one another. There were forty or fifty people there when the meeting began. Birch responded to a heckler by offering him the podium; he asked the man if he would come up and run the meeting, taking all questions as he had been doing. The heckling stopped.

A few hours into the session, Julia Carson, an African American woman from Indianapolis, approached the microphone. She spoke about the poverty of her family and how her mother was a household domestic who remained steadfast in her determination to educate her children. But the only school available to blacks in Indianapolis was Crispus

Attucks High, several miles from her home. She and her siblings walked the long distance to and from school in all kinds of weather, seeking the only education available to them. Her courage that day helped calm the crowd. Years later, Carson would be elected to Congress and serve with distinction for ten years, until her death in 2007. Birch continued to take questions for the rest of the day and into the night, finishing around 3:30 a.m. the next day.

The Juvenile Delinquency subcommittee held hearings on drug and barbiturate abuse, focusing on athletics at all levels, from Pop Warner football to junior college, college, and Olympic and pro sports. They also held hearings on methadone and how it had become a replacement drug for heroin, requiring its own regulations to prevent abuse. Birch introduced the Methadone Diversion Control Act of 1973, which became law, setting up the framework for practitioners who administered methadone and for those who regulated it.

Tornadoes struck Indiana once again in early April, killing 300 people and ruining the lives of many Hoosiers; there were 148 tornadoes throughout the United States and Canada, with the worst devastation in the Hanover-Madison area of the state. All but ten homes were wiped out in that area, and thirty-two of the thirty-three buildings at Hanover College were seriously damaged. The town of Monticello was virtually destroyed. The day after the tornadoes, Birch introduced legislation to provide $100 million in tornado disaster relief. It was the first step in a long and successful process to provide resources for rebuilding the devastated areas of the state. It was times like that when Birch recognized how he could make a difference as a senator. His stature in the Senate and membership on Appropriations put him in a position to be effective in making the lives of those most affected by the disaster a little easier.

On April 16, a subpoena was issued for sixty-four White House tapes by special prosecutor Jaworski, picking up where Cox had left off. On April 30, the White House responded by providing typed transcripts of the recordings, over 1,200 pages in all. The House Judiciary Committee insisted on the tapes themselves.

Birch continued his work on drug abuse, eventually seeing the Senate approve a measure placing federal control over methadone. He continued his efforts to secure disaster relief, again traveling to Indiana to witness the level of devastation his constituents had endured. The abortion and busing issues continued to simmer. Indianapolis was under a federal court order requiring busing, and the issue was red-hot. On May 22, the Disaster Relief Act of 1974 became law.

John Rector was the thirty-year-old chief counsel of the Subcommittee to Investigate Juvenile Delinquency, replacing Mathea Falco. Interviewed by *Roll Call* newspaper early in 1974, Rector said, "At least 50% of the serious crime in this country is committed by young people." Many learned their criminal trade after being incarcerated for lesser, juvenile violations. Incarceration actually created more serious crime and brought criticism from reformers and civil libertarians.[2] He was delighted that Chairman Bayh wanted to do something about this.

In May, Birch introduced the Juvenile Justice and Delinquency Prevention Act of 1974. He and Sen. Marlow Cook (R-KY) had introduced the same bill in November 1971 without success, and Birch reintroduced it the following February; it failed then too. Two years later, after holding hearings on juvenile justice and runaway youth, he tried again, on May 30, 1974. Despite the pressures of the campaign, he steered it to Senate passage on July 25. A House–Senate conference committee added Title III, the Runaway Youth Act, creating a federal role in providing alternatives for young people who feel compelled to run away from home. It passed Congress, and President Ford signed it into law on September 7, 1974.

Afterward, the National Council of Crime and Delinquency held a dinner in New York at which Birch was to be honored. He was campaigning in Indiana, so Rector substituted for him. President Ford and former attorney general Richardson were in attendance.[3] On Election Day, the *Indianapolis Star* editorialized against the act, saying it was going to turn criminals loose to terrorize their communities.

Marvella and actor Peter Graves became co-chairs of a public awareness campaign for the American Cancer Society. This effort curtailed her involvement with Birch's 1974 campaign, save for a few speaking engagements in Indiana. A video called *The Marvella Bayh Story* preceded her remarks to audiences around the country, seeking to help people understand and recognize cancer's seven warning signs. She also spoke about the ERA and urged its passage, but her public appearances were mostly about preventing cancer. That fall, she took a job at the NBC-TV affiliate in Washington, WRC-TV, as a reporter talking about events being planned to celebrate America's bicentennial in two years. Later, Birch described this period as one when he "reveled in his wife's success and admitted he had never seen her more fulfilled and happy in her professional life." Working for the American Cancer Society resulted in her receiving hundreds of heartfelt letters from across the country. She loved the one from a woman thanking her for alerting her to cancer's warning signs, adding that Marvella's information compelled her to see her doctor, and a malignant tumor was found. The warning may have saved her life, she wrote, adding, "Mrs. Bayh, you were my shining example."[4]

David Bochnowski remembered her appearance on *The Lawrence Welk Show* during this time, dancing on national TV with the famous bandleader. It was an appearance generated by her work fighting cancer, not because of her spouse. As a result of her working life, the strain on the family was lessened. Since her job made it problematic for her to be active in the campaign, she kept in daily touch with Birch by phone.

For the 1974 campaign, I became the "road show manager," traveling with the senator and managing his life on the road. Often referred to as the "body guy," I would travel constantly with the candidate and be at his beck and call.

The road show began when Birch arrived in Indianapolis on June 6. It lasted the next 153 days with only three days off, covering 90,000 miles without leaving the state except for events in Chicago, Louisville, and Cincinnati, all Indiana media hubs. In all, the road show included twenty-five county fairs in a state with ninety-two counties. Only a few times did Birch return to DC. Most of the miles were in a twin-engine airplane from Brown's Flying School in Terre Haute. Usually, the road show was met by local officials or coordinators.

Birch's opponent would be Indianapolis mayor Richard Lugar. The fact that he was widely advertised as "Nixon's favorite mayor" couldn't be a good thing for him.

The *Harvard Crimson* described Birch's opponent as follows: "Though Lugar is admittedly an underdog, Indiana Republicans believe their chances are a great deal more auspicious this year. . . . A former Rhodes Scholar and valedictorian of his Denison College class, Lugar has amassed an impressive record as two-term mayor.... [A]mong cities its size, citizens of Indianapolis suffered fewest homicides and robberies and second fewest rapes.... Lugar's innovative Uni-Gov program also merged metropolitan Indianapolis with the suburbs, thereby advancing Indianapolis from 26th to 11th largest U.S. city."[5]

Critics of Uni-Gov argued that the principle reason for consolidating the city and suburbs under the mayor's jurisdiction was not so that that the city could deliver services to the suburbs; it was meant to enable the Republican suburbs to vote for mayor to balance out urban Democrats. Many believed that both candidates had presidential aspirations. The election could determine which candidate might be able to pursue the larger ambition.

The *Harvard Crimson* went on to discuss Lugar's timing:

> Many Republicans are also skeptical of the timeliness of the Lugar campaign. Bayh is a youthful, handsome, down-home operator with a good number of accomplishments and a polished campaigning style. In his 12 years, the folksy junior senator has made numerous contacts, garnered seniority and made his activities known in his home state. Lugar, considered by many to be the best senatorial challenger, risks national obscurity if he doesn't run this year. He could also lose the Indianapolis mayoralty race in 1975, a defeat that would end his presidential aspirations. Though he cannot expect to match his opponent's recognition, Lugar can hope voters will reject incumbents this fall as a purging concomitant of Watergate. A win could put him in the vanguard of Republican presidential prospectives.
>
> That hope, however, could be counterbalanced by the white albatross of the Nixon administration. *The Washington Post* tagged Lugar "President Nixon's favorite mayor," an epithet which has stubbornly stuck. Lugar was the only big city mayor to serve as a surrogate speaker for the president in 1972 and his expertise in urban affairs as president of the National League of Cities made him a logical presidential consultant on urban issues.[6]

The article listed Bayh's accomplishments, adding,

> Aspects of his record, however, could work against Bayh this fall. Republicans are zeroing in on his support of forced school busing and his opposition to the Alaska pipeline—two positions which many of the Indiana constituency do not share. In the past, Bayh has also been an advocate of gun control and a major opponent of a proposed new Indianapolis area reservoir.... Lugar, a moderate Republican with views congruent with many in his conservative state, sees Bayh's voting record as his most exposed flank... coining the shibboleth of "the old politics of promise and spend, promise and spend."[7]

Years later, Lugar discussed the nickname "Nixon's favorite mayor" and asserted that it was not true. The veteran political reporter David Broder had written an article saying that Lugar was in and out of the White House so often that he must be Nixon's favorite mayor. The label stuck and hung around Lugar's neck with considerable weight in 1974. Years later and shortly before he died, Nixon was at an event, sitting next to Lugar. At one point he leaned over to him and whispered, "You really were my favorite mayor."[8]

The 1968 slogan had been the simple "Senator Bayh for Senator," reminding voters that he was the incumbent and taking advantage of the popularity that his service had created. This time, the slogan would be "One Man Who Makes a Difference," a phrase hoping to evoke pride in his accomplishments. A campaign brochure contained that slogan and sections highlighting a myriad of issues and related Bayh accomplishments.

As good as Birch was at remembering faces and names, he hated it when people came up to him and put him on the spot by saying something like, "You don't remember me, do you?" Often he knew the person's name, but sometimes he did not. We developed a routine that helped. If we were arriving by plane, for instance, I would ask him which people on the list of those meeting us he would recognize and which he would not. Leaving the plane first, I'd loudly introduce myself to those greeting us, so he could hear it. He'd then emerge from the plane and address the person by name. It worked pretty well.

This was before cell phones, but car phones existed, and one was installed in the campaign car. Often a sound truck playing "Hey, Look Him Over" met the road show when Birch was scheduled to walk door to door, announcing his presence in the community. He had over 90 percent name recognition, and most Hoosiers seemed to know how they felt about him—mostly positive, it seemed. One of the most difficult tasks was to change his reputation for tardiness, though he improved on that during the 1974 campaign. Our biggest obstacle was Birch's ice cream addiction. He seemed to know where every Dairy Queen in the state existed, and each time the car approached one, the schedule was threatened.

Indiana had multiple large and small media markets plus two time zones, Eastern and Central. The campaign would rotate in and out of each market to take advantage of as much free media as possible, also regularly visiting the most populous cities and counties.

For the first time, Birch and Larry Conrad were both on the Democratic ticket, campaigning together. Continually nagged by the "master plan" story, Conrad enjoyed seeing it overshadowed by Watergate. Few in Indiana were paying attention to his troubles when the larger national scandal was dominant. The two campaigns coordinated schedules, guaranteeing joint appearances at the most critical Democratic events. Conrad later recalled, "We were campaigning together . . . our last stop of the night, and after the dinner the owner of the restaurant grabs us and takes us through the kitchen and out the back door. He reaches over and grabs a jug of homemade brew and says: 'Want you two to have a little of this special after-dinner drink.' Birch, you grabbed and hoisted the jug and took a big swig. Your face turned red and your eyes started watering. You handed that thing to me, pointing to it and nodding, like it was French champagne. I grabbed it and took a pull. It felt like a hot poker was jammed down my throat. I fell to my knees on the ground, gasping for air. As I was on all fours . . . I heard the guy say to you, 'pretty good, huh?'"[9]

The Watergate scandal continued with the growing impression that the president could not last much longer. On July 24, the Supreme Court unanimously rejected claims of executive privilege by Nixon; the president was ordered to surrender sixty-four of the tape-recorded White House conversations. Far less attention was paid the next day when the Senate passed the Juvenile Justice and Delinquency Prevention Act.

On July 26, Birch faced a hot-button issue regarding the proposed Highland Lake reservoir, championed by the Indianapolis Water Company. His staff took the issue seriously and studied it carefully, concluding that Birch should oppose it. To build it meant taking away privately owned land through eminent domain and building an expensive reservoir. Expensive homes nearby would have a resort-like waterfront, all paid for by taxpayer money. Evidence was uncovered of underground wells that would allow any water shortages to be fulfilled by drilling, without taking privately owned land or funding major construction. Indianapolis water rates were already too high and would rise more if the expensive version of the project were approved. Also, the only private water company in a similar-sized city charging higher rates was in Philadelphia, which had overlapping boards of directors with the Indianapolis Water Company.[10]

At a luncheon in Indianapolis, Birch sat at the head table with the president of the Indianapolis Water Company, Tom Moses. While Moses sat there, Birch took to the podium and unloaded on him and the reservoir controversy, leading an audience member to later exclaim that Birch had "the balls of King Kong."

Fallout from Birch's effective Highland Lake opposition came from a prominent Democratic defection. The banker Frank McKinney, whose father had once been Truman's DNC chair, had a stake in the reservoir and became a vociferous opponent for the remainder of Birch's career. This was not the first time opposing a federal project cost him a friendship. Early in 1974, Bayh staffer Darry Sragow studied a proposal by the Army Corps of Engineers to build Wildcat Reservoir at Lafayette Lake, near Purdue. The proposed reservoir was supported by the business community, organized labor, and a Bayh mentor, Purdue dean David Pfendler. After considerable study, Sragow laid out the case to Birch, including the political risks. Birch interrupted him, saying he didn't want to hear about the politics; he wanted to know what Sragow thought was right. Convinced that the citizens opposed to it had a good case, he came down on the side of the opposition, as did Birch, and the project was killed.[11]

From July 27 to 30, the House Judiciary Committee did something that hadn't happened in a century: it cast votes on articles of impeachment, approving three of them that charged the president with obstructing the investigation, misuse of powers, and failure to comply with the House subpoenas.[12] It was becoming clear that Nixon was likely to be impeached by a vote of the full House of Representatives. Conviction in the Senate was also likely, and a delegation of Republican senators went to the White House to tell him that. Once again, it is easy to speculate that if Nixon's successor had been Speaker Albert instead of Vice President Ford, that delegation never would have made the trip to the White House. The impact of the Twenty-fifth Amendment in providing a process to fill a vice presidential vacancy for the first time, was felt again.

On August 8, Birch was walking in a parade when he learned that Nixon was going to resign the presidency that evening, and the schedule was being adjusted as a result. He was to go to an Indianapolis hotel where the press would be watching the televised resignation speech. Awaiting Birch at the hotel would be a prepared text for him to give to the press following the televised resignation.

That evening, Richard M. Nixon resigned the presidency before a national audience; it was indeed a sobering national event. Birch wrote changes, additions, and deletions on the typed statement, and let the assembled masses in, and the Bayh reaction to the Nixon resignation was given.

Elizabeth Drew wrote,

> Last fall, they had to invent a way of approving a Vice-President selected under the Twenty-fifth Amendment . . . things could be worse now. Agnew could be the Vice-President. And though the processes by which a new Vice-President was chosen under the Twenty-fifth Amendment are troubling, one must consider how much more complicated this would be if there had been no provision for selecting a new Vice-President and, with the Democrats in control of the Congress, Nixon's leaving office would have involved the accession of the Speaker of the House and a change of parties in power. Now people are speculating on who the *next* Vice-President will be.[13]

Watergate would have a long-lasting impact on American life. Nixon had broken the public trust, and Americans would no longer believe in their leaders as they once did. The public had been rocked by the revelations of lies about Vietnam and then Watergate, the scandal to beat all scandals.

The Bayh operation had reason to suspect that its office had been bugged and its copying machines subject to unauthorized users. It's impossible to know why a bigger deal wasn't made about this infraction at the time, but a sweep was made of Birch's suite, and he was told that listening devices had been detected there. It was also discovered that hundreds of unauthorized copies were made on his office Xerox copier. We later learned about other Watergate-related events directed at us by Nixon operatives. The FBI interviewed Larry Cummings, and he in turn directed the FBI questions to me. He said they had reason to believe that early in the campaign, a voluptuous redheaded woman was hired to get Birch into bed for the benefit of hidden cameras. They asked if I had any memory of a woman fitting that description making advances to Birch during that time frame. I did.

It was during the Indiana Democratic State Convention in June. I remembered the event because this stunning woman, a redhead with a fabulous figure, was sidling up to Birch during a reception and leaning in to him, whispering in his ear. Birch motioned to me and indicated to her that he needed a moment, ostensibly to confer with me. What he told me was a lesson that stuck with me and is largely why the incident was so memorable. He said that I should never let an attractive woman get a leg-lock on him as she was doing. The reasons were obvious, but it became another teachable moment, and I learned that protecting him from rumors was one of my tasks.

A humorous footnote to the FBI revelations was that this woman was hired to go after Congressman Andy Jacobs if she was unsuccessful with Birch. Jacobs, among the funniest of men, learned about it and responded, "Where was I?" expressing feigned shock that he had missed out on a great opportunity. He was unmarried at the time, so being photographed in a compromising position wouldn't have been nearly as scandalous as it would have been for Birch.

Another famous Jacobs story also focused on his romantic life. A longtime unmarried congressman, he was subject to rumors that he was gay, which he was not. When in a reception line, a constituent said to him, "Andy, I hear you're sleeping with everything in skirts in Washington." Jacobs responded, "At least the rumors are improving."

The night after Nixon resigned, there was a Marvin Gaye concert in Indianapolis. Gaye introduced Birch, brought him up to the stage, and the crowd went wild as they exchanged soul handshakes. As good as the experience was with Gaye, he had the opposite kind of experience with another African American celebrity in Indianapolis. The Indianapolis minority community held a "Black Expo" every year, and it was usually headlined by famous African American entertainers. The legendary jazz trumpeter Miles Davis performed at one of those events. Birch stood behind the stage in order to thank Davis after his performance. As Davis emerged from the stage, Birch thrust his hand forward, saying, "Miles, I just wanted to thank you for coming." Hardly hesitating, Davis responded, "I did it for the money." He did not return the handshake.

In August, the Juvenile Delinquency subcommittee reported the passage into law of several measures: the Methadone Diversion Control Act of 1973, reauthorization of the Drug Enforcement Administration, the Black Powder Bill (regulating the use of recreational firearms), the Runaway Youth Act, and the Juvenile Justice and Delinquency Prevention Act of 1974. When Birch completed his service as subcommittee chair, he had held seventy-five days of hearings on issues that included juvenile delinquency, black powder usage, drug abuse, school violence, runaways, vandalism, prisons, gun control, and a number of nominations to various federal offices. His bill to ratify a treaty to control the traffic of psychotropic drugs passed in 1978. Another he worked hard on, the Pharmacy Robbery Bill, became law after he left the Senate. The Juvenile Justice and Delinquency Prevention Act of 1974 attempted to change the way juveniles are treated when accused of a crime. All too often juveniles were unjustly confined to county jails or other adult facilities.[14] Forty years later, a *Washington Post* editorial stated, "It ended the practice of throwing convicted minors into adult prisons. It greatly curtailed the practice of locking up juveniles for status offenses (offenses adults would not be locked up for) such as truancy. It demanded that states report on racial disparities in the juvenile justice system."[15]

An editorial in the *New York Times* titled "Kids and Jails, a Bad Combination" begins, "There are few bright spots in America's four-decade-long incarceration boom, but one enduring success—amid all the wasted money and ruined lives—has been the Juvenile Justice and Delinquency Prevention Act, the landmark law passed by Congress in 1974."[16]

The Juvenile Justice and Delinquency Prevention Act, established with bipartisan support, "is based on a broad consensus that children, youth, and families involved with the juvenile and criminal courts should be guarded by federal standards for care and custody, while also upholding the interests of community safety and the prevention of victimization." It provides for

- A nationwide juvenile justice planning and advisory system spanning all states, territories, and the District of Columbia,
- Federal funding for delinquency prevention and improvements in state and local juvenile justice programs and practices; and

- The operation of a federal agency, the Office of Juvenile Justice and Delinquency Prevention, which is dedicated to training, technical assistance, model programs, and research and evaluation, to support state and local efforts.

The act's requirements are to deinstitutionalize status offenders, to remove juveniles from adult jails and detention facilities, to separate minors from adult inmates when they are detained, and to "reduce the disproportionate number of minorities who come into contact with the juvenile justice system."[17]

On August 20, President Ford announced that he was nominating Nelson Rockefeller, former governor of New York, for vice president. The grandson of the founder of Standard Oil, John D. Rockefeller, Nelson Rockefeller had been repeatedly elected governor but had unsuccessfully run for president several times. He represented a liberal wing of the Republican Party that seemed to shrink with each succeeding election. Ford considered selecting Donald Rumsfeld, US ambassador to NATO, and Republican Party chairman George Bush, but he settled on Rockefeller because of his executive expertise and because he represented a wing of the party that Ford did not. Rockefeller had stronger support from organized labor and minorities than most Republicans, and Ford assured him a role as full partner in his administration. This was the second nomination for vice president under the Twenty-Fifth Amendment. Rockefeller underwent contentious confirmation hearings in which Birch played no role; his time was spent in Indiana. Regardless, the country was tired from the interminable troubles of the Nixon administration, and it seemed as though Ford could choose almost anyone he wanted.

Much of that summer, the staff was joined by Birch's son, recent high school graduate Evan Bayh. Evan learned a lot about campaigning during those weeks. The time he spent with his father was invaluable to them both. We were under Marvella's strict orders never to let her two family members fly in a single-engine plane; she preferred them not to fly together at all. For the most part, we observed the twin-engine restriction, but they often flew together, and we protected those secrets like our lives depended on it. The most difficult task for Evan to master was to stay silent when we approached a Dairy Queen. Those abrupt stops for ice cream were wreaking havoc with the schedule. At one point, Birch was sleeping in the front seat and we were chatting in the back. When we saw a Dairy Queen coming up ahead, we all got quiet, which woke Birch up. He looked ahead, saying, "There's a Dairy Queen. Let's stop."

On August 22, Evan began his college career at Indiana University, and his time on the road show came to an end.

Tom Connaughton had left the Bayh staff in January 1970 to fulfill his military obligation and ultimately was sent to Vietnam. He returned in 1974 and found a senator who had been fully engaged in major national issues, with Carswell following on the heels of Haynsworth, then the ERA, Title IX, and the eighteen-year-old vote. But 1974 was also a reelection year, and Tom joined the campaign as the issues director.

Following Watergate, the important issues in the campaign were integrity in politics and the price of gasoline, which had gone way up because of OPEC reductions in the oil supply. It was also important to claim credit for federal funds that Birch was able to bring to Indiana. But one of Connaughton's tasks was to research Lugar's record. Lugar had blasted Birch with the classic Republican big-spender, tax-and-spend Democrat type of charge. But Lugar had been one of the progressive mayors who had taken advantage of many federal programs. While the mayor's budget had grown tremendously, that charge was complicated because of Uni-Gov. Connaughton's research revealed details on the bonded indebtedness and spending of the two jurisdictions before and after Lugar. The rise in both was dramatic. Under Lugar, bonded indebtedness rose over 200 percent and spending rose over 250 percent.

That's how Birch started the debate with Lugar. Talking about what he had accomplished in government, he said, "Now, here's what's happening in Indianapolis, where your budget has increased 250 percent and the bonded indebtedness has gone up." Lugar never recovered from that. He was unprepared for the charge and unarmed with facts to dispute the claim. Connaughton remembers thinking that Lugar's heart went out of the debate. Eventually, Lugar and his campaign manager, Mitch Daniels, argued that Birch was incorrect, but the numbers they came up with were not dramatically different from what Birch had cited. It's hard to mount an effective response when the facts say, "Oh no, it was only 220 percent" or something close to that.[18]

Lugar's staff had formed an anti-inflation task force, producing a lengthy document that itemized every Bayh vote and its related cost, arguing that he voted to outspend Congress's final appropriations in fiscal year 1973 by a whopping $25.1 billion. It further stated that he sponsored and cosponsored legislation in his second term in office that would have amounted to almost $236 billion. Unfortunately, claims like this generally made a typical voter's eyes glaze over and never seemed to hit home, whether accurate or not.

The statewide televised debate between Birch and Lugar was September 1. It was apparent to all that the debate could be a critical factor in the eventual outcome of the election. Staffers went to dinner early that evening, and campaign manager Jay Berman spent most of the evening in the bathroom, vomiting. The stakes and attendant anxiety were that high.

As hard as it may have been to be objective, the Bayh staff was convinced that Birch prevailed in the debate. He was personable and knowledgeable. Lugar was the latter but not the former; he was stiff and often boring. Birch arrived just before the debate was to begin. Lugar had been sitting at his podium, perspiring under the hot klieg lights. Walking in, saying hello to everyone, and shaking hands, Birch took the seat at his podium looking fresh and ready to go, in contrast to Lugar, who appeared hot and sweaty. Lugar charged that Bayh was a big spender and Indiana had not been receiving its fair share of federal money. Birch's response about Lugar's own spending muddied that issue sufficiently. Lugar also claimed that contributions to Birch's presidential campaign went unrecorded and that the senator obviously had something to hide. Birch took the populist stance at one point, jabbing Lugar for sitting in his office atop the City-County Building but caring not at all about the postal employee in his basement. As the "big spending" figures were bandied about, Birch responded with the old saying, "Figures don't lie, but liars can figure." Back at headquarters, he was greeted by cheering campaign staff. Larry Conrad spoke to the group and said, "He smoked him."

A *Harvard Crimson* article summarized the race at this point:

> Regardless of the record, recent events have made Lugar's defeat even more likely. In a well-publicized television debate, Lugar emerged with a bad make-up job, reminiscent of Nixon's debate debacle in 1960. The mayor, who generally speaks without notes, failed to win the victory for which he hoped and the debate (the only one to which Bayh has agreed). Bayh is also coming on strong with his country boy image. Once every six years, the liberal Indiana senator (ranked five points to the left of McGovern by the ADA) comes home with a hard-sell conservative act.... Bayh also loved to emphasize his rural origins (farm boy from Shirkieville, IN) and he has even scored Lugar, a Hoosier native, on his Oxford-Denison education, insisting that a senator with a Purdue education is best for Indiana.[19]

Ford's ascension to the presidency gave Birch an uneasy feeling about his own prospects. Watergate had been good for Democrats running in 1974, but Birch wondered if the era of good feelings created by Jerry Ford would remove that advantage in the remaining months of the campaign. He kept his concern private, though it created fears he couldn't shake. Would his opposition to Nixon over the years become a liability with Hoosiers who yearned to support the new president? Those fears disappeared when President Ford announced a pardon of the former president.

On September 7, President Ford signed the Juvenile Justice and Delinquency Prevention Act of 1974 into law. As proud as this made its author, it was completely overshadowed by the events the next day. On September 8, President Ford pardoned former president Nixon. History reveals that this was an honorable and courageous thing for Ford to do. He knew a Nixon trial would completely consume the nation and overshadow Ford's presidency. He became convinced that it was not in the country's interest to endure a trial of a former president. It was obvious that there would be a huge political price to pay for this, and his act probably doomed the Lugar campaign, if it wasn't already lost. But it was time to turn the page on Watergate and the Nixon scandals. The matter would be debated forever, but there has never been evidence of any untoward or sinister reason for the pardon. Ford did the honorable thing. He did what he thought was right, and he was determined to face the music. His was an act of statesmanship. The public may have wanted its pound of flesh from Nixon, but it wasn't to be.

When Nixon's pardon was announced, one of the most colorful reactions came from Sen. Bob Dole, who was running for reelection in Kansas: "I needed that like I needed a case of the clap."[20]

Lugar recalled that their internal polls showed the race growing extremely close, a virtual tie by the time of the resignation. When the pardon happened, the bottom dropped out for many Republican candidates, including Lugar, whose recovery was impossible.

On September 17, Birch and Lugar had a joint appearance in Indianapolis. Lugar took the hide off Birch in his speech, making points on the "big spender" charge, which convinced Birch of the necessity of going last in his engagements with Lugar. It was clear to him that Lugar was a formidable opponent.

Sen. Ted Kennedy's visit for Birch in Lake County was the next day. The appearance was in Merrillville for a rally and cocktail party, and a large crowd was in attendance. Later that evening was a Bayh fundraiser with Kennedy at St. Savas Hall in Hobart. It was a small-dollar event with three thousand people in attendance. A few demonstrators with signs referring to Chappaquiddick got a disproportionate amount of the press coverage, but the crowds were fantastic. Soon afterward, Kennedy announced that he would not be a candidate for president in 1976.

During a parade in late September, a note was handed to Birch telling him that First Lady Betty Ford had been operated on for breast cancer. When it was convenient to leave the parade, he made his way back to the car. An Associated Press (AP) reporter was traveling with the road show at the time. Birch asked the mobile operator to connect him with the White House switchboard, identified himself, and asked for the president's secretary. Instead of the secretary, President Ford got on the line, asking, "Birch, how in the world are you?" Birch expressed his concern for Mrs. Ford, and the president responded with appreciation. Their old relationship nurtured in the locker room before congressional baseball games was still there. Birch looked back on that greeting and mused, "Some things transcend politics."

The campaign was in full swing in September and October. Watergate was over, and the election loomed ahead. College campuses were great venues for Birch, and he experienced lots of support from students, particularly on September 25 at his alma mater, Purdue. In a large, packed auditorium, he gave a rousing speech that generated an enthusiastic response. Later, he met with the editorial board of the *Exponent,* Purdue's newspaper. Its editor, Bill Moreau, would later join the Bayh staff after volunteering in the 1976 presidential campaign. As an army brat, he had not been raised in Indiana and had never met the senator. A campus radical by Purdue standards, Moreau deeply admired Birch's progressive record. When he directed his editorial board that the paper would endorse Bayh for reelection, he encountered significant pushback from those colleagues who were from Indianapolis and admired its mayor, Richard Lugar. A compromise was struck: the editorial board would invite both candidates to be interviewed, and an endorsement vote would be taken after the interviews. Remarkably, both accepted the invitation.

Lugar was interviewed first. He arrived exactly on time and sat upright in his chair, every hair in place and his three-piece suit unruffled, while he fielded questions from the five young editors. He answered every question articulately. Precisely sixty minutes later, he excused himself and departed. Moreau was worried.

A few days later, Birch arrived at the editorial office after an energetic speech to the Purdue students. He was thirty minutes late, his shirt rumpled and sweat-stained, his coat over his shoulder, and his tie askew. When invited into the small office to be interviewed by the editorial board, he saw dozens of other staffers in the large, open newsroom. "How about we include everyone in our conversation?" he asked Moreau. Immediately, desks were shoved to the walls and a large open area was created. Then Birch said, "How about we all sit down on the floor and get acquainted?" For almost ninety minutes, Birch sat on the floor, got to know each student, and fielded every question.

At one point, Birch asked for a Tab, his favorite soft drink at the time. It was brought from the vending machine and handed to him. The soda cans in those days opened by snapping off a pop-top, and his habit was to drop it into the can before taking his first drink. This went unnoticed by one of the *Exponent* staffers, who politely reached out to take the pop-top and dispose of it for the Senator. Thinking the young man was thirsty, the senator reflexively said, "Here you go." People were struck that a United States senator would offer a drink of his soda to a perfect stranger.

Upon Birch's departure, an endorsement vote took place, and Birch was endorsed unanimously. Later that evening, the young staffers were regaling each other with what had transpired that day. Someone wondered what would have happened if, God forbid, the senator had accidentally ingested the pop-top. Another staffer quickly drafted a story announcing the untimely death of the senator—in the offices of the *Exponent*, no less—and Moreau added the headline: "Tab Tab Swallowed by Bayh: Bye-Bye, Bayh!"[21]

That phrase also showed up in a campaign song in the possession of the Lugar campaign. Beginning with the line, "Tonight we have a brand new hit—Listen to the name of it—Bye-Bye Birch Bayh," the song never gained traction.

At a parade the following week, Birch received yet another great response from the crowd. Birch was walking the parade, and as he passed the reviewing stand, there sat Lugar. They shook hands, and Birch walked on as everyone laughed and applauded.

Another parade was memorable, with Birch walking and those of us on staff walking behind the crowd but keeping pace. As we got ahead and waited for him to catch up, I heard two men talking, and one of them asked, "Has Bayh come by yet?" He was cradling something in a paper bag against his chest, trying to conceal it. It looked like it could be a weapon, and I was in a position to try to foil an assassination attempt. I walked up quietly behind the man with the bag, which also allowed me to see Birch approaching. As he came near, the man reached in the bag and pulled out . . . a bottle of beer. He waved aggressively, calling, "Hi, Birch," then unscrewed the lid and took a drink.

Throughout the campaign, instead of Watergate, Birch focused on his record and on Lugar. But he also spoke about the nation's lack of energy independence and our dependence on oil from foreign countries. He delighted in making it clear to people that we were talking about more than gasoline for our cars and shortening the lines at the pump. We were talking about how oil had seeped into many aspects of our lives, from the synthetics used to make our clothes, to the plastic frames on our glasses, to milk containers, wrappers on our bread, and on and on. Plastics, a major petroleum product, were everywhere in our daily lives, and we all depended on a steady diet of oil to continue our standard of living. The Republican administration had done little about it and electing another Republican wouldn't make the problem better. Lugar was also vulnerable on labor issues. His family's agricultural company had actually broken a strike, which was anathema to union members. For everyone else, we just kept referring to him as "Nixon's favorite mayor," a problematic designation for Lugar.

On the subject of energy, Birch promoted alcohol fuels generated from corn, known as gasohol. He was an early proponent of renewable energy, especially alcohol fuels from Indiana, providing a continuous source of revenue for farmers and other Hoosiers. Advocating renewable energy and ending our addiction to fossil fuels became a common theme during the campaign and for the rest of his career.

There was a gridiron dinner in Evansville in October. Birch developed a routine poking fun at his own image as a "limousine liberal," one who talks like a farm boy at home and mixes with the celebrities in Washington. He gave a hilarious speech, making fun of himself with pointed jabs at his own down-home reputation, a masterful display of self-deprecating humor. The crowd loved it, and Lugar, also in attendance, was compelled to watch his opponent on a roll.

Jules Witcover, a well-known national reporter and author, wrote an article for the *Washington Post* highlighting the event and provided an interesting assessment of the two competing candidates:

> Republican Mayor Richard Lugar of Indianapolis sat rather stiffly at the head table at the annual Evansville press gridiron dinner, listening to what he is up against as the underdog candidate in Indiana's U.S. Senate Race. Democratic Sen. Birch Bayh had just been ridiculed as a farm boy who had spent his childhood spreading manure on his family farm and since then broadened his scope. Bayh, at the microphone, turned his back to the audience, mussed his hair like some comic impersonator, swung around and began in his best barefoot manner: "Aw shucks, it sure is good to be back in Indiana. We were sitting in front of the stove the other day. I was reading the Constitution, Marvella was sewing a star on the flag and Evan was studying his Eagle Scout book, then Marvella looked up at me. "Blue Eyes," she said—she always calls me Blue Eyes—"I yearn for the farm." All right honey, I said, and we piled into our pickup truck, the one with the rifle rack up over the seat, and we headed out there."
>
> The large audience of press skeptics and politicians roared, and Birch Bayh had them. . . . But it took no insider to pick up the political vibes out of the fun poking: Dick Lugar, not long ago regarded a hot GOP presidential or

> vice-presidential prospect, was facing the distinct prospect of being engulfed by an irrepressible force—the personality of Birch Bayh.[22]

Because of the continuous, thorny gun issue, Birch took part in an event the next day that worked to his advantage. Knowing that the pro-gun people saw Birch as part of the Eastern elites who knew nothing about the glories of shooting and gun ownership, he was scheduled to shoot firearms in public. Some might forget that he was raised on a farm, had been in the army, and was comfortable around guns. Just because he felt that "Saturday night specials" shouldn't be manufactured and sold didn't mean he lacked empathy with gun owners. On October 27, in Jeffersonville, with an NBC film crew in attendance, Birch participated in the "Turkey Shoot at Utica Pike." He won the contest, which was covered extensively on TV news programs. Something similar took place years later when Birch visited a skeet shooting venue with Bob Novak, the syndicated columnist, in tow. Novak was as cynical as they come, and his sarcasm as we arrived at the facility was easy to detect. His amazement was just as serious when Birch hit every clay pigeon sent into the air.

Staff member Louis Mahern recalled how Birch prepared for events like this. He was as physically fit and as athletic as any senator, but he worked hard prior to competitions. Before attending a shooting contest sponsored by the NRA in Indiana, Birch went with Mahern to a shooting range to practice. He probably took a few hundred shots that day. When the event was over, Mahern asked him how he did. "It went great," was the reply. "They couldn't believe I hadn't shot a gun since I was in the army."

In the days before the election, the campaign implemented a voter protection strategy designed by staff member Gordon Alexander. Teams of young attorneys volunteered to man the polls for the Bayh campaign, to make sure that voters were neither intimidated nor prevented from voting. The strategy was a reaction to stories from previous Indianapolis elections. Whether factual or not, the stories were that police sat in their squad cars parked near the polling places in black neighborhoods. Any voters concerned about being seen by the police might be intimidated away from voting at those times. The assumption was that African Americans were far more likely to be voting Democratic and keeping them away could only help the Republicans. Volunteer lawyers would be present to make sure that any police hanging around were asked for their badge numbers, which was often enough to send them on their way. No one wants to lose an election because their voters can't cast their votes. Having friendly attorneys to help resolve any disputes on-site provides valuable insurance.

The campaign produced a short documentary called *Marvella's Last Minute Appeal*, in which she urged Hoosier voters to support Birch. Reminding them that she had "shared him with them for 12 years and over 4,000 roll call votes," she articulately evoked memories of the two previous campaigns in which she had played such a major role.[23]

During the final days of the campaign, Birch took a helicopter tour of the state, making multiple stops in each of Indiana's major and minor media markets. The day before the election was a flurry of activities in several cities, finishing in Terre Haute at an election eve rally, a fabulous and exciting event. Election Day was no less hectic. It started at the Chrysler foundry plant's gate at 4:30 a.m.; a second plant gate at Chrysler Motors followed at 6:15. After flying to Lake County for a rally and a plant tour, he returned to Terre Haute so he could cast his vote at the Goshen firehouse near Shirkieville. He had to confront a local opponent who challenged his voting residency. He won both the challenge and the election. That night was celebrated in Indianapolis at campaign headquarters.

Marvella spoke to Jay Berman that night and asked him if there was a campaign debt. When he replied in the negative, she expressed her gratitude.

Everyone in the campaign world knows that there is nothing like winning. The adrenalin that gets you to the end of the ordeal virtually explodes when victory results. Needless to say, election night 1974 was exciting for all involved. Birch, Marvella, and Evan stood on the stage together during the victory speech. While lots of factors go into every election decision, when you win, it's hard not to conclude that you have done something right. For whatever reason, Indiana voters spoke and decided to return Birch to the Senate for another six years. The final result was 889,269 for him with Lugar at 814,114, a margin of over 75,000 votes. There was also a third-party candidate, Don Lee; Bayh received 50.7 percent of the vote and Lugar 46.4 percent, and the balance, 2.9 percent, went to Lee. Larry Conrad was reelected secretary of state. One additional and satisfying result was that Marion County, Lugar's home, was won by all the Democrats on the statewide ticket.

This Watergate year was the best time for Birch to run for reelection, given Lugar's high quality as a candidate. Long after the campaign, Lugar described Birch as a "first-class campaigner." If he had been looking for a tutorial, running against Birch was the thing to do. "I learned a lot," he said.[24]

The election was successful for other Indiana Democrats. The congressional delegation had been 7–4 in favor of the Republicans. Now it was 9–2 Democratic with new Democrats in the 2nd (Floyd Fithian), 6th (Dave Evans), 8th (Phil Hayes), and 10th (Phil Sharp) congressional districts. Andy Jacobs won his seat back in the 11th, and Ray Madden (1st), John Brademas (3rd), Ed Roush (4th), and Lee Hamilton (9th) were reelected. Birch campaigned with each of those candidates during the previous five months and developed strong relationships with the new members and stronger relationships with those who were reelected.

The election was also good for Democrats nationally, unsurprising amid a scandal involving the Republican incumbent. The Democrats gained three seats in the Senate, adding another when a special election in New Hampshire resulted in Democrat John Durkin's election. The Democratic majority was 62–38, with a 291–144 majority in the House.

New Democrats included Dale Bumpers (AR), Gary Hart (CO), John Culver (IA), John Glenn (OH), and Patrick Leahy (VT). One Republican defeated was Marlow Cook (KY), who had played an important role during the Haynsworth and Carswell confirmation fights. Thirty years later in an op-ed, he wrote that he supported Democrat John Kerry for president in 2004, adding, "I have been, and will continue to be, a Republican. But when we as a party send the wrong person to the White House, then it is our responsibility to send him home if our nation suffers as a result of his actions."[25] This was a measure of how things changed over the next few decades, but ironically part of his legacy would be defined by former staff members who would serve in Congress, one of whom was Mitch McConnell, who was elected to the Senate and served as both minority and majority leader.

Looking back on his own career and discussing relationships with his colleagues, Birch described Leahy as "one of the best senators we have ever had" and someone he liked a lot. "He was probably the most decent human being of any I ever served with in the Senate."

Shortly after the election, President Ford issued a veto of amendments to the Freedom of Information Act, efforts to promote openness in government. Birch opposed this and urged his colleagues to override the veto. He supported a $4 million tax reform and relief package and introduced a bill exempting fraternities and sororities from Title IX, not wanting them to become unintended victims. The Senate passed it in December. Representative Brademas introduced the Presidential Recordings and Materials Preservation Act of 1974, nullifying Ford's agreement to give Nixon all the papers and tapes from his presidency. While Gaylord Nelson introduced it in the Senate, Birch helped lead the successful effort to pass it.[26] And on December 19, Nelson Rockefeller officially became vice president under the Twenty-Fifth Amendment.

Also in December, Birch was at Dulles Airport waiting for a flight to Indianapolis in order to take part in the swearing-in of state elected officials. Right before the plane was due to land at Dulles, it crashed, killing everyone aboard. Birch spent the rest of the day comforting loved ones who were awaiting the plane's arrival.[27]

Nineteen seventy-four will always be remembered as the year of Nixon's resignation. For much of the country, it was the year when *People* magazine was first released and when Hank Aaron passed Babe Ruth by slugging his 715th home run while the entire nation watched on television. Heiress Patty Hearst was kidnapped and surfaced as part of a bank robbery by the Symbionese Liberation Army; a photo of her wielding a gun was shown around the world. In Zaire, in a bout that became known as the "Rumble in the Jungle," Muhammad Ali knocked out George Foreman, regaining the heavyweight title he had lost. PepsiCo became the first American company to sell products in the Soviet Union. And an interesting and novel device, Rubik's Cube, was invented.

13

National Interests: 1975

Politics are almost as exciting as war, and quite as dangerous.
In war you can only be killed once, but in politics many times.

WINSTON CHURCHILL[1]

IN LATE 1974, BIRCH BAYH WAS FORTY-SIX YEARS OLD AND had been through an intense and grueling five-month campaign, an ordeal from which he found it hard to rebound. The work and responsibilities continued unabated, his role within the Senate now broader than ever before. Forty-first in seniority when Congress convened in January 1975, he was increasingly urged by political leaders across the country to seek the presidency once again. The Nixon pardon made President Ford vulnerable in a way that appeared impossible to overcome. There was no heir apparent in the Democratic Party, and while a campaign didn't need to start right away, it wasn't too early to begin thinking seriously about it. Birch understood that not taking certain steps toward a candidacy was tantamount to taking steps away from one. Some decisions would have to be made, if nothing simpler than a decision to consider running. But pulling the trigger so soon after the end of the last campaign was hard to do.

From one vantage point, a campaign made more sense in 1976 than in 1972. Marvella was now fully engaged in her work for the American Cancer Society, and Evan was safely ensconced in college. The tug and pull from the home front was greatly diminished in contrast to the family needs in 1972. Professionally and politically, his stature had grown. The way he departed from the race in late 1971 had brought him admirers. Having defeated a serious contender in 1974 and with his re-election behind him, he had additional legislative achievements to his credit. He needed to ponder how deeply he wanted the job of president. Few people realized or understood how dog-tired he was from the campaign and how much that would affect his ability to think carefully about the future.

Other politicians had their sights on 1976 well before 1975. While Birch was pursuing reelection, others had the luxury of time for planning and taking steps to organize for a possible candidacy. Nixon's demise strongly affected the atmosphere surrounding the next presidential campaign. Ford was crippled by the statesmanlike act he took to pardon Nixon, but he dearly wanted to redeem himself by being elected in his own right. The long odds were clear to him, at least as seen through the lens of early 1975, but optimism in politicians is not in short supply. Gerald Ford, the first president to serve without being chosen in a national election, wanted his chance at being elected in his own right. With the exception of Gov. Ronald Reagan of California, who would be leaving office in January after two terms, few Republicans appeared to be considering a presidential run. Reagan was known to be a formidable campaigner with broad national recognition resulting from his celebrity as an actor before becoming a politician. Since he no longer held elective office, he could be a full-time candidate.

Before 1974 ended, two potential candidates had already bowed out. Shortly after he campaigned with Birch in Indiana, Ted Kennedy had announced that he was not running. Kennedy would have been the front runner had he remained in the race. Also bowing out was Sen. Fritz Mondale, who had publicly questioned his own ambitions, admitting he lacked the burning desire to be president. You need "fire in your belly" to run for president, endure such an arduous campaign, and have a chance at being successful.

Birch felt he was as capable as any other Democrat who might seek the office, but given the circumstances, he knew it would be a large field in which to compete. Jules Witcover, a *Washington Post* political reporter who had traveled briefly with the 1974 Bayh campaign, wrote a book about the 1976 campaign and provided a list of those who were speculated to be candidates for the Democrats. He described Birch as "a shrewd and ambitious politician with strong labor support and a deceptive veneer of country-boyish looks and backslapping cordiality." There were several others, including senators Fred Harris (OK), Frank Church (ID), Scoop Jackson (WA), and Lloyd Bentsen (TX). Harris, considered a populist, was the youngest of the senators. Both Church and Jackson had distinguished records in the Senate, with the latter considered one of the more conservative Democrats running. The least experienced was Bentsen, who had been elected senator only two years before. Arizona's Morris Udall (AZ), a witty former professional basketball player, was the only House member throwing his hat in the ring. Former governors among the hopefuls included Terry Sanford (NC), Jimmy Carter (GA), and George Wallace (AL). As a Southerner, Sanford hoped to siphon off Wallace votes, while Carter was harder to pigeonhole. A bit of an evangelist, Carter portrayed himself as an efficient governor representing the "new South." Wallace could still claim a significant level of support in and out of the party. Sitting governors were Milton Shapp (PA) and Edmund "Jerry" Brown (CA). Shapp was liberal and Jewish, and though he represented a large Democratic state, he was probably taken less seriously than the other candidates. Only in his thirties, Brown was considered a comer, an attractive candidate who could represent the future of the party. Finally, there was Sargent Shriver, a Kennedy in-law, former ambassador, and head of the Peace Corps, also the vice presidential replacement candidate in 1972.[2]

That was a list of twelve candidates, and the longer Birch waited, the greater the chance that he would begin too far behind, that someone might grab the nation's attention and leap from the pack.

The year began with a Watergate verdict on January 1. Among those convicted in the Watergate cover-up were former attorney general and campaign manager John Mitchell plus H. R. Haldeman and John Ehrlichman, former top aides in the White House. Each was sentenced to prison.

For Birch, the year began with efforts to extend the Voting Rights Act and promoting a myriad of measures: direct election, the Family Farm Inheritance Act, increases for Social Security recipients, tax reform, public financing of congressional campaigns, and Medicare coverage of uterine cancer tests. Also in January, OPEC raised crude oil prices by 10 percent, continuing the energy crisis so prominent in the 1974 campaign plus adding economic pressures on an already struggling economy. Birch traveled to Los Angeles to give a speech on economic policy, thereby keeping his name before those paying attention to national presidential politics.

He was also learning about apartheid in South Africa, an issue important to Indiana's minority community, and he understood how little he knew about it. It was symbolized by the imprisonment of Nelson Mandela, leader of the African National Congress but a prisoner with a life sentence since 1962. Birch traveled to South Africa during this period, visited Mandela at Robben Island, and spent an entire evening at Mandela's home in Soweto with Mandela's wife, Winnie. He remembered how people came and went all evening long and that Winnie was truly the symbolic if not the actual leader of Soweto. He also remembered her as someone who couldn't be nicer.

Other events around the country included the Connecticut inauguration of the nation's first female governor who was not succeeding her husband, Ella Grasso. Vice President Rockefeller was appointed to head a commission to investigate alleged CIA abuses, an issue that would soon grow in importance. On January 26 was the first nationally televised women's basketball game, a contest between the University of Maryland and Immaculate University, something that would not have happened before Title IX.

On February 5, Birch was appointed chairman of the Transportation Subcommittee of the Appropriations Committee, a venue that involved him in the creation of the DC subway system as well as the federal funding of national rail systems Amtrak and Conrail. When he learned that money had been set aside for the building of another bridge over the Potomac linking Washington with Virginia, he redirected those funds toward a DC subway system, soon to be known as Metro. Birch was keenly aware that Washington was among the largest of American cities without a subway. He was familiar with subways in Moscow, London, and Paris, having visited those cities. Projects of this magnitude in the District of Columbia were made difficult because DC had only a single nonvoting delegate and therefore few voices lobbying on its behalf. He was pleased to be speaking up for DC.

The first 4.6 miles of Metro opened in March 1976. While it was being built, Evan Bayh took a summer job with an Indiana-owned company involved with its construction. Evan remembered how incredibly far underground the Metro was, with high scaffolding throughout the construction area. Having climbed up during one assignment, he was told to stretch out to help another worker with his welding. Afraid that he would fall to his death, he hesitated. The foreman told him to get out there "or I'll throw your ass off of here."

Amtrak was created to manage passenger trains, and Consolidated Rails Corporation, nicknamed Conrail, was to manage the bankrupt freight-carrying railroads. Both fell within Birch's jurisdiction.

He continued to pursue legislation on Foreign Service grievance procedures and participated in the Southern Conference on Correctional Juvenile Justice. Once again, he announced the participants for the Indiana High School Government Leadership Conference. The speakers would be US Transportation Secretary William T. Coleman; Under Secretary of the Treasury Edward C. Schmultz; Deputy Secretary of the Bureau of Consumer Protection (at the FTC) Joan Z. Bernstein; and a former astronaut, chair of the Nuclear Regulatory Commission (NRC) William A. Anders.

On March 1, Birch addressed the Democratic Conference of Mayors on juvenile justice and later that day brought attention to his growing presidential aspirations by traveling to Keene, New Hampshire, to give a political speech. A few days later, in recognition of his unique role in the fight for women's rights, he was appointed to the Commission on the Observance of International Women's Year. Also in March, he chaired hearings in the Juvenile Delinquency subcommittee on opium use, held Transportation hearings on highway funding, proposed tax cuts, and spoke out in opposition to increases in the cost of Medicare for seniors. He delivered speeches on strip mining and agricultural issues in Indiana and then traveled to California and Louisiana to give political speeches. On the last day of the month, he likely watched fellow Hoosier John

Wooden, a legendary UCLA basketball coach, coaching his final game. The UCLA Bruins defeated Kentucky to win the national championship for the tenth time in only twelve seasons.

The previous month, Margaret Thatcher was elected prime minister of the United Kingdom; she was the first woman to hold the position. The United Nations proclaimed March 8 International Women's Day. The role of women in America and around the world was undergoing seismic changes.

March 1975 was important in Senate history because of a decades-old battle to change the rule affecting filibusters. For more than a century, the Senate allowed unlimited debate. That practice led to filibusters during the Wilson presidency to prevent American involvement in the European conflict eventually known as World War I. Wilson, a congressional expert, pressured the Senate to institute a limit on debate that eventually passed as Rule 22. Two-thirds of the Senate could limit debate, a vote known as cloture. Over the next forty-six years, there were many attempts to invoke cloture, but only five were successful.[3]

Birch was one of a growing number of senators, particularly Democrats, who grew weary of the filibusters to prevent passage of the Civil Rights Act of 1964 and other progressive measures. After the 1974 election, there were sixty-one Democrats in the Senate and one independent organizing with them, so Rule 22 was in jeopardy. Birch was intimately involved in the effort to consider changing Rule 22 or eliminating altogether the right of unlimited debate. The new Democratic majority was successful in changing Rule 22, reducing the two-thirds requirement to three-fifths, or sixty senators. This was not a minor change for the Senate, and it would have a long-term impact on progressive legislation yet to come.

In April, Birch announced hearings on abortion, transportation, gun control, school violence, and the Juvenile Justice and Delinquency Prevention Act. He gave a political speech in New Jersey and spoke to the Americans for Democratic Action on a "new internationalism." Other speeches during this period were on railroad rehabilitation, the Voting Rights Act, and the Foreign Service Grievance Act.

Gun control once again came to the forefront of Senate debate. The 1968 law had been a bipartisan effort, and the same players remained on the Judiciary Committee and Birch's subcommittee when new gun legislation was introduced in 1971. The NRA at the time worked well with the senators and had not yet become a hard-right Republican organization. At the time, Mike Mansfield, a moderate to liberal Democrat from Montana, was an NRA supporter. After George Wallace was shot in 1972, the "Saturday night special" bill was passed in the Senate but failed in the House. A year or so later, Birch joined Sam Ervin to repeal a section of the crime bill passed in the late sixties. The bill allowed drug enforcement officials to enter premises without knocking if they had reason to believe that illegal drugs were inside. This "no-knock" provision was repealed by an amendment to the Controlled Substances Act, which became law in the 93rd Congress (1973–74).

In 1975, Birch once again pursued the prohibition of "Saturday night specials." His bill differentiated between the types of firearms used for sporting purposes and the easily concealable handguns most often used in crime. On May 17, the *Washington Post* printed his letter to the editor mentioning a Treasury Department study that found "70 per cent of the handguns used in crime would be covered by this specific factoring criteria" of the Bayh bill. "Similarly, the revolver fired at President Ford in San Francisco last September and the weapon used to cripple Governor Wallace in Maryland four years ago were non-sporting handguns covered by the factoring criteria. These handguns are not useful to the sport-shooter or target shooter. They are, however, the favorite side arm of the modern street criminal; the instrument which enables this relatively small group of individuals to turn our streets and neighborhoods into hostile territory for law abiding citizens."[4]

Later in 1975 a *Philadelphia Tribune* editorial ran entitled, "Give Victims of Crime a Break: Pass the Birch Bayh Crime Bill." It characterized the crime problem in America and how complicated the solutions seemed to be. "Nevertheless, U.S. Sen. Birch Bayh (D-Ind.) recently introduced a comprehensive piece of legislation—S. 1880 (the Violent Crime and Repeat Offender Act of 1975)—which could provide some of the answers." It listed the bill's features: "limits the availability of easily concealed handguns; provides mandatory penalties for the commission of crimes involving guns and for the illegal purchase of guns by criminals as well as gun dealers who knowingly sell guns to convicted felons; requires gun sellers to report all gun thefts with mandatory penalties for the sale of firearms for illicit inter-state purposes; requires mandatory prison sentences for the manufacture, distribution or sale of heroin and morphine. Finally, it establishes as a Federal crime the robbery of dangerous, addictive drugs."[5]

Unfortunately, only a few sections of this bill were enacted into law in 1976, though Birch would continue to pay a political price in Indiana because of his gun control sponsorship. A major accomplishment was the passage of the Maintenance of Effort Amendment to the Crime Bill in 1976. Birch heard testimony from the Justice Department that 19.15 percent of its budget was allocated to juvenile crime. His amendment required perpetual spending commitments on juvenile crime by the Law Enforcement Assistance Administration.

Americans paid great attention that month to the surrender of South Vietnam. Saigon had fallen, and there were televised scenes of the hurried evacuations of those Americans still in the city along with a number of South Vietnamese. Little attention was paid to the founding that month of Microsoft by Bill Gates and Paul Allen.

Birch was clearly leaning toward another run for president, though nothing formal had been announced. While growing increasingly critical of President Ford's leadership and the direction in which the country was heading, he vacillated about whether or not he should run. Vietnam disappearing as an issue was important, but it did not change his views about the administration. His areas of interest would be better served if he were in the White House, but in May he was still unready to pull the trigger. Marvella did not want him to run; she did not want the family threatened with possible new campaign debt, and she did not want another campaign to endanger her role with the American Cancer Society or the local television show she was a part of. The absence of her support for the effort he might be making was a burden to Birch.

One of the congressional rituals each spring takes place once baseball has begun. The annual congressional baseball game, Democrats against Republicans, was to be at the Baltimore Orioles' Memorial Stadium. Birch would be playing once again, and he agreed to take senators Joe Biden of Delaware and Dick Clark of Iowa with him to Baltimore. To make that easier on their schedules, because the Senate was in session, a helicopter was donated to transport the three senators to Baltimore before the game. I drove the senators in my small car, a Chevy Vega, to a vacant lot near the Capitol that would serve as a landing site for the chopper.

When we arrived at the helicopter, there was a disconcerted pilot walking around. Some kids had thrown rocks at the 'copter as it was landing and broken its windshield, limiting its capacity as a result. It could not withstand the air pressure with a full load, according to the pilot, and he offered to take Birch only. The other two senators and I sped away in the Vega, headed for Baltimore. We were crammed in, to say the least, but we made it to Memorial Stadium in enough time for them to get dressed and on the field. Birch hit a triple in the game, played third base, and won the Most Valuable Player award.

In May he generated more press in his Senate role than he did as a potential candidate, holding hearings on full DC voting rights and, in the Juvenile Delinquency subcommittee, on the problem of marijuana. He gave speeches that month concerning tax deductions for handicapped children, sex discrimination in the Social Security system, Title IX, direct election of the president, support for Israel, and the admission of women to the service academies.

The Ford administration faced an international crisis on May 12 with the seizure by Cambodia of the SS *Mayaguez* in international waters. A rescue mission was rapidly arranged, and three days after the American merchant ship was taken, American marine and naval forces took it back. Thirty-eight Americans lost their lives in this maneuver. As distressing as the loss of life was, the American public broadly supported the president's action. This was the kind of threatening situation that politicians dared not criticize, and Ford's popularity rose.

The Rockefeller Commission that was convened to investigate possible abuses by the CIA issued its report in June, recommending that Congress establish an oversight intelligence committee. The newly formed committee in the Senate would be chaired by Sen. Frank Church. Its hearings began in late July and continued until the following May. Many criticized the revelations resulting from its investigation, but there is no doubt that it not only made Church more famous but also turned him into a presidential contender. His committee ended up being replaced by the Senate Select Committee on Intelligence with Sen. Daniel Inouye as its first chair. That June and July, Birch gave speeches on banning polygraphs, land resource planning, foreign policy, gun control, defense spending, and crime. He held hearings on school violence, oil price increases, illegal drug use, and anti-discrimination enforcement provisions of government. He gave speeches on the supersonic transport (SST) and led the floor fight against the Concorde, the most prominent of the supersonic aircraft. He gave speeches on the Voting Rights Act, federal highway aid, the Foreign Service Grievance Act that passed the Senate on September 11, funds for runaway shelters and counseling centers, and Ford's mismanagement of the economy.

Birch was the last of the liberal candidates to make his intentions known about a presidential run. Reporter Jules Witcover described him as "the left-of-center Democrat with the best combination of political credentials—organizational ability and support and personal campaign magnetism." He wrote, "Bayh, blue-eyed, boyishly handsome with a Tom Sawyer style to go with his physical attributes, parlayed his gee-whiz, aw-gosh politicking into a position in the Senate from which he built a record of achievement that often was obscured by the very barefoot-boy image he projected." He went on to list Birch's accomplishments and talked about his first presidential effort that ended with Marvella's cancer surgery, noting that Bob Keefe was a prominent political force in that campaign but was running Scoop Jackson's campaign in 1975.

Birch remained reluctant about a presidential candidacy for most of 1975. His press secretary, Bill Wise, wrote him a memo in December 1974 saying that if he wanted to be President, 1976 was the year. With Marvella's reluctance, with Keefe running Jackson's campaign, and because of the sheer exhaustion he still felt from the reelection effort, he hardly responded to Wise's memo. It wasn't until he was persuaded that "nobody really was putting it together" that he agreed to begin organizing.[6] The Bayh Committee was registered with the Federal Election Commission in August.

He continued speaking publicly on a wide variety of subjects in August: oil price controls, Amtrak, and support of coal gasification and liquefaction plants to help fight the country's energy crisis. In the meantime, planning for the presidential campaign, with the necessary hiring and fundraising, had begun. Everyone involved was painfully aware that while there was no recognized frontrunner by August, Birch was still running woefully behind, if not in the polls, at least on the schedule. He continued the drumbeat of speeches into September, addressing the increase in university sports revenue because of Title IX, calling for the breakup of vertical integration of oil companies, and announcing new hearings on the problems of violence in our schools. Violence in the schools was a fitting topic for the country to address, and violence in other aspects of American life simply would not go away. There were two attempts to assassinate President Ford, both in California. One was on September 5 in Sacramento, when a member of the infamous Charles Manson family, Lynette "Squeaky" Fromme, fired a gun at him; later the same month, a deranged woman named Sara Jane Moore also tried to shoot him.

The Bayh campaign got a boost when Robert Abrams of New York endorsed Birch. A former state legislator and future New York attorney general, Abrams was then borough president of the Bronx and a major figure in New York's reform movement. He called for "early unity" around a Democratic candidate and described Birch as a "strong liberal candidate with broad-based appeal." Also joining the New York campaign was Harold Ickes, son of an FDR cabinet secretary and an active reform Democrat in the state. Ickes could remember a time when FDR came to his family home for dinner. He had been among the early leaders in the McCarthy campaign in 1968 and before that had been critically injured during a civil rights march in the deep South. An unusual family connection helped interest him in the Bayh candidacy. Harold's father-in-law ran a theater in New York and taught acting classes. One of his former students was Mary Alice Bayh, Birch's sister.

The next day, Justice William O. Douglas resigned from the Supreme Court. The Bayh staff hoped for an acceptable replacement, expecting better from Ford than from Nixon yet knowing that there would not be a Bayh campaign against a nomination this time. John Paul Stevens was nominated and confirmed without controversy in a vote of 98–0. He was replacing the longest-serving justice in American history and would become the third–longest serving Justice when he retired in 2010.

In early October, was a dinner thrown by the New Hampshire Democratic Committee at which six candidates were given ten minutes each to speak. It was an early opportunity for Birch to be on the same stage as others who already had been running. Joining him were Udall, Carter, Shriver, Harris, and Shapp.[7] On October 15, Birch gave a political speech in Claremont, New Hampshire; three days later spoke in Davenport and Vinton, Iowa.

The announcement of a candidacy, a ritual that all candidates go through and that is never a surprise to anyone, is almost never the actual start of a campaign. For the Bayh campaign, the announcement was closer to the actual commencement of the campaign than most people realized, which made his friends, staff, and associates very nervous. With Marvella and Evan by his side, Birch announced his candidacy at his farm in Shirkieville. After the speech, he repeated the announcement in cities around Indiana. Witcover described the event with a level of cynicism that only a Washington political reporter would write. According to Witcover, on October 21, "on his farm in Shirkieville, Indiana, Bayh officially took the plunge. For sheer corn, the scene rivaled anything Bayh had ever done in the past, and it was a perfect beginning for his heavy-on-the-trappings, light-on-the-substance campaign." Birch's remarks included the words "Those of you who know me longest here know I've never had a burning desire to be President of the United States. . . . I felt closer to my God and I felt more fulfilled out in these fields than anything else I've done," which caused Witcover to write, "The prime polyester candidate was now in the race and the campaign could go forward, assured that not a cliché would be left unspoken or an opportunity for the banal left untapped." He added that later on, after Birch had been needled by political reporters for those remarks, he "insisted it had been genuine, but said on reflection it probably had been a mistake to say it and he wouldn't say it again—at least to guys like you."[8]

For those who knew Birch well, we could admit that he was sappy at times and often corny to an extreme, but it was genuine. However, being corny with national political reporters was just not shrewd politics. Many years later, those qualities remained.

Announcement day was a long ordeal and included short speeches and huge crowds. As a baseball fan, it was a tough day for Birch to be away from television. It was the sixth game of the World Series between the Boston Red Sox and the Cincinnati Reds. At each stop, Birch gave his speech and then was told the score of the game. Each time it was checked, it seemed that the teams were alternating taking the lead. Birch wasn't able to see the twelfth-inning home run by Carlton Fisk that ended a game that many have described as the best game ever played in the World Series. The Reds defeated the Red Sox the following day, watched by the largest audience ever for a sporting event up to that point.

Soon, Birch was back in New York City for a press conference, and then he was in the Senate to announce hearings on gun control legislation. There couldn't be too many more commitments for hearings or other Senate business if he was to make up for lost time in the presidential sweepstakes. Nonetheless, the list of subjects he had spent time and effort on in 1975 was daunting.

The campaign was off and running in November, with Birch giving speeches on crime, revenue sharing, the SST, and food stamp reform. There were endorsements in November by a New Hampshire senior citizen's organization and a group of New York elected officials. The Nebraska Committee for Birch Bayh was formed. Over the last two weeks of the month, he received the endorsement of Massachusetts political leader Tom O'Neill, son of Rep. Tip O'Neill. In 1977, Tip O'Neill would become Speaker of the House of Representatives. Birch won the straw poll in Waterloo, Iowa, a state that for the first time figured prominently in the early presidential primary and caucus season. Finally, the campaign announced the hiring of Jim Friedman, a Cleveland attorney, to be its campaign manager, with Jay Berman as campaign director. The campaign chairman would be former governor Matt Welsh; Myer Feldman, a former counsel to presidents Kennedy and Johnson, was named finance chairman. Ann Lewis, a former aide to Boston mayor Kevin White and sister of future Massachusetts congressman Barney Frank, was named deputy campaign manager.

The reality was that the campaign was unprepared for an effort of such magnitude. When Jim Friedman arrived on the scene, Ann Lewis was the lone staff person. The operation had no headquarters, no finance plan, no field organization, little to no fundraising, and no system for financial reporting. Friedman first needed to find a headquarters but was afraid that a lengthy lease might be a mistake. Birch had not asked Friedman to serve as campaign manager until after his announcement to run.

At the time, Friedman was a thirty-four-year-old former chief of staff to Gov. Jack Gilligan in Ohio. Birch had visited Ohio often, especially while exploring his candidacy, and had met Friedman several times. Friedman didn't even begin his job in Washington until mid-December. He and Lewis found an office at 1801 K Street in northwest Washington, a few floors down from the Reagan campaign, which was also trying to unseat President Ford but in this case during the nomination process. Friedman felt the experience was "a great challenge, a great honor," but he was overwhelmed with the tasks before him that should have been implemented long before. He had to turn his attention to hiring staff and creating the necessary campaign functions, particularly involving fundraising and managing the campaign's financial resources.[9]

Campaigning in 1976 was different for one principal reason. The post-Watergate reforms created new contribution limits and a system of matching funds. Money had to be raised in certain denominations in a requisite number of states with extremely detailed reporting before the government would match the funds raised. This required a financial operation that hadn't yet begun. People didn't seem to understand the changes in politics or the increased cost of campaigning. Campaign infrastructure had taken on an importance that wasn't yet fully grasped by politicians and their staffs. For Democrats, this new reality was even more important because of the sheer number of candidates entering the race.

Jay Berman would be fundraising once again, heavily relying on money from the entertainment industry, which he and Birch assiduously courted over the years.[10]

Barbara Dixon specialized in women's issues in Bayh's Senate office, making her an important player in the women's movement. She joined the staff as a legislative assistant when the Bayh office, out front on women's issues, sorely needed a female legislative assistant. She quickly became engaged with women's groups as well as issues like Title IX regulations and worked against the anti-abortion, pro-life amendment in the Judiciary Committee.

She recalled how Birch struggled with the abortion issue, finding the practice repulsive but instead focusing on the question of who must ultimately decide the fate of a pregnancy: the woman with her doctor and possibly her church, or Congress. Barbara tried to get him to stop saying that he believed a fetus to be a life, arguing that it made his position harder to defend. He refused to change; he believed "we are talking about life" but also felt the decision had to be left up to the woman and not to politicians. During the 1976 presidential campaign, Dixon was able to get written endorsements from women activists. Gloria Steinem signed a fundraising letter for Birch, emphasizing his leadership on women's issues. Cathy Douglas, wife of the recently retired Supreme Court justice William O. Douglas, wrote a letter supporting Birch on the environment. Given Birch's achievements on behalf of women, it was overdue for leaders of the women's rights movement to support his career.[11]

Congressional Quarterly (CQ) produced a book on the coming campaign season and emphasized Birch's relationship with organized labor. He was the only contender to attend regional labor conferences in late 1975 and to be invited to an AFL-CIO meeting in October. The journal added that he had always managed the sometimes conflicting positions of labor and blacks. The Philadelphia Plan, a proposal to set black employment quotas in federal construction projects, was opposed by labor but supported by Birch. He also parted ways with unions when he supported Richard Gordon Hatcher to be mayor of Gary, Indiana. "Bayh's support of black causes has been consistent. He has supported the major civil rights bills throughout his Senate tenure, including voting rights, public accommodations and fair housing laws. He has been an advocate of federal funding of community and urban development programs and other efforts to relieve inner-city distress." The section on Birch also outlined his support for Hispanics, the poor, and especially women's rights.

Congressional Quarterly also pointed out his potential weaknesses, including his down-home style before sophisticated audiences, particularly pointing out his "closer to God" comment and his late entrance into the race. The book listed each candidate's rating by organizations that reflected differing philosophical views on politics. The liberal Americans for Democratic Action (ADA) ranked Birch at 62 percent, a lower score than in previous years due to missed votes while campaigning. He received a perfect 100 percent rating from the AFL-CIO's Committee on Political Education (COPE) and from the National Farmers Union

(NFU). The conservative Americans for Constitutional Action gave him 6 percent.[12]

Birch made an appearance at the National Press Club in Washington, DC, and was asked about a famous quote from Thomas Marshall, a Hoosier who had been Woodrow Wilson's vice president. Marshall was famous for the quip, "What this country needs is a really good five-cent cigar"— the only reason Marshall was remembered.[13] Still standing on the rostrum, Birch reacted with his own quip, suggesting, "What this country needs is a good five-dollar bag of groceries."

The Iowa caucuses were the first opportunities for voters to state their preferences for a candidate, and they were shaping up to be more important than ever. Different from primaries, caucuses resemble conventions where delegates are chosen. The first round in January elected delegates to the county conventions in March, followed by regional conventions in April, and the state convention in May, and finally delegates were elected to the national convention in July. Birch was encouraged by winning some early Iowa straw polls. A week before the caucuses, he shared the stage on *Meet the Press* with candidates Carter, Harris, and Shapp.

The genius of the Carter campaign was in the manner it made the Iowa caucuses important. They had been largely ignored in previous election years, but by luring top-tier candidates to compete with him, Carter created the perfect scenario in which he was best positioned to prevail. He had been virtually living in Iowa since he left the Georgia governorship in early 1975, though he began laying the groundwork prior to that.

The Bayh campaign would compete in the Iowa caucuses on January 19, followed by the New Hampshire primary on February 24 and the primary in Massachusetts a week later on March 2. Assuming he was still viable at that point, he could hope for a knockout punch in the big states of Florida on March 9 and New York on April 6. If anything made him feel confident, it was his organization in New York, and if he could remain in the race until then, it just might end there with him as the presumptive nominee of the Democratic Party.

The Carter campaign plan was to enter and win everywhere, expecting a snowball effect from the first win onward, with each win following on the heels of the publicity and notoriety from the previous victory.

Other campaigns seemed prepared to sit out Iowa, giving it little attention or none at all, waiting for the New Hampshire primary or, in the case of Jackson and Wallace, for the Massachusetts primary. For the Republicans, the principal news was Reagan's November 20 entrance into the race to oppose his own party's president.

On December 6, New York's New Democratic Coalition (NDC), "a collection of 112 liberal reform Democratic clubs all across the state," held a convention to determine whom to endorse in the April primary. The NDC nod would result in substantial Democratic manpower in the primary campaign, and Birch made it a priority. Because of the role of Bob Abrams, Birch was able to move quickly to the front of the pack seeking the endorsement. He made several visits to the state, arguing that he was the most electable of the aspirants, using the slogan, "Birch Bayh: The One Candidate for President Who Can Put It All Together." As things evolved, his main competition for the endorsement came from Fred Harris. The two candidates packed the galleries with their rowdy supporters at the NDC convention. When the votes were tallied, Birch had 59.974 percent to Harris's 30.21 percent with the rest voting "no endorsement," just shy of the 60 percent requirement for an endorsement. In fact, one delegate asked to switch to Bayh after the final vote was announced and was refused. Although Bayh trounced Harris almost two to one, being denied the endorsement was a painful defeat.[14]

On December 14, Bayh endorsements were announced from New Hampshire labor leaders and the following day from Iowa teachers and the Harvard Young Democrats. *Newsweek*, in an article summarizing the campaigns for the Democratic nod, described the Bayh campaign:

> In the little more than a month since he declared, Bayh has propelled himself into serious contention; he has won an impressive labor endorsement in New Hampshire and proven his liberal appeal by coming within less than one percentage point of claiming the New Democratic Coalition's endorsement in New York—a power play that deprived Udall and Harris of a boost they had hoped to earn. The Hoosier's aim is to knock fellow liberal Udall out of the early races, then beat Jackson—or whoever is the surviving opponent on the party's conservative side—for the nomination. To do that, the Bayh game plan now is to scrape up what he can in the Iowa precinct caucuses, then run ahead of Udall in New Hampshire and Massachusetts and concentrate on cutting down Jackson in New York and Pennsylvania. To stay in the running, Bayh must also soundly defeat Wallace back home in Indiana on May 4. Holding off on issues, Bayh's

> main pitch is his appeal to the old Democratic coalition: as a result of his fight against the Carswell and Haynsworth Supreme Court nominations and for the Equal Rights Amendment and the 18-year-old vote, Bayh appeals to big labor, blacks, women and young liberals. "He could be the strongest candidate of all because of his wide acceptability," says one hopeful staffer.[15]

Early in 1976, Birch took part in an event that the *Wall Street Journal* characterized as "the first serious contender for a major party's nomination to campaign in a gay bar." The Gay Political Union in New York City was an early organization to promote homosexual rights in American politics, and the organization arranged a reception for Birch. At the time, being identified with gay activists was at the very least unusual, if not altogether politically suicidal. The article said, "Accompanied by an entourage of Secret Service men (who didn't appear to be enjoying themselves), Sen. Bayh spoke and shook hands for about an hour."[16]

1970 Congressional Baseball Game.

Birch and Marvella.

Facing: Rally in the Dirksen Senate Office Building after the Kent State shootings, May 1970. *Courtesy of the Library of Congress.*

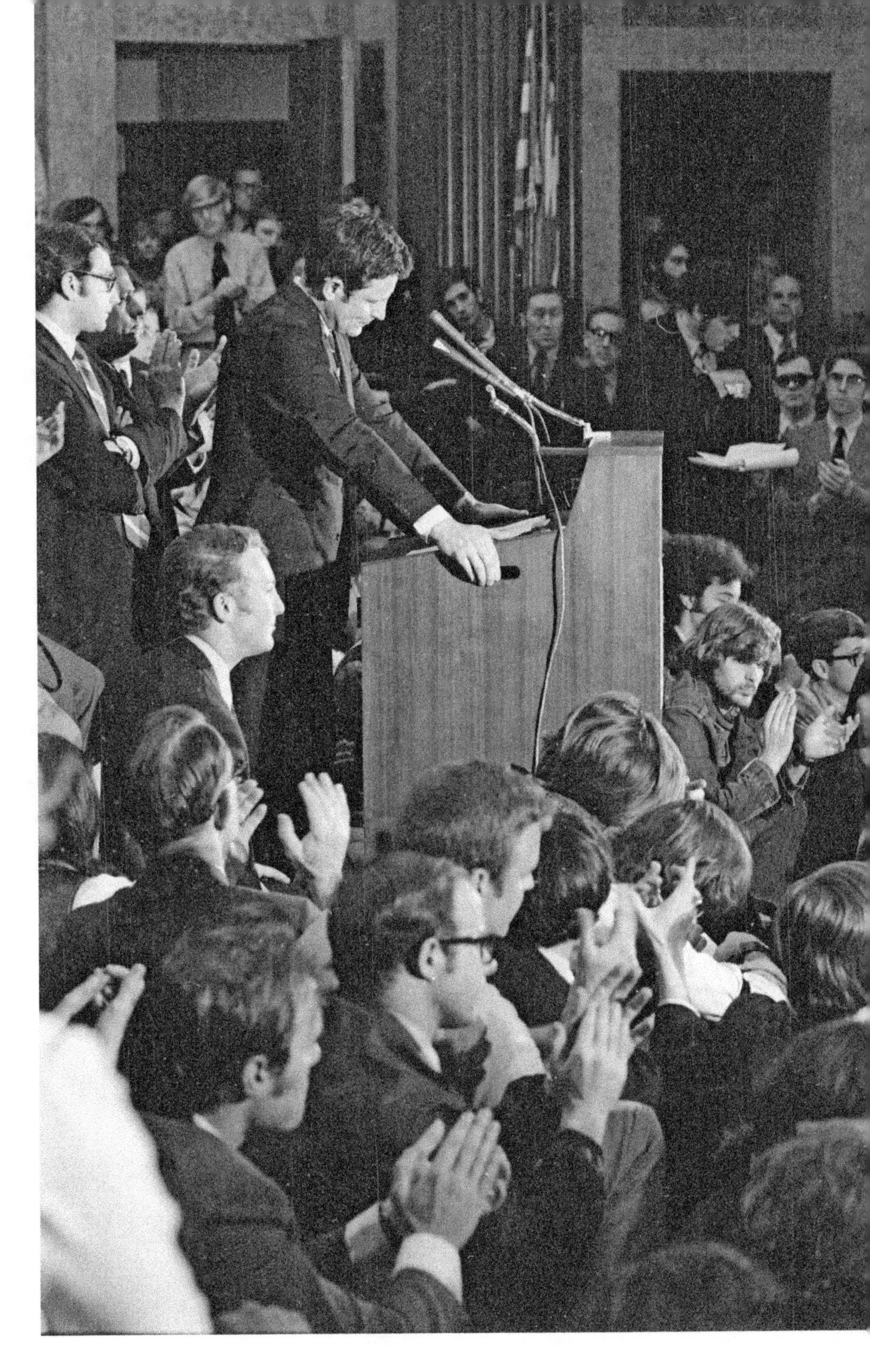

Campaigning in 1974: Author over Birch's right shoulder.
Facing: Brochure from 1974 Senate reelection campaign.

BIRCH BAYH
One man
who makes
a difference

Birch Bayh and Senator Ted Kennedy campaigning in Merrillville, IN,
September 18, 1974.

Birch and Marvella at the announcement of his candidacy for the 1976 presidential campaign at the farm in Shirkieville, October 21, 1975. *Photo from Marvella: A Personal Journey by Marvella Bayh with Mary Lynne Kotz.*

Senator
Birch Bayh
The Democratic
candidate for President
with a plan for
economic recovery...

Brochure from 1976 presidential campaign.

Facing top: 1978 Congressional Baseball Game.

Facing bottom: Birch Bayh and Arkansas governor Bill Clinton, 1979.

BAYH
22

With President Jimmy Carter in the Oval Office.

Birch Bayh with civil rights leaders (left to right) Jesse Jackson, Vernon Jordan, Andrew Young, and Gary mayor Richard Gordon Hatcher at a fundraising dinner for Hatcher, October 20, 1978. *Courtesy of Joe Smith.*

Birch Bayh jogging with female student athletes at Purdue, beneficiaries of Title IX, 1979.

Birch Bayh after the Senate.

Bottom: With President Bill Clinton and Hillary Clinton at the World Mine Property Agreement announcement at Yellowstone National Park, August 12, 1996. *Courtesy of William J. Clinton Presidential Library, NARA.*

Christopher, Kitty, and Birch Bayh at the dedication of the Birch Bayh Federal Building and US Courthouse, October 24, 2003.

With President Barack Obama.

Birch, Kitty, Jessica, and Christopher Bayh with Billie Jean King at White House ceremony on the fortieth anniversary of the passage of Title IX, June 20, 2012.

The author with Birch Bayh at the wedding of Jessica Pedley and Chris Bayh, 2010.

14

Bayh for President: 1976

All progress has resulted from people who took unpopular positions.
All change is the result of change in the contemporary state of mind.
Don't be afraid of being out of tune with your environment.

ADLAI E. STEVENSON[1]

A BAYH CAMPAIGN TACTIC WAS TO TAKE ADVANTAGE OF THE large number of enthusiastic supporters he had in Indiana, many of whom had surfaced during the previous campaign year. The Hoosier Travelers Program was designed for people to campaign for Birch in Iowa and New Hampshire at their own expense. Of the various forms of voter contact in political campaigns, the greatest impact is person-to-person. If the face-to-face contact is between persons who know one another, that is preferred, and using Bayh volunteers from his home state to talk with Iowa and New Hampshire voters was the next best method. Campaign coordinators in each state would find lodging for the travelers in the homes of supporters. Evan Bayh and his college roommate left school to join the travelers in Iowa.

One night in December, Birch called from Iowa, wanting to know how many people were coming to Iowa from Indiana. When given the update, he said, "No matter where I go in this state, that goddamned Jimmy Carter has been there four times before me." Despite his concern about Carter's coverage of Iowa, he didn't take Carter seriously up to that point. Carter was a one-term governor of Georgia who seemed overly pious, wearing his religion on his sleeve, and Birch probably felt the same about him as the other contenders did. They all knew he was clearly smart but didn't always come across as genuine. Carter was also a full-time politician who wanted to pretend that he wasn't.

Iowa caucuses presented a great unknown, given the structure of the meetings that would take place and the effort required for attendees to win in their respective caucuses. It was difficult to poll for caucus results because it wasn't simply a matter of candidate preference. If you wanted to be a Bayh delegate to the national convention, you needed to win the votes of your compatriots in the caucus meeting you were attending. It was important to understand that being for a certain candidate and getting elected a delegate were two very different things.

Now, after decades of presidential contests that were influenced by the Iowa caucuses, the results have become easier to predict. Knowing who had previously been elected to caucuses helps. Caucus watchers in the media have earned their stripes by covering multiple caucuses for both parties. But that was not true in 1976.

Carter's campaign was the first to recognize the potential importance of Iowa. Birch had visited many of Iowa's ninety-nine counties while campaigning in 1971 and felt that he had laid the groundwork for a successful 1976 campaign. Iowa's effort to cast the first votes before the fabled New Hampshire primary was successful and has remained a prominent fact of presidential campaigns ever since. Even the press took a while to appreciate the importance of the Iowa caucuses, and by then, Carter had been blanketing the state. When Birch started campaigning there, he was worried that his late start created a huge disadvantage but felt he needed to block Udall from becoming the leading liberal candidate. Neither the national press nor the other candidates took Jimmy Carter seriously, but his activities were largely responsible for attracting the interest of the other candidates. No one wanted an unknown candidate like former governor Jimmy Carter, a peanut farmer, to win simply because Bayh, Udall, and the other competitors were not paying attention.

When votes were tallied from the caucus sites around the state, the winner was "uncommitted," with 37 percent. Carter was first among the candidates with 27.6 percent, and Birch was next with 13.1 percent, followed by Harris, Udall, and Shriver in that order. Carter's success made everyone take notice.

Iowa was "really a blow to me," Birch recounted many years later. He had thought his 1971 efforts had created a good base of a support and a reason to feel optimistic. Plus, no one was yet taking Carter seriously. But Birch's campaign was only three months old; coming in second among the five men running seemed to be a substantial accomplishment. Carter, on the other hand, had been spending much of the previous two years in the state.

For the Republicans, the Reagan candidacy presented a serious threat to Ford's renomination. In Iowa, the president won the caucuses barely, by less than 3 percent.

Writing in *Rolling Stone* magazine about the caucuses, Joe Klein observed that no one was running away with a win in Iowa and asked why "none of the above" prevailed over all of the candidates. One activist was quoted as saying about his decision, "I did it by subtraction. I wanted to go with the guy who offended the least people. Birch Bayh offended the least people."[2] Nonetheless, the real story coming out of Iowa was the emergence of Carter.

James Earl "Jimmy" Carter ran as the "non-politician" during a period when, thanks to Richard Nixon, politics was scandal-ridden. When he came out of Iowa as the newly crowned frontrunner, the other candidates took aim at him. Birch saw Carter as a hypocrite. He asked, "How can a man who is a former governor and has been going around the country running for president say he isn't a politician?" To deal with that issue, the Bayh radio ads contrasted Birch's own political background with that of the candidate who confessed to being outside of the profession. The ads stated, "It takes a good politician to be a good president." There were various versions of television spots promoting the same theme, with different ads listing different accomplishments. The media campaign was managed by press secretary Bill Wise, who played a major role in Birch's communications strategy and hired media consultants Don Madden and Tony Isadore. But Marvella saw the ad and urged Birch not to run it.

To win the primary in New Hampshire, you need boots on the ground—a highly organized effort of people going door to door. The lateness of Birch's start made that difficult. While Birch's strategists felt that New Hampshire was the kind of state where he performed well, the development of an effective organization is time-consuming, and Birch was short on time. Nonetheless, the Hoosier Travelers program delivered several hundred people to the Granite State to join with hundreds of others going door to door for other candidates—probably more than a thousand canvassers in all. When Jim Friedman took over as campaign manager, he realized that Birch had no organization in New Hampshire and dispatched Mike Ford, an Ohio operative and friend, to create a campaign from scratch. Chairing the Bayh effort in New Hampshire was Chris Spiro, minority leader in the New Hampshire House of Representatives. Getting Spiro aboard had been a real coup and a shot in the arm to Birch and the rest of the campaign staff. Some polling was done, but not as much as would be preferred, because the budget was so tight. Two pollsters in their twenties, Dick Dresner and Dick Morris, the latter of whom would gain considerable notoriety in politics, conducted periodic polls.

The Bayh campaign handed an instruction sheet to each of its canvassers that said,

> This is it, folks. Effectively, this weekend represents our last large-scale contact with the New Hampshire electorate. . . . [I]n New Hampshire, more so than in most states, personal contact is by far the most effective campaign device. . . . New Hampshire is an organizer's dream—there are only 116,000 registered Democrats in the entire state, with 85% of those Democrats living in the population centers in which you will be canvassing. The people in this state are political pros in terms of campaign techniques, as most every technique known to the world of politics has been tried on these people at one time or another. The voters know precisely what we are doing. So does everyone else. We must tell you that the techniques we are using are also being used by every other campaign as well. This is natural, as many of us had the same training ground—the McGovern campaign.

Jules Witcover wrote cynically on the political activities leading up to the New Hampshire primary but provided a useful description of Birch's New Hampshire campaign:

> Of all the others, Bayh seemed to be doing the best. Trying to make up for lost time, he grimly set about cleaning up his own act. That is, he tried to put his aw-shucks style on the back burner and come on as the Experienced Legislator, which his record substantiated but his cornball manner and boyish looks always blurred. To achieve this reformation, his campaign cooked up a special routine: a documentary film emphasizing Bayh's record in the Senate, which was shown at evening gatherings in high-school cafeterias and the like. When it was over

> and the lights were turned on, a local supporter holding a hand microphone with a long portable cord would stand in front of the screen and, like some Yankee Ed McMahon, proclaim: "Ladies and gentlemen, the next President of the United States, United States Senator Birch Bayh!" From the rear, wearing his dark-blue sincerity suit and a serious look, would stride the senator from Indiana. He would take the mike, let some of the cord out behind him, and then start roaming around the room, a political Johnny Carson on the "Tonight Show," talking up his experience. "Why Birch Bayh?" he would ask, in that maddening habit of calling himself by his full name, in the third person, and then proceed to tick off the "tough battles" he had fought in the Senate: opposition to the Haynsworth and Carswell Supreme Court nominations; extension of the voting-rights act; passage of constitutional amendments lowering the voting age to eighteen and revamping presidential succession. (One problem with this whole approach was that it reinforced Bayh's image as a Washington political creature, in a year when Jimmy Carter was making hay running against Washington. But he had no choice; if his experience was vulnerability, it was also his underutilized strength.)[3]

As primary day approached, Marvella told Birch that she wanted to come up to New Hampshire and campaign with him. He reminded her that they had an agreement—that she shouldn't jeopardize her job with the American Cancer Society and therefore should not spend time campaigning. On the other hand, even though she had been largely absent during the 1974 campaign, it was still strange that their long-standing partnership was not in action as it had been.

While the focus was clearly on New Hampshire, efforts were needed to get on the ballot and have slates of delegates in other states. Harold Ickes was dispatched to manage Pennsylvania. At one point, he remembered sitting in a phone booth in Harrisburg's train station, holding a cup full of dimes so he could make a series of phone calls to Pennsylvania supporters in order to fill the delegate slate. Ickes was more successful in getting delegates onto the ballot than was the sitting governor, another Democratic contender, Milton Shapp.[4] It was an unusually daunting task. Pennsylvania requires delegates to circulate petitions to get themselves on the ballot and a second set of petitions to get their candidate on the ballot. Only forms produced by the Secretary of State's office could be used, and after completion they had to be notarized by two notaries. In Connecticut there was a Bayh effort as well, with Joe Lieberman among his supporters. Lieberman was then a state senator who would go on to become a US senator and a Democratic nominee for vice president.

The New Hampshire primary was February 24, and Carter was the frontrunner. The outcome would determine which candidate would wear the liberal mantle in opposition to Carter, something that never made Birch totally comfortable. He had entered Iowa to combat Udall but found that it became a contest against Carter, and in the first primary, he found himself having to lower expectations. In this instance his instincts were right, and when the votes came in, he ran third, with Udall becoming the "liberal leader." Carter had won 30 percent in the crowded field, with Udall second at 24 percent and Birch at 16 percent. Fred Harris's campaign often seemed to be the most fervent and emotional. His fourth-place finish found him sending strong signals that he was about finished, as was Shriver.

Not long after the primary, post-election polls showed that despite the expectation that the second choice of Bayh's supporters would be Udall, they were largely for Carter instead. This raises the question whether Birch should have competed in New Hampshire at all, given his late start. If it can be assumed that the lion's share of the Bayh votes would have gone to Carter, the Georgian might have beaten Udall so badly that it would have ended the Arizonan's race. In that scenario, Birch could have picked up the liberal mantle. At the same time, it can be argued that if he had avoided the contest with Carter in Iowa and begun in New Hampshire, Carter might have knocked out Udall in the caucuses, changing everything later on. Instead, he moved on to the Massachusetts primary the following week, after a disappointing showing in New Hampshire.

For the Republicans, New Hampshire was almost a carbon copy of Iowa, with Ford defeating Reagan by a margin of less than 2 percent. But Ford went on to win by considerably larger margins in Massachusetts, Vermont, Florida, and Illinois.

A brochure produced for the March 2 primary focused on Birch's accomplishments and succinctly summarized his career up to that point. One page was headlined, "This year, when the Presidential candidates tell you what they're going to do, ask them what they've done." Each page highlighted key parts of Birch's Senate career. One said, "When no one else would take on Nixon, Birch Bayh beat him . . . twice," going on to describe the Haynsworth and Carswell nominations. The brochure also said, "Birch Bayh wants to create jobs by throwing two guys out of work," specifically referring to Treasury Secretary William Simon and Federal Reserve Board chairman Arthur Burns, two architects of the Nixon–Ford economic policies that were a major campaign issue. Another matter affecting the economy was oil prices and the lack of competition in the oil industry; in the brochure, this was highlighted with the Bayh quote, "Let's give the oil companies a dose of free enterprise." On the subject of women's rights was a page labeled "Susan B. Anthony would probably vote for Birch Bayh" describing his efforts on the ERA and Title IX and as the first senator to introduce major child care legislation in the Senate. Columnist Marianne Means was quoted as saying, "Birch Bayh has done more to advance the cause of women's rights than any other single Senator." Another key issue was crime; a page in the brochure stating, "Birch Bayh wants to stop giving out advanced degrees in crime. That's why he wrote and passed the Juvenile Justice and Delinquency Prevention Act of 1974." And finally, another page in the brochure said, "Birch Bayh has written more words in the Constitution than anyone since James Madison."

Carter's strategy to win early meant that he would also compete in Massachusetts. Carter had long felt that he would be the nominee and that he had a decent chance to win. He was not intimidated by a possible Ted Kennedy candidacy. In fact, his attitude was "Bring him on." Hamilton Jordan, Carter's campaign manager, wrote a memo in August 1974 laying out the plan for obtaining the nomination. Witcover writes, "Of all the 1976 Democratic hopefuls, only one—Jimmy Carter—could be said to have been somewhat let down by Kennedy's decision not to be a candidate. He alone had predicated a 1976 campaign not on the possibility that Kennedy would stay out, but—brazen as it seemed at the time—in the hope that he would get in, keeping all the other ambitious Democrats on the side lines."[5]

Udall needed to wound or defeat Birch by the time Massachusetts was over, as next up was New York, where Birch was particularly strong. For the Bayh strategists, there were unrealistic beliefs that the plane crash with Kennedy would translate into support. But that had been twelve years earlier—an eternity in politics. Massachusetts followed closely on the heels of New Hampshire; without a win there, the unraveling would begin. The "I'm a politician" ads were being replaced, but trouble loomed large on the horizon, partly because there were insufficient funds to hardly run ads at all. Harold Ickes recalled objecting to an ad that showed Birch on a tractor, looking over his shoulder. Ickes knew that anyone plowing a field would look straight ahead to avoid running over his crops.

Labor leaders wanted to defeat the more conservative challengers, Jackson and Wallace, and many concluded that Birch was not the candidate to fill that bill. Carter began to look acceptable to many; others were looking more closely at Udall.

Carter appeared vulnerable after stating that he would do away with the home interest deduction, something several candidates jumped on at once. Birch's reaction was to suggest that Carter's move "would undercut the chances for millions of Americans to buy better homes." His criticism was echoed by Jackson and Udall, causing Carter to lash out at it as being unfair, treating it as if it were coming from Joe McCarthy or as though it were similar to Nixon's dirty tricks. Birch responded by criticizing the former governor on issues of importance to labor. Whether the series of attacks and responses had an impact is hard to say, but the major effort mounted by the Jackson campaign was the bigger story. As the frontrunner, it was not at all surprising that Carter was attacked by most or all of his opponents, nor that those attacks would have an effect. Nonetheless, Carter's finish in Massachusetts was a disappointment, but not as disappointing as Birch's. Jackson with 23 percent of the vote was the big winner, 5 percent ahead of the second-place finisher, Udall. Wallace was a close third, and Carter finished fourth with only 14 percent. Birch received only 5 percent of the vote and knew that his wounds were fatal.[6]

Birch Bayh won only a single convention delegate, sarcastically referred to as "the million-dollar delegate." The delegate was Tom O'Neill, son of the future Speaker Tip O'Neill. Mark Shields, then a campaign operative and later a national political pundit, referred to Tom as "Tiplet." Bayh's staff knew that the money would start drying up and that the momentum that is so important in primaries was not with their candidate. The day after the Massachusetts primary, Birch withdrew from the presidential campaign.

In Indiana, preparations were on for the May primary. It was likely that most candidates would not oppose Birch on his home ground but that George Wallace, who had run so strongly there in the past, just might. If that ended up happening, a Bayh–Wallace debate might be staged that could end up with Birch being the Wallace-killer. The prospect may have been far-fetched, but regardless, the campaign was over.

The campaign had been crippled from the start. It started too late and was made harder by the competition for Birch's time between the campaign, his Senate duties, and Marvella. For instance, Marvella made it clear that she and Birch had agreed that he would be home every Monday night. David Rubenstein, a former Bayh staffer who joined the Carter campaign, looked back on Bayh's campaign in later years and concluded that to run for president meant doing only that, including missing all the Senate votes, something Birch did not do. "Carter was not qualified to be president in the traditional sense," commented Rubenstein. Carter hadn't climbed the ladder of politics, but he had concluded that an early effort in Iowa was the key to success. It was clear that with an earlier start and a greater effort in Iowa, Birch could have won there, and then it would have been, in Rubenstein's words, "off to the races." It was also clear that everyone underestimated Carter, who had, according to Rubenstein, for more than a year devoted six days a week to campaigning. But he always returned to Plains, Georgia on the weekend to teach Sunday school.[7]

Elizabeth Drew wrote that Birch's strategy "seemed so sound on the drawing boards . . . to build a candidacy on a collection of blocs, including labor (traditional and liberal), farmers, blacks, women, and Jews. The strategy so impressed Party professionals that in January some were predicting Bayh's nomination. But it collapsed utterly at the hands of the voters."[8]

Following Birch's withdrawal, Democratic activists who had supported him largely gravitated to Udall, particularly in the New York primary, which was coming up. But Udall shifted his efforts to Wisconsin, not New York. Carter won Wisconsin anyway. In New York, Jackson won but hurt himself by predicting a landslide victory, which didn't materialize. The expectations game often dominates presidential primary politics, and Jackson set a trap for himself. His victory was characterized as disappointing, as reported by Witcover.[9]

Three weeks later, Carter won a resounding victory in Pennsylvania. Rumors abounded that Hubert Humphrey would get into the race, and there were a few potential candidates not yet running: Sen. Frank Church, Gov. Jerry Brown of California, and possibly Humphrey. Democrats wondered whether an "Anybody but Carter" campaign had any chance at all.[10]

In late April that year, I ran into Peter Emerson, who had traveled with Birch throughout the presidential campaign as I had done in 1974 and was disappointed not to be doing so in 1976. When Birch dropped out, Emerson had joined the Carter campaign. He told me that Bayh was the candidate that most concerned Carter before Birch ended his effort.

Carter was amassing so many delegates that it appeared impossible to stop him, despite new candidates getting into the race and the unwillingness of Mo Udall to get out. The Georgian won the next three states and Indiana on May 4. Birch endorsed Carter the day before the Indiana primary, acknowledging that he was "philosophically closer" to Udall but adding, "One thing I learned long ago is how to count."[11] No one expected a Bayh endorsement to change the results of the primary, but it seemed sensible to get on the good side of the likely nominee. A side issue was created, though, that raised questions about Carter's truthfulness. At the endorsement, Birch said that Carter phoned him the week before and asked for his help. The following day, Carter stated that he had "never gone to anyone yet and asked them to endorse me." In response to a reporter who pointed out the contradiction, Carter said, "My point was that I have never depended on endorsements to put me in office," and he said that he had asked for Bayh's help, not specifically for an endorsement.[12]

The Indiana primary was dramatic for other reasons entirely. Congressman Phil Hayes, first elected in 1974, challenged Vance Hartke for the US Senate nomination. Hartke narrowly won and was heading into a difficult general election against Richard Lugar, the man Birch had defeated. Larry Conrad won the Democratic nomination for governor to face the incumbent, Otis Bowen. Congressional nominations included state senator Adam Benjamin, defeating longtime incumbent Ray Madden in the 1st District. Madden was first elected in 1942 during the Roosevelt administration and was someone who always referred to that era when discussing a current event. In the 4th District in the northeastern part of the state, the Republican nominee for Congress was a young, attractive candidate named J. Danforth Quayle. Dan Quayle was the scion of the Pulliam family, publishers of newspapers in Indianapolis and Huntington, among other cities. He presented a solid challenge to Ed Roush, the incumbent. Because of his challenge to Hartke, Hayes did not seek reelection to Congress from the 8th District in the southwest corner of the state. The new nominee for his seat was David Cornwell. In the Republican primary, Indiana's conservative bent showed itself by the defeat of President Ford at the hands of Reagan. Wallace voters likely realized that requesting a Republican primary ballot and supporting Reagan was a more comfortable way to participate.

Shortly after the Indiana primary, Hartke's wife, Martha, confronted Birch, saying, "Well, you didn't get us, did you?" Birch told her he wasn't involved in that race at all, which should have seemed obvious to anyone observing him during that period. He said that if he had gotten involved in that race to oppose Vance, he "would have been a fool to do so." After all, they shared many of the same supporters in Indiana, particularly among Democrats. It was clear to Birch that the Hartkes perceived his real feelings about Vance, but there was no way Birch would take any steps to hurt Hartke politically.

The presidential primary season continued with Brown and Church getting into the race. Humphrey eventually announced that he was not running. This meant that it took Carter longer to secure the delegates he needed to get the Democratic nomination, but it was still resolved well in advance of the convention. The Republican primaries remained competitive, with Ford and Reagan trading victories.

In his book *The Last Great Senate,* Ira Shapiro described the Democrats who had thrown their hats into the presidential campaign ring:

> Bayh, Church, Jackson and Harris quickly discovered what many senators had found out before them: you could be a very good senator and still be an awful presidential candidate. Bayh's liberal supporters split in too many directions, and he never won a primary. Church started much too late. Jackson, extraordinarily effective on Capitol Hill, was simply unable to communicate on television or in speeches. As one humorist noted, "If Scoop Jackson gave a

> fireside chat, the fire would go out." After Harris was crushed in the crucial New Hampshire primary, receiving less than 1 percent of the vote, he explained: "The little people couldn't reach the levers."[13]

Once out of the campaign, Birch had to deal with a health issue of his own. A checkup revealed a growth on his lung, and it was quickly determined that he would need surgery. After the surgery, it was revealed that the tumor was benign and the result of histoplasmosis, a condition common among people who live in Indiana's Wabash Valley. Marvella was thrilled to tell Birch that the doctors claimed it a complete success: they had removed the noncancerous growth.

As Carter began considering possible candidates for vice president, his head of congressional relations, Frank Moore, talked with Birch and his staff about the process. Moore strongly intimated that Birch was on Carter's short list, but many on the Bayh staff argued that he shouldn't be believed. While it made sense for Carter to have a senator on the ticket because he lacked Washington experience, there was no evidence that he felt charitable toward any of those who had run against him in the primaries and criticized him so roundly in the process. Additionally, Birch could not guarantee any additional electoral votes for Carter from Indiana, which had only voted for a Democratic presidential nominee four times in the twentieth century. Even with Birch on the ticket, there was little reason to expect that to change in 1976. History has since shown that none of Carter's primary competitors, with the possible exception of Frank Church, was on the short list for vice president.

Carter was initially opposed to using other Democrats as surrogates in his presidential campaign, but he ultimately relented. Carter's campaign manager, Hamilton Jordan, came up with a list of fourteen names of people who would be asked to devote time to the campaign. Birch was recruited to campaign four days. Others asked for two to six days of campaigning were Udall, Brown, Church, Humphrey, Jackson, Kennedy, Wallace, representatives Peter Rodino and Barbara Jordan, senators Biden and Glenn, and governors Colon of Puerto Rico and Apodaca of New Mexico.[14]

Birch inadvertently made a contribution to the Carter campaign when the chief counsel of the Constitutional Amendments subcommittee,

David Rubenstein, left the Senate to join the Carter campaign. He was eager to work in a presidential campaign, later saying that his goal was "to play tennis on the White House tennis courts." When Rubenstein joined the Carter campaign, he found no paid staff concentrating on issues and policy; everything was focused on politics. He was "amazed how thin the Carter staff was." That was the area where he devoted himself, and he filled that void admirably. He also admired a Carter technique for writing thank-you notes. Rather than worry about how to address people, with titles or first names or nicknames, Carter addressed all notes "To first name, last name" in handwriting that was not only legible but almost a style a calligrapher would admire.[15] Replacing Rubenstein in the Senate was Nels Ackerson, an Indiana attorney who had once been president of Future Farmers of America and had met Birch in 1960 when he was the minority leader in the legislature.

The Democratic Party held its convention in New York City at Madison Square Garden beginning on July 12. Like most conventions since 1960, the Democratic choice for vice president was announced well before the delegates arrived. Sen. Walter "Fritz" Mondale was nominated as Carter's running mate. He was neither a surprising nor a controversial choice. Few issues needed to be resolved, and the party seemed united in its desire to end eight years of Republican rule. The keynote speech was given by Congresswoman Barbara Jordan of Texas, the first African American to keynote a party convention.

Birch was elected chair of the Indiana delegation. The only controversy facing the delegates had nothing to do with the nomination but was created by a shortage of guest passes for family and friends also in New York. It took an enormous amount of wheeling and dealing to acquire the necessary credentials, and more people were taken care of than anticipated.

One memory from convention week involves Birch and Paul Newman. Birch was chairing a platform session on energy, and the actor was on his panel. As Birch spoke, he leaned forward dramatically and looked into Newman's eyes, exclaiming, "They really are that blue!" Birch's blue eyes were pretty striking but couldn't compare with Newman's.

Once the convention was over, the campaign began in earnest. We knew that Indiana would not be in play, but Birch was given the names of the people who would run the Indiana effort. In charge would be Doug Coulter of New Hampshire and two women from Arkansas, Ruth Hargraves and Hillary Rodham. Bayh staffer David Bochnowski was friendly with Bill Clinton, Hillary's husband, who had been his classmate at Georgetown University and was running for attorney general in Arkansas. Clinton had impressed Bochnowski when they were in law school together.

Larry Conrad's campaign was the highest priority that year for Birch and his staff, even though they recognized what an uphill battle he faced. Birch tried to do as much as he could for Conrad and others on the ticket, and he spent a good deal of time on the road around the state.

The unofficial kickoff of the fall campaign was the weekend before Labor Day, held at the IDEA convention in French Lick, Indiana. The guest speaker at the 1976 event was California governor Edmund "Jerry" Brown. Brown had been a contender for the nomination and was an enormously attractive and charismatic candidate. Should Carter lose, Brown would be among the immediate front runners for 1980. The Bayh staff was told that he was to be picked up at the French Lick airport in an old blue Plymouth. After considerable effort, one was located, and the car was driven to the small airport in the countryside, several miles from the French Lick Sheraton where the event was held. The assembled Democrats would see Governor Brown arrive in this common auto, and the image he was attempting to portray would be maintained. For those who saw him land at the airport in a Lear jet, the entire affair appeared artificial and phony.

The GOP convention was held in Kansas City in late August. Reagan's effort faltered after he announced his choice for running mate, Richard Schweiker of Pennsylvania, a senator considered too liberal for Reagan supporters. As a result, Ford narrowly beat Reagan, 1,187 to 1,070. The Republican ticket would be Ford and Kansas senator Bob Dole. Vice President Rockefeller had announced his decision to leave the ticket the year before, something no one thought was voluntary. As has become customary, both the presidential and vice presidential candidates debated each other on national television. Dole's dark visage and sarcastic nature didn't do him any good in his debate with Mondale, and when President Ford characterized countries in the Soviet bloc as nations that didn't believe they were under the thumb of the Soviet Union, Democrats across the country rejoiced.

While it may have felt like there was no way Ford could be elected in his own right after pardoning Nixon and debating as he did, the race tightened in the final weeks. Carter made a few missteps, one of which was an interview he gave to *Playboy* magazine a few weeks before the election. Many Americans were shocked at the evangelical Christian deigning to be interviewed by *Playboy*. As strange as that seemed, the interview itself was worse. Carter liked to project a holier-than-thou image, and the negative optics of a *Playboy* interview could be politically fatal.

In an effort to quiet the uproar, Carter characterized his comments as "just part of being a human being" and described *Playboy* as "just another forum." But his remarks shocked voters across the country. Carter's comments would be remembered as singularly unpresidential:

> Anyone who looks on a woman with lust has in his heart already committed adultery. . . . I've looked on a lot of women with lust. I've committed adultery in my heart many times. This is something that God recognizes I will do—and I have done it—and God forgives me for it. But that doesn't mean that I condemn someone who not only looks on a woman with lust but who leaves his wife and shacks up with somebody out of wedlock. Christ says, don't consider yourself better than someone else because one guy screws a whole bunch of women while the other guy is loyal to his wife. The guy who's loyal to his wife ought not to be condescending or proud because of the relative degree of sinfulness.[16]

There was serious concern, so late in the game, that he might lose the race after all. One memory sticks out from when Birch traveled to La-Porte, Indiana. Joe Farina, the county chair, who looked like he stepped out of central casting for *The Godfather*, was running for county office. To associate himself with the Democratic standard-bearer, he borrowed the green and white colors of the Carter campaign; his Carter green bumper stickers with white lettering said "Carter–Farina." When the *Playboy* interview came out, Farina collected every bumper sticker that hadn't been distributed and cut off Carter's name. Seeing a car with a green and white "Farina" bumper sticker was hilarious.

Indiana is rarely a priority state in presidential politics, and 1976 was not any different. But the Hartke race, along with a few contested congressional races, gave it greater importance. As a result, Hubert

Humphrey came to Indiana to stump for the ticket and gave a barnburner of a speech. Reacting to the perception of Jerry Ford as such a nice guy, Humphrey agreed that he was, "but so is my Uncle Fred, and he shouldn't be president either."

Election night 1976 was the first time television election coverage used the colors red and blue to indicate Republican or Democratic results in a given state, now customary in American politics. Indiana was the first state to be shown in red that night. Ford defeated Carter in Indiana by a margin of about 169,000 votes out of 2.15 million cast. Carter won only twenty-six of Indiana's ninety-two counties. Nationally, however, he squeaked by Ford with 50.08 percent of the popular vote. Lugar defeated Hartke, setting a record margin of 384,000 votes (roughly 59%–41%). Hartke won only ten counties. The composition of the Indiana delegation changed, with Adam Benjamin winning in the 1st District, Dave Cornwell in the 8th, and Dan Quayle in the 4th.

Serving in the Senate with Lugar would be unusual, to be sure. It was also uncommon. Senators Glenn and Metzenbaum had competed in an Ohio primary and ultimately ended up serving in the Senate together, but we knew of no others. It would be interesting to see the former opponents Bayh and Lugar serving as colleagues.

Carter might not have become president if either Bayh or Udall had stayed out of the early battle for the nomination. It can be argued that the liberal activist vote was split between the two of them and Carter was the beneficiary. It is conceivable that if both stayed out of Iowa, others might have as well, giving Carter a caucus victory that would hardly be news. And he needed the media bounce that helped him win New Hampshire. Without winning both Iowa and New Hampshire, he might have not been able to mount a successful campaign later on. But we'll never know.

"If you're going to run for president of the United States," Birch later said, "it ought to be the most important thing in the world to you. It wasn't." This may be a rationalization, but his late entry into the race seems to substantiate that recollection. Also, he seemed reluctant to ignore his Senate responsibilities, showing far more involvement in legislation than would be expected from someone who should be consumed by campaigning with a "fire in his belly."

David Rubenstein discussed how Carter made claims during the campaign that were hard to support but he escaped serious damage because of them. Rubenstein met with Georgia budget director Jim McIntyre to discuss the Carter claim that he used "zero-based budgeting" as governor. McIntyre said he told Carter to stop saying that; it doesn't work. When Rubenstein asked McIntyre about Carter's claim that he had taken 1,200 agencies and "squished" them into 15, he was told that didn't happen either. But Rubenstein said Carter was a "very focused guy, very disciplined."[17] Those qualities served him well.

Carter's election was without coattails and did not change the partisan makeup of the Senate, though a number of new senators were elected. One new Democratic senator was Paul Sarbanes in Maryland. Birch described Sarbanes, a Rhodes scholar like Lugar, as "a good man, one of the smartest members of the Senate I ever served with."

New Republican senators included John Heinz III (PA), who replaced the retired Hugh Scott, and Orrin Hatch (UT), who defeated Frank Moss. Birch said of the minority leader, "I liked Hugh Scott. He was a progressive figure heading the Republicans." Orrin Hatch and Birch would develop a close relationship while standing on different sides of most issues, a relationship that continued after Birch left the Senate. They teamed up on a bill to protect the rights of the institutionalized. Cosponsoring the Bayh legislation, Hatch complained to him, "Bayh, this thing is killing me [at home]. It's terrible!"

Birch said, "But it's the right thing to do, wasn't it, Orrin?"

Hatch responded, "Yes, it was."

Hatch said that Birch was "one of the real powerhouses on the Judiciary Committee." He considered Birch "an unrequited liberal, but [he] was there to try to do what was right." And Hatch said it was hard for him to dislike colleagues who were sincere. "I could never get mad at those who believed in what they were doing, and Birch believed in what he was doing," Hatch recalled. But his favorite memory of Birch was when a witness at a committee hearing collapsed with a heart attack. Birch immediately rushed to the man and gave him mouth-to-mouth resuscitation while Hatch pressed down on the man's chest.[18] Northern Illinois University professor Martin Diamond, age 57, died later that day at Capitol Hill Hospital.[19]

Democrat Howard Metzenbaum was elected in Ohio and had an interesting story. He had filled the unexpired term of Sen. Bill Saxbe in 1974 when Saxbe became Nixon's attorney general. That same year, Metzenbaum was defeated in the Democratic primary by John Glenn. Two years later, he was elected in his own right and became Glenn's junior colleague. Bayh and Metzenbaum got along well, but there was an incident in 1980 when they went head to head in the Senate. A visit by seventy-six soft drink bottlers from Indiana convinced Birch to offer legislation that may appear inconsistent with his legislative history. They sought legislation to protect their regional monopolies that had been created by the large soft drink manufacturers. While it might appear to be supporting monopolies in restraint of trade, it actually prevented the larger bottling companies from swallowing up the smaller ones, small enterprises that remained in business only because of the regions set up for them by the manufacturers. Birch introduced the Soft Drink Interbrand Competition Act and soon had seventy-nine cosponsors in the Senate. Metzenbaum objected to the bill and introduced amendments aimed at killing it. The Supreme Court had ruled in a case known as *Illinois Brick* that indirect purchasers did not have standing to sue in federal court. If Metzenbaum's amendment were to pass, the number of plaintiffs in anti-trust lawsuits would rise exponentially. The business community vigorously opposed the amendment that, if passed, would kill Birch's bill.

In an unusual alliance, Birch and Sen. Strom Thurmond devised a novel approach to avoid dealing with the amendment: they filibustered their own bill. A successful cloture vote would bring the unamended bill up for a vote. A filibuster could prevent Metzenbaum's amendment from being considered; once cloture was invoked and debate was ended, the bill as it stood at the time must be voted on. That was exactly what happened, and the measure passed. At one point, Metzenbaum confronted Birch, saying, "I thought you and I were friends." Birch told him that they were friends, but so were several others among the seventy-nine cosponsors of the legislation.

Both party leaders, senators Robert Byrd (D-WV) and Howard Baker (R-TN), praised the duo of Thurmond and Bayh for their creative effort to pass a major piece of legislation. Baker, speaking immediately after the bill was passed, noted the unusual combination of the two senators who rarely voted alike, if ever. He said he had never seen a preemptive or preventative filibuster and didn't know if it was a good or a bad thing, "but it worked and worked very well." Byrd stated, "I think that in the future, I might be well advised, when attempting to get cloture, to just turn to Senators Bayh and Thurmond and leave the job to them. If we can get cloture on the first vote, we can save the time of the Senate."[20]

While the year was consumed with campaigning, Senate responsibilities were not ignored. Birch picked up where he left off before the presidential campaigning began, speaking out on a wide variety of issues. He also assumed leadership in the Constitutional Amendments subcommittee to prevent the passage of an amendment designed to overturn *Roe v. Wade*. The Appropriations Committee approved Birch's proposal to fund a national campaign to fight maladies found among many of the underprivileged: lead poisoning and rat-borne or other communicable diseases.

In reaction to the revelations of the Church Committee, he also supported establishing a new Senate Committee on Intelligence and opened hearings to promote the Foreign Intelligence Surveillance Act (FISA). Birch was appointed to the Intelligence Committee in May. His legislative activities for the remainder of 1976 were centered on FISA and his desire to see the oil companies broken up to diminish their dominance of the American economy. As the year neared its end, Birch announced that he would once again introduce his direct election constitutional amendment; the issue seemed more pertinent after another close election. President Ford named him to a presidential mission that would travel to the People's Republic of China.

Shortly before the end of the year, Senator-Elect Lugar and staff arrived in the Senate for orientation sessions and to be allocated office space. Birch and members of his staff trooped down there to meet them, offering to help get them settled and engaging in pleasantries and small talk. Lugar appeared rather stiff and formal; the experience was a bit surreal. Lugar's administrative assistant was Mitch Daniels, who had managed both of Lugar's Senate campaigns. Mitch went on to have a distinguished career in Republican politics. He would become executive director of the Republican Senatorial Campaign Committee, a top political adviser to President Reagan, head of the Office of Management and Budget under President George W. Bush, governor of Indiana, and president of Purdue University.

The previous summer, the Senate finally had passed legislation restoring citizenship to Eugene V. Debs, something that had been percolating on the Hill since Birch's first year in the Senate. In August, Birch toured the new courthouse and downtown structures in tornado-ravaged Monticello, Indiana. One staffer remarked that Birch's role in obtaining federal funding for the city might result in the citizens of Monticello throwing rose petals in his path when he visited—an exaggeration, to be sure, but he was treated like a conquering hero when he arrived there.

Now that the boss was no longer running for president, life in the Senate office was settling down. Jay Berman and Bill Wise left the office, and P. A. Mack became administrative assistant. Staffer Abby Saffold remembered a going-away dinner held for Wise, at which Birch was asked about his philosophy of staff management. Saffold quoted the words she remembered from Birch:

> "I know more about Indiana and its politics than anyone on my staff. I also trust that everyone who works for me has my best interest at heart. When a staffer comes to me with a suggestion about how I should vote or something I should do regarding Indiana politics, I do not tell them that I may think the suggestion is lousy and/or stupid, and I am not going to scream at the staffer. I will explain why I think a different approach is better.
>
> "If you scream at a person and tell him or her that they do not know what they are talking about, they might come to you a second or third time with suggestions . . . but the odds are that if you have screamed at them every time they have talked to you . . . they will not come to see you a fourth time. And the fourth time may be THE time I really needed to hear what they had to say . . . but they had been beaten down too often to risk being yelled at again."[21]

I was promoted to head up the Indiana Department with responsibilities for constituent service, casework, and special projects involving federal funding for Indiana. The job included managing the staff in Indiana, including the field staff, called regional coordinators, as well as the ninety-two volunteer county coordinators. Heading up the Indiana office was Diane Meyer, finance director in the 1974 campaign. In many respects, removing Birch's national ambitions allowed us to better focus on Senate responsibilities and the needs of our state. We had four years to prepare for reelection, so while the election wasn't yet a high priority, improving his standing at home was. Like the aftermath of the first presidential campaign, much of Indiana's population seemed to sour on the ambitions of their senator. It was important for Hoosiers to know that they were his first priority. In December was established a "Hotline for Help" in the Indianapolis office, a toll-free phone number that people could call when they needed assistance. That same month, Birch urged the Transportation Department to approve a feasibility study to fund a "Downtown People Mover" in Indianapolis, a futuristic way of moving people more efficiently around the capital city. Just before Christmas, he announced that Indiana would receive more than $22 million for a variety of public works projects.

Gordon Alexander also left the staff in 1976, but not before playing a part in the passage of a major piece of legislation. Commonly known as the 4R Act, the Railroad Revitalization and Regulatory Reform Act approved a final system plan for Conrail and authorized the acquisition of railroad tracks and facilities by Amtrak. Alexander prevailed on Birch to introduce an amendment guaranteeing minorities a share in the contracting that would result from the act's implementation. Alexander proudly recalled visits to the railroad stations that had been rehabilitated as a result of its passage, many planned by black architects and constructed by minority-owned companies, much of which would not have occurred had it not been for the Bayh amendment.

The nation's bicentennial, 1976, would be remembered for a number of things. Celebrating 200 years since the Declaration of Independence, Washington experienced the grandest fireworks display ever seen. Less recognized at the time was the creation of Apple Computer that year by two unknown young entrepreneurs, Steve Jobs and Steve Wozniak.

Internationally, the Socialist Republic of Vietnam was formally established. On the same day Americans celebrated the bicentennial, Israeli commandos raided Uganda's Entebbe Airport to free 103 hostages held by Palestinian terrorists. That summer, the US Naval Academy inducted its first class of women, while in New York City there began a year of terror with the first of several murders by David Berkowitz, better known as "Son of Sam."

In August a dispersal draft was held in professional basketball, a step toward finalizing the merger of the National Basketball Association (NBA) with the American Basketball Association (ABA). Birch had been involved in helping pass the legislation enabling the merger to take place. Also in August was the construction of a third Senate office building. Named for Sen. Philip Hart of Michigan, a man of sterling reputation and unparalleled integrity, it was one of the only buildings named for a living person. Hart died only four months later. The Old and New Senate office buildings had already been named for the late senators Richard Russell and Everett Dirksen.

In September, Chinese leader Chairman Mao Zedong died of a heart attack. On September 21, a car blew up while driving around Sheridan Circle in northwest DC. Among its passengers were Orlando Letelier, the former Chilean ambassador under Salvador Allende, and his American assistant, Ronni Moffitt. Living in exile in the United States, Letelier was an outspoken opponent of Chilean dictator Augusto Pinochet, whose agents were likely responsible for the assassination.[22]

While 1976 may have been the year when Birch Bayh's presidential hopes ended, America would have its 200th birthday and elect a president from the Deep South for the first time since the Civil War, not counting LBJ, who ascended to the presidency following an assassination. It clearly signified the end of an era when Richard J. Daley died on December 20. He had been the mayor of Chicago for twenty-one years. For the Bayh family and staff, all were looking forward to the inauguration in January and the arrival in town of a Democratic administration. It had been a long eight years under Republican rule, and Birch anticipated an enhanced ability to get things done for the state as well as the nation. With the loss of Hartke, Birch was Indiana's only statewide-elected Democrat and the unquestioned leader of the Indiana congressional delegation. This new status was both exciting and sobering.

15

The Carter Administration: 1977

We need to draw upon America's entire reservoir of talent and skill to help conduct our generation's most important business—the public business.

JOHN F. KENNEDY[1]

A PRESIDENTIAL INAUGURATION IS A UNIQUE TIME IN THE history of the United States of America. It is a moment when we reflect on the importance of our democracy and what it took to build it and maintain it all of these many years. The swearing-in of a new president marks a peaceful transition of power, rare in many parts of the world. When George Washington decided not to seek a third term as president, even King George III of Britain remarked that he must be a great man indeed, to give up power voluntarily. While giving up power is not always voluntary, the fact that we in the United States have had a peaceful inauguration every four years since Washington was first elected should be a reminder of what a unique nation we are.

As 1977 dawned, the country was preparing for a new president and another inauguration ceremony. For those living in the nation's capital, inaugurations are busy times. Since the days of Jefferson, most inaugural ceremonies have taken place on the east front of the Capitol. In 1981 President Reagan moved the ceremony to the west front, where they have been held ever since. On January 20, 1977, the Carter inauguration took place on the east front on a very cold day in the nation's capital.

Like all inaugurations, there was a scramble for tickets. As large as the Capitol grounds might appear, there is a strict limit on the number of people allowed in the seats or standing close enough to see and hear the president well. That ticketing competition was one of many issues facing the Bayh office in the weeks prior to January 20. Birch was completely engaged in the beginning of a new Congress, preparing once again to try passing the direct election amendment along with other legislation. His seniority had now risen to thirty-first. As the head of the Indiana congressional delegation, he took seriously the responsibility to make the group more effective for Indiana by working together more closely than in the past. The Democratic delegation was made up of Birch plus Adam Benjamin, Floyd Fithian, John Brademas, Dave Evans, Dave Cornwell, Phil Sharp, Lee Hamilton, and Andy Jacobs Jr. Democrats held eight of the eleven congressional seats, an absolute rarity in Indiana political history but one seat fewer than the prior year, with Dan Quayle replacing Ed Roush. As Democratic as the federal office holders may have been, Birch occupied the only statewide office held by a Democrat. Indiana's Republican reputation is well earned, and most governors and presidential winners over the years have been from the GOP. The state legislature was mostly run by Republicans, but there were more local Democratic officeholders in the counties, a fact that sometimes seemed hard to believe.

The concern about federal appointments in 1977 was substantial. Birch reminded the staff that for each appointment an elected official has, he makes "ten enemies and one ingrate." The pressure for federal appointments was a never-ending task, reminding one of the myth of Sisyphus—eternally pushing a rock up a hill, only to have it roll back down again. The more prominent the appointment, the greater the pressure from politicians and donors to help their friends get the job. Among the appointments at stake were two US attorneys, two federal marshals, farm home loan administrator, chair of the Agriculture Stabilization and Conservation Service, and dozens of lesser posts. Traditionally, the administration makes appointments in a given state based on the recommendations of that state's United States senators of the president's party. While not always true, it was largely the practice for more years than anyone could remember. Out of office for eight years, Democrats were hungry for these last vestiges of patronage. Dealing with the question of federal jobs and appointments consumed much of 1977.

January was hectic. In addition to Birch's activities related to the inauguration and federal jobs, the Indiana state legislature was considering passage of the Equal Rights Amendment.

Congress passed the Equal Rights Amendment (ERA) in 1972; if ratified, it would be the twenty-seventh amendment to the Constitution and the third amendment with Birch Bayh as the principle author. Congress passed enabling legislation giving the amendment seven years to be ratified by thirty-eight states, the required three-fourths. The seven-year period would end on March 22, 1979. If Indiana ratified the amendment, it would be the thirty-fifth state to do so, with only three more states needed, with seemingly ample time to accomplish that goal. On January 12, 1977, the Indiana General Assembly approved the measure by a vote of 54–45. The Indiana Senate was the remaining obstacle.

I headed up the Bayh effort in Washington; Mary Scifres and Diane Meyer were the key staffers in charge in Indianapolis. Scifres had deep personal relationships with many Indiana legislators and was a terrific politician. I had the ability to speak for the senator and to generate outside pressure on members when necessary. Since I had the senator's ear, my role was to manage the process and make judgments about how to land votes that might not have been secured by the Indianapolis office. Scifres provided an assessment of those senators who were either opposed or on the fence. Of the former, there were a few who said their constituents were opposed to the ERA but whose seats were sufficiently safe to vote with us if they were persuaded to do so. One of those was Chester Dobis, a legislator from Lake County who held a very safe seat and could safely vote as he chose on the issue. I asked the staff to give me a report on how often Dobis asked us for favors and learned, to my surprise, that he had done so many times over several years. He was in fact the legislator who asked for more favors than any other. With this in mind, I decided to call him.

Dobis took the call and began whining about how many people in his district were opposed to the ERA and how he just couldn't see any way to vote for ratification. With several people in my office listening to the call, I told Dobis that if he didn't support us on this, he could no longer call on our office for help. There was a pause while I prepared myself for anger, threats to complain to my boss about what I said, or an abrupt disconnection. Instead, Dobis replied, "Now you're talking my language." To my relief, the conversation lightened up a bit, and I told him a vote against ratification would be fine if we didn't need his vote. Otherwise, he had to vote for it. He agreed. I hung up, and my coworkers breathed an audible sigh of relief and showered me with hugs and pats on the back.

I had met State Senator Lindel Hume in 1974, and he told me he would vote with us if we could get the Princeton High School Band included in the inaugural parade. That was not something I could easily promise, but I called a friend on the Inaugural Committee and told her what we needed. She made it happen, Lindel was ecstatic, and we got his vote.

A few years later, the actor Alan Alda spent time in our Senate office researching his role for the film *The Seduction of Joe Tynan*. He had contacted Birch, saying that he was his model for the senator that Alda wanted to portray in the film. He came to the office, spoke to the entire staff about his reasons for being there, and then met several of us one on one for background conversations about what it was like to work in a Senate office. He had an assistant with him, and both assured us that anything we told them would be held in the strictest confidence. Foolishly, I told them the Lindel Hume story, which they loved. Unfortunately, it was retold in the *New York Times* by the Alda assistant after the movie came out. Senator Hume was none too pleased with me.

As the vote neared, we were getting close to the twenty-six votes we needed but not quite there. Among those lobbying Republicans to support ratification were Maureen Reagan, daughter of Ronald Reagan, and George H. W. Bush, former congressman, chairman of the Republican National Committee, and ambassador to China and the United Nations. Bush was preparing to depart his current job as director of the Central Intelligence Agency, once Carter took office.[2] The Democratic lobbying was led by Birch and his staff. He made every call he was asked to, often to the legislators directly but sometimes to others in Indiana who might be able to influence a legislator's vote. As the vote drew near, we learned that one of his best friends and political allies, State Senator Wayne Townsend, was wavering in his support for the ERA. Townsend and Bayh had been fellow students at Purdue University. Calls to him didn't bring us the necessary assurances, so we ratcheted up the pressure. Birch called Rosalynn Carter, only days away from becoming First Lady, and asked her to call Townsend. She did and offered to campaign for him in Indiana, getting a commitment from him at the same time. He was quite embarrassed when the press got wind of the call and gave it substantial media coverage shortly before the vote.

On January 18, the Indiana Senate ratified the Equal Rights Amendment by a vote of 26–24. Sadly, it was the last state to ratify the amendment. Even after Congress passed a three-year extension for the ratification process, pushing the final date back to June 30, 1982, no other state voted to ratify. A few even voted to rescind ratification, an action many constitutional experts consider invalid. Birch was adamant about the futility of rescission. The US Constitution says that a state may ratify but is silent on the issue of withdrawing ratification once it is given. In other words, a strong case can be made that once a state ratifies an amendment, its role has concluded.

With the inauguration of a new Democratic administration, Democrats in the Capitol felt that a new day had dawned. There was relief and palpable excitement throughout the halls of Congress. Watching the Carters walking in the inaugural parade was memorable. We searched for a sighting of the Princeton High School Band, to no avail. President Carter's post-inaugural honeymoon ended the day after he was sworn in, when he issued a pardon to Vietnam War draft evaders.

Carter clearly understood that the furor resulting from his action would blow over in time but not right away. He believed he had the four years of his presidency in which to weather the tempest caused by this decision. He was right. As much as veterans' groups decried the action, it was largely forgotten in subsequent years. About 100,000 Americans moved to Canada and other places to avoid the draft. Following the president's pardon, many of those who feared conviction for draft dodging returned to the United States; more than half of those who went to Canada remained in that country.[3]

In addition to lobbying for votes for ERA passage, fielding demands from those seeking federal jobs, and inauguration-related activities, Birch prepared to reintroduce the direct election amendment. He also put forth a bill to eliminate sex discrimination in the Social Security system and supported a resolution condemning Soviet harassment of Jews and others who were trying to emigrate from the Soviet Union. On January 12, he cosponsored a resolution designating January 15 as Martin Luther King Day, which eventually passed and created a new national holiday.

After the inauguration but early in the appointments process, Birch had lunch in the senators' dining room with Frank Moore, the head of Carter's Congressional Relations office. Moore instructed us to forward all job requests to him, describing the manner by which we could indicate when we were serious about an appointment versus those being passed along as a courtesy. But it never really worked like he suggested. In fact, everything we really wanted had to be fought for, and far too often we were unsuccessful. This became one of the early rifts between the Carter White House and Congress. The White House was not responsive to congressional requests, even from fellow Democrats.

One appointment important to Birch was the post of United States–Canada border commissioner. Bill Schreiber was a prominent Marion County politician and an early Carter supporter who paid a huge political price for that support in the early days of the Carter campaign. A US–Canadian treaty establishing the post meant that if there were any change at the boundary between the nations—whether by a river changing course, an earthquake, a volcano, or some other natural alteration—the lines could be redrawn and agreed to with the Canadian counterpart, and a new treaty would not be necessary. On this one Birch was successful, and Schreiber was appointed to the post.

Stories started circulating quickly, though, about Republicans being appointed in states with Democratic senators. Often, the senators learned about these in the press. And the Carter folks were appointing people across the country who were political unknowns. This does not mean that they weren't qualified, but it did create a large number of ruffled feathers when people were chosen who were not deemed politically worthy. To the dismay of many, unhappiness among their Democratic congressional counterparts did not seem to bother the Carter people. A story involving Speaker of the House Tip O'Neill is legendary. The House Speaker's tickets for the Inaugural Gala at the Kennedy Center landed him near the back of the auditorium. He called Hamilton Jordan, soon to be Carter's chief of staff, and asked him if there was a message being sent to him with the tickets' location. Jordan responded that if he didn't like the tickets, he could return them. Whether true or not, this was one of many stories circulating around Capitol Hill like wildfire. Later on, O'Neill would famously refer to Jordan as "Hannibal Jerkin."

Carter was perceived by many to be a holier-than-thou candidate, and now as president he seemed to be bending over backward to avoid "playing ball" with Democrats in Congress. This led to rocky relationships with Congress, a body already irritated by the pardons issued on Carter's first day. The stage was set, and relations between Carter and Congress seemed destined for difficulty. Ira Shapiro wrote,

> It would become clear soon enough: the new president disliked political small talk, disliked politicians, and liked politics least of all. He thought that the interest groups who had worked to elect him were selfishly pursuing their own narrow agendas at the expense of the national interest. He saw members of Congress as either complete captives of interest groups or too political. The only politician he had faith in was Jimmy Carter. Principled compromise was the heart and soul of the legislative process, but Carter was not a believer in compromise.[4]

President Carter consistently showed a lack of sensitivity to those things congressmen saw as prerogatives. Congressman Brademas remembered a time when Carter had given a speech at Notre Dame after a Brademas invitation. He and Birch had shared the stage with the president; Carter neglected to introduce either one of them. Introductions such as these were expected, considered gracious, and one of the basic rules of getting along in politics. The men and women of Congress frowned upon all who did otherwise.

An absurd example of how bad the relations were between the Carter people and Congress was when Indiana congressman Adam Benjamin called the Bayh office. He had been unable to get White House tour passes for his brother's family, and he asked for help. While the Bayh staff was able to secure those visitors' passes, this was an irksome situation that never should have occurred. The White House staff should have just made this happen. Carter was not your typical politician. He seemed more comfortable dealing with the minutiae of policy matters than chatting with a member of Congress. One unusual story was recounted by Lee Hamilton, who was invited to the White House for lunch with the president. It was just the two of them in the Rose Garden, and Hamilton found himself wondering why he was there. As they engaged in small talk about the weather and the New York Yankees, Carter seemed to be wondering the same thing. The lunch was obviously the idea of a staffer, and Carter had not been properly briefed. Trying to raise the level of seriousness in the meeting, Hamilton asked Carter what they were going to do about the high inflation rate at the time. Carter thought for a moment and responded, "Lee, it's a bitch." The small talk continued.[5]

On the jobs front, Birch kept at it and did his best to help out as many deserving people as he could. There were all kinds of outrageous requests, including a realtor from an Indianapolis suburb who asked Birch to promote him for secretary of state. A process was developed for filling the vacancies that would soon exist in Indiana for two US attorneys and two federal marshals, one that probably ended up creating more than ten enemies per job and certainly at least a few ingrates. Birch wanted a woman to be appointed US attorney for the Southern District of Indiana, headquartered in Indianapolis, and soon determined that Virginia Dill McCarty would be appointed. She was formally recommended in February and would eventually become the first woman to hold the post of US attorney, another badge of pride for the senator.

Of those Carter appointed to his administration, Birch praised the selection of the mayor of Atlanta, Andrew Young, to be ambassador to the United Nations. Honoring the civil rights leader by bestowing this office on him underscored the American commitment to majority rule in South Africa, he proclaimed. Lee Hamilton joined Birch in praising the naming of two Hoosiers, James Joseph and Leo Krulitz, to high posts in the Interior Department. There was, however, no prior consultation on these appointments with either Bayh or Hamilton.

Birch's profile was growing, both in the Senate and in Indiana politics. The future looked promising for a number of his legislative initiatives. In February, calling for a comprehensive energy policy, he introduced the Petroleum Industry Competition Act of 1977, an effort to break up the vertical integration in the oil industry. He also released a report on school violence and vandalism, made statements concerning the Alaskan pipeline, and began hearings on the direct election amendment, which

President Carter endorsed shortly afterward. In March, he introduced the Family Farm Antitrust Act to limit takeovers of family farms by large non-farm interests, a bill to include Pap smears in Medicare coverage, and the Privacy Protection for Rape Victims Act, and he recommended Frank Anderson for US marshal. Anderson was formally nominated in June and confirmed in July.

In March, Birch began a practice that he had not done in previous years but that had become very popular with members of the House. He started hosting town meetings in communities across the state. In recognition of the ill will created by his presidential candidacy, the meetings were designed to bring him closer to everyday Hoosiers. With franked mail, we blanketed each community with letters announcing the meeting, and as a result, the meetings proved wildly successful. He ended up drawing large crowds at these events, and with staff support, he was able to handle more constituent problems.

Once Virginia Dill McCarty's nomination was approved, Birch's attention turned to the other vacancy in the Northern District of Indiana. This nomination proved far more problematic. Lake County, one of the largest counties in Indiana, is also its most Democratic, with a longtime reputation for corruption and machine-like politics. Similar to its neighbor Chicago, strong political leaders and a vibrant patronage system dominated the politics of northwest Indiana. In Lake County were several cities, each with a different Democratic mayor, and there was also a strong county political organization. The people of Lake County had voted overwhelmingly for Birch, and he felt indebted to them for their support.

The cities of Gary, Hammond, and East Chicago were all run by Democrats: Dick Hatcher, Ed Raskosky, and Bob Pastrick, respectively. Pastrick was also Democratic county chairman. A number of Lake County officials were under indictment at the time, and Birch worried about those among them who were his friends. It was natural to hope that your friends hadn't broken any laws, but political prudence required steering clear of those being accused, something not always possible. Immediately after the election, Birch began receiving letters and calls urging him to nominate John Stanish, a Lake County attorney who was obviously close to prominent elected officials and party leaders. Birch was inclined to support Stanish, assuming he could pass the qualifying tests.

During this period, however, Birch received a call from a judge on the 7th Circuit US Court of Appeals. He was told that some of the indictments of Lake County politicians by the Republican US attorney John Wilkes had been examined closely and that they didn't have merit. It was his view that the next US attorney, deciding on the merits, would dismiss these indictments. It was obvious that this was something valuable for Birch to know.

Birch discussed this with me as the staff person heading up the appointments process. It was clear to both of us that appointing someone out of the Lake County political structure who then dismissed indictments of the very people recommending him would appear as if by design. No one would believe that we hadn't taken that step to help out our friends. The solution appeared fairly simple: The appointment had to come from some other county in the Northern District. The information Birch had received was in confidence and had to remain so. This meant we would be denying the requests from our friends and key supporters in Lake County and could not tell them the real reason why. So we began looking at other names that had been recommended and from whom those recommendations came. Eventually, we settled on Dave Ready of South Bend. As difficult as this seemed at the time, we knew it would blow over and we'd have ample time to recover.

Around this time, it became clear that there would also be federal judgeships opening up in Indiana. The Senate Judiciary Committee had formally recommended two new judgeships, one in the Northern District and one in the Southern. Birch announced a merit selection system for the recommendations he would eventually make to the president. While this system proved very useful in providing a forum for people to express their views on a nominee, it also created a dilemma. How could he possibly ignore its recommendation once made? Over the next four years, large amounts of time would be invested in the process of nominating two federal judicial appointments.

In March, the Senate passed a Bayh resolution disapproving of the Soviet expulsion of Associated Press correspondent George Krimsky.

In April, Birch introduced the Civil Rights Commission Authorization Act and a bill on behalf of the Carter administration to extend the Drug Enforcement Administration, also expressing his determination to propose an amendment to decriminalize marijuana. He initiated legislation forbidding discrimination on the basis of pregnancy and childbirth, responding to the Supreme Court decision *General Electric Company v. Gilbert*. Similarly, he announced his intention to introduce legislation to protect the nation's handicapped from job discrimination and to provide elderly people with sufficient access to polling places. Finally, he introduced a bill giving the Justice Department the right to initiate civil actions on behalf of institutionalized persons. It seemed there was a full-court press to protect those segments of our society that needed a spokesman, from women seeking Pap smears, to victims of rape, to the elderly, to those institutionalized with mental illness, to farmers trying to protect the future of their farms. These efforts were taking place at the same time we were working to install our people in coveted federal jobs as well as trying to pass the direct election amendment, secure the best means for transporting Alaskan oil, and break up the oil companies.

The variety of issues that Birch and his staff handled seemed to grow with each passing week. In May, he introduced legislation involving Social Security, animal fighting, and cigarette smoking and proposed funding for research into genetic diseases, for improving American highways, and for the National Cancer Institute. He spoke out in support of the Foreign Intelligence Surveillance Act (FISA), an important effort by the administration to better regulate intelligence activities. Before the month ended, he formally recommended David Ready for US attorney, which was approved in September.

Before announcing Ready's recommendation, Birch spent time with John Stanish to figure out what he could do for him. It was made clear to the White House that Stanish was important to Birch and that making Stanish happy would go a long way toward easing the unhappiness from his Lake County allies. Working at the Justice Department in Washington held a great deal of interest for Stanish, and Birch was able to secure him an appointment to become US pardon attorney. He was formally recommended to the president in June.

The McCarty nomination sailed through the Judiciary Committee, and she was sworn in by the president in a Rose Garden ceremony in June. Others sworn in at that same time included Patricia Derian, Carter's choice for assistant secretary of state for human rights and humanitarian affairs, and Eleanor Holmes Norton, chairperson of the Equal Opportunity Employment Commission (EEOC). Norton went on to be elected DC's non-voting delegate in Congress, serving there for many years. Had Birch's efforts for full voting representation in the District of Columbia succeeded, the many votes she had cast over those years would have counted.

Also in June, Birch's efforts included hearings on the rights of the institutionalized. While he was working on advancing women's rights and protecting the institutionalized, other cultural changes were taking place in the country. Concerns about the rights of homosexuals made the news when the singer Anita Bryant brought to Miami her anti-gay crusade, which she called "Save Our Children." San Franciscans marched in the streets, protesting her anti-gay remarks. A local murder, the result of an anti-gay attack, had taken place. The Supreme Court ruled in favor of states that chose not to allow their Medicaid funds to be used for abortions, further exacerbating an already intense issue. At month's end, the Women Marines were discontinued as a separate organization and made part of the regular Marine Corps.

The summer found Birch introducing a bill to prevent abuses of civil liberties through polygraph tests, joining the supporters of a solar energy program to drive down the price of solar cells, and calling once again for full congressional representation for Washington, DC. In late July, Alaskan oil started moving through the Trans-Alaska Pipeline System, and the Department of Energy was established. In October, Joseph Novotny was recommended to the Justice Department as Indiana's second US marshal.

Author Ira Shapiro described the large role the Senate played during the sixties and seventies. It was a period of major innovation with a number of senators many referred to as giants. The prominent role played by the Senate in the country's affairs felt appropriate because of the large number of social issues fermenting in the nation. As president, Jimmy Carter was often successful because of the Democratic majorities in each house of Congress, with the Senate often leading the charge. The Vietnam War might have been over as an issue, but its absence unleashed energy for many other things: fighting for environmental protections and arms control; working to halt nuclear proliferation, to solve the Arab–Israeli conflict, to protect the rights of Soviet dissidents, and to promote urban renewal; and work advancing civil rights, including women's rights and gay rights—measures to protect those segments of society left behind and most needing protection. The country yearned for energy independence, an issue of both foreign and domestic concern, but also cared about human rights around the world, an issue that Carter made his own.[6] Birch seemed to be involved in all of these issues during this period.

After spending fourteen years in the Senate and finding himself working with an activist administration promoting major issues and forcing Congress to act, Birch had become the kind of senator he wanted to be. There are various components of the job, emphasized differently by different senators. There was the role to represent one's constituents, to serve their needs, often individual problems needing solutions from the government or bureaucracy. On a larger scale, constituent assistance might mean helping a community get an overpass built or a dam authorized or disaster relief. Then there was the role each senator plays to remain in office, to tend to politics in the home state and do everything possible to keep his or her name before the public and grab credit for accomplishments whenever possible. For some, there are ambitions to higher office, to be able to operate on a larger stage and have a greater impact on the country's affairs. Finally, there is the legislative role: to enact laws that will improve the lives of Americans, to solve problems facing the country, or to promote an issue that deserves attention. There are senators who develop areas of expertise such as foreign affairs and focus their time and energy on solving problems with countries around the world; others just spend time keeping themselves in office. Some are better at "bringing home the bacon" than others, and for many, constituent service is the highest priority. More senators than not seem to be maneuvering toward higher office, seeing themselves as future presidents.

Birch's staff managed constituent service skillfully, solving individual case problems and helping to deliver federal projects for Indiana.

He described casework by saying, "You could know that you touched the life of one person." The operation worked well, requiring his attention only when necessary. Other staff members were adept at politics, doing those daily things that might contribute to Birch's reelection without allowing politics to dominate his schedule. Having abandoned his presidential ambitions, he was free to work on the issues he cared most about rather than aspirational issues that would buttress a presidential vision. His priority was the passage of important legislation. He felt he had hit his stride, having become an effective legislator who could make important things happen—things that affected the country. His ambition was to have the greatest impact on as wide an array of issues as possible; he felt that half a victory was better than no victory at all and that sometimes one had to accept the necessity of taking baby steps toward an ultimate legislative solution of a problem.

It seemed like Birch was in many places at the same time, moving legislation, dealing with individual constituent problems and requests from municipalities, and tending to the needs of Indiana Democrats as the only statewide-elected leader. His staff could help him perform this juggling act, largely because he empowered them to do so. And he encouraged dissent in the office. He never cut off the head of the messenger giving him bad news. His was an environment where staff people were encouraged to speak their minds; it was never sinful to let the senator know when they thought him wrong. This ingredient served as an elixir for the staff and fueled aggressive activity in support of the common agenda.

The staff learned how to operate in close quarters on a tight timetable, how to get along with coworkers as well as constituents, and how to develop and implement solutions to problems. It was an exhilarating time when each person felt he or she was contributing to the public welfare and possibly some history in the making. We learned how to manage our time and efforts while often sublimating our own interests to those of the group or the boss.

In September, Birch announced his support for import quotas on specialty steel. He gave a speech that cited the potential for abuse in the IRS proposals to computerize the agency. That month, the United States signed a nuclear non-proliferation treaty with fourteen other countries, including the Soviet Union. The modern food stamp program began with the enactment of the Food Stamp Act of 1977. The Senate Judiciary Committee favorably reported Birch's direct election amendment to the full Senate. Also in September, the United States and Panama signed treaties concerning the status of the Panama Canal. It was agreed that control of the canal would fully become Panama's responsibility at the end of the century. This underscores a prominent feature of the Carter presidency. Regardless of how you felt about President Carter or these issues, he took stands that he felt were right for the country though politically unpopular. Prior to his negotiations in Panama, the canal was not an issue that people seemed to care about, and when it was raised, Democrats were none too pleased to have to deal with it. Unfortunately, Carter also took several positions that were unpopular on both sides and not always successful. There is no question that he paid a price for these matters, as did his supporters in Congress.

Difficulties with Carter's energy policies and his relationship with Democratic senators coalesced in October 1977. Carter opposed legislation to deregulate natural gas rather than retaining the existing price controls, a view shared by many in the liberal wing of Senate Democrats. Senators Abourezk (D-SD) and Metzenbaum (D-OH) led a filibuster to stall passage of the deregulation bill. Abourezk and Metzenbaum were considered an unusual pair. Abourezk was viewed as a bomb thrower, not one to compromise with his peers. Metzenbaum was considered an obstructionist, someone who seemed to be opposed to most bills on many issues. Nonetheless Birch, along with Ted Kennedy, participated in the filibuster, which was out of character for both. They were members of the Northeast–Midwest coalition, a group organized to assess the regional impact of legislation. They were wary of deregulation and its impact on their states and wanted to support the president.

The filibuster went on for weeks, angering Majority Leader Robert Byrd, who often was more concerned about the Senate running smoothly than about the possible impact of a proposed piece of legislation. He got so upset about the filibuster that he started keeping the Senate in session all night long, calling votes every twenty minutes. He hoped that enough senators would be annoyed by the all-night process of legislating that it would be easier to win a cloture vote. Those participating in the filibuster organized relay teams. Cots were set up outside the chamber each night, allowing the senators a resting place when they were not participating in the ongoing debate. Eve Lubalin was Birch's staff person on this effort. She recalled the drama of that time and noted the differences in the Senate of the 1970s versus the Senate of the twenty-first century. At one point while Birch was in charge of the filibuster, he almost lost control of it but did not, due to a level of Senate courtesy that has become rare in the twenty-first century.

> So there was a time when Birch Bayh was supposedly holding the floor and Jesse Helms (R-NC) walked onto the floor. It was a slow period in the afternoon, not much going on, and Bayh really wanted to make a phone call. He was getting restless and bored and I just had this feeling that Helms was circling around us. . . . He'd walk around, sit down, chat with someone on the floor, go into the cloakroom, come out of the cloakroom. And Bayh kept saying, "There's no one here. We're in a quorum call. I need to make this phone call." And I was like, "No!" Now, I had known him for all of maybe three months at this time and kept saying, "You are going to lose the floor. You are going to lose this whole thing." And he said, "I've been here 15 years. You've been here three months. Just trust me on this. I am going to go make this call, and no one is going to call off the quorum call." So he stands up from his desk, walks to the cloakroom and disappears.
>
> And out the other door comes Jesse Helms and picks up his microphone—and I'm telling you my heart stopped—and says, "Mr. President, I ask unanimous consent that the quorum call be rescinded." And as the gavel is going up to give him the motion, I run up the stairs to the cloakroom—which I was not allowed to enter myself without a Senator, but did—and he is sitting in one of the telephone booths making his phone call. And I am motioning and banging on the doors and trying to get his attention, and he is deeply involved in his discussion. I am just about in tears, and I am like, my whole life, I see it going up in smoke. So he comes sauntering out of the phone booth, always calm, cool and collected and I tell him, "Jesse Helms has the floor. He asked unanimous consent for the quorum call to be called off and that's when I came to get you." And now he looks concerned and he goes out of the cloakroom, walks down the stairs, and sure enough, Helms is standing there and has the floor. I don't remember what Senator Bayh said. Something like "point of inquiry" or something. And Helms turned around and said something like, "I see the Senator from Indiana on the floor. I know that the Senator from Indiana, my colleague, has been conducting a lengthy discussion of the natural gas bill on the Senate floor and because I am a gentleman and don't want to take advantage of the fact that the Senator left the floor for a moment, I will yield the floor back to the Senator from Indiana," which was amazing. And it just captures everything about working in the Senate at the time. That would never happen now.[7]

Ironically, the filibuster was broken by rulings from the chair by Vice President Mondale. The senators thought they were supporting the president, and his own vice president took actions that made them feel betrayed. The anger was palpable.

In October, Birch supported a bill to prohibit the use of children in the production of sexually explicit materials and introduced the Alcohol Fuel Incentive Act, an effort to jumpstart the prospects of gasohol and other renewable fuel sources. In November, the Judiciary Committee recommended a criminal code containing nine Bayh amendments, after which he cosponsored a resolution denouncing South Africa over its human rights violations and introduced the Buy American Act and the Trade Procedures Improvement Act. In December, he held hearings on the National Initiative Amendment, a measure that would enact laws by popular vote.

I traveled to Indiana with him during the first two weeks of December and wrote a memo summarizing the trip for the entire staff. Many years later, it is a reminder of how aggressively he tended to his political responsibilities, especially during a time when he was neither running for reelection nor for president. This was our longest trip to the state in 1977; these are my recollections on what was accomplished.

> First the quantity—thirteen days in Indiana, visited 24 counties, each Congressional District was visited and a total of 31 cities and towns with multiple stops in several. More than 2,000 miles in the state and saw more than 18,000 people. Sixteen Democratic party events plus meetings, breakfasts, lunches and dinners. The Senator's appearances included 14 high schools, 4 labor meetings, 4 media events, 3 plant gates, 10 newspapers and/or interviews, 7 scheduled radio interviews, 4 scheduled television shows and a variety of scheduled and unscheduled meetings, events and interviews.
>
> Second the quality—It was a fine trip because of the type of time we spent at most places. The Senator was relaxed and asking for nothing. Time was well spent because many people, especially those most important to us, knew he was in the state and many commented about the good job he is doing and how well we are keeping in touch with Indiana. At party events, it was clear that Birch Bayh is the leader and many Hoosiers care about his future. More trips like this in the next year will make the subsequent staff tasks much easier.

> The perception of being in touch was there. We encountered less grief on issues than on any trip I've been on. Gasohol may be one that turns out big for us according to the response we had with it. The trip went a ways toward solidifying our base, and extending ourselves to groups not normally for us—as well as the public at large.

After reading it, Birch scratched across the top of the memo, "I'm tired! B2."

Since the end of the presidential campaign, Birch's efforts were focused on business at home, to shore up his popularity and demonstrate to his constituents that his presence in the Senate was valuable to them. Constituent services were beefed up, and town meetings were added to the senator's agenda. He conducted eight town meetings that year and visited fifty-three counties and sixty-two individual cities and towns; he spoke at twenty-three high schools and attended thirty-eight Democratic Party functions. With an eye to his 1980 reelection campaign, eighteen "Special Friends Receptions" were held, events to which donors, elected officials, and other opinion leaders were invited.

Congress in 1977 passed a number of pieces of key legislation. The Urban Development Action Grant (UDAG) program was an inspired piece of legislation. Created to deal with the growing decay in many American cities, the program provided federal matching funds to dollars raised locally by businesses and individuals. It encouraged businesspeople and philanthropists to invest in their home communities, stipulating that every dollar spent really meant two dollars for the projects included in the UDAG application. This would eventually transform downtown Indianapolis and other cities that put together successful cooperative projects.

The year was one of change outside Congress as well, with cultural and historical changes in the world. In May, while Menachem Begin was elected prime minister in Israel, *Star Wars* opened in cinemas across the United States. It would become the highest-grossing film ever, a distinction it held for many years. In South Africa in September, activist Steve Biko was killed by police while in police custody. In October was another memorable World Series when New York Yankee Reggie Jackson hit three home runs in a game against the Los Angeles Dodgers. In San Francisco, the first openly gay official of any major American city was elected when Harvey Milk became city supervisor. Also that October, Egyptian president Anwar Sadat traveled to Israel to meet with Israeli prime minister Menachem Begin, making him the first Arab leader to do so. On Christmas Day, legendary actor and comedian Charlie Chaplin died in his sleep at his home in Vevey, Switzerland, at the age of eighty-eight.[8]

16

Foreign Intelligence: 1978

The object of government is the welfare of the people. The material progress and prosperity of a nation are desirable chiefly so far as they lead to the moral and material welfare of all good citizens.

THEODORE ROOSEVELT[1]

CARTER HIT THE GROUND RUNNING IN HIS FIRST YEAR, AND the energy expended in 1978 would be no less. Despite the ill will created among fellow Democrats, Carter successfully engaged Congress with his initiatives and vice versa.

In 1978, Birch was engaged in a wide variety of issues: energy, gasohol, Social Security, dumping of foreign steel, government reorganization, direct election, defense of Soviet dissidents, economic assistance to Vietnam, rights of the institutionalized, anti-smoking campaigns, and the transportation budget. There were environmental needs at the Indiana Dunes, the Humphrey–Hawkins bill, the Panama Canal treaties, assistance for the employment of women, the Israeli–Palestinian conflict, disaster relief, and the Buy American Act. As chair of the Appropriations Subcommittee on Transportation, Birch was the principal sponsor for creating and funding the Amtrak passenger rail system, and the work done in 1978 was key to its success.

Birch's increased stature in the Senate was highlighted by an enhanced ability to bring home the bacon. The Bayh archives are filled with announcements of federal monies being directed to Indiana projects for several airports, an amusement park, a National Auto Safety Center at

Purdue, and a variety of highway and transportation projects. He helped obtain federal support for a study to convert waste products to energy, for food and nutrition programs in Indiana, for five Indiana art museums, and he obtained Amtrak funds to support jobs at minority-owned contracting firms in Indianapolis.

The federal projects illustrate the ongoing challenge by those who might consider misspending public dollars. There certainly are politicians who profit from public expenditures in their jurisdictions, and any official who strives to be ethical must always remain on his guard from them, many who might be valuable political allies. There were occasions when Birch had private discussions with the politicians in charge of particular projects to ensure that these funds were handled as intended. On at least one occasion, he told a mayor that if he heard even a whisper that any of the federal funds weren't being spent as intended, he would make it his personal mission to see the mayor behind bars.

There was a sad passing on January 13 when Hubert Humphrey died. A giant political figure who played a major role in the previous three decades of American life, he burst onto the national scene when, as mayor of Minneapolis, he championed the civil rights platform at the 1948 Democratic Convention, leading to the walkout by Strom Thurmond and his eventual independent candidacy opposing Truman's reelection. Humphrey was Senate majority whip when Birch arrived, and he became the colleague who impressed Birch most. He served as vice president with President Johnson and was the Democratic presidential nominee in 1968, losing by a hair's breadth to Richard Nixon. Returning to the Senate in 1971, he was one of its stars and universally respected on both sides of the aisle. Not long before Humphrey died, Birch was walking from the Senate to an elevator and saw him, a frail man with thinning white hair, coming his way. Birch called out, "Going down, Hubert?" to which he replied, "Faster than I want, Birch." His wife, Muriel, would be appointed to his seat until a special election was held to replace him.

Jay Berman described Humphrey as the nicest and most decent senator he had known in his decade working in the Senate. Berman recalled giving a Capitol Hill tour to his in-laws. They ran into Humphrey, and he introduced them to the legendary senator by saying, "They worship you," to which Humphrey replied, "Let me worship them back."

Humphrey was truly a giant in the Senate, a liberal lion and a legend who walked the halls as a true celebrity. I remember one time when he got into a Senate subway car with me. He seemed genuinely interested in what I was studying. When I told him we were studying organized labor, he talked about his role in the passage of the Landrum–Griffin Act of 1959, a landmark of labor law. Political consultant Bob Shrum reflected on a time when Michael Dukakis, the Democratic presidential nominee in 1988, "criticized Humphrey as an outworn relic of the old politics. Humphrey's response tumbled out: 'I tell you the difference between Dukakis and me. He wants the pipeline to be nice and clean and shiny, and as long as it is, he doesn't care if shit comes out the other end. I don't care if the pipeline's messy and even shitty at times as long as the right result comes out the end.' That's the best description I've ever heard of the dividing line between process liberals—reformers—and results-oriented progressives."[2]

Harder on Birch than the passing of Humphrey was a diagnosis Marvella received on January 25. The cancer had reappeared, this time in the bones of her lower body, and it was deemed terminal. The doctors estimated that she had one to five years to live. Birch took it hard. Marvella was resolved to fight this development as she had fought the original bout of cancer in 1971, but this time it was to be in private. She was adamant about not becoming a public victim once again, and those few who knew about the diagnosis were sworn to secrecy. It was immediately clear that Birch would be doing little to no traveling. It became the staff's responsibility to substitute for him when appropriate, but it was imperative that we turn down scheduling requests without telling people the real reason why. His Washington schedule appeared unaffected. The ordeal brought the two of them closer. Marvella said she would no longer hold him to a "promise made after the collapse of his 1976 campaign—that he would not run for political office again. 'He has been my anchor,' she told a reporter. 'If I started to be blue, he was there to buoy me up. He never took any kind of attitude except we are going to lick this thing.'"[3] Geoff Paddock wrote an article about Marvella in 2013 and added,

> Marvella's faith grew. She asked people to pray for her and many did, including President Carter and Vice President Mondale. She wanted to spend her remaining years nurturing a closer relationship with God and concentrating

> on issues she cared deeply about. Less than a month after the diagnosis, she kept her commitment to debate the ERA with conservative activist Phyllis Schlafly at a forum in Lynchburg, Virginia. Marvella continued to champion women's rights in various forums around the country while continuing her work for the American Cancer Society. She felt the "positive pressure" of work and commitment could help her heal, along with prayer. Birch and Evan joined prayer circles as well and her renewed strength allowed her to decrease her reliance on tranquilizers. That spring Reverend Robert Schuller invited her to appear on and give testimony about her life on his popular television program, *Hour of Power*.[4]

That January, Birch appointed Joseph Smith to the Indianapolis staff as special assistant for urban and minority affairs. Joe was prominent in Indianapolis politics but particularly in minority affairs statewide. His appointment was the first of several steps taken to prepare for a tough reelection in two years, but it also underscored the growing role that African Americans were playing in Indiana life. The next month, Birch introduced a resolution designating February as Black History Month.

On January 27, Birch was appointed to be the next chair of the Senate Select Committee on Intelligence. Sen. Daniel Inouye (D-HI) was its first chair, serving from 1975 until 1979. Birch would be its second chairman when Inouye's term expired in a year.

The demands relating to legislative activities continued unabated. On January 30, Birch announced the creation of the National Commission on Alcohol Fuels to help make gasohol a reality. Those early efforts helped generate public support for gasohol, sowing the seeds for the major industry gasohol is in the United States today. As a member of the Intelligence Committee, he issued a proposal for a criminal standard for the electronic surveillance of Americans in the United States. The Foreign Intelligence Surveillance Act (FISA) was under consideration, and he was determined to play a prominent role in its passage. The committee approved it on February 27.

The Panama Canal treaties dominated the Senate. Many senators grumbled that this was an issue that didn't need to happen, but Carter brought it front and center nonetheless. President Theodore Roosevelt had accomplished what was considered impossible at the time, pushing through the passage of the original Panama Canal treaty in 1903 and overseeing the massive construction of the canal across Panama's isthmus. The treaty provided to the United States the right to build the canal and control a ten-mile-wide zone through which the canal was built. In 1964 there were riots in Panama over the American role in the canal, and a joint treaty was renegotiated. In 1977, President Carter and Panamanian strongman Omar Torrijos Herrera signed two treaties. The first was a neutrality treaty to ensure that the canal would be neutral and open to all vessels. The second provided for joint U.S.–Panamanian control of the canal until it reverted to full Panamanian control on December 31, 1999.

The proposed treaties quickly became a heated issue. Most Americans saw no reason to change the status quo and could not care less about the concerns of Panamanians. The issue's importance was shown on February 8 when the Senate debate on the treaties was broadcast on the radio—the first time that had ever happened. There were allegations of drug trafficking by General Torrijos and his family, a topic Birch had to address on the Senate floor speaking for the Intelligence Committee. Opponents of the treaties also helped generate a new industry in politics, one that has continued to exist and that has reshaped political campaigns since.

Conservative activist and direct mail entrepreneur Richard Viguerie put his expertise to work helping generate opposition to the treaties. Creating a coalition of like-minded organizations seeking to defeat the treaties, Viguerie sent millions of pieces of mail to people who were thought to share his views. He raised millions of dollars for the organizations but also generated an avalanche of mail to Senate offices calling for the treaties' defeat. The Bayh office was no exception. Thousands of letters and postcards arriving in mailbags stunned the senators and their staff. The exercise helped Viguerie develop valuable mailing lists of people who would give money and take action in support of conservative causes. Nonetheless, the first treaty was approved by the Senate on March 16, a courageous vote for many senators who understood the likely political price. The issue dominated the Senate for another month with constituent pressure building every day until April 18, when the Senate voted 68–32 to approve the second treaty, returning the canal's control to the Panamanians at the end of the century.

That same day, a congressional conference committee approved the creation of a federal judgeship for the southern district of Indiana. Another responsibility had fallen into Birch's lap.

While the treaties were being debated, Congress passed the Nuclear Non-Proliferation Act of 1978. Birch announced his strong support for its passage, emphasizing its importance to the future of the United States; we could not afford for the nuclear club to grow in size. At that time, there were six countries with nuclear weapons and there was little desire to see a seventh. During this period, President Carter decided that the United States would postpone plans to produce a neutron bomb, a weapon that would kill people by its radiation alone while leaving behind less physical destruction.

The growing movement for homosexual rights increasingly gained national attention. America's most comprehensive bill to protect homosexual rights was enacted in San Francisco, while leaders in St. Paul, Minnesota, repealed a gay rights ordinance, both actions in reaction to Anita Bryant's well-publicized anti-gay campaign. Later in the year, California voters defeated a prohibition of gay school teachers. On November 27, however, were the assassinations in San Francisco of its mayor and Supervisor Harvey Milk, the latter a prominent and gay elected official.

Birch was honored on March 7 with an award from the Association for Children with Learning Disabilities. In April, he announced the resumption of the Indiana High School Leadership Conference, which had been interrupted by the presidential campaign. The April leadership conference would be headlined by United Nations ambassador Andrew Young. The other speakers were author, columnist, and former *CBS News* reporter Daniel Schorr; the director of ACTION (a federal agency created to administer federal volunteer programs like Peace Corps, VISTA, etc.), Sam Brown; and Grace Olivarez, director of the Community Services Administration.

In December 1976, the Supreme Court decided the case of *General Electric Company v. Gilbert,* ruling that a company denying non-occupational sickness or disability benefits to pregnant workers did not "constitute sex discrimination in violation of Title VII" of the Civil Rights Act of 1964.[5] It stated that illnesses covered under company health plans were involuntary while pregnancy was not. Birch disagreed.

The remedy seemed clear. If the Court felt the law did not address occupational benefits for pregnant workers, the law should be changed. He wrote the bill and the Pregnancy Discrimination Act of 1978 was passed, writing into law that "pregnancy discrimination is a form of sex discrimination under the Civil Rights Act of 1964." Rep. Augustus Hawkins of California sponsored the bill in the House. Under the law, employers must treat "women affected by pregnancy . . . the same for all employment-related purposes . . . as other persons not so affected but similar in their ability or inability to work."[6]

On May 18, Marvella had surgery to remove swollen nodes that had appeared on her neck. Around the same time, a fuller body examination detected more malignant tumors. Her involvement with religion took on a new urgency. Geoff Paddock wrote about Marvella, "She prayed that God would comfort her in her final months and allow her to understand his plan for her. 'I have a wonderful new relationship with God. We believe in God's plan, although we do not always understand it.' That summer, she was well enough to travel with Evan to Europe. She felt particularly blessed to spend time alone with her only child and fortunate to return to the work she loved."[7]

During this time of sadness at home, the drama of who would be in charge of running the Indiana Democratic Party was playing out. It had become increasingly clear that the Indiana Democratic Party was an albatross around Birch's neck. It provided no services to campaigns, and often its chairman, Bill Trisler, made public statements in direct contradiction to public stands that Birch had taken. Don Michael, a regional coordinator on Birch's staff who was also Cass County Democratic chairman, wanted to challenge Trisler in 1977, but Birch prevented him from doing so. Beating a sitting chair in an off-year election required a two-thirds vote of the state committee, and that seemed unlikely. As the year went on and Michael maintained his desire to be elected state chair, Birch began to take another look at that possibility for 1978.

With Marvella's illness, there was no way he could adequately plan a campaign for reelection in 1980. If he wanted to have any sort of campaign operation, it would best be housed at the state committee, but not with its current leadership. Also, under federal election law, the state party could spend two cents per registered voter on behalf of a Bayh race, about $400,000 in spending capability. That was money the state committee did not have, and even if it had, Birch could not assume it would be spent as desired. The Bayh operation could help raise that money for the state committee but not unless it was in control of it. All of this nudged Birch toward making a change in the chairmanship of the Indiana Democratic Party. To do so meant winning the election at its reorganization meeting to take place after the primary in May.

Despite this, Birch was more than a little reluctant. His previous entry into state party reorganization had ended up with his candidate getting a single vote. Ironically, that vote was Trisler's. Diane Meyer, Don Michael, and I spoke with Birch over a period of months about getting involved in the contest. He came around, but that only happened after he became convinced we would win. His role in the party was more prominent than it had ever been, which made the difference. Once in, Birch was fully committed; after an effort lasting ten weeks, he prevailed, and Don Michael was the new state chair.

Birch was clearly in control of the Indiana Democratic Party and promised, if the chairmanship race was successful, to try to get President Carter as the speaker at the Jefferson–Jackson dinner that summer, the state party's major fundraising event. Carter agreed to attend, and the successful fundraising event was a great kickoff for Michael's tenure as state chair.

In April, Birch held hearings on his proposed amendment on DC representation. Its objective was for the District of Columbia to be treated like a congressional district, served by an elected representative with the same privileges as other congressmen and congresswomen. In May, there were floods in three Indiana counties that Birch successfully had declared as disaster areas, opening up the flow of federal funds for disaster relief to the areas most affected. In June, the Senate approved his proposal to establish a National Commission on Alcohol Fuels.

On July 20, Birch introduced the Consumer Privacy Protection Act. Looking out for the rights of consumers was another example of his long reach into affecting change in the United States. On August 16, he endorsed the House version of the amendment for full DC representation. The Senate passed it on August 22, and it went to the states for ratification, another opportunity for Birch to be the author of three constitutional amendments. Unfortunately, in 1985, the seven-year ratification period ended after only sixteen states had signed on. The amendment would have provided full congressional rights for Washington, at least one House member and two senators, full electoral vote participation, and an equal role in future deliberations on constitutional amendments. This was the fourth constitutional amendment Birch authored that was passed by Congress; two were ratified by the states, and two were not.

Catholics all over the world mourned the death on August 6 of Pope Paul VI. He was replaced on August 26 by Pope John Paul I as the 263rd pope, a papacy lasting only 33 days when he also died. It would be mid-October before another papal election resulted in John Paul II becoming pope. Not only was he the first pope from Poland; he was also the first pope from outside Italy in more than five centuries. He would be the pope for nearly three decades until his death in 2005.

Quickly on the heels of the Gilbert decision and the Pregnancy Discrimination Act came the effort to extend the period for ratification of the ERA, due to expire in 1979 and still three states shy of the thirty-eight required. Congresswoman Elizabeth Holtzman of New York introduced a joint resolution in the House extending the process for another three years. Birch introduced the same legislation in the Senate. On October 2, he appeared on *The Today Show* to make his case for an ERA extension.

One concern was the nature of the Republican opposition. Birch didn't want Sen. Orrin Hatch (R-UT)—whom he considered smart and also extremely conservative—to lead the opposition. He was aware that Hatch wanted to be nominated to the Supreme Court, and Birch involved women's groups in the process. They approached Hatch, promising not to oppose his nomination if it took place, providing he not lead the opposition to the Bayh ERA legislation. He did not. Also, Birch led the extension effort against the wishes of Senate Majority Leader Robert Byrd. Byrd relented after he was lobbied by a few senatorial wives, telling Birch that he would allow the effort to proceed, but Birch or another manager of the effort must be on the floor at all times while it was under consideration.

Birch faced two dilemmas. One was Marvella's health, often requiring him to leave the Senate at a moment's notice. Then another senator would be needed to replace him on the Senate floor when he left. The second was an amendment by Sen. Bill Scott of Virginia. Scott's amendment would allow states to rescind a previous vote to ratify, something Birch always felt would be overturned by the Supreme Court. Sen. Tom Eagleton was supportive of the extension but also supported Scott's rescission amendment. Birch maneuvered to have Eagleton sent to the Vatican to represent the United States at the funeral of Pope John Paul I. With Eagleton gone, the amendment was defeated, setting the stage for the Senate to pass the extension, which it did. His and Rep. Holtzman's bills were successful, and the amendment would still have a chance to be ratified until June 30, 1982. Marvella was one of Birch's allies in the effort, getting personally involved in helping to seek passage of the extension.[8]

In September, Birch introduced legislation to ban the use of polygraphs in employee hiring. In October, FISA became law. Also in October, Birch announced that gasohol would be for sale to the public throughout Indiana.

The biggest news riveting the nation that fall concerned the Middle East. Egyptian president Sadat had visited Israel the previous year, a bold and dramatic step toward seeking a permanent solution to the Arab–Israeli conflict. On September 5, President Carter stepped up to the plate, thrusting himself into the middle of the peace process. He invited Sadat and Israeli prime minister Begin to Camp David for negotiations that he conducted with a passion and skill that amazed much of the nation. On September 17 in a dramatic White House ceremony, the Camp David Accords were signed. Its signers, Egypt's Sadat and Israel's Begin, would be honored with the Nobel Peace Prize for their efforts in seeking Middle East peace. Carter's peacemaking skills were widely applauded, and his popularity soared.

Measures important to women were passed in October, including the Susan B. Anthony Dollar Coin Act, the Pregnancy Discrimination Act, and the ERA extension. Government reformers were delighted by the passage of the Civil Service Reform Act and the Ethics in Government Act. Also passed were the Drug Abuse Prevention, Treatment and Rehabilitation Act, the Humphrey–Hawkins Full Employment Act, the Contract Disputes Act, the Airline Deregulation Act, the Bankruptcy Act, and the National Energy Conservation Policy Act.[9] Birch's efforts were rewarded by the passage of the Foreign Intelligence Surveillance Act.

Since Marvella's diagnosis and treatment for cancer, Birch rarely traveled to Indiana. Instead, they both went to religious retreats, about which Marvella had become enthusiastic. They visited faith healers, some well known and others not, people like Oral Roberts and the Reverend Robert Schuller. Adding to these personal problems were rumblings of a Birch Bayh ethics problem, something antithetical to the image he had so successfully crafted. His name began appearing in newspaper stories relating to a possible influence-peddling scandal known as Koreagate.

In 1976, federal charges were brought against South Korean businessman Tongsun Park. It was alleged that he had used money from his government to bribe members of Congress in an effort to force the United States to keep its troops in South Korea. The following year, Park was indicted on thirty-six counts but avoided going to trial in exchange for an offer of testimony. One of the members of Congress he testified against was House Speaker Carl Albert, who soon announced he would not run for reelection.

Early in 1978, Park appeared before the Senate Ethics Committee to respond to stories that he had made illegal campaign contributions to a number of senators, including Birch. When press reports started occurring in March, Birch and Jay Berman fervently denied any illegality. Only campaign contributions from American citizens are legal. Furthermore, it is illegal to accept contributions at the Capitol or on its grounds. Articles reported that Park told the Ethics Committee that he gave $1,500 to $1,800 to the Bayh campaign in 1974. Birch told reporters, "It's for damn sure I didn't know about it," and indicated that he didn't believe that it had happened at all. Park had said that he believed he gave the money to Berman at the time, who categorically denied taking a contribution from Park.

Birch was despondent about the charges. Added to the distress he felt over Marvella's illness, he began to wonder, probably for the first time, if it all was worth it. He told a reporter, "Given the total context of things, that is just about as damaging in the minds of the public today as if I had stuck eighteen $100 bills in my coat pocket."[10] Birch recalled years later, "He [Tongsun Park] never asked me for anything."

The issue seemed to go away for several months. But in October, it exploded back onto the scene when the report of the Ethics Committee was released. Days prior to that, however, committee leaks led to a spate of entirely inaccurate stories implying that Birch was dishonest and had broken the law in several ways.

The story dogged him all year. Beginning in March, it seemed that there were articles monthly, many about the rumors printed elsewhere. Rumors are impossible to put down, and this one was extremely harmful. Birch and Berman had received a contribution from DC lawyer Ed Merrigan that was later revealed to be given at Park's request. Merrigan first contended that it had been handed over inside the Capitol, which would be illegal. Birch and Berman denied that, and Merrigan backtracked later on. There was never evidence that Park had actually contributed. No report of a Tongsun Park contribution appeared on the extensive campaign contribution filings with the Federal Elections Commission (FEC) by the Bayh campaign in 1974. No canceled check from Park's account was ever produced. In other words, there was no physical proof of illegality, and regardless of what one may have believed at the time, the dispute was largely one person's word against another's. In this case, it was one person (Park) against two (Bayh and Berman).[11] No one was charged with a crime, but that didn't keep the story from being recirculated and, in the case of several Indiana newspapers, misreported.

With regard to campaign contributions, I recalled an incident that took place in Birch's Indianapolis office. A prominent Bayh donor visited and made a request but was unsubtle in his offer of campaign money in return. While I don't recall what he requested, the conversation felt completely inappropriate, and I asked him to leave. Being new to running the Indiana operation, I began to feel nervous about it. When I next met with Birch, I told him about it and was relieved by his response, which was that I had done the right thing and that I should never question myself on a matter like that. There was no further contact with the donor.

As the year moved along, Marvella started traveling again, and the need for Birch to be home with her on a regular basis lessened. He decided to take an extended trip to Indiana. I accompanied him on a lengthy Indiana sojourn in mid-October, and it was memorable. The first crisis faced was when the Ethics Committee report was leaked to the press. The Fort Wayne *Journal Gazette,* normally a friendly newspaper, editorialized that Birch had broken the law and characterized the charges in the worst possible way. It was incredible and included fourteen factual errors. Eventually, the paper actually apologized for the editorial and printed the entire rebuttal that Birch and his press staff wrote. But the damage was done, and for many days as we traveled around Indiana, the Ethics Committee report was the first topic brought up by reporters.

The next development on the trip was troubling news from the Associated Press that Marvella had written an article for *Good Housekeeping* that stated she only had one year to live. Birch was asked the question and could only confirm that she had written an article for the magazine. Then we were informed of a death threat against Birch, and he was to be given protection by federal marshals. Both these matters were troubling, but it got worse when a local TV station, also in Fort Wayne, broadcast, "Senator Birch Bayh told reporters in Fort Wayne last night that his wife Marvella was dying of cancer and had one year to live." This was absolutely untrue.

The next several days were spent crisscrossing the state, accompanied by federal marshals as well as news crews. Eventually, Marvella's article hit the newsstands, and the *Good Housekeeping* headline read, "One Year to Live." This personal crisis involving her health, preceded by the death threat and the Ethics Committee report, made for a stressful tour of the state. Then Birch was faced with another crisis that threatened his political health.

We traveled to Gary, where he was to be honored at a peculiar affair. Billed as "A Night to Remember," it was an attempt to highlight Gary as America's civil rights capital but was, at $100 per person, a fundraiser for Mayor Hatcher's 1979 campaign. Among the guests were Andrew Young, Coretta Scott King, Vernon Jordan, Jesse Jackson, Dick Benjamin, and Rep. Adam Benjamin. But a police and firefighter picket line was planned around the event. Birch had never crossed a picket line. He received mailgrams from AFL-CIO Indiana leader Willis Zagrovich, who, like other local labor leaders, refused to participate in the event. If Birch had only been among the guests, he could have stayed away, but the event was in his honor.

The leaders of the strike, who also considered themselves Bayh allies, agreed to meet with him to see if a solution could be reached. Several firefighters and policemen came, all union leaders, and discussed their grievances, of which there were many. They also expressed extreme enmity for the mayor. They didn't want to hurt Birch, and they understood Hatcher's political importance to the senator. The problem was that a pro-labor senator was about to be embarrassed—he could either cross the line and anger labor unions around the state and country, demonstrating hypocrisy on labor issues, or miss the dinner, humiliating one of the most powerful and influential politicians in the party. Beyond that, he also might avoid an assassin.

Since Birch was Hatcher's friend and Hatcher wouldn't listen to them, we suggested, why couldn't Birch be escorted across the line as the union's emissary to the mayor? The discussion absolutely halted. Will Smith of the International Association of Fire Fighters (IAFF) suggested they depart and call us back with their answer. We agreed and offered any assistance we could provide, whatever their response might be. Two minutes later they called us with the answer. They agreed to our proposal, and a meeting place was appointed for the walk across the line. The press would be notified of the agreement, and pictures would be taken.

We cleaned up and drove to the dinner. A black-tie affair, it was in stark contrast to the Gary neighborhoods we passed through and the economic woes of the pickets we approached. At each corner were police cars with protest signs leaning against the cars. All off-duty cops were forced to work that night under threat of dismissal, and they were unhappy. We arrived at the picket line, had photos taken, and were accompanied across the long yard by the firemen, police and now four federal marshals. It was a long walk.

Inside the venue, the celebrities were gathering with the political dignitaries. They were introduced and led to their seats. To our surprise, all the dignitaries crossed the picket lines, including civil rights leaders Young, Jackson, and King. There were many tributes to Birch from the crowd of two thousand people. The evening was a huge success. Discussing the death threat, Birch said I was standing a bit far away from him and must be worried that the assassin was a bad shot. I acknowledged that it had occurred to me. After his speech, he sat down and said to me, "The asshole missed his best shot."

Birch spoke to Hatcher about the police and fire dispute and would follow up the next day. We left in a police escort, stopping for ice cream at a shop advertising "groceries and auto parts." Once at the airport, we considered the threat over and flew away with considerably less anxiety.

Here is another memorable episode during this Indiana trip from notes I wrote at the time:

> Learning that Marvella would be on the 8:30 segment of ABC's "Good Morning America" the next day, we found a television we could watch in the car while driving to Terre Haute. At 7:30 AM, we left and watched the TV on the way. At 8:30, and 15 miles before Terre Haute, the TV lost frequency and went blank. We took the next exit and headed toward the home of a Bayh friend.
>
> Realizing we'd never make it in time, we pulled into a yard where a house had its lights on. Senator ran to the door in the rain and asked the astonished woman in the doorway if he could watch her television saying, "Hi, I'm Birch Bayh and my wife's on TV. Do you mind if we watch it?" He then called me in and we watched it.
>
> Marvella did well in her interview with David Hartman of ABC-TV, discussing living with cancer. Mrs. Ramos, who had answered the door, was quite enthralled by the circumstances. We soon left with the Senator inviting her to his home if she gets to D.C.
>
> At subsequent events, we were continually impressed with Birch's recognition factor and repeated questions about Marvella. It was one of the most gratifying parts of this or any trip, the genuine affection demonstrated by so many people. There is no way to put a value on the warmth and support we felt from so many human beings. The experience was priceless and will remain memorable to Birch and to me.
>
> Flying to Elkhart, Senator called Governor Bowen about an effort to persuade Pepsi-Cola to move its manufacturing plant to Indiana. Bowen told Birch he had defended him on the Ethics Committee charges in a press conference. We were surprised by this and glad to see that the state's main Republican chose not to turn our troubles into partisan gains for himself.
>
> We proceeded to fly to Valparaiso. The county chair met us. While waiting for us, a young man hitch-hiking from California had bummed

> spare change for cigarettes and was told who would soon be arriving. The hitch-hiker remarked, "He's having problems with his woman, isn't he?"

The trip continued in this vein until a few days later when, while preparing for a live TV interview, Birch called home and had an upsetting conversation. Marvella's cancer was in her lungs, and she wanted him with her. It was a difficult and challenging moment.

The election came, and Indiana Democrats were beaten badly, losing the statewide ticket, both houses of the legislature, and most county offices. All congressional incumbents won, with one exception, David Cornwell. Cornwell's loss was not surprising, but the overall effect of the 1978 election was a warning about the next election in two years.

That fall, after rioters in Tehran sacked the British embassy, two million Iranians demonstrated against the shah in Iran, the first stirrings of events yet to happen.

In the Senate elections, thirteen seats changed hands, with a net GOP gain of three seats, though it remained Democratic 58–41, with one independent. The House had a net loss of fifteen Democrats but retained a 277–158 majority. The ominous signs, however, came with the rise of the New Right, a combination of organizations seeking to defeat liberal Democrats, mostly in the Senate. Fueled by the fundraising machine of Richard Viguerie, groups like the National Conservative Political Action Committee (NCPAC) and the Moral Majority burst onto the scene. Viguerie had helped create the movement with his direct mail expertise. The Panama Canal treaties presented the first opportunity for Viguerie, who used that expertise to generate massive amounts of mail to Congress against the treaties. Groups started popping up everywhere, largely funded by direct mail contributions, each with its own conservative slant but focusing on single hot-button issues that could rouse those in the country who seemed to be waiting to be aroused. Jerry Falwell, a Lynchburg minister, launched a political group known as the Moral Majority and quickly became among the most strident of New Right spokesmen. A close second was Terry Dolan, head of NCPAC, showing considerable competence in using single issues to target political opponents for defeat while raising money to fuel conservative efforts. The emotional hot-button issues were always best—abortion, gun control, the Panama Canal.

Many decried their slash-and-burn tactics, a demagoguery that threatened the civility necessary to healthy political dialogue. Exemplifying a win-at-all-costs philosophy, they received substantial press because of the senators they had targeted for defeat, with five Democrats—Haskell, Clark, Hathaway, Anderson, and McIntyre—losing their seats. Much was made of these Senate targets in the national news, and their defeats bolstered the reputations of these groups. It didn't take a crystal ball to know who they would target in 1980.

Among Senate Republicans, Jesse Helms (NC) personified the far right, a leader among the New Right forces so active in that election. He served in the Senate with Birch, and Birch remembered the two of them to be "as far apart on the issues as could be." Yet when news of Marvella's health issues arose, Helms was among the first colleagues to call Birch and express his concern. Alan Simpson of Wyoming was another and also became a close Bayh friend. "We have a very good relationship to this day," Birch said almost four decades later.

A sign of the times was the defeat in California of a ballot measure prohibiting gay school teachers. The nascent gay rights movement was having an impact, while organizations in the New Right, hardly supporters of gay rights, were on the rise.

Also in November was the Jonestown massacre in Guyana. Jim Jones, a cult leader from Indiana who presided over what he called the Peoples Temple, was under investigation by California congressman Leo J. Ryan because of stories circulating about the cult. He led his followers to commit mass suicide, claiming 918 lives, including over 270 children. While on his mission, Congressman Ryan was assassinated by Jones's supporters.

Here at home, the office of Mayor Dennis Kucinich announced that Cleveland was going into default, something no other major American city had gone through since the Depression. In China, the Communist Party and its leader, Deng Xiaoping, announced a reversal of Mao-era policies in order to pursue economic reforms. And America learned in 1978 that its nuclear weapons stockpile was exceeded by that of the Soviet Union.[12]

17

The Death of Marvella: 1979–80

There is no sea more dangerous than the ocean of practical politics—none in which there is more need of good pilots and of a single, unfaltering purpose when the waves rise high.

THOMAS HUXLEY[1]

IN 1979, BIRCH BAYH BEGAN HIS SEVENTEENTH YEAR AS A United States senator and the final two years of his third term. Now twenty-third in seniority, he could justifiably feel that his role in that august body was more prominent than the ranking implied. A key member of the Democratic majority and the chair of active and important subcommittees, he would soon also assume the chairmanship of a full committee for the first time, the Senate Select Committee on Intelligence. His Senate activities reflected interests far wider than his committee or subcommittee roles might suggest. Throughout the year, the Bayh office issued over 375 press releases displaying an astonishingly large number of interests on a myriad of issues. Other changes in the Senate took place in early 1979, namely the ascension to the chairmanship of the Judiciary Committee by Ted Kennedy, while Frank Church took over the Foreign Relations Committee.

When Kennedy took over Judiciary, he reorganized it, abolishing some of the smaller subcommittees and combining them into fewer, larger ones. Birch would chair the Subcommittee on the Constitution, combining the Constitutional Amendments, Constitutional Rights, and Juvenile Delinquency subcommittees. The first order of business was revising the federal criminal justice code, a priority of Kennedy's in which Birch played a major role.

Birch was a member of the Intelligence Committee during its first two years, and becoming chair brought him into more direct contact with President Carter and other leaders in the intelligence community. Not long after assuming the chairmanship, he recognized how important it was to identify those members of the committee he could trust. When President Carter shared a secret with him, Birch felt it appropriate to share that secret with the handful of senators who could be trusted. Sharing a secret with a senator who couldn't be trusted would undermine his own role as chair and would also make the country less secure. Those secrets were shared with senators Daniel Inouye (D-HI), Walter Huddleston (D-KY), John Chafee (R-RI), and Indiana's Dick Lugar. They all proved trustworthy, and secure information was never revealed. Each knew that a leak from the committee would result in an administration that would stop sharing secrets with them. The fact that Birch's relationship with Lugar had evolved to this point was no small thing.

Time was spent on the direct election proposal, a possible balanced budget amendment, the welfare of Russian dissident Anatoly Scharansky, promotion of alcohol fuels, drilling of Alaskan oil, juvenile justice matters, and the SALT II Treaty. There was also the Marble Hill nuclear plant in Indiana, oil divestiture, anti-cigarette smoking, child abuse issues, the Family Farm Antitrust Act, the proposed Bayh–Dole Act, the welfare of Vietnam veterans, chairmanship of the National Alcohol Fuels Commission, a proposed Department of Education, the Chrysler bailout, and improper steel imports. President Carter had begun deregulating airlines, and Birch warned about the discontinuation of service to many Indiana cities, an accurate prediction.

There were a number of staff promotions in the Bayh office, with Tom Connaughton becoming administrative assistant, Kevin Faley replacing Nels Ackerson as chief counsel of the Subcommittee on the Constitution, and Eve Lubalin becoming legislative director. Connaughton and Faley, both Hoosiers, worked themselves up the ladder to reach those upper-level jobs. Lubalin was not from Indiana but typified those staffers who had gravitated to the Bayh office because of certain issues.

Eve Lubalin joined the office in the mid-seventies as a Congressional Fellow, eventually leaving in order to write her dissertation. She returned to the Bayh office in May 1977 to work on energy and environmental issues. Her commitment to women's issues resulted in part from being denied a teaching position at the University of Virginia (UVA) because she was pregnant. Birch had written Title IX to ensure that university women receive the same treatment as men with regard to teaching posts, sports, and campus life. Because of Title IX, she wanted to work for him. The fact that UVA was the same institution that had denied Marvella Bayh entrance because of a males-only policy was unknown to her at the time. She found the office environment inviting, with other women holding similar, important legislative roles. Barbara Dixon, Mary Jane Checchi, and Abby Saffold all held key jobs with significant responsibilities in major legislative areas, including managing Senate floor activities for Birch.

As Birch looked toward a 1980 reelection campaign, there was much to be concerned about from the 1978 election. Several key Democratic senators had lost, and there was the rise of the "New Right," a panoply of groups that became anti-liberal attack dogs in the political landscape. The next election seemed right around the corner, and the lesson of 1978 was that these groups had to be taken seriously.

"The 1980 Senate campaigns started almost immediately," wrote Ira Shapiro about early 1979. "Paul Brown, the leader of the Life Amendment PAC (LAPAC), took aim at what his group called the 'deadly dozen.' Boasting that his group had been instrumental in defeating pro-choice senators . . . Brown said: 'We've proven our point. There's a pro-life vote. We've come of age as a political force.' Birch Bayh headed the group's 'Deadly Dozen' list, followed by George McGovern, Frank Church, and Representative Morris Udall. Also targeted were John Culver and Patrick Leahy, and, for good measure, one Republican—Bob Packwood, who had been the most outspoken Senate advocate of abortion rights."[2]

Birch would face reelection, and these organized efforts against him after having been the deciding vote against the pro-life, anti-abortion constitutional amendment in his subcommittee not once but five times. Those votes played right into their hands.

This flurry of activity took place during a time of intense worry about Marvella's health. Her diagnosis and ongoing medical procedures cast a pall over Birch's life, making it increasingly difficult to be effective on all the political issues he cared about. His staff faced the challenge of how to plan for a reelection campaign without knowing whether or not he would run. We knew that if Marvella lived, he would not run for reelection. There was no way he could justify being away from an ailing wife while pursuing his own reelection. This was the unspoken reality that was fully understood by those of us with the responsibility to plan for the campaign.

The year began with the reintroduction of the amendment to abolish the Electoral College in favor of direct election of the president. Soon, Birch announced support for a $41.3 million gasohol research project at Purdue. On January 15, he introduced a bill to honor Martin Luther King by making his birthday a national holiday. A few days later he introduced legislation to include Pap smears under Medicare coverage. Had Pap smears been more widely available when Birch was a child, his mother's premature death might have been avoided. Shifting gears the following day, he introduced legislation requiring divestiture of the oil companies.

Also on January 15, he introduced a bill to guarantee civil rights to the institutionalized. His involvement with the rights of those who were incarcerated or committed to mental hospitals or other state institutions began a few years earlier when Patricia Wald, President Carter's assistant attorney general for legislative affairs, asked Birch to sponsor the legislation. The measure was a far-reaching effort to protect those in jail, prisons, pretrial detention facilities, nursing homes, mental institutions—any such facility or institution that was owned, operated, or managed on behalf of a state or municipality. There was no conceivable political benefit from investing time and effort in the matter, and opposition arose from Southern senators who interpreted it as an attack on the way Southern prisons were operated. Birch agreed to sponsor the bill and got personally involved in drafting it. Having been unsuccessful during the previous session, he resolved to bring it up again at the beginning of the new Congress and did so.

Hearings were held, and it would be an understatement to call them contentious. Many senators were contacted by state officials lobbying to keep the federal "busybodies" out of their business and allow them to run those institutions within their borders unshackled by regulations related to federally approved legislation. Strom Thurmond was a prominent opponent. The Judiciary Committee was among the most sharply divided of committees in Congress, and its jurisdiction included many of the most hotly debated issues before the country, especially civil rights. Birch was hard pressed to find the necessary votes to get the bill out of his own subcommittee, much less the full committee.

Then, seemingly out of nowhere, the ranking member of the subcommittee, Republican Orrin Hatch, changed his mind, supported the bill, and helped Birch get it passed out of the subcommittee. Hatch endured the wrath of his own ally Thurmond, who never forgave him.[3] Because of this unusual alliance, the bill passed out of Judiciary, passed the House and Senate, and became law with President Carter's signature. The Civil Rights of Institutionalized Persons Act became the law of the land in 1980. Twenty-two years later, Nels Ackerson ran into Hatch at the World Economic Forum and reminded him who he was and of their work together on the bill. Hatch told him, "It was one of the best votes I ever made."[4]

At the same time, much was going on internationally. On New Year's Day, the United States and China established full diplomatic relations. Also in January, the Pol Pot regime in Cambodia fell to Vietnamese troops. In February, the American ambassador to Afghanistan was kidnapped and would later be killed. As important as these developments were, events in Iran would come to dominate them.

It is unclear how many Americans in 1979 knew anything about Iran. But that would change, probably forever, with an impact on American history that would resonate for decades to come. On January 16, after a year of turmoil, the shah of Iran and his family fled Iran and traveled to Egypt. The shah had been a close ally of the United States, and his opponents turned much of their anger against his regime toward those Americans still working and living in Iran. On February 1, Muslim cleric Ayatollah Ruhollah Khomeini, in exile for almost fifteen years, returned to Iran. Greeted by a measure of enthusiasm unfamiliar to most Americans, he took control of the country, creating the Council of the Islamic Revolution. By the end of his first week, his supporters had taken over all legal and governmental institutions. Soon, the military mutinied and joined the Islamic Revolution. On April 1, a national election was held, and Iran voted overwhelmingly to become an Islamic republic. The shah's overthrow was made official.[5] It was not yet clear what this would mean to the average American.

In February, Birch introduced the Family Farm Antitrust Act. He also introduced a bill that would have far-reaching implications, what became known as Bayh–Dole. Formally the University and Small Business Patent Procedures Act, it acknowledged that the ability to market patents and inventions to which the taxpayers had contributed money was being stifled by government bureaucracy. Birch and Sen. Bob Dole worked to change this. Bayh–Dole's intent was to transfer those rights to the universities and small businesses involved instead of the federal government.

The matter of federal appointments continued as well. Knowing there would be judgeship appointments in all the judicial districts in Indiana, Birch instituted a judicial selection process to make a concerted effort not only to find the best persons to recommend to the president but also to involve the best legal minds in the state. It was critical to screen each person thoroughly to ensure that a man like Carswell would not benefit from a Bayh nod. On February 16, James Kimbrough and Eugene Brooks were recommended for judgeships representing the northern and southern districts, respectively. While Brooks was a close friend of Birch's, Kimbrough was not well known to him. An African American attorney from Lake County, the scrutiny of Kimbrough was intense and the process of choosing him far more complex than it was for Brooks. Making these recommendations to the president was a time-consuming process.

In March was an event important to many people around the country but particularly to Indiana. The NCAA basketball finals, now known as "March Madness," took place with Indiana State University (ISU) playing Michigan State University (MSU). Larry Bird of the ISU Sycamores and Earvin "Magic" Johnson of the MSU Spartans were the stars, and both would go on to dominate professional basketball for years to come. Michigan State won the game, but Bird, known as "the hick from French Lick," would become one of the most recognizable Hoosiers. When Bayh friend Joe Anderson in Terre Haute proclaimed the existence there of the greatest basketball player who ever lived, scant attention was paid to him. Birch attended a game during that championship year and was taken to the locker room by ISU president Richard Landini. When he was introduced to Larry Bird, the player proceeded to pick up the senator and carry him fully clothed into the showers. Birch was soaked. Before meeting with the press afterward, he was asked if he wanted to change into clean clothes, but he declined, understanding that only good feelings could emanate from a politician dampened by the exuberance of Indiana's greatest star.

As Marvella's health deteriorated and Birch was unavailable to attend events he otherwise would, the American Bar Association held a convention to which he was invited to speak. Staff member Bill Moreau represented him and was acknowledged from the podium by Attorney General Benjamin Civiletti, who proceeded to talk about Birch, pointing out that he was involved in an effort to protect the rights of the institutionalized. It was obvious to those present that this was not an issue that would gain the Indiana senator votes but was one of those issues that he felt needed to be addressed. The group was reminded of how rare it was for a senator to spend political capital on behalf of those who could not represent themselves. "He just always felt very, very strongly to protect those who can't be protected and to help those who need help the most" was Tom Connaughton's characterization.[6]

In early 1979 Marvella's health problems continued unabated. The previous fall, she had to be hospitalized after promoting the American Cancer Society at a town hall meeting. Doctors found that her cancer had spread and detected fluid in her left lung. That night, her struggle with cancer was featured by Walter Cronkite on the *CBS Evening News*. He described her as "someone who is living with cancer and walking with God every step of the way." On February 14, she observed her forty-sixth birthday, and a few weeks later, on March 2, she gave the last speech she would ever give. It was to an audience in Charlottesville at the University of Virginia. Watching in the audience was law student Evan Bayh.[7]

Marvella was honored with the Hubert H. Humphrey Inspirational Award for Courage on March 28 but was too ill to accept it in person.

Her written words were read at the ceremony. "Courage? If I have shown courage it has been saying, I alone am not strong enough, and I need help from God and others. I am blessed for they have been and always are, there."[8]

Birch described the next several days as a time of holding hands and reminiscing. They talked about Evan's future and wondered about whom he might marry. "Marvella hoped to be a grandmother, but she was not afraid of dying, yet she wanted very much to live," he said. The last time Evan would see his mother was Easter Sunday. Birch and Evan attended services at Washington National Cathedral and upon leaving saw a woman selling daffodils. Birch bought a few for Evan to give to his mother. When they got to her hospital room, she was not hooked up to machines but was unable to be heard when she tried to speak. Evan could see her mouthing the words "I love you."[9]

Marvella Hern Bayh passed away on April 24. Birch talked about those last days, saying, "She was at peace and secure in her faith in God. Those of us who loved her cried. But somehow I have the feeling that those tears expressed only our grief, our own personal loss. Marvella must have been smiling down on us." At her memorial service at the Washington National Cathedral, attended by President Carter, Vice President and Mrs. Mondale, and many senators and members of Congress, "Birch read from a speech she never had the opportunity to deliver. 'I have been blessed by the help of others reaching out to me with encouragement and hope.'"[10]

Evan's memories of his mother brought to mind the Bayh family vacations, special experiences for Evan as an only child whose father was often away. At the age of eight, he learned to ski, probably at the same time as Birch, and they had many skiing trips together. He retained fond memories of their trip abroad during the Christmas season in 1977, to Israel, Switzerland, and Morocco. The following summer, he went on a cruise with his mother, which turned out to be the last opportunity for them to have concentrated time together. It was on that cruise, he felt, that he first looked at his mother through the eyes of an adult. Evan remembered his mother's last speech, when he had sat with her on the dais at an American Cancer Society event in Virginia. For most of the time since she had fallen ill in 1971 and up until that point, he had not accepted her possible death from cancer. Hearing her labored breathing as they sat next to each other was a shock and a wake-up call to him.[11]

A few days after Marvella's death, I wrote several pages worth of memories in order to remember this period in our lives. An excerpt follows:

> In March 1979 Marvella took a turn for the worse. She was admitted to Columbia Hospital for Women. A few days later, she was home feeling fine and Evan came home for a visit. The latest illness had drawn the family closer together and Evan grieved over his mother's condition, yet he endured his first year of law school at the University of Virginia and intended to take his final exams as she urged him to do.
>
> After Evan's visit, she went downhill again and we had to rush oxygen out to the house. She got increasingly worse and the doctors urged chemotherapy for her. From the pain of her early cancer treatment, she had vowed never to take those treatments again. This time it was clear. If she didn't agree to at least two rounds of chemotherapy at the National Cancer Institute (NCI), she would soon die.
>
> By the end of the first week in April, she was placed on life support. Evan was not told this in the hope that it would be only temporary sustenance for her and he could finish his finals unimpeded. Most of the staff was kept in the dark for the first several days. Birch and Marvella had earlier agreed that neither would keep the other alive artificially and he agonized over the decision to do so at this time. The doctors, however, insisted that unless they revived her strength and sustained her vital functions, they could not attempt chemotherapy, it would surely kill her. If they didn't try it, on the other hand, the cancer would either kill her outright or sufficiently weaken her to the point that a germ or infection would mortally inflict her. Neither were pretty alternatives but Birch agreed to two weeks of life support. If she could not tolerate the chemotherapy, they would pull the plug.
>
> On April 11, at 6:30 that evening, Senator called me. He rambled on about how many times she had almost died that week and Evan had to know, but he couldn't tell him by phone. I suggested we drive to Charlottesville so he could tell him face-to-face and he agreed.
>
> We arrived at Evan's about 10:00 pm and he went in to tell him. Evan was crushed and remarked that he was shocked as if his Dad was telling him she had been hit by a car. She had looked so good when he last saw her. We stayed with Evan until 1 AM, both he and Birch crying on each other's

shoulders. Birch persuaded him to remain at school until he was really needed at home. The Senator and I drove back to NCI in pell-mell fashion, getting stopped for speeding once. The cop knew about Marvella's condition and let us go. We talked about Evan and Senator decided that Evan couldn't be protected; as an adult he had to make up his own mind about being there. He would probably rather see her alive again, even hooked up to machines, than never see her alive again. Birch agreed to have Evan make the choice and knew they'd be together soon.

Monday, April 23 at the office, I was accosted by Mrs. Koutsoumpas with tears in her eyes. "Bob, she's gone. Marvella died."

The events following her death moved rapidly. We put a news clamp on her death so Birch could tell Evan personally. P. A., Mrs. K and I immediately left the office that afternoon for NCI. Arriving there, we found the Senator composed and exhausted. He had already begun making calls to family and friends. Our press secretary had also received press inquiries asking for confirmation of her death. It seems that some news organ paid an informant in the hospital and it was on the wires within 30 minutes. He was directed to confirm her death but to request an embargo until Evan had been told.

We left for another quick ride to Charlottesville. Senator was relaxed, almost relieved, and we talked through the steps that would need to take place over the next three to four days. Arriving at Evan's, we found him crushed with grief. During our drive, the news was on the air and the office called Jane Sinnenberg to let her know. She called Evan so he would not hear it on radio or television. Soon afterwards, Evan's friends began calling. We left Charlottesville almost immediately.

We got to the Bayh home quite late to find it filled with neighbors and friends. The next day we began making the necessary arrangements. While trying to plan all details, we were besieged by visitors, dignitaries and the not-so-famous; staff, family and friends; political allies and adversaries. We hosted Senators, Congressmen, Ambassadors; handled many phone calls and telegrams. The crush of people became tremendous. The Senator made dozens of calls to his and Marvella's family, friends around the country, old political associates, and those who would be asked to participate in the services. The staff called political friends in Indiana who the boss couldn't call himself.

On Wednesday night, the Senator, Evan, Tom, Lynne and I rested in his study. Marvella's presence was everywhere but less so there than in other rooms of the house. We watched the NBC Nightly News that ended with a tribute to Marvella and her war on cancer. It got to us for two reasons. The first was hearing her voice. No one had heard it for nearly a month. Then there was the tagline by correspondent Tom Pettit. "While not universally loved, Marvella was an inspiration to many." God knows why that disclaimer was necessary for anyone's benefit. I know it hurt her son and husband.

The memorial service was scheduled for Friday noon at the Cathedral. Taking part were Abigail Phillips, better known as Abigail Van Buren, or "Dear Abby," Pat and Shirley Boone, among others. Senator Bayh spoke as well. It was emotional and the sobs could be heard throughout the Cathedral. Among the ushers were Senators Pryor, McGovern and Church. Among those attending were nearly the entire Senate, Vice President and Mrs. Mondale and President Carter. Abby took part because of her long friendship with Marvella. She struck the most appropriate chords for what was to be an emotional but upbeat occasion. She had traveled with Marvella in our 1974 campaign and gave remembrances of those times. The Boones sang "Amazing Grace" with the Cathedral Choir. The memorial service at Washington's National Cathedral was to be described as the largest memorial to any woman since Eleanor Roosevelt.

Despite the obsession with Marvella's health in those months prior to her death, Birch and his staff remained intensely busy. In March, he introduced the Buy American Act of 1979, to incentivize the purchasing of products made in the United States. He introduced legislation to expand the Indiana Dunes, announced a schedule for hearings on the proposed balanced budget amendment, and held hearings on the direct election amendment, featuring testimony by author James Michener. Michener had written a book called *Presidential Lottery* about the Electoral College, and his criticism of the manner by which Americans elect their presidents generated a great deal of attention.

Also that month, Three Mile Island, Pennsylvania, experienced the worst nuclear power plant accident in the United States. In many ways, this accident was another blow to the president. His support of nuclear power came into question, and the accident seemed to underscore an attitude that nothing was working these days. When people feel that way, it is rarely good for the person in charge. Notably, March 26 was the date when the peace treaty was signed in a White House ceremony between

Egypt's Sadat and Israel's Begin. This would not have happened without the intense negotiating skill of President Carter, who brokered the deal.

Days before Marvella's death, there was a bizarre incident involving the president that seemed to typify the kind of problem he would encounter in the next year and a half. He was in his hometown of Plains, Georgia, when he was attacked by a swamp rabbit while fishing from a boat, a humorous image that would haunt him as other problems piled up about him. Around this same time, the news was filled with photos of the president nearly collapsing while trying to run in a race. He appeared snakebitten, unable to dominate events around him and in many ways victimized by them.

Seemingly a harbinger of things to come, May elections brought Conservative Party leader Margaret Thatcher to power as the United Kingdom's first female prime minister. Was there a conservative trend in the United States as well? The 1978 Senate elections seemed to indicate that there was. Despite that, some events indicated that the progressive agenda was alive and well. In June the Los Angeles City Council passed the city's first bill to protect homosexual rights. Its signing by Mayor Tom Bradley was done with little fanfare, possibly because of the riots taking place in San Francisco, dominated by gay men who were outraged by the lenient sentence Dan White received after his conviction for assassinating Mayor George Moscone and Supervisor Harvey Milk.[12]

Birch's sister Mary Alice had gone back to school to get her nursing degree in the 1970s, having given up her professional acting career outside of stints in community theater around Maryland. She was married to a man named Larry Feather. One of the memorable events for Birch in 1979 was giving the commencement address at Mary Alice's graduation. He remembered looking into the audience and feeling the influence of Title IX. The satisfaction was on a personal as well as a professional level.

If his career was to continue, he needed to spend time in Indiana, shoring up support for another reelection campaign. With Marvella's passing, it seemed that pursuing reelection was the obvious move for his future. He wanted to continue in the Senate and had only contemplated retiring when it seemed he should be staying home with Marvella, who would need his support and presence. It was also true that Birch's national reputation was exacting a cost in Indiana. He hosted town meetings in several cities, and the staff beefed up the franked mail operation. He became convinced that his relationship with the Hoosier press would improve if he had a press secretary who came from among them. His instincts told him that his pursuit of national office and the recent Tongsun Park controversy strained that relationship, which had always been fairly comfortable. We began a quiet search to identify someone in the Indiana press community whom we could hire to be the Bayh press secretary.

Fred Nation was a reporter with the *Terre Haute Star* whom Birch and the staff had come to know. Born and raised in Terre Haute, Fred had observed Birch's career since Fred was in high school, always aware of the local politician who was rising through the ranks of Indiana politics. He was energized by the presidency of JFK and felt favorably toward Birch because of his affiliation with the Kennedys. When the group Bipartisans Against Bayh was active in the 1974 campaign, Fred investigated to find out who was involved, since federal law did not require that kind of disclosure at the time. When he had a list of names of those in the group, he shared it with Birch and the staff. Birch first approached Fred to feel him out about becoming press secretary in 1978, but the effort to bring him aboard intensified after Marvella's death as campaign planning also got more serious. Fred was hired in late 1979 and moved to Washington early in January 1980.

In the wake of Marvella's death, Birch refocused his energies on Indiana politics, but he remained engaged with the national issues he cared so much about. A Carter executive order in July 1979 created the Federal Emergency Management Agency (FEMA) in response to the federal responsibilities required by the Disaster Relief Act of 1974, which Birch cosponsored. The creation of FEMA was an attempt to merge the multiagency responsibilities for disaster relief, a satisfying conclusion to a long-term effort that began in Birch's first Senate term. His first success in the effort to bring federal assistance to those affected by natural disasters was the passage of his bill, the Disaster Relief Act of 1970. For years, he had felt like a lone wolf baying at the moon—until Hurricane Camille hit the South and Southwest in 1969. Only then was he able to find enough colleagues to cosponsor his legislation and secure its passage into law. By 1974, amendments to the existing law ended up becoming the Disaster Relief Act of 1974, introduced by Sen. Quentin Burdick with Birch as cosponsor. The House legislation set up a mechanism for disaster relief determinations to be made only after a disaster actually occurred. Birch argued that solutions needed to be in place in advance of a disaster. What if the crisis happened while Congress was out of session or on a holiday? When that did happen, he secured support from those who previously had endorsed the House's approach. Victims of a disaster would not have to wait for Congress to act; a process would be in place to provide relief quickly.

After a tornado hit southern Indiana, Birch toured the area, assuring constituents affected by the tornado that he would make sure they got help. When there were bureaucratic difficulties, a meeting was scheduled at the White House. Staffer Lew Borman said, "After listening to a series of excuses from the administration's Jack Watson, Birch ran out of patience and said, 'Listen Jack, don't tell me about what FEMA can and can't do, I wrote the bill. Now I'd appreciate your consideration to do ALL you can for Southern Indiana.' The red-faced Watson said he would revisit the earlier decision."[13]

In June 1979, President Carter and Soviet leader Brezhnev signed the SALT II Treaty. At the same time and unknown to Brezhnev, Carter also signed an authorization for covertly aiding the Soviet opposition in Afghanistan. Birch found himself embroiled in proposals for a constitutional amendment to require a balanced budget. He publicized his cautions about the matter, concerned that such a measure might handcuff the government at inopportune times and ought to stay out of the Constitution.

On July 15, on the third anniversary of President Carter's acceptance of the Democratic nomination for president, he gave a speech that will forever be known as his "malaise" speech. While the word described the tone of the speech, it actually was never used. Announced as a speech about the nation's energy crisis, with gas lines across the country and an increased need to import foreign oil, it was instead a speech about what Carter termed a "crisis of confidence." He spoke like the preacher he was and was widely criticized for blaming the American people, not the government or his policies, for the problems the country faced. The speech landed with a thud and did little to buttress Carter's waning fortunes.

Birch was driving across Indiana and listened to the speech on the car radio. He commented, "Pat Caddell finally got someone to listen to him." Caddell had been the wunderkind of politics in 1972, a young pollster who became a key member of McGovern's campaign and rebounded after that defeat to become one of the Democratic Party's leading pollsters. He was hired by the Bayh campaign in 1974 and eventually fired. The message Carter gave the nation in July 1976 was the same message Caddell had been giving the Bayh campaign three years earlier: people had lost faith in their government; there was a crisis of confidence in America; people felt helpless. Carter began his speech talking about the number of Americans he had been listening to at Camp David and around the country over the previous ten days, but it struck us that while he had listened to many, he only heard Caddell.

Before the speech, Carter's approval ratings had fallen into the mid-twenties, competing with the lowest levels reached by Nixon during Watergate. Following the speech, he fired several cabinet secretaries, and his ratings fell even more.

In late July, responding to Birch's recommendation, the president nominated Gene Brooks to be federal district court judge. As summer turned to fall, Birch turned his attention once again to the Indiana High Leadership Conference. The speakers were Frank Reynolds from Munster, Indiana, anchorman for *ABC World News*; Wade McCree Jr., Solicitor General of the United States; Dr. Harold Denton of the Nuclear Regulatory Commission; and Joan Claybrook of the National Highway Traffic Safety Administration.

In October, Birch introduced legislation to help the Chrysler Corporation avoid bankruptcy. It was clear that Chrysler going under not only would cost thousands of employees their jobs but also would have a deleterious ripple effect throughout the economy. On November 2, the Senate Judiciary Committee approved Birch's legislation aimed at protecting the constitutional rights of the institutionalized.

Staffing Birch on trips to Indiana during extended recesses was an exhausting yet coveted duty. During these recesses, it was not uncommon to have three male staffers spend a different week apiece on the road, always returning completely drained, while Birch seemed to gain energy with each passing day. Each day was packed with events and held promise for the staffer that there would be one-on-one time with the senator. This was an era before cell phones, so the quiet car ride was an ideal place for a staffer to get to know the real Birch Bayh. These could barely be called conversations, because the staffer's goal was always the same: get the boss talking and just absorb every word.

During one of those road trips, Bill Moreau recalled asking him, "Are you sure you are from and represent Indiana?" How could such a progressive legislator be elected three times by a largely conservative state?

Birch said, "I didn't really expect to win when we ran the first time in 1962, and I took away a very important lesson. If I get one more vote than the other guy, I get to continue doing the job I love. Sure, it'd be great to have an easy reelection, but I guess I'm not willing to give up working on the issues that are important to me in order to be popular with everyone."[14]

That fall was also a time to raise money for the coming campaign. Knowing that Birch would accumulate significant labor money, always an issue in Indiana, the campaign planned to raise a substantial amount from business-oriented political action committees (PACs). After securing an agreement from Ethel Kennedy to use her home, Hickory Hill, for a fundraiser that fall, the Bayh office asked each Democratic member of the Senate Appropriations Committee to permit the use of his name on a fundraising invitation. As a result, every PAC director in DC was invited to an event sponsored by every Democratic Appropriations Committee member, at the home of the late Robert Kennedy. As expected, it was a huge success, and Birch had the ammunition he desired to combat the inevitable stories about the amount of labor money contributed to his campaign.

November was a difficult month for the United States and especially for Jimmy Carter. Iranian turmoil continued with Ayatollah Khomeini urging his people to demonstrate and expand their attacks on the interests of the United States and Israel. As a result, thousands of Iranians, mostly students, invaded the U.S. embassy in Tehran. Ninety hostages were taken, of which fifty-two were American. The hostage-takers demanded that the shah be returned to Iran to stand trial. Carter responded by halting the imports of Iranian oil and shortly afterward freezing all Iranian assets within US control. Americans old enough to remember the hostage crisis recall that *ABC News*, with Ted Koppel reporting, featured a nightly update on the crisis called "America Held Hostage." The program would continue after the hostage crisis was resolved, later known as *Nightline*.

While the president was consumed by the hostage crisis and working to protect the lives of the hostages, he found opposition of another kind arising within his own party. There were many hints that it was about to happen, and on November 7, Ted Kennedy announced that he would seek the Democratic presidential nomination, a direct challenge to Carter. This put Birch in a difficult position. His relationship with Kennedy was well known, and Kennedy had already agreed to attend Bayh's birthday celebration in Indiana in January. But many close Bayh associates and former staffers were working in the Carter White House, and several Bayh initiatives would be severely jeopardized without presidential support. Remaining neutral seemed to be the prudent course and consistent with Birch's previous habits. Once again, choosing sides, regardless of his relationships with both men, would cause a rift among his own supporters and threaten his survival in the coming election. While Kennedy and he were closer personally and philosophically, Birch had forged a good working relationship with the president and knew full well how difficult and divisive a challenge to a sitting incumbent would be.

He had read that Democratic senators up for reelection in 1980 preferred Kennedy on the ballot rather than Carter. As a result, Birch left the Senate one day with Kennedy and, walking down the Capitol steps together, told him that he hoped Kennedy would do what was right for him and not to "worry about the rest of us."

While the country wrestled with the hostage crisis, Birch remained busy in Washington. With Kennedy on the road campaigning much of the time, Birch became the acting Judiciary Committee chair. On November 20, the committee approved the Bayh–Dole bill. Later, hearings began in the Constitution subcommittee on procedures that should be in place if a constitutional convention were to be called, an issue that could have momentous results affecting all Americans. Other major pieces of legislation coming out of the committee during this period were Civil Rights of the Institutionalized, the Soft Drink Interbrand Competition Act, the Privacy Protection Act of 1980 (requiring federal agents to subpoena reporters' notes rather than using search warrants to search newsrooms), the Family Farm Anti-Trust Act (preventing family farms from being broken up by the anti-trust laws), the Fair Housing Act of 1979, and several Juvenile Justice amendments. The Carter UDAG program, Urban Development Action Grants, declared Indianapolis to be eligible in December. Birch had worked with his close friend Herb Simon to help raise private money for urban development projects that the federal government would match. Birch and Republican mayor Bill Hudnut jointly announced the government's endorsement of eligibility.

Before the year was out, Chrysler received the government loan guarantees requested by popular Chrysler CEO Lee Iacocca and supported by Birch and many of his Democratic colleagues. The day before the Senate vote on the bailout, Birch was at a Chrysler plant in Indiana, shaking hands with the workers. From many workers, he faced hostility on the gun issue rather than gratitude for helping save their jobs. Once again, the intensity of the gun issue intruded on a matter that would have a greater impact on those workers' lives.

On Christmas Day, the Soviet Union invaded Afghanistan. Soviet puppet Babrak Karmal replaced the overthrown and executed President Hafizullah Amin.

The year 1979 will also be remembered for the launch of cable television's first full-time sports channel, the Entertainment Sports Programming Network, known as ESPN. In October, Pope John Paul II visited the United States, and there was a major gay rights march in DC involving tens of thousands of people.

As Americans woke up to a new year and a new decade in 1980, they faced a variety of problems. Inflation and unemployment were at dangerously high levels, and gas shortages were causing automobile lines at the pumps. Iran held Americans hostage, and the Soviets had just invaded Afghanistan. The president seemed besieged by these difficulties as well as a challenge from Kennedy within his own party. Carter's poor relations with the Senate were being tested more than ever, because of the thirty-four Senate seats up in 1980, twenty-four were occupied by Democrats. It was hard to imagine that having him on the ticket would be helpful in most of those races.

The year started off with a bang. On January 4, President Carter authorized an embargo of grain shipments to the USSR because of its Afghanistan invasion. Several of his allies, particularly those from farm states, criticized this action as punishment of American farmers for Soviet misdeeds. Birch felt that such an action could not be effective unless other countries joined the embargo, and he was concerned about the heavy price that Hoosier farmers might be paying for this policy. The grain embargo was one of several issues he wrestled with in 1980 while also planning a campaign. He had been greatly invested in the Chrysler bailout issue, which concluded during that first week in the year as Carter signed the legislation giving the automobile manufacturer $1.5 billion in loan guarantees.[15]

Birch was seated next to Katherine "Kitty" Halpin at Washington's annual Press Club Dinner in January 1980. Normally he would not have attended the dinner by himself; he had always accompanied Marvella at the dinner previously, because she loved those events. But he had just hired Fred Nation of Terre Haute to be his press secretary and felt the dinner might be a good opportunity for him to "get some of the rough edges taken off." He did not know that Kitty, whose job involved her in the planning of the dinner, had arranged for their seats to be side by side. Birch had briefly met Kitty a few years earlier at a meeting in his Senate office when she was a staffer of the DNC. The meeting was to discuss the DNC's upcoming telethon. The dinner conversation would be the first real conversation they ever had.

They proceeded to talk throughout the evening, and eventually Birch told her that he wouldn't want to do or say anything that would hurt their new friendship. She replied that she couldn't imagine that he would. To that he said, "Then may I make a proposition?" She was clearly on her guard when he followed with, "Can we go to Howard Johnson's . . . and get some ice cream?"

She agreed and left with Fred and Birch, who was driving. Fred remembered being asked if he wanted to go out with them after the dinner and not knowing what to expect. He could only imagine how a United States senator might entertain himself after one of the big political dinners in Washington, DC. Going to Howard Johnson's for ice cream was hardly what he anticipated. They all went to the Howard Johnson's across the street from the Watergate, made famous as the location where the burglars were being monitored by Nixon campaign operatives. After the ice cream, Birch drove Fred to his home on Capitol Hill and proceeded to drive to the Capitol steps, where he and Kitty sat and talked until past midnight. When they left, he drove her home, walked her to her doorstep, and didn't try to kiss her, concerned that he might blow it. When he got home, he called 411 for her phone number and dialed her up. When he spoke, she responded, "How did you get my number?"

"There is such a thing as Information. I just wanted to let you know what a good time I had tonight," he said and hung up.

Kitty was working at ABC-TV at the time and was very busy with the approaching primary season. They worked it out that Wednesdays would be the safest evenings to get together. Since primaries were on Tuesdays, Wednesdays were a safe bet, and he could be sure not to be home campaigning in that middle part of the week, at least for a while. When Birch was able to spend time in DC during 1980, he spent as much of it as possible with Kitty. Their romance blossomed.

She could hardly suspect that they would spend the rest of their lives together.

18

The Last Campaign: 1980

The hardest thing about any political campaign is how to win without proving that you are unworthy of winning.

ADLAI STEVENSON[1]

JANUARY BROUGHT ANOTHER BAYH BIRTHDAY GALA. JIMMY Durante, who had appeared at the 1968 gala, passed away the same week as the 1980 gala. This one was headlined by Sen. Edward M. Kennedy, now a declared candidate for president. Bob Shrum, a media consultant, wrote about Kennedy's participation at the birthday dinner, recalling that they "flew through an ice storm to get there." According to Shrum,

> Kennedy had to sit patiently and apparently unperturbed at the head table as Bayh delivered a speech that all but endorsed Carter. It was so graceless that some of the more boisterous members of the press corps, who couldn't believe we'd knocked around the sky for this, retreated to a corner and popped balloons while Bayh was droning on. Back on the plane, in the front compartment with Kennedy, I said, "What a son of a bitch." No one on the staff disagreed. But Kennedy couldn't get mad at Bayh. Sixteen years before, when the small plane carrying him, Bayh, and Bayh's wife to the Massachusetts State Convention had crashed, Bayh had climbed back into the burning fuselage, pulled out a paralyzed Kennedy, and saved his life. Kennedy was constantly reminded of the crash by the continuing back pain that plagued him every campaign day . . . Kennedy was disappointed in Bayh but he didn't want to hear anyone bitching about him. Bayh, he said, had a pass and always would.[2]

Shrum is allowed his recollections, to be sure, but they don't square with the memories of others who were there. Nor would pro-Carter remarks have been consistent with Birch's personal feelings or the actions he had already taken to help Kennedy. If Shrum perceived the remarks as pro-Carter, it may have been because Birch was concerned that Kennedy's presence at the event constituted a Bayh endorsement. He wanted to appear neutral and clearly did not endorse his Massachusetts colleague. Neutrality still made sense; many former Bayh staff worked for the Carter White House, and so much of Birch's Senate work could be adversely affected by endorsing Kennedy. Paul Kirk, for many years one of Kennedy's top lieutenants, met with Birch and me in the Senate office. He understood the policy of neutrality but asked if we could be a resource for information about Indiana. Who could they count on for support? Who was trustworthy? What made sense, and what did not? Birch agreed, and I became the sole contact for Kirk whenever he needed Birch's point of view or to discuss anything about Indiana or its politics. It was a weird limbo to be in. The Kennedy people, unaware of the back-channel communication, thought we should be helping them, and the Carter people assumed that we were.

Fred Nation remembered writing Birch's speech for the event with Kennedy. It was written longhand on the plane to Indiana and handed to a staffer to type up just before the dinner began. Nation was giving pages of the speech to Kennedy on the stage to hand to Birch after Birch had already begun speaking. The challenge for Fred was to write a speech appropriate to the occasion, as this would be the first time Birch would be addressing statewide Democrats since Marvella's death. Intended as a short speech, it was too long and rambling and was largely ignored, but in no way did it endorse Carter.[3]

On the final day of January, the Senate approved a Bayh amendment to eliminate the disability benefit waiting period for the terminally ill. Other issues of 1980 continued to dominate most of Birch's time: the grain embargo, US intelligence, the balanced budget amendment, alcohol fuels, SALT II, the steel industry, coal production, runaway youth, taxing Social Security benefits, auto imports from Japan, juvenile crime, "Buy American" proposals, gun control, and more.

The Soviet Union seemed impervious to the views of the West. The Afghanistan invasion was one way of thumbing its nose at the United States, an attitude reinforced when noted Soviet scientist and human rights activist Andrei Sakharov was arrested in Moscow. The fact that many American political leaders were keeping up a drumbeat to free Soviet dissidents seemed irrelevant. For years, Birch had been engaged in the effort to free dissidents in the Soviet Union. His exposure to the issue was personal and went back to 1969, when he had first spent time in the USSR with students in Moscow. The issue grew in importance again after Woodford McClellan visited Bayh's Senate office. A professor at the University of Virginia, McClellan had not seen his wife, Irina, a Soviet citizen, since 1974, when they were married in the Soviet Union. His cause was kept alive with an aggressive petition campaign coordinated by Rabbi Gedallyah Engel from Lafayette, Indiana. Birch joined Congressman Floyd Fithian in forwarding the petitions to the Soviet embassy and requesting an exit visa for Irina McClellan.[4] This effort, coupled with the invasion of Afghanistan, led Birch to support Carter's decision to delay consideration of SALT II. Americans rejoiced in February during the Winter Olympics when America defeated the Soviet Union in hockey, an event known as the "Miracle on Ice." *Sports Illustrated* later described it as the greatest moment in sports history.[5] In March, responding further to the Soviet invasion, President Carter announced that the United States would boycott the upcoming Summer Olympics in Moscow—another unpopular decision. Later in the year, Birch took gymnast Kurt Thomas, an Olympic hopeful from Indiana State University, to the White House to meet President Carter as partial compensation for missing the games.

As dismal as American–Soviet relations were, a bright spot in American foreign policy was in the Middle East, where Israel and Egypt had established diplomatic relations. But the Middle East was also the site of the cartel controlling the price of crude oil, and gas prices in the United States continued to rise. The price of oil and our dependence on foreign leaders provided incentive for using alternative fuels. Birch promoted gasohol proposals and announced a coal gasification project in Lake County. Then he proposed legislation to create a grain reserve earmarked for alcohol fuel production, supporting gasohol while also creating an alternative demand for embargoed grain. Birch chaired the National Commission on Alcohol Fuels, established by Congress in 1978 under the Surface Transportation Act. It gave him a public forum to continue promoting gasohol, and he supported a proposal to exempt gasohol from excise taxes. The proposed exemption passed before the end of February.

During his first month in Washington, DC, Fred Nation was fully engaged at work when the *Wall Street Journal* printed an article implying that Birch had taken a bribe. Birch's name was mentioned in an FBI wiretap as part of a Teamsters investigation. Not long after, the Justice Department said that he was not a target of the investigation, and the Senate Ethics Committee cleared him.[6] But it was an example of how hard it is to dismiss charges like this. The wiretap discussion seemed to be name-dropping and bragging by a person targeted by the investigation, but that hardly mattered once the charge was made.[7] Birch knew that charges like this were hard to weather politically and a blight on one's reputation, but he could do little about it. As troubling as the story was to Birch and to his staff, it never became an issue in the campaign. Despite his innocence, no political candidate ever wants to be forced to deny illegal activity. The lack of any damning facts in the matter allowed it to go away.

Birch remained busy with his Senate activities, but this competed with the coming campaign. Travel to Indiana increased; with Evan away in law school, Birch could afford to be away more often. He knew his relationship with Indiana voters to be his greatest asset. The state of affairs in the country wasn't helping his prospects, and he hoped the dismal national outlook would not prove fatal to him. There was little he could do about that beyond his role as a senator, but he would do whatever was necessary to shore up his political base and his standing with his constituents.

The rumor mill is always an uncomfortable part of campaigning in Indiana. Birch's staff needed to protect him from rumors. People in politics love gossip, and many thrive on passing along stories regardless of their truth. Marvella was a revered figure in the state, and any rumors connecting Birch with another woman were to be avoided. The upcoming election would be the first time he would campaign as an unmarried man. Therefore, the staff was told that under no circumstances should a woman ever drop him off at night, pick him up from a hotel, or be seen in public alone with him. This proved to be an effective strategy.

The campaign team was set. David Bochnowski, political director in 1974, moved into the role of campaign manager. I became the political director with responsibility for directing the field operation. The Senate press secretary, Fred Nation, became the director of communications; at age thirty-six, he was among the oldest on the staff. From the Indianapolis office, Ann Latscha assumed the role of fundraiser-in-chief. Lynne Mann left the DC Senate office to manage the campaign's finance operation. Tom Connaughton and Eve Lubalin made up the issues team.

An interesting addition to the campaign leadership team was the former Indiana Democratic chair Manfred Core. Many staffers were unhappy with his hiring, feeling that Birch had so little faith in them that he felt compelled to bring in an old hand. But attitudes quickly changed as Manfred showed himself to be a fount of wisdom and cool-headed advice, always ready to reach back into the past to make a cogent point about a matter before us. Rev. Gary Kornell also joined the team. He was a young Presbyterian minister who would help the campaign deal effectively with the abortion issue and the right-wing attacks.

Herb Simon became the principal fundraiser for the Bayh campaign, a job he had held in Larry Conrad's 1976 campaign. When asked about his memories of Birch, Simon described a meeting in Israel with Menachem Begin just after Carter was elected president. He said that Begin hated Carter and loved Birch Bayh, an opinion that had a marked impact on Simon at that time. Of course, Begin may have felt differently following the Camp David Accords, which Carter managed expertly and for which he was given much credit. Born in Brooklyn, Simon immigrated to Indiana in 1960 to work for his brother Mel, who had been drafted into the army and stationed at Indianapolis's Fort Benjamin Harrison. Mel later worked for Albert Frankel, a principal creator of Eastgate, the first shopping center in Indiana and among the first anywhere in the United States. When Mel left Frankel to pursue his own development ideas, he asked his brothers Fred and Herb to move to Indiana to help him at $100 per week each. The Simon business became enormously successful, developing shopping malls across the United

States including the Mall of America in Minneapolis and Pentagon City outside of Washington, DC.

When Simon helped Larry Conrad run for governor, he became friendly with Diane Meyer and, because of her, began raising money for Birch. He described Birch as incredibly kind and open. Simon eventually would buy the Indiana Pacers NBA team, the Reno Aces, a minor league baseball team, and *Kirkus Reviews*, among other holdings. He and Diane Meyer were later married.[8]

The professional team hired by the campaign included Bob Squier for media, Bill Hamilton for polling, and Matt Reese for voter contact. The consultants collaborated on a unique form of targeting that had been tried only once and never for a candidate. It was used by Reese in the successful Missouri right-to-work campaign in 1978. The AFL-CIO hired Reese to run the voter contact effort in the hope of defeating the anti-labor referendum on the ballot that year. Reese understood that he was starting out far behind, but Hamilton was able to show him that the more attention the referendum got, the farther behind they became. Voters didn't understand that the term "right to work" really meant the right not to be represented by a union. But polling demonstrated that when it was understood, the referendum suffered. Therefore, it made sense to run a stealth campaign to educate voters by mail and phone and very little television or radio.

The social scientist Jonathan Robbins had developed a form of targeting known as Claritas, taking decades of census data to create forty clusters that everyone in the country fell into. A cluster was defined by a combination of sociodemographic characteristics, and groupings could be defined geographically. Citing the proverb "Birds of a feather flock together," Robbins described how each zip code in the country was dominated by people that fell into a unique cluster. Americans tend to reside in neighborhoods with people very much like themselves, and once a campaign decided which clusters it needed to target, all adults within the cluster-defined zip codes would be contacted. He cleverly labeled these clusters with names like "Pools and Patios," "Hardscrabble," and "Shotguns and Pickups." For example, he marketed the Claritas cluster system to *Field and Stream* magazine, contending that they could improve their subscription solicitations if they confined their efforts to the "Shotguns and Pickups" cluster. A test proved successful, and he was hired. Hamilton polled by cluster in Missouri, identifying clusters that were most "fair-minded" about right-to-work. Reese sent persuasive messages to the targeted households, varying them by cluster, and it proved effective. A campaign deemed hopeless in the beginning was successful, and Reese became an evangelist for Claritas targeting. We became convinced that it made sense for us to use Claritas in our campaign.

During late 1979, Birch made quiet efforts toward neutralizing the increasingly powerful National Rifle Association (NRA). While there was no way to erase his support of gun control legislation and the perception that he was "anti-gun," if there was a way to keep the NRA from being active in the campaign, it might help. Based on information provided by the NRA, he wrote a letter to the director of the Bureau of Alcohol, Tobacco and Firearms pointing out serious problems involving accuracy in the record-keeping of the firearms registration system, raising questions about the efficacy of the Federal Gun Control Act and its implementation. The NRA distributed the letter on letterhead from its Gun Owners Foundation.[9] Birch held hearings to help define who was allowed to be a gun dealer and who would be required to have a license under the Federal Gun Control Act. National Rifle Association spokesman Wayne LaPierre testified and expressed support for the National Instant Check System for gun dealers. Birch agreed to hold hearings concerning "alleged government abuses of power against gun owners and dealers."[10] The hoped-for result was accomplished: The NRA remained neutral in Birch's race.

We learned a disappointing lesson about interest group politics early in 1980. A representative of the pro-choice organization the National Abortion Rights Action League (NARAL), met with Bayh's staff to discuss a contribution of PAC money to his campaign, insisting that in 1980 Birch was their highest priority for support. Knowing the sensitivity of the abortion issue in Indiana, the Bayh staff suggested that the best way they could help would be to give the designated funds to the Indiana Democratic Party rather than directly to the campaign. The NARAL representative replied that she thought it a bad precedent; they didn't give contributions to state parties. It was reiterated that if NARAL wanted to help Birch, this was the best way to do it. She declined, and the Indiana Democratic Party never received a penny from NARAL.

Throughout the year, abortion continued to be a troublesome issue. One unique approach was to have Birch photographed with the pope in the hope of muting antagonism generated by Catholics in Indiana. Interrupting his campaign schedule, Birch flew to Rome to appear with Pope John Paul II for a "photo opportunity." The picture was taken, and he flew back to Indianapolis on the next plane.

Politics received another black eye on February 2 with revelations about an FBI sting operation later known as ABSCAM. Several members of Congress were targeted; thirty were under investigation. Eventually, six congressmen and one senator, Harrison "Pete" Williams (D-NJ), were convicted of taking bribes from fictitious Arab businessmen. It was a scandal that dominated the US news cycles for months. The guilt by association could not improve the image of incumbents seeking reelection that year.

Birch also did what he could to tout his leadership of the Senate Intelligence Committee. In February, CIA director Stansfield Turner traveled to Indianapolis to address a public meeting with Chairman Bayh at his side. That same month, Sen. Bill Bradley of New Jersey, former star of the New York Knicks NBA team, was the special guest at a Bayh fundraising dinner in South Bend.

At the end of February, Birch spoke in support of his legislation on the civil rights of the institutionalized, and a few days later, the Senate passed it. He also promoted the sale of American cars to Japan, urged the president to adopt measures to protect specialty steel manufacturing jobs in Indiana cities, and urged Ford Motors to build its new engines in the United States rather than in Mexico as planned. "Buy American" was clearly in the air, bolstered by a surge of patriotism fueled by the Iran hostage-taking.

In March, the balanced budget issue continued to demand Birch's time and attention. He held a news conference to announce that he was recommending balancing the budget without a constitutional amendment, emphasizing his confidence in the ability of Congress to balance the budget. Before the month was out, the Muskie–Bayh balanced budget resolution passed the Senate.

International events again dominated the news. On February 4, Iran's Ayatollah Khomeini named Abolhassan Banisadr as the new Iranian president. Khomeini later stated that the fate of the American hostages would be determined by Iran's parliament. On April 7, the United States cut off all diplomatic relations with Iran and, in an effort to apply pressure for the release of the hostages, imposed new economic sanctions. On April 24, Americans awoke to the news of Operation Eagle Claw, later known as Desert One.[11] Desert One was a failed attempt by American soldiers to rescue the hostages. The combination of a sandstorm and mechanical problems caused the helicopters to crash in the desert, killing eight Americans in a mid-air collision. When the Desert One helicopters crashed—an event that might have altered the electoral results in 1980 had it turned out differently—CIA Director Turner called Birch at 1:00 a.m. to tell him the bad news. Birch had no advance notice of the event, something the Intelligence Committee chair might have expected. Secretary of State Cyrus Vance resigned because of his opposition to the rescue attempt. Had the operation succeeded, Carter would have been hailed as a hero. Instead, he appeared snakebitten and inept.

In 2015 Carter was asked what he might have done differently during his presidency, and he responded by saying that he should have had an additional helicopter take part in the rescue attempt. It could have meant his reelection. Not only did the failed rescue operation make Carter's reelection unlikely; it had the same effect on several other campaigns and had a huge impact on the direction of the country.

In April, due to an economic downturn in Cuba, its leader Fidel Castro announced that people who wanted to leave the island could do so, and an exodus to Florida began. Castro also emptied his prisons; thousands of Cubans ended up in Florida in a mass emigration known as the Mariel boatlift. The American government appeared incompetent and helpless. It could not prevent an onslaught of needy immigrants to our shores—another blow to the Carter image.

The president's image was also under assault in the primaries with Kennedy's entrance into the race along with that of California governor Jerry Brown. Nonetheless, Carter defeated Kennedy in the Iowa and Maine caucuses in swift succession, followed by the New Hampshire primary. In Massachusetts, Kennedy defeated Carter for the first time, while the president won a beauty contest vote in Vermont the same day. Carter then ran off a string of victories with Kennedy bouncing back with two wins. After a dozen contests, Brown withdrew, having won none. Carter and Kennedy continued alternating victories in the remaining primaries and caucuses.

Sensing the president's vulnerability, several candidates entered the race for the Republican nomination. Former California governor Ronald Reagan was running for the third time. A candidate who once said that trees cause pollution, he was also generally considered too right-wing to be viable. He had been an actor in a string of B movies; if elected, he would be almost seventy years old on Inauguration Day, the oldest president in history. Many Democrats wanted Reagan to be the nominee, figuring there was no way he would be elected. George H. W. Bush also joined the race. The son of a Connecticut US senator, Bush had a distinguished résumé, having served in a number of high-profile positions: member of Congress, CIA director, chair of the Republican National Committee, ambassador to China, and United Nations ambassador. Other candidates included senators Bob Dole of Kansas and Howard Baker of Tennessee, Illinois congressmen John Anderson and Phil Crane, and former governor John Connally of Texas, formerly a Democrat and best known for being wounded in the Kennedy assassination in Dallas. By the time the primaries were over, only Reagan and Bush had won any of the contests. John Anderson eventually left the primaries and ran as an independent.

The Bayh campaign was warming up. Evan Bayh was named the chair of the Birch Bayh for Senate Committee, a largely honorary designation since he had never run a campaign. Winning the primary was not a concern, because there was no opposition, but the campaign engaged with potential candidates for other slots on the ballot. The greatest drama before the primary was one involving John Hillenbrand, a millionaire businessman from Batesville and odds-on favorite to win the primary for governor. Democrats liked his deep pockets, assuming that the Democratic nominee would not be outspent, as was the norm. But a challenge by Birch's close friend State Senator Wayne Townsend was the choice of the more progressive wing of the party, particularly labor. Birch remained neutral. After a spirited campaign, Hillenbrand barely prevailed.

The relationship with Gary's Mayor Hatcher remained important, and that area of the state always yielded the highest percentage of

Democratic votes. But times had changed, as had the law. Hatcher was known as someone who could deliver votes on Election Day, but the normal processes wouldn't work any longer. Street money, or walking-around money, was usually distributed to precinct workers in cash. Large amounts of cash were traditionally handed to political leaders to be doled out to precinct workers. In 1980, it had to be handled differently. Federal election law prohibited that kind of cash transfer, and Hatcher was told that all of his precinct workers would be put on the payroll for one day and compensation would be based on results. No automatic payments would be made to anyone. Because of federal appointments yet to be made and Birch's prominence in the Senate, Hatcher reluctantly agreed to this arrangement.

After the primary, there was a fundraiser featuring a concert by the rock group Fleetwood Mac on May 16. It was set up consistent with federal finance laws, defining each ticket as a contribution with the appropriate information collected from the ticket buyer. Almost 16,000 donors were listed afterward. Fleetwood Mac had previously played a concert to help pay off debts from Birch's 1976 presidential campaign.

In May, Birch denounced the huge profits of the oil companies at a time when the price of gas was rising sharply amid shortages everywhere. At the same time, he took the first steps toward killing the development of the Marble Hill Nuclear Plant in southeastern Indiana, something many observers found hard to understand, seeing nuclear power as an alternative to fossil fuels. Yet concerns about nuclear waste were paramount, and no safe or reasonable plan to dispose of them appeared on the horizon. Marble Hill began construction in 1977 and almost immediately ran into difficulties. Environmentalists raised questions about the safety of nuclear power plants, and many in the scientific community were sounding alarms about nuclear waste and the inadequacy of procedures for handling it while also keeping it out of the hands of terrorists. President Carter had worked on nuclear submarines while in the military and was an advocate of nuclear power as part of the answer to America's energy crisis. The Public Service Company of Indiana (PSI) was spending millions of dollars building the plant, and many people living in the area wanted the jobs that would result. Both Birch and Lee Hamilton were conflicted, noting the need for new energy sources coupled with the desire to create more jobs versus the concerns of the environmental and scientific communities. The controversy seemed unwinnable.

The March 1979 incident at Pennsylvania's Three Mile Island nuclear power plant changed everything. Support for Marble Hill began to evaporate quickly, but PSI struggled to move forward rather than lose its huge financial investment. As a political issue, what seemed to be a potential and significant liability in 1980 had fizzled substantially by the time the campaign began. It ended up being a non-issue. In 1984, after spending $2.5 billion, PSI abandoned the project.

Birch continued to call for alternatives to fossil fuels, suggesting that renewable energy and grain-based alcohol fuels were the answer. In May he proposed federal money for Fort Wayne to improve its transportation system. Later, he announced federal loans that would create 200 new jobs in the town of Salem. While crisscrossing the state, Birch was stunned to learn that his friend, the civil rights leader Vernon Jordan, had been shot in Fort Wayne in an attempted assassination. He traveled to Fort Wayne to visit Jordan in the hospital; Vernon recovered. His shooting was the first major news story for CNN, the new twenty-four-hour news network that launched on June 1. Jordan was president of the National Urban League and had been a prominent African American leader for many years. He had visited Indianapolis during the 1974 campaign to speak at the Children's Museum, quoting Martin Luther King in his remarks: "We may have all come on different ships, but we're in the same boat now."

On the national political landscape, the Bayh campaign was a blip on the radar screen. Carter and Kennedy continued to trade primary wins until Carter ran off a string of victories that left no doubt about who the nominee would be. Kennedy won several of the largest states, but Carter won thirty-six states plus Puerto Rico, finishing the primaries with 60 percent of the pledged delegates.

The GOP campaign continued as a two-man race in May with Reagan winning six states to Bush's one plus Washington, DC. After that, Bush did not win again, and Reagan won the rest. Bush formally dropped out of the race.

The Bayh opponent would be the Republican congressman from the 4th Congressional District, J. Danforth Quayle. Earlier speculation was that incumbent Republican governor Otis Bowen would be the nominee, but he declined to run. Birch hoped that it would be Bowen. He saw polling that had them neck and neck, which told him that if Bowen wasn't considerably ahead and was in the state every day, he was unlikely to improve his numbers by campaigning, which Birch was confident he could do. Quayle was an unknown, a matter of some concern. Birch's challenge was as an eighteen-year incumbent Democrat on the ticket with an unpopular Democratic president and an economy facing "double-digit inflation, double-digit interest rates and double-digit unemployment,"[12] along with gas lines at the pump and hostages in Iran. Reagan referred to the poor economic indicators as the "misery index." The term was coined by University of Chicago economist Robert Barro "to measure the combined effect of inflation and unemployment." It simply added the two percentages together and they rose to an "unbearable 19" in 1979 and to 20 in the election year.[13] Was it possible to escape responsibility for that state of affairs as a Democratic senator? At one point, the race was described as a tidal wave with Birch on a surfboard. How good was he? The Bayh argument was that yes, he had been in the Senate for the last eighteen years; no, he was not responsible for our troubles; and things would get worse if he lost. It was a pretty hard argument to make.

The Bayh campaign polled early in 1980 in a way to adequately sample all of the Claritas clusters. The poll results became the lynchpin for all targeting throughout the campaign. One polling item was described as the "Jesus Christ question." When asking the head-to-head question of Bayh versus Quayle, Birch ran ahead 72–10, largely because no one knew who Quayle was. But when asking voters whom they would vote for with Birch Bayh as the Democrat and describing the Republican as a young, attractive conservative who agreed with you on the major issues of 1980, Birch trailed by a margin of 46–40. In other words, running against the perfect candidate, aka Jesus Christ, Birch would likely lose. Forty percent of the voters were with him in the direst of circumstances. The campaign's task was to keep Quayle from becoming the perfect candidate and to target those undecided and persuadable voters who could make the difference for Birch.

It soon became clear that the two-term, thirty-three-year-old Quayle was far from perfect. His assets included his youth and good looks plus the fact that he was the grandson of Eugene Pulliam, owner and publisher of the *Indianapolis Star* and other papers. The conservative *Star* had always treated Birch like he was the devil incarnate and would now have a greater incentive to defeat him. Once when Quayle and his grandfather were golfing a few days after Pulliam had written an editorial that seemed to be favorable toward Birch, Quayle recalled that he and his grandfather "got into a big altercation on the first tee." Clearly, Quayle shared his grandfather's negative views about the senator.[14] But shortly after it became clear that Quayle would be the candidate, Birch was called by a friend in the Indiana National Guard. His friend told him that years earlier, Quayle had slipped into the Guard when it was virtually impossible to do so because of the number of draft-eligible young people who were seeking an alternative to fighting in Vietnam. A number of people, including officials in the governor's office, had taken steps to ensure that Quayle would be accepted into the Guard rather than drafted. Major General Alfred Ahner said that he was directed to hold a place open for Quayle by Wendell Phillippi, a retired National Guard commander and then a high-ranking official at one of the Pulliam newspapers. Quayle went into the Guard a mere six days before he became draft-eligible. Later, it was reported that he did poorly on the required tests, scoring a 56 on the written test, considerably below the average score of 75.[15] The Republican mantra against liberal Democrats included the charge that they were soft on defense, so the situation was a perfect example of hypocrisy, which wouldn't go down well with Hoosier voters.

Tom Buis, the Bayh staffer responsible for agricultural affairs, was also the former Democratic county chairman in Putnam County. Quayle was a graduate of DePauw University in Greencastle, the Putnam county seat. Buis was contacted by Bob Sedlack, a precinct committeeman and a professor at DePauw. Sedlack had learned from an official at DePauw that Quayle committed plagiarism while in college. Quayle had failed a course due to plagiarism, and the professor who failed him, Bill Morrow, was then teaching at William and Mary College in Williamsburg, Virginia. The official at DePauw who had seen the transcripts, which were considered private, would lose her job if the plagiarism were revealed. It was far better to hear it from the professor involved. Bochnowski traveled to Williamsburg, where he met with Professor Morrow, who confirmed the story but would not agree to go on the record.

Despite what were felt to be two explosive stories, Birch would not use either one, insisting that it wasn't the way he campaigned and that Hoosier voters expected better than that from him. Nonetheless, the staff was determined to find a way to bring these stories to light, even if only surreptitiously. Later the campaign learned of a third story that was making its way around DC. The rumor was that Quayle was one of a few congressmen who had gone away for a weekend of sex with a beautiful blonde female lobbyist. If the story were true and became public, Quayle would be done. But it was only a rumor, and the hope was that the rumor would be published with his name included. The *Washington Evening Star* published a piece about three congressmen and a "love nest" retreat they shared with a female lobbyist, but no names were given. And like the other stories, Birch would not allow it to be used.

The campaign continued to emphasize Birch's accomplishments for Indiana, downplaying national issues with which he was identified. The "Birch Bayh—Fighting for Indiana" and "Birch Bayh—Fighting for You" slogans were displayed everywhere. Also appearing were prominent plays on his name, bumper stickers saying "Bayh American," "Bayh Coal," "Bayh Gasohol," or "Bayh Steel," identifying him with these issues to the greatest extent possible. Photos showed him shooting rifles, produced on literature aimed at people against gun control.

In June, while traveling throughout the state in full campaign mode, Birch returned to Washington to cast a vote to override Carter's veto of an anti–oil import bill, another blow to the president and the first successful override vote of a Democratic president since 1952. Congress had passed legislation to negate a Carter executive order assessing an import fee on foreign oil. Arguing that the fee would do little to conserve energy and added to the inflationary pressures already existing in the economy, Carter nonetheless vetoed the bill. The override vote in both Houses far exceeded the two-thirds vote required, a clear rebuke to Carter that almost appeared like piling on. But it is easy to understand why Congress acted as it did.[16]

During the primaries, Carter insisted that the hostage crisis was his first priority, leaving him little time to campaign or to debate Kennedy. Being presidential was the right strategy, but by late spring he gave in and went on the road.

In July, an unwelcome development interrupted Birch's campaigning: A new task fell to him as chair of the Intelligence Committee. Author Ira Shapiro writes,

> Billy Carter, the president's brother, had accepted $220,000 from Libyan friends and agreed to register as a foreign agent. This wasn't the first time Billy Carter's connection to the Libyan government had come to light. The year before, he had taken a trip to Libya and returned to help set up a Libya–Arab–Georgia Friendship Society. When criticized for his pro-Libyan activities, the president's brother had responded: "There's a helluva lot more Arabians than there is Jews," and referring to his Jewish critics: "They can kiss my ass."
>
> Birch Bayh sat in his office in August 1980, ruefully contemplating his prospects. Bayh had intensely enjoyed chairing the new Senate Intelligence Committee, but no one could have predicted this assignment. With only a few months left before the election . . . it looked as though he would have to take on Jimmy Carter in a most awkward and embarrassing way: by investigating the activities of the president's brother . . . Bayh understood that great senators did not always get to pick and choose their defining issues.
>
> Bayh had no illusions about the job: "It's going to be like walking through a minefield." He said that his aides had urged him to turn it down and concentrate on his campaign. He deserved praise for stepping up to a tough assignment like the distinguished senator he was. Nevertheless, events would prove that his staff was probably right.[17]

Early on, Birch suggested that what became known as Billygate "may not amount to a hill of peanuts." For Birch, the Billygate hearings required an inordinate amount of time away from his campaign. The hearings ended with a whimper, not a bang. No connection between Billy Carter's activities and the president was found. The hearings had been underwhelming because there was nothing there. Birch told reporters, "I'll wager that 90 percent of everything we will hear, you have already written about." But, he said, "until you stir the pot, you can't say whether you have a mouse or a dinosaur."

The Republicans nominated Ronald Reagan for president with George Bush as his running mate. To the dismay of moderate Republicans, the influence of the religious right was felt when their convention dropped its long-standing support for the Equal Rights Amendment. Naturally, the Republican convention focused on the reasons why Americans should not support President Carter for reelection. With the opening of the Summer Olympics in Moscow, Americans were reminded why they didn't like the president who had taken us out of the games.

On August 14 at the Democratic convention, President Carter was formally nominated. The convention was held at Madison Square Garden in New York City, just as it was four years earlier. Birch did not attend, feeling that identifying himself with the national ticket was unhelpful. The campaign was on, and if you were a Democratic candidate there was much to be concerned about.

Ira Shapiro described the 1980 campaign in the following manner:

> Perhaps busy with Billygate, perhaps caught up in the insulated environment of the Senate, Bayh had failed to get ahead of the country's changing politics. Republican staffers on the Appropriations Committee had begun keeping track of his absences from committee meetings, using them as campaign fodder. Moreover, the single-issue groups, which had not existed six years earlier, had become forcefully and stridently involved in the Indiana Senate race. "It's been vitriolic," Bayh observed in October. "The outsiders have come here in force."[18]

Birch stressed his seniority and his accomplishments for agriculture, the steel and coal industries, national security, and inflation, and he was shown target shooting and talking with National Guard troops. Quayle's consultants came up with a clever line: "Bayh suffers from the two Georges syndrome. He sounds like McGovern in Washington and like Wallace in Indiana."[19]

Late in the campaign, Bochnowski and I met with Birch to discuss ways he might fine-tune his message to voters. He listened patiently to our thoughts and told us firmly that he had come to the Senate with certain ideals and wasn't about to change them simply because the wind was blowing in a different direction. He knew who he was and was comfortable in his own skin. Although we remained concerned, we both felt proud of him at that moment.

Right-wing single-issue groups had a potent effect on the 1980 campaign, just as they did in 1978. Organizations like the Moral Majority, the National Conservative Political Action Committee (NCPAC), the Conservative Caucus, and a number of pro-life groups spent large amounts of money targeting liberal senators on a few hot-button issues, mainly abortion, guns, and the Panama Canal. These negative efforts by outside groups were relatively new to American politics, and those affected were unclear about ways to counter the attacks. Could they be kept off the air? How should a targeted senator respond?

The National Conservative Political Action Committee was the most notorious of the right-wing groups. It raised $1.2 million in 1980 to spend against six Democratic senators and was one of the first organizations to take advantage of the independent expenditure rules. Terry Dolan, head of NCPAC, said, "A group like ours can lie through its teeth and the candidate it helps stays clean." He labeled the Civil Rights Act "irrelevant" and called the Voting Rights Act "absolutely silly."[20] In addition to the scurrilous things he said about the Democrats he was targeting, he also lambasted Ronald Reagan's daughter Maureen for her liberalism. "Maureen Reagan is the type of person who, in the middle of a war, would go out and shoot our wounded."[21] Jerry Falwell's Moral Majority joined NCPAC in their "Declaration of War against Homosexuality," writing in a piece of direct mail, "Our nation's moral fiber is being weakened by the growing homosexual movement and the fanatical E.R.A. pushers (many of whom publicly brag they are lesbians)."[22] Ironically, Dolan was a closet homosexual who would die of AIDS in 1986.[23]

The American Enterprise Institute authored a book on the 1980 election that aptly characterized the issue facing the Bayh campaign. "The NCPAC attacks, combined with those of antiabortion groups and of Bayh's opponent, Republican Congressman Dan Quayle, shrank Bayh's lead in polls from a 58 to 34 percent margin in fall 1979 to a mere ten percentage points by June 1980. Bayh, too, began to fight back, attacking the 'outside' interests and their ties to Quayle."[24]

Birch campaigned at full throttle in 1980. He bristled at the suggestion that he was getting older and shouldn't begin as many days with plant gate visits at 5:00 a.m. as he had in 1974. Having done those plant gates with him when I was twenty-five and he was forty-six, I was convinced that he would be a better candidate if he scaled back on those kinds of energy-draining events. Birch had to face the fact that he was fifty-two; though physically fit, he couldn't pretend that he was as young as he once was.

The *New York Times* ran an article in September about the Billy Carter investigation and how it was hurting the Bayh campaign. Birch was described as frustrated by the responsibility. "Believe me, folks, I'd rather be out here campaigning . . . But somehow or other . . . somebody hung a thing called Billy Carter around my neck. Now we've got to finish this crazy investigation and we're going to do it right—impartially, judiciously." Quayle's campaign manager Mark Miles commented, "Bayh's political strength in Indiana is his personality rather than his record . . . To the extent that the hearings have kept him out of Indiana, it helped us."[25] Columnist Mary McGrory wrote about the same subject, characterizing Birch's feelings. "'Someone had to do it,' he sighs with a martyr's air that is totally believable."[26]

The two candidates locked horns in a televised debate at the Indianapolis Children's Museum on September 14. Quayle focused in the debate on Birch's eighteen years in office as a "very long time," stressing that "leaders of the 1960s and 1970s won't get us through the 1980s."[27] He pressed the issue of Birch being too liberal for the state, while Birch criticized Quayle for his support of the Kemp–Roth tax cut plan and for being too cozy with the oil companies. Finding out that Quayle was attending a fundraiser sponsored by oil company executives at their headquarters in Dallas, the campaign dispatched Squier to Texas to produce an ad showing the young congressman at the event. During debate prep, Squier had been asked how to know who won the debate. His comment was that the losing-side candidate and staff usually made a quick exit from the venue. Quayle and his staff departed first.

The nastiness of the independent campaign against Birch mirrored that in other states where liberal Democratic incumbents were fighting for their political lives. The Moral Majority targeted Birch on abortion, its Indianapolis leader, Rev. Greg Dixon, calling him a "baby killer" on TV. NCPAC had organized large quantities of mail going into the state and half-page ads in newspapers around the state were arranged by the Ship Out Bayh Committee. They said, "We've looked him over and here is what we've found," following with criticism of his positions on the ERA, abortion, guns, defense, and an accusation that he was "feathering the nest," based on his support of legislation that increased congressional salaries. The mailers said Bayh supported measures that "build up our Marxist enemies while at the same time voting to undermine our anti-Communist allies."

The Quayle for Senate Committee reprinted literature from an organization called FaithAmerica summarizing the "major issues of concern to the Christian community." Eleven "Key Moral and Religious Liberty Issues" were listed, with Birch voting against "voluntary prayer in the public schools, right to life, the Family Protection Act, support of Christian education and new tax deductions for church and charitable giving." He was shown as voting in favor of "support for the religion of secular humanism, forced school busing and a constitutional amendment to eliminate all legal differences between men and women," a characterization of the ERA. Listed as "unknown" were his views on "parental approval of sex education for elementary students," the "promotion of homosexuality," and "federal control of all church youth camps and conference grounds." The flyers were placed under the windshield wipers of cars in church parking lots the Sunday before the election throughout the state.

The Quayle campaign produced a flyer on the abortion issue. "Bayh continues to insist he is 'personally opposed to abortion on moral and religious grounds.' Millions of unborn children would disagree, but then their voices will never be heard," it read, going on to say, "Not only has Mr. Bayh voted pro-abortion, he has also become a *leading* spokesman for those who believe in *abortion on demand*."

Years later, when Quayle had become vice president, he responded to questions about the abortion issue by saying, "I think people make a mistake trying to exploit it in the political realm."[28]

The *Indianapolis Star* had kept up an anti-Bayh drumbeat all year, even removing its own political reporter Joe Gelarden from the beat because he was rumored to be close to the Bayh staff. On October 14, the *Star* ran an article with the huge headline "Poll Gives Quayle Commanding Lead: Victory Over Bayh 'Almost Certain.'" While this differed substantially from the Bayh campaign polling, it was troubling nonetheless. Indianapolis professor Brian Vargus managed the poll and had a reputation for accuracy. He cited a volatile electorate that perceived Birch as a "big spender with an undesirable voting record." Some evidence of the impact of NCPAC's direct mail campaign was also noted, wounding the campaign with its theme of "If Bayh Wins, You Lose." The poll had Quayle at 50.1 and Birch at only 34.8, with the rest undecided or refused.[29]

In late October, the Bayh campaign sponsored a fun and fascinating fundraising event. The special guests for the Indianapolis event were actors Angie Dickenson, Marlo Thomas, Hal Holbrook, Ed Asner, Martin Sheen, and Robert Walden, author Kurt Vonnegut, and sports stars Oscar Robinson, Rosie Grier, and Tim Richmond. In an earlier fundraiser taking advantage of the celebrities supporting Birch, actor Kirk Douglas was the featured guest. Another celebrity, Robert Redford, sent a telegram of support to Birch objecting to the description of Dan Quayle as looking like Redford.

Nationally, all attention was on the Carter/Reagan debate scheduled one week before the election. Years later, David Rubenstein, a staffer in the Carter White House and a former Bayh staffer, said that Carter didn't take Reagan seriously. Carter took part in a single debate preparation session at which a university professor who had studied Reagan for years portrayed him on the stage. A president can usually talk as long as he wants about any subject, but a debate requires short, concise statements and responses. Carter did poorly in the debate prep and stormed off the stage, angry at the experience.[30] Reagan's task was fairly straightforward: to show the American people that he was not as flaky as some of his comments implied. Showing substance and an ability to go toe to toe with an incumbent president was more important than whatever policies might come out of his mouth. But Reagan did far better than that, turning aside some of Carter's comments by saying, "There you go again," and famously asking a question to the audience in his closing remarks: "Are you better off now than you were four years ago?" Carter needed to put the former actor away if he were to regain the momentum. Instead, Carter's debate performance will be remembered for the moment when he told the audience that he had asked his twelve-year-old daughter to name the most important policy issue in the campaign. The fact that she replied "nuclear proliferation" was not nearly as important as the image of a president consulting a child.

The *Indianapolis Star* endorsed Quayle on October 26, following with two editorials against Birch in the ensuing three days—not at all surprising to anyone associated with the Bayh campaign. One of the editorials was called "Bye-Bye Bayh," but it and the others were filled with factual inaccuracies about his record, outraging Birch. An apology was demanded, and the newspaper admitted its mistakes in an article on November 1, printed on page 27 near the sports and comics section. On November 2, despite the earlier poll it had printed, the *Star* called the race "too close to call."[31]

Election Day arrived, and things were looking bleak. A race that was neck and neck for so long began to slip away over the final ten days. It appeared that Carter would be demolished by Reagan. The Bayh campaign remained nervously optimistic. By early afternoon, AFL-CIO political director John Perkins, also a Hoosier, reported that he was shown exit polls predicting a Bayh defeat. In the ensuing hours, nothing happened to change that prediction.

The 1980 road show was managed by Tim Minor, who experienced the same type of grueling experience that I had six years earlier. He had a bird's-eye view of the end of the election while on the road with Birch and Evan:

> Most of Indiana was on Eastern Time but small portions in the northwest and southwest were in Central Time, creating scheduling challenges and opportunities. On Election Day, it was decided that the Senator and Evan would do one final plant gate stop at a factory in Evansville before the polls closed. After shaking hands and talking with the factory employees as they got off their shift that evening, we flew to Indianapolis.
>
> On our way, the polls closed and the networks predicted that Quayle would win. Campaign staff decided to give the Senator and Evan a heads-up while in the air. They were sitting in the rear of the aircraft. Someone radioed the pilot, who turned to me and whispered: "All three networks have predicted a Quayle victory." We were nearing the airport and I turned to the Bayhs and said, "Senator, we just heard from the staff and all three TV networks are predicting that Quayle will win." Evan looked at his Dad; the Senator just turned and looked out the window. We landed at the Indianapolis airport in a very quiet plane.
>
> After landing, as we taxied to the end of the runway, I saw a large number of bright TV lights at the airport. I felt sad and miserable but was most disappointed that the Senator would not be able to continue the successful work he so loved and thrived on. I felt horrible for him and did not want to see him

> rushed into a media frenzy so soon after receiving the news of his defeat. I said, "Senator, there appears to be a ton of media on the tarmac; would you prefer that we go to another area of the airport to deplane?" He looked at me, smiled and in that positive, upbeat tone that he was so well known for said, "No, let's go talk to them and see what's on their minds."[32]

Fred Nation was given a heads-up about the election outcome from one of the networks and made the call to the airplane. When Birch arrived at the headquarters, a number of us met him outside to make sure he knew what we knew before entering the building filled with press, staff, and a large gathering of voters.

On November 4, 1980, Americans went to the polls, and Reagan defeated Carter badly. Attracting a large number of Democrats to his side, Reagan won in a landslide, 50.75 to 41.1 percent, a margin of over eight and a half million votes. Carter won only six states and the District of Columbia, making the electoral vote margin even more lopsided at 489–49. In Indiana, Reagan walloped Carter by over 411,000 votes out of 2.2 million votes cast, a 56.1 to 37.7 percent victory.

In the Senate race, Quayle beat Bayh 54–46, winning by 166,492 votes. Hillenbrand was defeated by Bob Orr by almost 350,000 votes. Election night was also the one-year anniversary of the hostage-taking in Iran. Birch took comfort that he had run far ahead of the rest of the ticket; his margin in defeat was much smaller than Hillenbrand's and Carter's. Throughout the year, those of us in the campaign speculated that Birch could afford a quarter-million drag from the top of the ticket. His margin of defeat was about 245,000 votes less than Carter's, thus proving that prediction to be on the mark. The results in Indiana left so few Democrats in the General Assembly that they could not prevent a quorum from meeting. In the congressional delegation, John Brademas was also defeated, ending his dream of becoming Speaker. After a friendship of over forty years, Brademas would say, "I cannot think of a single matter, political or policy, on which Birch and I ever disagreed."[33]

Tom Connaughton looked back on 1980 this way: "Birch was trying to present an image that he needed to be returned to the Senate. For years it was the Birch Bayh personality, much more than issues, which allowed him to prevail. The personality was still there, but when the country has 18 percent unemployment, that's a tough time to be seeking reelection.

While Quayle may have been a lightweight, he didn't make any major mistakes. Birch may have won in the debate, but Quayle didn't make a fool of himself."[34]

Late in 1980, while he was shaking hands at the plant gates, Birch thought the workers seemed reticent. "There were an awful lot of those guys who weren't looking him in the eye," recalled Connaughton, and Birch said, "If I've lost those votes, I'm in big trouble." Evan had been traveling the state independently and had showed that he had inherited his father's political instincts with his own capacity for reading the political tea leaves. Davis–Bacon was a popular piece of labor law, but after being briefed on labor issues and having recently experienced labor picnics and plant gate visits, Evan's comment was, "Working men and women weren't worried about Davis–Bacon; they were worried about eggs and bacon."

Former Bayh staffer Jerry Udell was in DC on Election Day and called the Bayh campaign and spoke to Connaughton, saying that he had talked to people at the White House and was told they had been polling over the weekend and "the Senate is going to change hands. Everybody's gone. You're gone. Church is gone. Culver is gone . . . they said the president's going to lose by 20 points and this is going to be a catastrophe."[35]

Knowing what we know now, it seems that a Bayh victory was impossible. One factor that may have had an impact was Indiana's election technology. Most counties still used machines in which pulling a single lever could vote the entire party ticket. The Bayh campaign had to educate people on ticket splitting in those areas where he could run significantly ahead but might suffer because of the voting machine design. Bayh voters intending to vote a mostly Republican ticket would need to complete a complicated task: pulling the party lever down, flipping Quayle's lever up, and pulling Birch's lever down.

Birch looked back at that election and concluded that another million dollars spent on television ads would not have made a difference. "People had stopped listening." He remembered something he felt the last day while shaking hands at a plant gate. "There was not the fervent positive response I was used to . . . something, I sensed, was missing." Nonetheless, many years later he would describe the effort as "the best campaign we ever ran." He had been heartened that he wasn't sent to events that didn't make sense, as in every other campaign, and that the campaign accomplished everything it set out to do, apart from winning.

Democratic pollster Peter Hart remembered a call he received on election night from Doug Bailey, a prominent Republican consultant. Bailey described Birch as "the best retail politician" he had ever seen.[36]

The Republicans gained twelve seats and won control of the Senate, 53–46, the largest turnaround in the Senate in twenty-two years and the first time the Republicans had controlled either house of Congress since 1954. Senators Robert Byrd and Howard Baker exchanged places with each other as majority and minority leaders. Twenty-two Democratic senators had stood for reelection. Three lost their primaries; each of those seats went Republican in the fall. Nine others lost; twelve of the twenty-two would no longer be senators. The Republicans, on the other hand, had six incumbents running in the fall; all were reelected. Their only casualty was Jacob Javits, perhaps the smartest senator then in office, who lost the primary in New York to Al D'Amato, who also won in November. None of the five open seats of retiring senators changed parties. The other eight Democratic incumbents who were unseated included senior Democrats Herman Talmadge (GA) and Warren Magnuson (WA) plus four others who, like Birch, had been targets of the New Right groups: Frank Church (ID), John Culver (IA), George McGovern, (SD) and Gaylord Nelson (WI). Only Democratic senators Gary Hart (CO), Patrick Leahy (VT), and Tom Eagleton (MO) survived the election.

The fact that the voters in Washington would turn away Magnuson, the chairman of the Senate Appropriations Committee, always struck Birch as "unbelievable."

Months afterward, Birch was told that a poll of Hoosiers ranked him as the most popular political figure in the state. Former staffer Louis Mahern's view was that the voters wanted to send Birch a message about the problems in the country and they had no intention of firing him. Birch added that his presidential campaign made him a national figure, which made it much harder for him to separate himself from national problems.

Quayle would go on to greater fame and some ignominy, being chosen by George H. W. Bush to be his running mate in 1988. He would be elected vice president in a campaign in which he was constantly under assault for some of the same issues that the Bayh campaign was aware of but never developed. The National Guard issue was huge in the immediate aftermath of the Republican convention in 1988, but the plagiarism issue died away when the professor who had acknowledged the facts to us changed his story. Quayle was also plagued by allegations of drug usage. As vice president, he earned a reputation for gaffes that often made him the butt of late-night comedy shows. When speaking to a United Negro College Fund event, whose motto is "a mind is a terrible thing to waste,"[37] he mangled the motto to say, "What a terrible thing it is to lose one's mind."

There were many others. While vice president, Quayle was asked by reporter Diane Sawyer about some of the quotations, and he responded, "I stand by my misstatements."

Nineteen-eighty was a watershed year, one that saw so many prominent politicians defeated, the beginning of the "Reagan Revolution," and the influence of the New Right. Many would argue that the comity and civility that typified our politics was lost, perhaps for a long time—perhaps forever.

The United States in 1980 was a country very different from the way it was when Birch first came to the Senate. The role of women and of African Americans in American life looked very different in 1980. An innocence that seemed to exist in 1962 was lost after the failure in Vietnam and the scandal of Watergate. And the role of the Senate in public life seemed different as well. Maybe it was due to the growing impact of television in American society, but it may have been caused by the presence of certain men and women in the Senate during those years.

When Birch arrived in the Senate, his salary was $22,500 per year. This small salary presented a challenge to many senators who wanted residences in their home states as well as in the Capitol. That salary would tick upward during Birch's tenure, reaching $42,500 by the beginning of his second term and $60,662.50 at the end of his service. Also, few senators would have been recognized on the streets in 1962, while by 1980 many had obtained celebrity status.

For Birch Bayh, it was the end of his Senate career. When he spoke to the campaign staff, press, and supporters on election night, he was upbeat. How could he feel too badly about losing? Instead, he said, he was grateful to the people of Indiana for giving him eighteen years to do precisely what he wanted to do in his life. He also thanked the people of Indiana for giving him an opportunity to practice law, which he would soon begin to do.

19

Capstone

Birch Bayh was a senator's senator.

SEN. PATRICK LEAHY[1]

AN ELECTION DEFEAT REASONABLY SIGNALS AN END TO A career. Birch Bayh could not have known that between the election and the end of the year he would add a crowning achievement to his career. A piece of legislation with his name would have an enormous impact on the US economy and millions of lives for years to come.

In early 1978, while struggling with Marvella's illness, the Bayhs learned about an innovative technology that could better determine how a patient might react to chemotherapy. It was not available to the public, however, because it was snared in bureaucratic restrictions concerning patent rights for federally sponsored research discoveries. Further investigation revealed that it was part of a larger issue involving inventions with great potential, a number of which were created with federal funding. At the time, patent rights were being governed by as many as twenty-two agencies, each in its own way.

Birch convened a meeting with officials from Purdue University, who explained their concerns. Professor Ralph Davis from Purdue's patent office brought Howard Bremer, his counterpart from the University of Wisconsin, and Norman Latker, patent counsel at the National Institutes of Health. They described several promising government-funded inventions that Purdue could not bring to market because of existing federal patent policies. It seemed logical to Birch that it was to the country's advantage to have those inventions brought to market and made accessible; it made no sense for the government to stand in their way. He learned that after World War II, the policy on handling government-funded research was to make it available to the public under non-exclusive licenses that negated patent protection needed for development. But no business could afford to invest in the commercialization of a product unless it knew that it had the rights to that invention, rights that were not provided by the government. Two champions of the existing patent policy were Hyman Rickover, famously known as the "father of the nuclear navy,"[2] and Sen. Russell Long, son of the famous politician Huey Long and chairman of the Senate Finance Committee. Rickover was important because of his close relationship with President Carter. They had served together on a nuclear submarine during Carter's navy service. Long insisted that if the government was contributing money, it had first rights to the innovation. With Birch at the Purdue meeting was another member of his Judiciary Committee staff, Joe Allen, who would become a champion of changing this patent policy.

Allen was also friendly with Latker, whose former office was part of the Department of Health, Education and Welfare (HEW). Joseph Califano, the secretary of HEW, fired Latker because of his strong advocacy for changing the way federal patent policy worked. Latker believed that those people seeking patents for innovations partly funded by taxpayers should have the ability to bring their new products to market. Califano wanted publicly funded inventions to be freely available to all, but Latker argued that doing so meant they would never be developed into useful products benefiting taxpayers. Latker worked closely with Allen to draft the legislation.

An image comes to mind. At the end of the movie *Raiders of the Lost Ark*, the Ark of the Covenant that had been captured by Indiana Jones is turned over to the government and gets stored in a huge warehouse with many corridors of shelves stacked with all manner of artifacts. That was what was happening with American innovation. Inventions were happening, but no one was benefiting from them. In 1978 it was learned that no medicines invented with federal funding had ever been brought to market when the government took them away from the universities that made them. None. And government funding had led to approximately 28,000 inventions, of which less than 5 percent were licensed. The United States, once the land of innovation and new inventions, from Bell's telephone to Edison's lightbulb, was no longer the world leader in new patents.

Serving on the Judiciary Committee with Birch was conservative Republican Bob Dole, who was also looking into federal patent policies. The two combined forces to work together on legislation that would alter these policies, an alliance that proved formidable.

Hearings were held, and in Birch's statement to the committee on September 13, 1978, he said,

> A wealth of scientific talent at American colleges and universities—talent responsible for the development of numerous innovative scientific breakthroughs each year—is going to waste as a result of bureaucratic red tape and illogical government regulations . . . Unless private industry has the protection of some exclusive use under patent or license agreement, they cannot afford the risk of commercialization expenditures. As a result, many new developments resulting from government research are left idle.[3]

The hearings illustrated that there was no incentive for universities and labs to commercialize research and development, because the government owned the rights—and there were thirty-five different federal patent policies. Occasionally an individual inventor fought within the system and won, but that process was cumbersome and lengthy, usually taking a minimum of two years and requiring substantial legal fees. It seemed that American taxpayers would be well served if government would get out of the way. Also, small, innovative companies were not incentivized to participate in government-funded projects, since any inventions they developed would be taken from them, becoming available through the government to larger rivals. The US patent system was created to protect inventors and their inventions, to give them exclusive use of the fruits of their own creations. Yet government policy was standing in the way.

Bayh, Dole, and their staffs worked together to deal with this. The result was the University and Small Business Patent Procedures Act. Now known as Bayh–Dole, it allowed "universities, small businesses, and non-profit organizations to retain intellectual property rights of inventions developed from federal government-funded research."[4]

- Universities and small businesses would be permitted to own the inventions they had developed with federal financial assistance
- It was mandated that universities share the royalties from those inventions with their inventors
- Licensing preference would be extended to small businesses and companies proposing substantial manufacturing within the U.S.
- The federal government would have the right to use the inventions for its own purposes without royalty, including for treaty obligations[5]

The Senate Judiciary Committee unanimously passed Bayh–Dole in December 1979. Debate began in February 1980.

Birch led the successful effort to defeat amendments to the bill, but the biggest obstacle remained Russell Long, who exclaimed, "This is the worst bill I've seen in my life."[6] Nonetheless, the Bayh–Dole bill passed 91–4 in the Senate, with Long voting against it. The House bill, however, contained provisions that Birch could not support, and it was not passed by the time Congress adjourned for the election. At this point, everyone thought the measure was dead.

When Congress recessed before the election, the budget had not been passed; as a result, a lame duck session was convened to deal with that. Birch was among the legislators who would finish their careers with that session. Bayh–Dole was still on his agenda, but he didn't think he had time to deal with the House. Rep. Robert Kastenmeier of Wisconsin came to the rescue. Hearing from people at the University of Wisconsin about their support for Bayh–Dole, Kastenmeier offered to accept the legislation if Birch would accept Kastenmeier's amendment to improve a system of review for issued patents, minimizing subsequent litigation. The deal was struck. The House passed an omnibus patent reform bill containing Bayh–Dole and sent it to the Senate.

As the clock ticked toward the end of the session, time was running out to bring up the House-passed legislation that included Bayh–Dole.

On the final day, November 21, Sen. Long called Birch to say, "Birch, you can pass your damn patent bill—and I'm really going to miss working with you."[7] In the final minutes that day, Congress unanimously passed the Bayh-Dole Act.

It was a parting gift from Russell Long, an act of friendship, a testament to collegiality in the Senate. Had Birch won re-election, it might never have happened. The lame-duck president had to sign the bill within ten days, or it would be deader than dead as the result of a "pocket veto." He faced opposition from within his administration, particularly from the Department of Energy, which had amassed a large staff that reviewed patent petitions dealing with energy and was very sensitive about how the bill might affect their work. Concerns about Admiral Rickover and his influence on Carter were substantial. Nonetheless, Bayh–Dole became law when President Carter affixed his signature on December 12, 1980, the last day before it would have died.

In 2002, an *Economist* editorial said, "Possibly the most inspired piece of legislation to be enacted in America over the past half-century was the Bayh–Dole act of 1980. Together with amendments in 1984 and augmentation in 1986, this unlocked all the inventions and discoveries that had been made in laboratories throughout the United States with the help of taxpayers' money. More than anything, this single policy measure helped to reverse America's precipitous slide into industrial irrelevance."[8]

Ira Shapiro wrote, "Once Long lifted his hold, the Bayh–Dole Act became law . . . The president of NASDAQ estimated that 30 percent of the value of companies listed came from university research that was commercialized only because of Bayh–Dole."[9]

Norman Latker wrote, "Before the enactment of Bayh–Dole, an enormous amount of government-sponsored research and innovation went to waste, as there were no clear mechanisms in existence to transfer the resultant inventions to the marketplace . . . the drafters of the act wanted to ensure that adequate incentives were in place to facilitate invention and to attract corporate investment into their development and distribution . . . Our answer to the problem was that intellectual property rights should be accorded in full to the innovators, rather than to the government agency that financed their research."[10]

The National Academy of Science reported that the United States was "once again a leader in basic and applied research and development to improve economic performance."

A survey published in 2013 by the Association of University Technology Managers (AUTM) said, "818 start-up companies were formed around academic patents, up 16% from 2012."[11] Overall, there were 4,200 startups operating, usually in the state where the parent research institution was located, indicating substantial regional development. "$22.8 million in sales from commercialized academic inventions"[12] were directly tied to policies in place because of Bayh–Dole. The National Science Foundation reported, "The U.S. dominates international scientific publications in the Bayh–Dole era."[13]

Among the drugs developed because of Bayh–Dole are the hepatitis B vaccine, the nicotine patch, avian flu vaccine, and a once-a-day human immunodeficiency virus (HIV) medication. One of the companies created by the partnership between the government (National Science Foundation) and universities (Stanford) since the passage of Bayh–Dole is Google.

Joe Allen recounted a story that came out of this effort concerning Leland Clark of the University of Cincinnati Hospital. Clark had developed, with EPA funding, a technology for detecting the oxygen levels in rivers. This technology helped the fight against water pollution, serving the public and giving the EPA what it wanted. Because Clark was able to own the rights and bring it to market, his technology evolved and is now used to detect oxygen levels in newborn babies.[14]

Birch has received a number of honors and recognitions for this through the years, as has Joe Allen. Allen went on to become executive director of Intellectual Property Owners; he would serve as director of technology commercialization at the US Department of Commerce and work at the National Technology Transfer Center, where he was able to continue his efforts to foster innovation around the country. In 1999, AUTM awarded him the Bayh–Dole Award for the key role he played in securing passage of the landmark legislation. Birch often referred to it as the Bayh–Dole–Allen Act.

It became apparent, as Congress convened in 1981, that the election represented a sea change in the Senate. Not only were there changes of personalities, but there was also a change in mentality. The level of partisanship ratcheted up as the level of civility decreased. Many have argued that comity was lost; the idea of "go along to get along" became obsolete. Throughout the decade, events would reshape the US Senate, and for most Senate watchers, it was becomingly increasingly difficult to imagine how it could ever return to working the way it once did, what author Ira Shapiro called "The Last Great Senate." The first twenty-four-hour news channel, CNN, was created in 1980, and it has changed our cultural dialogue forever. A requirement to fill all available airtime, whether there is breaking news or not, has meant the amplification of news that is not news, creating an atmosphere where public officials need to be "on" at all times. Small indiscretions or misstatements become prominent news items, creating a culture of caution among elected politicians. The "gotcha" mentality that seems to typify the current twenty-four-hour news cycle may be the principle reason why it seems there are few politicians writing their own chapters in "Profiles in Courage." When Sen. John Tower (R-TX) was rejected for a cabinet appointment in 1989, it was the first time in thirty years that a cabinet appointment had failed confirmation by the Senate. The Senate rejecting one of its own would have been impossible only a few years before.

It was not unusual in that every time Birch wanted to introduce legislation, a Republican cosponsor was the first order of business. He understood that bipartisanship was the key to effectiveness, the key to success. This was the message he received from Everett Dirksen early in his career, a message that seemed to commence a slow death after the 1980 election. In 1995, a story circulated around Washington that when Sen. Joe Biden met his new colleague on the Senate floor, Rick Santorum of Pennsylvania, he said, "Nice to meet you Rick. Congratulations on your election. I look forward to working with you." To which Santorum replied, "I don't know why. We won't get along on anything."

After leaving office, Birch established a law firm, joining attorneys Don Tabbert and Jim Capehart to form the Bayh, Tabbert & Capehart firm with offices in Indianapolis and DC. Evan would join his father's firm briefly before turning his attention to pursuing his own political career. Having graduated from law school in 1981, Evan clerked for a federal judge before joining the firm. Later, the firm was dissolved and

Birch transitioned to different law firms with various partners, each firm with his name on the letterhead, including a few resulting from mergers with other firms. Eventually, he accepted the invitation of former attorney general Benjamin Civiletti to join his law firm, then known as Venable Baetjer Howard & Civiletti, eventually as Venable LLP. Former staffers Tom Connaughton and Kevin Faley would join him in the practice of law.

Life after the Senate was almost bizarre for him. Not only did he begin the year unemployed and without a staff to take care of him, he was a widower living alone. He was also seriously in debt. During the better economic times early in recent years, he bought a group of six houses on Pennsylvania Avenue in DC as investments. He took out six mortgages on the properties and a substantial housing rehabilitation loan in order to make the houses more attractive on the seller's market. When the economy tanked in the late 70s, he had houses without tenants paying rent and a daunting financial obligation. Added to that, he assumed responsibility for the campaign debt of almost a quarter of a million dollars. It would take him many years to unload the houses and to pay off the campaign debt, the latter accomplished through a direct mail campaign managed by several formers staffers.

The good news was that Birch's romance with Kitty Halpin blossomed. They were married on Christmas Eve, 1981. The following year they became parents when Christopher John Bayh was born. This marked an incredible four years of important life events for Birch Bayh. Marvella died in 1979, Birch lost his Senate seat in 1980, Birch and Kitty were married in 1981, and Christopher was born in 1982. Christopher attended Sidwell Friends in Washington, DC and become another Bayh to attend Indiana University for his undergraduate degree.

After I.U., Chris attended law school at Washington University in St. Louis, where he got his degree. Later, he worked as a research assistant to former Indiana congressman Lee Hamilton at the Woodrow Wilson International Center for Scholars. Following that, he clerked for Judge David F. Hamilton of the Seventh Circuit. After working in two prominent law firms in Washington, DC, he moved to Indianapolis to join the firm of Barnes & Thornburg LLP, which made it easier to have a home and start a family. When Kitty discussed her son's work with

Judge Hamilton, she was told that he worked hard and had "no sense of entitlement," something that made both of his parents very proud. In 2010, Chris married Jessica Pedley, and in 2015 Jessica gave birth to their daughter, Georgia. They would have a second daughter, Josephine, in 2017. Anticipating the arrival of Georgia, Birch commented that he had to "come to grips with living with a grandmother."

Several years earlier, Bill Arceneaux, who had headed up Birch's presidential effort in Louisiana, recalled having lunch with him. After the lunch, Birch was on his way to a Little League baseball game. He told Arceneaux, "I'm the only guy in the world who has one son who is the governor of Indiana and another who's playing Little League baseball."[15]

In 1985, Evan married Susan Breshears, the daughter of one of Birch's California supporters. He would run for and win the office of Indiana secretary of state in 1986, defeating the son of former governor Bowen, an election that was by no means a slam dunk. He was twice elected governor, in 1988 and 1992, serving two terms as Indiana chief executive. During his second term in 1995, Evan made Birch a grandfather with the birth of twin boys, Nicholas and Birch Evans Bayh IV, known as Nick and Beau, respectively. Talking about his grandsons as they prepared to enter Harvard University, Birch said they were great guys who seemed totally unaffected by the career of their father, something that made him proud. In 1996 was the reelection campaign of President Bill Clinton, and Evan was the keynote speaker at the Democratic convention in Chicago. Two years after leaving the governor's mansion, Evan was elected United States senator and claimed the seat his father once held. In 1999, *People* magazine featured him in its issue on the "Most Beautiful People" in the country.[16] In 2004, he was reelected overwhelmingly, having won more than 60 percent of the vote in both elections. In 2008, he began to seek the Democratic nomination for president. Once Barack Obama entered the fray and became the leading competitor, Evan dropped out of the race. Despite his support for Hillary Clinton in the primaries, he had hoped to be Obama's choice as the vice-presidential running mate and was disappointed when Obama chose Sen. Joe Biden.

Evan talked about the legacy of his father and his own years in the Senate. He laughed when recalling how he had bonded with Iowa's senator Tom Harkin over Harkin's authorship of a book about Dairy Queens in Iowa. When campaigning in Indiana, Evan repeatedly met women who claimed they had changed his diaper when he was a baby. It made him muse that he was probably the "driest baby anywhere."

The 1980 campaign, in which Evan was the chairman and its principle surrogate, gave him invaluable political experience. Taking off the fall semester from law school and travelling to more than 65 counties during that summer and fall, he received a training that would be hard to experience any other way. He had poignant memories of the last stop with his father, at the Alcoa plant in Evansville. When the election defeat happened, he went into a shell for a period, leaving the country to travel around Europe and get away from all he had recently experienced. In retrospect, he wished that he had spent time instead with his father, not appreciating how difficult life probably was for him. When asked why he chose to pursue a political career, he talked about the example his father had provided, describing him as a "heroic figure."[17]

In 2010, Evan announced his retirement from the Senate, decrying the nonstop partisan bickering there; it was no longer the institution his father had known. He talked about the dysfunction, the overt partisanship, and the inability to be effective and get substantive work done. Birch was sorry to see Evan leave the US Senate as he did. Six years later, in 2016, Democratic leaders prevailed on Evan to run for the Senate. That effort failed in a surprising election year that brought Donald Trump to the presidency. Sadly, Susan Bayh was diagnosed with brain cancer in 2018 and operated on the same week that the Bayh twins graduated from Harvard. It means the third generation of Bayh men faced with their mother's cancer.

A private citizen living in the Maryland suburbs, Birch was appointed chair of the National Commission Against Prejudice and Violence by Maryland governor Harry Hughes, the first ever national commission to fight hate crimes. He served from 1984 through 1994 and helped write the hate crimes law that Congress passed. In 1995, President Clinton appointed him to the Fulbright Foreign Scholarship Board. In 2003, the US Courthouse and Post Office in Indianapolis where his Senate office had been housed was officially dedicated as the Birch Bayh Federal Building and US Courthouse. Birch attended the ceremony with Kitty,

Christopher, and Evan present. The legislation was sponsored by his former rival and colleague Richard Lugar.

In 2004, Birch took part in one of many events that honored him for the authorship of Title IX. During halftime of the University of Connecticut and Rutgers women's basketball game, Sen. Christopher Dodd (D-CT) spoke, saying, "Thanks to Title IX, women have taken their rightful place in American education—as students, teachers, administrators, and athletes. Sen. Bayh's leadership as original author of this legislation has directly impacted millions of young women whose lives have been touched and bettered through equality in education, collegiate athletics, and opportunities for success in virtually every aspect of American life." UConn president Philip Austin said, "Title IX represented a major advance not just for women, but for all Americans and for higher education." He noted that "although UConn began admitting women in 1893, many publicly funded universities did not admit women, and many women who did enter universities were discouraged from studying math, science, law or medicine, before Title IX became law in 1972."[18]

In 2006, Birch was appointed to the C. V. Starr Center for the Study of the American Experience at Washington College in Chestertown, Maryland, as a senior fellow.[19] That experience would turn out to be one he loved, inspiring the Bayhs to move to Easton, Maryland, a much shorter commute to Chestertown. At one point when discussing retirement, Birch said he would like to move to Bloomington, Indiana. Kitty wanted to move to Easton. "So," Birch said, "we compromised and moved to Easton." He would later institute a program at the college called the Senatorial Colloquy on American History and Politics, bringing former and current senators to Washington College to talk with Birch in front of a live audience and take questions from them. The first participant was former senator Gary Hart (D-CO), followed by Paul Laxalt (R-NV), Dale Bumpers (D-AR), and Richard Lugar. That same year, Birch was feted at an exhibition of his Senate papers, donated in 1981 to Indiana University in Bloomington, inaugurating a five-month exhibit called "The Art of Leadership." Also in 2006, because of the impact of Title IX on women's sports, he was awarded the NCAA Gerald R. Ford Award for leadership. Birch was given the honorary degree of Doctor of Laws from Washington College in 2008. The following year, he was awarded the Distinguished Alumni Service Award by Indiana University.

When Title IX was being challenged in the Supreme Court, Birch was invited to appear in court as the act's founder, leading to comments from the bench about how unusual it was for the justices to have a law's author present for its deliberations. In June 2012, he was honored at the White House in a ceremony recognizing the fortieth anniversary of Title IX. Later that evening at a Women's National Basketball Association (WNBA) game in Indianapolis, he was recognized at courtside for the same reason. That same year, the Women's Sport Foundation awarded him the Billie Jean King Contribution award. During this fortieth anniversary period, Melissa Isaacson, writing columns for ESPNW.com, ESPN Chicago, and ESPN.com and formerly with the *Chicago Tribune*, wrote a tribute entitled "Birch Bayh: A Senator Who Changed Lives." While researching a book in 2005, she recalled a conversation with Birch. "There was one more thing I wanted to say to Birch Bayh, one more thing I had to say. But I was afraid it would sound corny—or worse, insincere. And I wasn't sure I could spit it out. 'I know you must hear this all the time,' I said, not such a great start. 'But you changed our lives. The job I chose, the man I married, the mother I've tried to be . . . You affected the women we became.'"[20]

After that fortieth anniversary celebration of Title IX, Birch arrived at the Indianapolis Airport for the flight home. A middle-aged African American man helped with his luggage. He told Birch that he knew who he was and remembered him from a parade when he was a kid; Birch had shaken his hand and spoke with him that day, something he never forgot. That moment gave Birch satisfactions large and small.

For many years after leaving the Senate, he campaigned for candidates he supported, particularly in Indiana and often for Evan. In 2008, at the age of eighty, he campaigned aggressively across Indiana for Barack Obama; when Obama won the state, it was one of the few times in the previous century that a Democratic presidential candidate had won there and the first time since 1964.

Also in 2012, the Robert H. McKinney School of Law at Indiana University inaugurated the Birch Bayh lecture. The lecture has since become a major event for the Indiana legal profession, attracting scholars from across the country.

In May 2013, Indiana State University dedicated its new Bayh College of Education, a family-wide honor because of the number of Bayhs associated with the institution. Leah Hollingsworth Bayh, Birch's mother, had met Birch Bayh Sr. while she attended Indiana State Normal School, which became Indiana State. The school's name changed as a result of a bill written by State Representative Birch Bayh. When Birch brought Marvella to Terre Haute, she took classes there. Birch as senator and Evan as governor and senator had both proved themselves valuable friends of the university over four decades. All of this is depicted in the Bayh Legacy Wall, an interactive display at Indiana State that illustrates the Bayh careers. The display is regularly updated.

On May 10, 2017, to celebrate the fiftieth anniversary of the ratification of the Twenty-Fifth Amendment, Birch was honored at an event sponsored by the Bipartisan Policy Center in cooperation with the American Bar Association and Fordham University School of Law. Among those paying tribute was former Fordham Law School dean John Feerick, a key player during the process to pass the Twenty-Fifth Amendment while a young lawyer. Feerick spoke about Bayh the legislator but also focused on Bayh the man, whose collegiality was the necessary ingredient in making the bipartisan effort to pass the measure successful. Jay Berman accepted the award on Birch's behalf.

When his career was over, Birch described his greatest legislative disappointment as the failure to pass direct election of the presidency. He was reminded that some of those opposing direct election were longtime allies from other battles, people from the American Civil Liberties Union and Jewish and minority groups. He recalled an argument with one of them, civil rights stalwart Eddie Williams. Williams argued that they didn't want to give up the advantages they had in places like New York and California, where they could make a difference in where the electoral votes went. Birch's anger at the experience remained fresh almost forty years later. In his retirement, he has contributed his advice and writings to organizations that are working to get states to approve measures committing their electoral votes to the results of the actual votes in the election, still trying to achieve a direct popular vote for the nation.

Recounting the years of his career, he talked about how lucky he was that he was able to be married for twenty-six and a half years to Marvella, "one of the most exceptional people I ever met," and married even longer to Kitty. About Kitty he often repeated, "I don't know what I would do without her."[21]

He served with senators whose fathers had also served: Robert Taft Jr. (OH), Harry Byrd Jr. (VA), and Joe Tydings (MD). His son, Evan, served in the Senate with the sons of Birch's former colleagues, Mark Pryor (AR), Bob Bennett (UT), and Lincoln Chafee (RI).

Indiana politics can make a deep and lasting impression on you. How can you ever forget a county chairman named Peanut Courtney, who was a hunchback and wore a gun? Or names like Greasy Willis, Knute Dobkins, Casey Pajakowski, and Fanny May Hummer? There was a candidate in Delaware County handing out campaign literature highlighting his name, Scott A. Hole. It was in Indiana where Birch developed his view of politics, something he described as less an interstate highway than a country road wending its way around. One has to be prepared to change course and to stop, slow down, and speed up to get where you are heading.

Birch talked many years later about his own education in politics and how he learned to get along well with people from different walks of life, people with ideas different from his. Being elected president of his fraternity at Purdue was a major first step; it was a body with students still in their teens and others who were war veterans, one of whom had flown twenty-seven missions in World War II. Birch's major regret had nothing to do with his political career; it was that he wasn't born at least two years earlier so he could have joined the military in World War II, at least near the end of the conflict.

His reminiscences are largely good ones when looking back on his life and career, always somewhat amazed by the breadth of issues in which he was actively engaged. Several measures Birch pursued in the Senate had no possible political benefit for him, something that made his Senate service memorable. One of them was voting rights for Washington, DC. Creation of the DC subway was another, along with key measures to protect the rights of the institutionalized. There were those issues in which he assumed a leadership role that hurt him politically, particularly gun control. The legacy for the country is substantial:

- **Bayh–Dole:** University innovations created almost two startup companies per day by 2005 and two new commercial products per day by 2013; by 2010, 6,500 new companies had been founded as a result of these innovations and 153 drugs, vaccines, and in-vitro devices had been brought to market; campus patents grew to 3,278 by 2005 as well as 4,932 academic licenses and 28,349 active licenses overall. The act caused the creation of 279,000 new jobs, and support was provided for another 3.8 million jobs. By 2015, the academic–industry licensing created by Bayh–Dole had "contributed up to $1.18 trillion to the U.S. economy" with $22.8 million in directly reported sales from commercialized academic inventions. The biotechnology industry that is so large today is rooted in this academic research, as is nanotechnology, all of this directly tied to the policies in place because of Bayh–Dole.[22]
- **Title IX:** This federal civil rights law increased women's "access to higher education, career education, education for pregnant and parenting students; employment; enhancing the learning environment, increased access to math and science; standardized testing and technology."[23] There is now an increasing number of remedies for sexual harassment. Fifty-seven percent of college enrollments by 2003 were by women. Title IX virtually created organized women's sports; the athletic participation of girls went from 294,015 before Title IX to over three million in its first forty years—before passage, only one in twenty-seven girls participated in high school sports, two in five by the time of its fortieth anniversary.[24]
- **The Twenty-Fifth Amendment:** The amendment created the path for the resignation of President Richard Nixon by providing a mechanism to fill a vacancy in the office of the vice presidency for the first time, possibly heading off an unimaginable constitutional crisis. It ensures that the country has a way for

a successful transfer of power should a presidential disability ever again occur.

- **The Twenty-Sixth Amendment:** Birch Bayh's second constitutional amendment led to millions of eighteen-to-twenty-one-year-old Americans having the right to vote.
- **The Federal Emergency Management Agency (FEMA):** This agency created the mechanism for the government to manage natural disasters.
- **The Foreign Intelligence Surveillance Act (FISA):** This act provided the means to obtain foreign intelligence while protecting individual rights with FISA courts.
- **Juvenile justice:** The Juvenile Justice and Delinquency Prevention Act has stood up as a pillar of legal protections for juveniles, while the Runaway Youth Act has prevented runaways from being treated like criminals.
- **Civil rights of the institutionalized:** This measure put in place federal rules protecting institutionalized persons from mistreatment, whether they are mentally ill, incarcerated, or otherwise resident in institutions.
- **Drug abuse:** The Methadone Diversion Control Act of 1973 set up the framework for practitioners who administer methadone and for its regulation.
- **DC Metro:** Metro, the country's second-busiest transit system, has fundamentally changed neighborhoods, bringing thousands of new residents and jobs to places that were starving for people and development.[25]
- **Public works:** In addition to the DC subway system is a long list of public works in Indiana that would not have happened without Birch Bayh, including the rebuilding of tornado-ravaged communities like Monticello, but perhaps the most notable being the transformation of the city of Indianapolis because of the UDAG (Urban Development Action Grant) money he championed, a development that he was only able to appreciate after he left the Senate. Nationally, he was a key participant in decisions that created the federal criteria for clean air and clean water regulations.

- **Alcohol fuels:** The alcohol fuels industry was virtually created by the efforts of Birch Bayh, who promoted gasohol production as a way to deal with our dependence on foreign oil while helping American farmers, an industry that has grown to huge proportions since.
- **Women's rights:** Promotion of the Equal Rights Amendment (ERA) and the Pregnancy Discrimination Act, the appointment of the country's first female US attorney and first female chief counsel of a Senate subcommittee, changed the status of women in government and the environment of women's rights in America.
- **Protecting the Constitution:** While adding more to the Constitution than any person since the founding fathers, he also stood in the way of several efforts to amend it, preventing amendments he considered harmful to the country, measures dealing with school prayer, a woman's right to choose, busing, guns, and a balanced budget.

The 1962 campaign song, "Hey, Look Him Over," had a line in it that has proven to be true: "For Indiana he will do more than anyone has done before." It can be easily argued that he did.

A number of former staff people provided memories from their years with Birch; many remarked about how different politics is now than it was then, and how different the Senate was. One consistent recollection is that whenever a staffer mentioned the political impact of a pending vote or policy dispute, Birch always replied that he would worry about the politics. "Just tell me what you feel is the right thing to do." He once said that what he accomplished "was the result of people and circumstances," but history will not assess his accomplishments so passively.

A number of those staff memories provide a more three-dimensional view of the person. Lynne Mann wrote that he "never passed a Dairy Queen he didn't like, he could write lovely poetry... loved homegrown tomatoes and grew plants at his house . . . loved music, had a nice voice... had a good sense of humor."[26] John Rector remembered how Birch had a pull-up bar across the door of the bathroom in his Senate suite and would do one-armed pull-ups there, often while being briefed early in the morning. And Rector remembered seeing a Valentine's Day ad in his local California newspaper that included a quotation: "We should take the time to hold hands, to say a kind word now and then, to be polite and giving and sharing. More importantly, we should take the time to get back to the basics of life—love. Birch Bayh."[27] Abby Saffold told stories about staffing Birch on the Appropriations Committee that demonstrated a lot about "the Senator's humor and lack of pretense."[28] Louis Mahern fondly remembered him as someone who never lost his temper and never treated staff members with anything but respect. "It is said that no man is a hero to his valet," Mahern commented, "but we were close to being his valets and Birch Bayh is my hero."[29] Jeff Smulyan recalled the softball games with other Senate staffs, one in particular with Sen. Bob Packwood and staff. Here was a Democratic office playing a Republican one with all the participants going to a pizza place afterward for beer and pizza. That camaraderie seems like something from a bygone era.[30] David Rubenstein described Birch as a senator who "had the ability to connect with people" and said that Birch would put aside his own ego when dealing with his older colleagues.[31] Allan Rachles expressed a hope that Birch would be remembered for his collegiality with his Senate colleagues, how he worked hard on his relationships with them. When he felt strongly about some matter before the Senate, he would have detailed and serious discussions with senators, regardless of how much or how little they may have agreed with his views.[32] Diane Meyer characterized Birch as always humble and kind.[33]

Speaking about the Senate, Tom Connaughton said,

> It was different then. We worked very closely with Republican staff, and if there were big issues, we'd try to work out the things that were there . . . In those days, the Senate had the reputation of truly being a club, and you didn't insult other members of that club. The fights could be terribly emotional . . . people really felt passionately. Yet still it would not be personal, and I don't think the issues are any more passionate today than they were then. But even on minor issues, it gets very personal now . . . You worked with all members and tried to get as much of a consensus as you could. Now and then an issue would come down to a party vote . . . but you walked away from it still friends.[34]

Ray Scheele, Larry Conrad's biographer, said, "I think the guy will go down in history as not only a great senator from Indiana, I think he will go down in history as a great senator, a great U.S. Senator who served in a time of giants, and Birch was a giant among giants."[35]

Lee Hamilton, who served in Congress from 1965 to 1999, had an incredibly distinguished career capped off by receiving the Presidential Medal of Freedom from President Obama in 2015. Lee talked about how Birch mastered both the inside and outside power of the Senate. Inside power requires expertise in drafting and passing legislation. Outside power is the court of public opinion, where one addresses the larger public beyond the Senate chamber. He described Birch as someone who was great at networking, someone who liked the Senate and who was liked by fellow senators. He was, described Hamilton, "one of the finest legislators of the twentieth century."[36]

Sen. Richard Lugar, whom Birch defeated in 1974, became a close ally, colleague, and friend. They became friendly during their work together on the Senate Intelligence Committee, and he was someone Birch learned he could trust completely. Lugar would later recount his feelings about seeking the Senate in 1976, saying that he had superb training as a candidate by first running against Birch, something he had no idea about when he initially decided to run.[37] After Lugar lost his reelection in the 2012 primary, Birch called him to offer his regrets, telling him, "The country has lost a real resource."

Another former colleague, Sen. Patrick Leahy of Vermont, fondly remembered his own arrival in the Senate in 1975 and how the two senators who made him feel most welcome were Humphrey and Bayh. He spoke about how he admired Birch's persistence and the way he deftly dealt with conservatives and liberals in order to get things done. Who got the credit was not important to Birch, whom Leahy described as "the go-to guy to get things done . . . Birch Bayh was a senator's senator." Expounding on what was required to be effective at the time, Leahy emphasized how critical it was to keep your word and said everyone knew that Birch could be relied upon to keep his. According to Leahy, older senators considered it "inconceivable to break your word." He underlined his point with a story that illustrated the values that existed in the Senate. Ted Kennedy had asked Chairman Eastland to schedule a Kennedy priority in the Judiciary Committee. Eastland wouldn't agree until Kennedy showed him that he had the votes necessary to win. When

Kennedy provided his list of commitments, Eastland scheduled the matter and a vote was taken, with Kennedy's side losing by a single vote. Eastland asked the senator voting no whose name was on Kennedy's list of commitments if he had pledged his vote to Kennedy. The senator confirmed that he had but since changed his mind. Turning to the clerk, Eastland said, "Change my vote to aye."[38]

In many ways, Birch was a Renaissance man. He wrote poetry and played guitar; he had been a farmer, soldier, pilot, boxer, baseball player, and author. He was also a marksman with firearms and an expert horseshoe player. But it was as a politician and legislator that he made his mark.

At the dedication of the Birch Bayh Federal Building and US Courthouse in 2003, Birch was summoned to speak with an elderly woman in a wheelchair. He learned that for her hundredth birthday, she was asked what she wanted. Her reply was, "I want to meet Birch Bayh." It was momentous for both of them.

Congressional gridlock, as it typified the first decades of the twenty-first century, resulted from a conflict between those who feel it most important to prevail and those who are willing to compromise. The nation was built on compromise: taking disparate views and philosophies and weaving them into a fabric of a system that allowed this country to prosper. History reminds us that the Civil War was the result of a conflict between those who insisted on preserving a particular way of life and those who believed the union must be preserved. The first would tear the nation asunder to protect their customs, while the other would resist that effort in order keep the union intact.

The era in which Birch served was one when the Senate made a real difference in American life. The commitment to compromise was absolute; winning half the battle was better than winning none of it. There was also a recognition that today's ally could be tomorrow's adversary and vice versa. One had to go along to get along, and the main goal was to move the ball forward, to improve the lives of all Americans, to move our history forward. Members of the Senate revered the institution and treated their colleagues as fellow travelers in a special place, each committed to pursuing his or her own goals but ultimately to making the system work. Henry Clay, speaking about the Compromise of 1850, rhetorically asked what a compromise was. "It is a work of mutual concession," he answered, and "a measure of mutual sacrifice."[39] In a 2015 biography of George H. W. Bush, it was said that "he embraced compromise as a necessary element of public life."[40]

Birch Bayh was fortunate to be serving in an incredibly important time in American history. From the luster of Camelot until the dawn of Reaganism, it was a time of growth, change, and turmoil. The eighteen years he was a senator included four assassinations of major public figures in American history: Medgar Evers, Martin Luther King Jr., John Kennedy, and Robert Kennedy. It was a time of war and an antiwar movement that embroiled the nation, of civil rights, women's rights and gay rights, of political scandal in Washington and American hostages in Iran. There were wars in the Middle East and Jewish dissidents in the Soviet Union, a Sino–Soviet conflict while Americans sought détente with the Soviets and made an opening to China. These major issues led to controversies over busing, guns, crime, prayer, abortion, the energy crisis, civil disobedience and protest. In this era, the country lost one president to assassination and another to scandal, with a vice president also resigning in disgrace. The Vietnam War was the catalyst of massive student protests as well as the killings at Kent State and Jackson State. Taken along with Watergate, the American willingness to believe its leaders may have changed forever.

This was the environment in which the career of Birch Bayh flourished, creating the basis for his many accomplishments: the twenty-fifth and twenty-sixth amendments to the Constitution, successfully leading the opposition to two Supreme Court nominees, the authorship of Title IX, and his many efforts to pass a constitutional amendment for direct election of the president as well as the Equal Rights Amendment. It can be argued that what he kept out of the Constitution may have been more important than what he put into it. He played a role in professionalizing the nation's intelligence community while protecting individual rights. He played a major role in bringing a subway system to the nation's capital. His career would end during the period of hostage-taking, double-digit inflation and unemployment, gas lines at the pump, and a growing sense of disillusion with government, each a factor in his defeat. Issues like the Panama Canal treaty exacted a price, contributing to his career's demise. Ironically, the end of that career became the reason for the success of Bayh–Dole and its deep and lasting impact on American innovation and the economic health of those centers of academia participating in that innovation.

On a personal level, he aged from thirty-four to fifty-two and suffered the death of his father, the murder-suicide by his father-in-law, and the death of his wife and found himself starting a new life as an unemployed widower who needed to learn, in his sixth decade, the routines of life that most people consider normal. And he would see his eldest son elected both governor and US senator from the state of Indiana that he loved. His life was a life of value. "I wanted to make a difference," he once said. "If you get into politics to make a difference, I think we did."

For those who want to read more about the life of Birch Bayh, bonus features are available online at www.blaemire.us. They are:

Birch Bayh and the Indiana Democratic Party
Senate Colleagues
Haynsworth and Carswell
Bayh Staff Stories
Birch Bayh Campaigns
The Birch Bayh Legacy

Notes

1. Farmer, Soldier, Legislator

1. Roosevelt quote from radio address, Selective Service Registration Day, October 16, 1940.

2. Early Indiana history from "Prehistoric Indians of Indiana," Division of Historic Preservation and Archeology, Indiana Department of Natural Resources, March 10, 2004.

3. Indiana's growth and expansion from "Indiana History Part 4," Northern Indiana Center for History, May 11, 2008.

4. Discussion of German immigration from Theodore H. White, *The Making of the President 1960*, page 245.

5. Events of the year 1928 are found on the website Onthisday.com.

6. Quotations and stories about Birch Bayh's early life are taken from the author's interviews with him in twelve sessions in 2012–2015.

7. Description of Terre Haute by Michael Coakley, "Bayh: A Name to Watch in '72," *Chicago Today Magazine*, August 23, 1970, pages 8–9, 15.

8. Story about Birch Bayh helping young people in Germany to grow crops for themselves: Karl Detzer, "GI Ambassador," *Reader's Digest*, November 1948, page 36.

9. Ibid.

10. The story about Birch Bayh capsizing is from an interview with the author and Wayne Townsend on May 25, 2015.

11. Quote about precinct committeemen as "the backbone of the system" from Katie Wolf, in the film *Hey Look Him Over* (Sautter Films, 2005).

12. On early accomplishments in the General Assembly: Kate Cruikshank, *The Art of Leadership: A Companion to an Exhibition from the Senatorial Papers of Birch Bayh*, Lilly Library, Indiana University, January 29–May 5, 2007, page 21.

13. Ibid.

14. "Bayh Heads Jaycees 10 Young Men of '63," Fort Wayne *Journal Gazette*, January 14, 1964.

2. US Senator

1. John F. Kennedy, speech to Massachusetts State Legislature, January 9, 1961.
2. Discussions of the 1962 campaign are taken from the author's interviews with Birch Bayh, in twelve sessions in 2012–2015.
3. Description of the 1960 Kennedy campaign in Indiana in Michael Coakley, "Bayh: A Name To Watch In '72," *Chicago Today Magazine*, August 23, 1970, pages 8–9, 15.
4. https://www.alaforveterans.org/ALA-Girls-State/
5. Description of Kennedy's relationship with Homer Capehart in the Senate in Ted Sorenson, *Counselor*, pages 150–51.
6. Scott Crass, "Surprise '62 Upsets Gave Us Bayh Wonder and Glorious Gaylord," TheModerateVoice.com (blog), June 6, 2013 (accessed October 2015).
7. Capehart's role in the Senate: William B. Pickett, *Homer E. Capehart: A Senator's Life, 1897–1979*, pages 151–58.
8. Description of the Indiana Republicans: William B. Pickett, *Homer E. Capehart: A Senator's Life, 1897–1979*, pages 165–66.
9. Ed Ziegner quotation: Raymond H. Scheele, *Larry Conrad of Indiana*, page 18.
10. Description of the Indiana Republicans: William B. Pickett, *Homer E. Capehart: A Senator's Life, 1897–1979*, pages 165–66.
11. Description of the battle for the Senate nomination in 1962: "Voters Size Up Birch Bayh," *South Bend Record*, September 22, 1961.
12. Birch Bayh and Larry Conrad first meet: Raymond H. Scheele, *Larry Conrad of Indiana*, page 16.
13. Bob Hinshaw describes his role in the campaign in an interview with the author on May 27, 2015.
14. Homer Capehart's background description: William B. Pickett, *Homer E. Capehart: A Senator's Life, 1897–1979*, page 148.
15. Rumors about Governor Welsh supporting Boswell: Edward Ziegner, *Indianapolis News*, April 26, 1962.
16. Capehart scolding Boswell: James Carroll, *South Bend Tribune*, "Welsh May Face Balky Convention."
17. Description of Welsh endorsement of Bayh: Irwin J. Miller, "Gov. Welsh Endorses Bayh," *Terre Haute Tribune*, May 10, 1962.
18. Results in the 1962 Democratic Convention: Edward Ziegner, *Indianapolis News*, June 23, 1962.
19. Democratic Party's budget for the fall campaign: Raymond H. Scheele, *Larry Conrad of Indiana*, page 25.
20. The genesis of "Hey, Look Him Over" as told by Bob Hinshaw to the author on May 27, 2015.
21. The memory about realizing the song's impact was taken from the author's interviews with Birch Bayh, in twelve sessions from 2012 to 2015.
22. Evan Bayh, interview with the author on July 16, 2015.
23. Description of the impact of the Bayh song: Associated Press, "Bayh Jingle Called Best Commercial," May 25, 1963.

24. Description of Capehart as a candidate: William B. Pickett, *Homer E. Capehart: A Senator's Life, 1897–1979*, pages 167, 172–73.

25. The issues between Bayh and Capehart: William B. Pickett, *Homer E. Capehart: A Senator's Life, 1897–1979*, pages 169–70.

26. Kennedy quotation at Indianapolis rally: Theodore C. Sorensen, *Kennedy*, page 674.

27. Bob Boxell, interview with the author on May 29, 2015.

28. Quotation describing JFK's feelings about the impact of the Cuban Missile Crisis on the Indiana Senate campaign: Hugh Sidey, *John F. Kennedy, President*, page 278.

29. "Would you believe it? Homer Capehart is the Winston Churchill of our time!" Ted Sorenson, *Kennedy*, page 2.

30. Evan Bayh, interview with the author on July 16, 2015.

31. Bob Hinshaw, interview with the author on May 27, 2015.

32. Lee Hamilton, interview with the author on May 28, 2015.

33. Description of the reasons for Bayh's victory: Hugh Sidey, *John F. Kennedy, President*, page 301.

34. Kennedy's approval rating after the Cuban Missile Crisis: Robert Dallek, *Camelot's Court*, page 335.

35. Description of the Senate when Bayh arrived: Ira Shapiro, *The Last Great Senate*, page 13.

3. Assassination and Amendment: 1963

1. Theodore Roosevelt quote from speech to the Sorbonne in Paris, France, April 23, 1910.

2. Scott Crass, "Surprise '62 Upsets Gave Us Bayh Wonder and Glorious Gaylord," The ModerateVoice.com (blog), June 6, 2013 (accessed October 2015).

3. Chris Sautter quote from Scott Crass, TheModerateVoice.com blog.

4. Stories from the early years of Birch Bayh's career are taken from the author's interviews with Birch Bayh, in twelve sessions from 2012 to 2015.

5. Congressional baseball game described in the *Washington Post*, July 31, 1963, page B7.

6. Letter from Harry Kingman to Edward Strong, chancellor of Berkeley, August 26, 1963, provided by Clay Risen to Adam Goodheart.

7. Lee Hamilton, interview with the author on May 28, 2015.

8. Description of celebrity and politics from Robert A. Caro, *The Passage of Power*, pages 52–53.

9. Story about Bayh taking Bobby Baker to Indianapolis 500 from Bill Wildhack, "Baker at 500 Race as Guest of Bayh," *Indianapolis News*, January 23, 1964.

10. Bobby Baker hearings described in Robert A. Caro, *The Passage of Power*, page 287.

11. Bobby Baker, *Wheeling and Dealing*, pages 68–69, 94–95, 100.

12. Bob Keefe, interview with the author on February 13, 2015.

13. Problems in the Senate office as described in Raymond H. Scheele, *Larry Conrad of Indiana*, page 31.

14. Description of Marvella's first year in Washington from Jane Howard, "Imagine Me Here in D.C.!" *Life*, July 26, 1963.

15. Experiences with the Kennedys described in Marvella Bayh and Mary Lynn Kotz, *Marvella*, pages 112–14.

16. Bob Keefe, interview with the author on February 13, 2015.

4. Crash and Constitution: 1964

1. "Bayh Heads Jaycees 10 Young Men of '63," *Fort Wayne Journal Gazette*, January 14, 1964.

2. History.com, "Slain Civil Rights Workers Found," http://www.history.com/this-day-in-history/slain-civil-rights-workers-found (accessed August 19, 2018).

3. Details of the Kennedy plane crash from Check-Six.Com, "The Luck of the Kennedys," http://www.check-six.com/Crash_Sites/Kennedy-N344S.htm (accessed June 23, 2014).

4. Ted Kennedy's recounting of the plane crash in Edward M. Kennedy, *True Compass*, pages 220–21.

5. Birch Bayh's memories of the crash, often different from Kennedy's, as well as events throughout 1964 are taken from the author's interviews with Bayh in twelve sessions from 2012 to 2015.

6. Bob Keefe, interview with the author on February 13, 2015.

7. Description of Bob Boxell and Larry Conrad experience at the hospital: Raymond H. Scheele, *Larry Conrad of Indiana*, page 37.

8. Patty Rees description of the aftermath of the plane crash as told in an email to the author.

9. Clay Risen, *The Bill of the Century*, pages 1–2.

10. "Memorandum for the President," June 29, 1964, provided to author by Bill Moyers.

11. Indiana primary detail from Theodore H. White, *The Making of the President 1964*, page 24.

12. Sen. Ernest Gruening quote from "Senate Debate on the Gulf of Tonkin Resolution (1964)," Alpha History, https://alphahistory.com/vietnamwar/senate-debate-gulf-of-tonkin-re/ (accessed October 26, 2017).

13. Birch Bayh, *One Heartbeat Away*, foreword and preface.

14. Birch Bayh, *One Heartbeat Away*, page 83.

15. Birch Bayh's description of Vice President Nixon during the Twenty-Fifth Amendment deliberations is taken from the author's interviews with Bayh in twelve sessions from 2012 to 2015.

16. Kenneth Crawford, "Blessings on Bayh," *Newsweek*, October 26, 1964.

17. Don Bacon, "Freshman Senator Succeeds Where Predecessors Failed, Writes Presidential Succession Amendment," *Jersey Journal*, February 10, 1965.

18. "A Feather for Sen. Bayh," *Kokomo Tribune*, February 13, 1967, page 4.

19. Association for Diplomatic Studies and Training, "Al Haig and the Reagan Assassination Attempt—'I'm In Charge Here,'" March 2014, http://adst.org/2014/03/al-haig-and-the-reagan-assassination-attempt-im-in-charge-here.

5. Civil Rights, Guns, and Vietnam: 1965–68

1. Lincoln's message to Congress in special session, July 4, 1861.

2. Birch Bayh stories about the prayer amendment and the events of 1965–68 are taken from the author's interviews with him in twelve sessions from 2012 to 2015.

3. Birch Bayh telephone conversation with President Johnson from Michael Beschloss, *Reaching for Glory: Lyndon Johnson's Secret White House Tapes, 1964–1965*, pages 313–14.

4. Jay Berman, interview with the author on March 12, 2015.

5. LBJ conversation with President Eisenhower as told by Andrew Goodpaster in the KPBS-TV documentary *25 Years of the Presidency*, 1986.

6. Tom Littlewood, "Indiana's Bayh Hits Peak on Capital Prestige Scale," *Chicago Sun-Times*, October 10, 1965, page 26.

7. Robert N. Branson, "Freshman Sen. Bayh 'Like a Son' to President Johnson," *Lafayette Journal and Courier*, October 2, 1965.

8. "Congress: A Crop of Bright Young Men," *Newsweek*, November 8, 1965.

9. Events of the years 1965–66 were found on Onthisday.com.

10. Robert P. Mooney, "Bayh's Future Is Rosy: Pastore," *Indianapolis Star*, June 5, 1966.

11. Thomas Paine and Abigail Adams quotations in Jill Lepore, *Book of Ages*, pages 182–84.

12. Ibid.

13. Abraham Lincoln's support for female suffrage in Harold Holzer and Norton Garfinkle, *A Just and Generous Nation*, page 23.

14. http://msmagazine.com/blog/2014/05/07/wheels-of-change-the-bicycle-and-womens-rights/comment-page-1.

15. Senator Dirksen as quoted in E. W. Kenworthy, "Dirksen Dead in Capital at 73," *New York Times*, September 8, 1969, http://movies2.nytimes.com/learning/general/onthisday/bday/0104.html (accessed August 20, 2015).

16. Description of LBJ's job offer to Marvella from Marvella Bayh and Mary Lynn Kotz, *Marvella*, pages 159–62.

17. Allan Rachle, interview with the author on May 27, 2015.

18. Birch Bayh comment on *Miranda* decision: Marguerite Davis, "Protect the Law-Abiding Citizens," *News American*, March 19, 1967, page 21A.

19. John Glenn's comments about Gus Grissom from a conversation Glenn had with the author in 1978.

20. Discussion about Gary and election of Mayor Hatcher from Rich James, "Gary Should Honor Richard Hatcher," Howey Politics Indiana blog, September 19, 2012.

21. Events of the year 1967 were found on Onthisday.com.

22. Antiwar demonstrations at University of Wisconsin in David Maraniss, *They Marched into Sunlight*, page 477.

6. 1968

1. Lincoln's desire as expressed in Joshua Wolf Shenk, *Lincoln's Melancholy* (Houghton Mifflin Harcourt, 2005), page 190.

2. Stories and Birch Bayh quotations from 1968 are taken from the author's interviews with Birch Bayh in twelve sessions from 2012 to 2015.

3. Robert Vaughn's endorsement: Jack Doyle, "1968 Presidential Race," Pop History Dig, August 14, 2008, http://www.pophistorydig.com/topics/1968-presidential-race-democrats.

4. Tony Podesta quote about McCarthy: Jack Doyle, "1968 Presidential Race," Pop History Dig, August 14, 2008, http://www.pophistorydig.com/topics/1968-presidential-race-democrats.

5. Gordon Alexander, interview with the author on July 1, 2015.

6. Celebrities taking part in 1968 presidential campaign: Jack Doyle, "1968 Presidential Race," Pop History Dig, August 14, 2008, http://www.pophistorydig.com/topics/1968-presidential-race-democrats.

7. Description of Robert Kennedy's task in the Indiana primary: Ray E. Boomhower, *Robert F. Kennedy and the 1968 Indiana Primary*, page 6.

8. McCarthy comment from *St. Louis Post-Dispatch*, June 3, 1968, page 23.

9. "Statement by Senator Robert F. Kennedy on the Death of the Reverend Martin Luther King, Rally in Indianapolis, April 4, 1968," John Fitzgerald Kennedy Library transcript as quoted in Raymond H. Scheele, *Larry Conrad of Indiana*, pages 44–45.

10. Diane Meyer, interview with the author on August 6, 2015.

11. Ruckelshaus positions in Senate race: Brian Haycock, "Ruckelshaus Visits BSU," *Ball State University Daily News*, October 23, 1968, page 1.

12. "Brief History of Chicago's 1968 Democratic Convention," CNN, 1997, http://www.cnn.com/ALLPOLITICS/1996/conventions/chicago/facts/chicago68/index.shtml.

13. LBJ's announcement of the Great Society: Michael Beschloss, *Taking Charge: The Johnson White House Tapes, 1963–1964*.

14. Events of the year 1968 were found on Onthisday.com.

15. Allan Rachles, interview with the author on May 27, 2015.

16. Gordon Alexander, interview with the author on July 1, 2015.

17. Kenneth Ikenberry, "Bayh Emerges Only Victor in Bitter Court Fights," *Washington Star*, April 19, 1970, page B-3.

18. Events of the year 1968 were found on Onthisday.com.

7. Haynsworth: 1969–70

1. Message to New York State Legislature, February 28, 1929.

2. Richard M. Nixon, *RN: The Memoirs of Richard Nixon*, page 245

3. Theodore H. White, *The Making of the President 1968*, page 474.

4. Andrew Jackson's feelings about the Electoral College: Robert V. Remini, *The Life of Andrew Jackson*, page 209.

5. "The Nixon Doctrine Is Announced," History.com, http://www.history.com/this-day-in-history/the-nixon-doctrine-is-announced (accessed August 26, 2018).

6. The Fortas nomination: Laura Kalman, *Abe Fortas* (Yale University Press, 1990).

7. Ibid.

8. John P. Frank, *Clement Haynsworth, the Senate, and the Supreme Court,* pages 4 and 6.

9. Kevin Phillips quotation: James Boyd, "Nixon's Southern Strategy: 'It's All in the Charts,'" *New York Times,* May 17, 1970.

10. Events of the year 1969 were found on the Onthisday.com.

11. Judge Parker's nomination rejection: Stephen L. Wasby and Joel B. Grossman, "Judge Clement F. Haynsworth, Jr.: New Perspective on his Nomination to the Supreme Court," *Duke Law Journal,* February 1990, page 75.

12. Haynsworth quotation about corporate directorships in Kenneth Auchincloss, "The Nomination is Rejected," *Newsweek,* December 1, 1969, page 22.

13. Haynsworth decision on collective bargaining: John P. Frank, *Clement Haynsworth, the Senate, and the Supreme Court,* page 20.

14. Stephen L. Wasby and Joel B. Grossman, "Judge Clement F. Haynsworth, Jr.: New Perspective on his Nomination to the Supreme Court," *Duke Law Journal,* February 1990, page 78.

15. Alfonso A. Narvaez, "Clement Haynsworth Dies at 77: Lost Struggle for High Court Seat," *New York Times,* November 23, 1989, http://www.nytimes.com/1989/11/23/obituaries/clement-haynsworth-dies-at-77-lost-struggle-for-high-court-seat.html, accessed November 2, 2017.

16. "A sort of laundered segregationist" in ibid.

17. Grace Lines case and Sen. Robert Griffin's opposition: John P. Frank, *Clement Haynsworth, the Senate, and the Supreme Court,* pages 70–71.

18. Richard Harwood, "Haynsworth's Defeat," *Washington Post,* November 23, 1969, pages A1 and A7.

19. Alfonso A. Narvaez, Clement Haynsworth Dies at 77; Lost Struggle for High Court Seat," *New York Times,* November 23, 1989, http://www.nytimes.com/1989/11/23/obituaries/clement-haynsworth-dies-at-77-lost-struggle-for-high-court-seat.html, accessed November 2, 2017.

20. Quotations by and about Sen. Everett Dirksen from E. W. Kenworthy, "Dirksen Dead in Capital at 73," *New York Times,* September 8, 1969, http://movies2.nytimes.com/learning/general/onthisday/bday/0104.html, accessed on August 20, 2015.

21. AZ Quotes, "Everett Dirksen Quotes," http://www.azquotes.com/author/3997-Everett_Dirksen (accessed August 26, 2018).

22. Sen. Robert Griffin's opposition: John P. Frank, *Clement Haynsworth, the Senate, and the Supreme Court,* pages 70–71.

23. Sen. James Eastland's pessimism: ibid., page 35.

24. George Meany joining the alliance against Haynsworth: ibid., page 30.

25. Birch Bayh's bill of particulars: ibid., pages 70–71.

26. Ibid.

27. Characterization of Sen. Strom Thurmond's support: ibid., pages 23 and 32.

28. Tom Connaughton, interview with the author on August 27, 2015.

29. "Top of the Week," *Newsweek,* December 1, 1969, page 3.

30. *Washington Post* urges withdrawal of nomination: John P. Frank, *Clement Haynsworth, the Senate, and the Supreme Court,* page 61.

31. Ibid., p. x.

32. Ibid., pages 92–93.
33. Tom Connaughton, interview with the author on August 27, 2015.
34. Agnew quotes from AZ Quotes (azquotes.com).

8. Carswell: 1970

1. Speech to the Institute of Journalists Dinner in London, England, November 17, 1906.
2. Bayh quotation about difficulty of taking on Carswell: Richard Harris, *Decision*, pages 12–13.
3. Jay Berman, interview with the author on March 12, 2015.
4. Bob Keefe, interview with the author on February 13, 2015.
5. Bayh as quoted in Richard Harris, *Decision*, pages 7–9.
6. Carswell as quoted in ibid., pages 14–16.
7. Republican senator describing the White House attitude: ibid., pages 11–12.
8. Carswell drawing up papers for whites-only covenant: ibid., page 18.
9. Tom Connaughton, interview with the author on August 27, 2015.
10. Bayh's speech at Statler Hilton to the Leadership Conference on Civil Rights: Richard Harris, *Decision*, pages 23–24.
11. "Let's crank it up," Richard Harris, *Decision*, page 24
12. First senators in opposition: Richard Harris, *Decision*, pages 23–24, 60.
13. Sen. Edward Brooke's deliberations: ibid., pages 14–18.
14. Horsky memo, ibid., page 134.
15. Carswell's racist remarks: John P. Frank, *Clement Haynsworth, the Senate, and the Supreme Court*, pages 102–103.
16. Sen. Roman Hruska comment: Richard Harris, *Decision*, pages 110–11.
17. News about Marvella's father: ibid., pages 168–69.
18. Background on Marvella's father: Geoff Paddock, "A Shining Example: The Journey of Marvella Hern Bayh," *Traces*, Fall 2013, page 49.
19. What Marvella and Birch experienced in Oklahoma: Marvella Bayh and Mary Lynn Kotz, *Marvella*, pages 122–23.
20. Final Carswell strategy: Richard Harris, *Decision*, pages 177–80, 189, 195–96.
21. Final Bayh speech before vote: Robert Barr, "Bayh—Could Be Household Word in White House Race," *Women's Wear Daily*, April 10, 1970, pages 1, 11.
22. Final drama in the Senate: Richard Harris, *Decision*, pages 200–202.
23. John P. Frank, *Clement Haynsworth, the Senate, and the Supreme Court*, pages 116–117, 135.
24. Sen. Marlow Cook and Sen. Bob Dole after the vote: Richard Harris, *Decision*, page 100, 173, 204.
25. Address of Waffle Shop: John De Ferrari, *Historic Restaurants of Washington, DC*, page 135.
26. Richard M. Nixon, *RN: The Memoirs of Richard Nixon*, pages 422–23.
27. Carswell's arrest in *Baltimore Sun* obituary, "G. Harrold Carswell, Nixon Nominee, Tallahassee, Florida," August 1, 1992; Joyce Murdoch and Deb Price, *Courting Justice: Gay Men and Lesbians v. the Supreme Court* (2002), page 187; UPI report reprinted in the *Sarasota Herald-Tribune*, September 12, 1979, page 23.

28. Chet Huntley statement provided by NBC news, April 17, 1970
29. Eric Severeid statement provided by CBS news, April 8, 1970
30. Kenneth Ikenberry, "Bayh Emerges Only Victor in Bitter Court Fights," *Washington Star*, April 19, 1970, page B-3.
31. Ibid.
32. Hamilton E. Davis, "Birch Bayh 'a Midwest JFK,'" *Providence (RI) Journal-Bulletin*, July 1, 1970, pages 1, 18.
33. "Killings at Jackson State University," *African American Registry*, May 14, 1970.
34. Leonard Shapiro, "GOP Continues Superiority in Off-Year Baseball, 6–4," *Washington Post*, June 25, 1970.
35. Sanford J. Ungar, "3 Living in Hut 'Found' Near Hill," *Washington Post*, July 3, 1970, pages B1, B6.
36. Women's rights timeline from Thoughtco.com.
37. Henry Giniger, "Nixon Peace Plan Assailed by Reds at Talks in Paris," *New York Times*, October 9, 1970, page 1.
38. Bayh effort to abolish Electoral College: John D. Morris, "Articulate Popular-Vote Advocate," *New York Times*, September 25, 1970.
39. Bayh effort to abolish Electoral College: "A Welcome Showdown in the Senate," *Washington Post*, September 23, 1970.
40. Bayh traveling across the country: Ken W. Clawson, "Bayh Is Busy Running 'Availability' Campaign," *Washington Post*, 1970 (n.d.).
41. Events in 1970 from "This Day in History," History.com, www.history.com/this-day-in-history.
42. "Who vs. Nixon in '72," *U.S. News & World Report*, November 16, 1970, pages 37–39.
43. "World Trade Center," History.com, http://www.history.com/topics/world-trade-center.

9. Campaign and Cancer: 1971

1. Speech at State Capitol, Albany, New York, September 29, 1960.
2. Marvella speaking about day care: Linda Citro, "How Mrs. Bayh Sees Her Job," *Philadelphia Bulletin*, February 10, 1971.
3. Elizabeth Bennett, "Protect a Tot, Save a Teen, Bayh Says," *Houston Post*, February 22, 1971.
4. Juvenile delinquency statistics from "Bayh at Panel Helm, Vows Aid for Youths," *Miami Herald*, January 2, 1971.
5. Democratic Party and quotas: Theodore H. White, *The Making of the President 1972*.
6. Birch Bayh comments about Agnew: Richard Rodda, "Sen. Bayh Puts Last Touches on California Democrats' Plans to Defeat Nixon," *Sacramento Bee*, January 25, page A4.
7. Events in 1971 and about Lt. Calley from This Day in History, www.history.com/this-day-in-history.
8. Bayh comments on Calley and tally of the mail received: William Chapman, "Bayh: Nixon Plays Politics on Calley," *Washington Post*, April 8, 1971, page A1.
9. Description of Bayh stance on Lt. Calley: United Press International, "Bayh Gambles Anew in Blasting Nixon Calley Case Action," *Fort Wayne Journal Gazette*, April 12, 1971.

10. Comments about Birch Bayh's candidacy and his comments to Richard Stout: "Birch Bayh's Bid," *Newsweek*, April 19, 1971.

11. David Broder, "Hustle, Detail, Money: Bayh Picking up Steam," *Washington Post*, May 10, 1971, pages A1, A8.

12. John Bartlett, "Martin Niemoeller," in *Bartlett's Familiar Quotations*, page 824.

13. Lee Hamilton, interview with the author on May 28, 2015.

14. Allan Rachles, interview with the author on May 27, 2015.

15. Bayh trip to Paris, Marvella's involvement with a presidential campaign, description of her cancer checkup, diagnosis, and surgery: Marvella Bayh and Mary Lynn Kotz, *Marvella*, pages 217–20.

16. Allan Rachles, interview with the author on May 27, 2015.

17. Gail Alexander, email sent to the author.

18. Allan Rachles, interview with the author on May 27, 2015.

19. Richard M. Nixon, *RN: The Memoirs of Richard Nixon*, page 423.

20. Description of the Gun Control Act of 1968: William Vizzard, "The Gun Control Act of 1968," newsletter of Jews for the Preservation of Firearms Ownership, pages 4–8.

21. S. 2507, Hearings before the Subcommittee to Investigate Juvenile Delinquency of the Committee on the Judiciary, United States Senate, Ninety-Second Congress, First Session, September 13 and 14, October 5 and 27, November 1, 1971, pages 26–29.

10. Title IX: 1972

1. From Winston Churchill, *The Story of the Malakand Field Force*, 1898, p. 41.

2. Alice Paul Institute, The Equal Rights Amendment, www.equalrightsamendment.org.

3. Rep. Howard Smith's actions on the Civil Rights Act: Louis Menand, "The Sex Amendment," *New Yorker*, July 21, 2014, pages 74–81.

4. The ERA being introduced in every congressional session: Clay Risen, *The Bill of the Century*, page 160.

5. Rep. Howard Smith's actions on the Civil Rights Act: Louis Menand, "The Sex Amendment," *New Yorker*, July 21, 2014, pages 74–81.

6. Clay Risen, *The Bill of the Century*, page 161.

7. Franklin Roosevelt Jr.'s comment from Louis Menand, "The Sex Amendment," *New Yorker*, July 21, 2014, pages 74–81.

8. Rep. Martha Griffiths and the ERA: ibid.

9. UVA founders from Medlibrary.org.

10. Wording of Title IX found on KnowYourIX.Org.

11. Lee Hamilton, interview with the author on May 28, 2015.

12. "History of Title IX," TitleIX.info, www.titleix.info/History/History-Overview.aspx (accessed August 30, 2018).

13. Mark Emmert and Bernice Resnick Sandler quotations from NCAA, *Sporting Chance: The Lasting Legacy of Title IX*, 2012.

14. Senate report on Kleindienst nomination: "ITT Dispute Fails to Block Kleindienst Confirmation," *CQ Almanac*, 1972, https://library.cqpress.com/cqalmanac/document.php?id=cqal72-1250367

15. Mathea Falco, "The Equal Rights Amendment: You've Come a Long Way, Baby," *Radcliffe Quarterly*, 1972, page 39.
16. Eastland story after the ERA as told to author by former staff member Joe Rees.
17. Frank Mankiewicz quotation from Bob Shrum, *No Excuses: Confessions of a Serial Campaigner*, page 36.
18. Sen. Ed Muskie incident in New Hampshire: Elisabeth Goodridge, "Front-Runner Ed Muskie's Tears (or Melted Snow) Hurt His Presidential Bid," *US News & World Report*, January 17, 2008.
19. Bayh advice to Larry Conrad on governor's race: Raymond H. Scheele, *Larry Conrad of Indiana*, pages 95–96.
20. Conversation between former senator Gary Hart and the author, October 17, 2007.
21. Events of the year 1972 were found on History.com.

11. Watergate: 1973

1. Roosevelt quotation from speech at the Groton School, Groton, MA, May 24, 1904.
2. "Timeline 1973," Timelines of History, http://timelines.ws/20thcent/1973.HTML (accessed August 31, 2018).
3. Birch Bayh, conversation with the author on a flight to Indiana.
4. McCord's letter as described in John J. Sirica, *To Set the Record Straight*, pages 88 and 96.
5. https://www.senate.gov/artandhistory/history/common/investigations/Watergate.htm
6. Ibid.
7. Sam Ervin story as told to the author by Bill Wise, who wrote a book about Ervin, *The Wisdom of Sam Ervin*.
8. Elizabeth Drew, *Washington Journal: The Events of 1973–1974*, pages 158, 160, 180.
9. Truman quotation from Merle Miller, *Plain Speaking*, pages 135 and 178.
10. John Dean conversation with Barry Goldwater: John Dean, *Blind Ambition*, page 299.
11. Media coverage as described in Facts on File, *Watergate and the White House*, volume 1, pages 96–97.
12. Peter Grier, "Howard Baker: The Real Story of His Famous Watergate Question," *Christian Science Monitor*, June 26, 2014, https://www.csmonitor.com/USA/Politics/Decoder/2014/0626/Howard-Baker-the-real-story-of-his-famous-Watergate-question (accessed August 31, 2018).
13. Enemies list quotation from Dredd Blog, blogdredd.blogspot.com, https://www.colorado.edu/AmStudies/lewis/film/enemies.htm (accessed April 5, 2016).
14. Elizabeth Drew, *Washington Journal: The Events of 1973–1974*, pages 50–51.
15. Ibid., pages 92–93.
16. Library of Congress, "War Powers," https://www.loc.gov/law/help/war-powers.php (accessed August 31, 2018).
17. David Kopel, "The Missing 18½ Minutes: Presidential Destruction of Incriminating Evidence," *Washington Post*, June 16, 2014, https://www.washingtonpost.com

/news/volokh-conspiracy/wp/2014/06/16/the-missing-18-12-minutes-presidential-destruction-of-incriminating-evidence/?utm_term=.012361fd7ef0

18. Diane Meyer Simon, interview with the author on August 6, 2015.

12. Bayh versus Lugar: 1974

1. Adlai Stevenson quotation from a 1948 speech in Bloomington, IL.
2. Cade Ware, "Rector Involved with the Young," *Roll Call*, March 1974, page 5.
3. John Rector, in an email to the author.
4. Discussion of Marvella Bayh's work with the American Cancer Society: Geoff Paddock, "A Shining Example: The Journey of Marvella Hern Bayh," *Traces*, Fall 2013, page 52.
5. Anne D. Neal, "Hot and Heavy Hoosiers," *Harvard Crimson*, October 5, 1974, https://www.thecrimson.com/article/1974/10/5/hot-and-heavy-hoosiers-pbpbeople-on/ (Accessed October 30, 2014).
6. Ibid.
7. Ibid.
8. Richard Lugar, interview with the author on June 3, 2015.
9. Larry Conrad story from Raymond H. Scheele, *Larry Conrad of Indiana*, page 171.
10. Louis Mahern, interview with the author on May 26, 2015.
11. Darry Sragow, interview with the author on August 7, 2015.
12. Elizabeth Drew, *Washington Journal: The Events of 1973–1974*, pages 358–82.
13. Ibid., pages 406–408.
14. Juvenile Justice Act description in email from John Rector, December 30, 2014.
15. https://www.washingtonpost.com/opinions/stop-mistreating-youth-at-rikers-island-and-other-institutions-around-the-country/2014/12/21/3b50cc1e-87c6-11e4-b9b7-b8632ae73d25_story.html
16. "Kids and Jails a Bad Combination," *New York Times*, December 29, 2014, page A16.
17. "Juvenile Justice and Delinquency Prevention Act," Coalition for Juvenile Justice, http://www.juvjustice.org/federal-policy/juvenile-justice-and-delinquency-prevention-act (accessed August 31, 2018).
18. Tom Connaughton, interview with the author on August 27, 2015.
19. Anne D. Neal, "Hot and Heavy Hoosiers," *Harvard Crimson*, October 5, 1974, https://www.thecrimson.com/article/1974/10/5/hot-and-heavy-hoosiers-pbpbeople-on/ (accessed October 30, 2014).
20. Bob Dole comment provided by Tom Connaughton in an interview with the author on August 27, 2015.
21. Bill Moreau, interview with the author on May 26, 2015.
22. Jules Witcover, "Sen. Bayh's Style Frustrates Underdog Challenger," *Washington Post*, October 23, 1974.
23. "Marvella's Last Minute Appeal" from Geoff Paddock, "A Shining Example: The Journey of Marvella Hern Bayh," *Traces*, Fall 2013, page 52.
24. Lessons from the 1974 campaign provided by Richard Lugar in an interview with the author on June 3, 2015.

25. Marlow Cook, "A Former Republican Senator for Kerry," *Louisville Courier-Journal*, October 20, 2004.

26. Presidential Recordings and Materials Preservation Act of 1974 described by Frank Sullivan; interview with John Brademas, Vol. II, September 2–3, 2002, pages 94–95. Frank Sullivan was Chief of Staff to Brademas for many years, interviewed him extensively, and provided the author with the transcript of the interview.

27. Airplane crash near Dulles as described in Marvella Bayh and Mary Lynn Kotz, *Marvella*, page 24.

13. National Interests: 1975

1. From a conversational exchange with Harold Begbie as cited in Harold Begbie, *Master Workers* (Begbie, Methuen & Co., 1906), page 177.

2. Jules Witcover, *Marathon: The Pursuit of the Presidency 1972–1976*, pages 127–28.

3. United States Senate, "Filibuster and Cloture," https://www.senate.gov/artandhistory/history/common/briefing/Filibuster_Cloture.htm (accessed September 1, 2018).

4. Letter to the editor written by Birch Bayh, "The Gun Bill: 'A Workable Compromise,'" *Washington Post*, May 17, 1976, page A20.

5. "Give Victims of Crime a Break: Pass the Birch Bayh Crime Bill," *Philadelphia Tribune*, November 8, 1975, pages 1-A and 14-A.

6. The beginning of the presidential campaign as described by Jules Witcover, *Marathon: The Pursuit of the Presidency 1972–1976*, pages 152–54.

7. Ibid., page 177.

8. Ibid., pages 152–54.

9. Jim Friedman, interview with the author on July 5, 2015.

10. Jay Berman, interview with the author on March 12, 2015.

11. Barbara Dixon, telephone conversation with the author on September 28, 2015.

12. Barry Hager, "Appealing to the Old Democratic Coalition," *Congressional Quarterly*, November 1975, pages 17–21.

13. "Thomas R. Marshall Quotes," AZ Quotes, www.azquotes.com/author/43654-Thomas_R_Marshall (accessed September 1, 2018).

14. NDC convention as described by Jules Witcover, *Marathon: The Pursuit of the Presidency 1972–1976*, pages 187–88.

15. Characterization of Bayh's presidential campaign: David Alpern and Hal Bruno, "Free-For-All," *Newsweek*, January 12, 1976, pages 20–21.

16. Attending gay event: Roger Ricklefs, "A New Constituency: Political Candidates Seek Out Gay Votes," *Wall Street Journal*, October 20, 1976, page 35.

14. Bayh for President: 1976

1. Adlai E. Stevenson, address at Princeton University, "The Educated Citizen," March 22, 1954.

2. Joe Klein, "Indecision 76," *Rolling Stone*, March 11, 1976, page 7.

3. The Bayh effort in New Hampshire as described by Jules Witcover, *Marathon: The Pursuit of the Presidency 1972–1976*, pages 228–31, 234–35.

4. Harold Ickes, interview with the author on November 20, 2015.

5. Kennedy's decision not to run as described by Jules Witcover, *Marathon: The Pursuit of the Presidency 1972–1976*, page 132.

6. Ibid., pages 249–50.

7. David Rubenstein, interview with the author on July 24, 2015.

8. Bayh strategy as described by Elizabeth Drew, *American Journal: The Events of 1976*, page 67.

9. Jackson's New York victory as described by Jules Witcover, *Marathon: The Pursuit of the Presidency 1972–1976*, page 295.

10. Efforts to stop Carter as described by Jules Witcover, *Marathon: The Pursuit of the Presidency 1972–1976*, pages 286–309.

11. Ibid., page 327.

12. Carter's claim that he never asked for an endorsement: Elizabeth Drew, *American Journal: The Events of 1976*, page 174.

13. Description of sitting senators running for president by Ira Shapiro, *The Last Great Senate*, page 40.

14. Surrogates for Carter as described by Jules Witcover, *Marathon: The Pursuit of the Presidency 1972–1976*, page 521.

15. David Rubenstein, interview with the author on July 24, 2015.

16. Robert Scheer, "The Playboy Interview: Jimmy Carter," *Playboy*, November 1976, Vol. 23, Issue. 11, pages 63–86.

17. David Rubenstein, interview with the author on July 24, 2015.

18. Orrin Hatch talked about his memories of Birch Bayh in an interview with the author on December 14, 2015.

19. Jack Eisen, "2 Senators Fail in Attempt To Revive Dying Witness," *Washington Post*, July 22, 1977.

20. Sen. Howard Baker's and Sen. Robert Byrd's remarks: Congressional Record, May 15, 1980, page 11360.

21. Birch Bayh as quoted by Abby Saffold in an email message to the author.

22. Events of 1976 from History.com.

15. The Carter Administration: 1977

1. John F. Kennedy, "Special Message to the Congress on Conflict-of-Interest Legislation and on Problems of Ethics in Government," April 27, 1961.

2. Description of the Indiana ERA battle by Beth Van Vorst Gray, "Equal Rights Amendment Ratification in Indiana," Broad Ripples in the Water (blog), March 19, 2010, http://broadripplesinthewater.blogspot.com/2010/03/equal-rights-amendment-ratification-in.htm (accessed April 5, 2015).

3. Draft dodging described by Andrew Glass, "Carter Pardons Draft Dodgers Jan. 21, 1977," Politico, January 21, 2008, https://www.politico.com/story/2008/01/carter-pardons-draft-dodgers-jan-21-1977-007974 (accessed April 10, 2015).

4. Carter versus Congress from Ira Shapiro, *The Last Great Senate*, pages 43–44.

5. Lee Hamilton, interview with the author on May 28, 2015.
6. The importance of the Senate in the sixties and seventies as described by Ira Shapiro, *The Last Great Senate*, pages 43–44.
7. Eve Lubalin's account of the filibuster provided in writing to the author.
8. Events of 1977 from History.com.

16. Foreign Intelligence: 1978

1. Theodore Roosevelt quotation from "The New Nationalism" speech, delivered at the dedication of the John Brown Memorial Park in Osawatomie, Kansas, on August 31, 1910.
2. Hubert Humphrey's reaction to Gov. Dukakis's comments about him as described by Bob Shrum, *No Excuses: Confessions of a Serial Campaigner*, page 60.
3. Description of Marvella's new medical crisis early in 1978 by Geoff Paddock, "A Shining Example: The Journey of Marvella Hern Bayh," *Traces*, Fall 2013, pages 53–55.
4. Ibid.
5. https://www.oyez.org/cases/1975/74-1589
6. "The Pregnancy Discrimination Act of 1978," US Equal Employment Opportunity Commission, https://www.eeoc.gov/laws/statutes/pregnancy.cfm (accessed September 3, 2018).
7. Marvella's new health crisis in May 1978 as described by Geoff Paddock, "A Shining Example; The Journey of Marvella Hern Bayh," *Traces*, Fall 2013, page 55.
8. Barbara Dixon's role in the ERA extension effort as described in a telephone conversation with the author on September 28, 2015.
9. Congressional action in 1978 from Congress.gov.
10. Birch Bayh's first reactions to Tongsun Park allegations in "Bayh Denies Park Aided Campaign," *Chicago Tribune*, March 30, 1978, page 14.
11. Associated Press, "Senate Ethics Panel Suspects Birch Bayh of Taking Illegal Gifts," *Cornell Daily Sun*, Volume 95, Number 34, October 17, 1978.
12. Events of 1978 from History.com.

17. The Death of Marvella: 1979–80

1. Thomas H. Huxley, "On the Natural Inequality of Men," in *Collected Essays*, vol. 1 (1893).
2. Emergence of the New Right from Ira Shapiro, *The Last Great Senate*, page 225.
3. Sen. Orrin Hatch's characterization of Strom Thurmond's reaction to the Civil Rights of the Institutionalized legislation as told to the author in an interview on December 14, 2015.
4. Nels Ackerson's description of Pat Wald's request and the quote about Sen. Orrin Hatch as told by him to the author.
5. Events of 1979 from History.com.
6. Tom Connaughton's description of Civil Rights of the Institutionalized from an interview with the author on August 27, 2015.

7. The last days of Marvella Bayh's life as described by Geoff Paddock, "A Shining Example: The Journey of Marvella Hern Bayh," *Traces*, Fall 2013, page 55.
8. Ibid.
9. Ibid.
10. Ibid.
11. Evan Bayh's memories about his mother and her death as told to the author in an interview on July 16, 2015.
12. Events of 1979 from History.com.
13. Conversation with Jack Watson recounted by Lew Borman in a memo written for the author about his staff experiences.
14. Bill Moreau, interview with the author on May 26, 2015.
15. Events of 1979 and 1980 from History.com.

18. The Last Campaign: 1980

1. Adlai Stevenson, speech in Fresno, California, October 11, 1956.
2. Bob Shrum, *No Excuses: Confessions of a Serial Campaigner*, page 89.
3. Fred Nation, interview with the author, May 26, 2015.
4. Chris Aldridge's description of the Irina McClellan case in an email message to the author.
5. "Miracle on Ice" as described in James S. Hirsch, *Willie Mays: The Life, the Legend*, page 134.
6. Teamsters investigation turned over to Ethics Committee, but Justice Department tells Bayh and Cannon they are not under investigation: Frank Reynolds, "Cannon and Bayh Investigation," *ABC Evening News*, February 6, 1980.
7. First description of charges relating to senators Cannon and Bayh: Del Tartikoff, "FBI Wiretaps Brought Down Nevada's Senator Cannon," Nevada Journal Online, 1996, http://www.newsnet1.com/electricnevada.com/pages96/mob3.htm (accessed October 31, 2015).
8. Herb Simon, interview with the author on August 6, 2015.
9. Gun Owners Foundation, "Letter to Hon. G. R. Dickerson from Sen. Bayh," October 15, 1979.
10. Bayh hearings on instant background checks for gun ownership as described by Howard Kohn, "Inside the Gun Lobby," *Rolling Stone*, May 14, 1981, page 9.
11. Events of 1980 from History.com.
12. Russ, "The Monetary Problem with the Fiscal Problem," Eagle Watch, October 20, 2010, http://www.eaglewatchonline.net/2010/10/20/the-monetary-problem-with-the-fiscal-problem
13. Description of the misery index by Harold Holzer and Norton Garfinkle, *A Just and Generous Nation*, page 250.
14. David S. Broder and Bob Woodward, *The Man Who Would Be President: Dan Quayle*, page 48.
15. Dan Quayle's score on the National Guard test from "Quayle and Paula Parkinson," *Orlando Sentinel*, August 24, 1988, http://articles.orlandosentinel.com/1988-08-24/news/0060270264_1_quayle-parkinson-guard (accessed on July 8, 2015).

16. "Congress Overrides Veto on Oil Import Fee," *Congressional Quarterly Almanac*, 1980.

17. Description of Bayh's role in Billy Carter affair by Ira Shapiro, *The Last Great Senate*, pages 327–37.

18. Ibid.

19. Ibid., page 339.

20. Terry Dolan as quoted in Michael Vernetti, *Senator Howard Cannon of Nevada: A Biography*, page 202.

21. Ibid.

22. Terry Dolan as quoted in Paul Houston, "Terry Dolan, 36, Chairman of Conservative Lobby, Dies," *New York Times*, December 31, 1986, http://articles.latimes.com/1986-12-31/news/mn-1478_1_terry-dolan (accessed November 1, 2015).

23. Terry Dolan's homosexuality as described by Frank Rich, "Just How Gay Is the Right?" *New York Times*, May 15, 2005, https://www.nytimes.com/2005/05/15/opinion/just-how-gay-is-the-right.html (accessed November 1, 2015).

24. American Enterprise Institute, *The American Elections of 1980*, edited by Austin Ranney, page 271.

25. Mark Miles quotation in Judith Miller, "Billy Carter Investigation Hobbles Bayh Re-Election Drive in Indiana," *New York Times*, September 17, 1980, page B14.

26. Mary McGrory, "Damage Control Officer," *Boston Globe*, August 3, 1980, page A7.

27. Quayle debate quotes in Edward Wills Jr., "Bayh, Quayle Lock Horns in Debate," *Indianapolis Star*, September 15, 1980, page 1.

28. David S. Broder and Bob Woodward, *The Man Who Would Be President: Dan Quayle*, page 193.

29. Poll showing Bayh behind in James G. Newland Jr., "Poll Gives Quayle Commanding Lead," *Indianapolis Star*, October 14, 1980, page 1.

30. David Rubenstein, interview with the author on July 24, 2015.

31. Patrick T. Morrison and Patrick J. Traub, "Senate Campaign Strategies Leave Bayh–Quayle Race Too Close to Call," *Indianapolis Star*, November 2, 1980, section 5, page 1.

32. Tim Minor's memories provided in writing to the author.

33. John Brademas comment from Frank Sullivan, made during Sullivan's remarks on the occasion of the Ball State University 2002 Stephen J. Senior and Beatrice Brademas Memorial Lecture, November 13, 2002, page 3.

34. Tom Connaughton, interview with the author on August 27, 2015.

35. Ibid.

36. Peter Hart, conversation with the author shortly after the 1980 election.

37. "United Negro College Fund," Ad Council, https://www.adcouncil.org/Our-Campaigns/The-Classics/United-Negro-College-Fund (accessed September 3, 2018).

19. Capstone

1. Patrick Leahy, interview with the author, December 8, 2015.

2. "Hyman G. Rickover," Atomic Heritage Foundation, https://www.atomicheritage.org/profile/hyman-g-rickover (accessed September 3, 2018).

3. Bayh statement on the introduction of the Bayh-Dole act from a press release, "News from Birch Bayh," Washington, DC, September 13, 1978.

4. Steve Clemons, "Birch Bayh: What Leadership Ought to Look Like," Huffington Post, August 7, 2008, http://www.huffingtonpost.com/steve-clemons/birch-bayh-what-leadershi_b_117513.html.

5. "Bayh-Dole Act," Association of University Technology Managers (AUTM), June 19, 2015. https://autm.net/about-tech-transfer/advocacy/legislation/bayh-dole-act/ (accessed June 10, 2015).

6. Joseph Allen, "The Enactment of Bayh–Dole, an Inside Perspective," IP Watchdog, November 28, 2010, http://ipwatchdog.com/2010/11/28/the-enactment-of-bayh-dole-an-inside-perspective.

7. Ibid.

8. "Innovation's Golden Goose," *Economist Technology Quarterly*, December 12, 2002.

9. Ira Shapiro, *The Last Great Senate*, page 357.

10. Norman J. Latker, "Founding Fathers—Senator Robert Dole—Biography," Bayh DoleCentral.Org, University of New Hampshire Law School, https://ipmall.law.unh.edu/content/bayh-dole-act-research-history-central-founding-fathers-norman-j-latker (accessed October 29, 2015).

11. A. S. Rao, "The Economic Contribution of University/Nonprofit Inventions in the United States: 1996–2013," India Invents, March 20, 2015, http://indiainvents.blogspot.com/2015/03/the-economic-contribution-of.html.

12. Ibid.

13. Statistics provided by Biotech-now.org and by Joseph Allen in an article he authored with Sen. Birch Bayh, "The Bayh-Dole Act: 35 Years of Public Benefit," and from "Products and Drugs: A 2012 Report," Association of University Technology Managers (AUTM).

14. Joseph Allen, "Bayh–Dole at 35: Lauded in Kazakhstan, Dissed in Boston," IP Watchdog, December 21, 2015, http://www.ipwatchdog.com/2015/12/21/bayh-dole-at-35-lauded-in-kazakhstan-dissed-in-boston.

15. Bill Arceneaux quotation from an email he sent to the author.

16. "Evan Bayh: Senator," *People*, May 10, 1999, https://people.com/archive/evan-bayh-senator-vol-51-no-17/ (accessed November 5, 2015).

17. Evan Bayh, interview with the author on July 16, 2015.

18. Bayh's visit to the University of Connecticut as described by Richard Veilleux, "'Father' of Title IX Honored," *UConn Advance*, January 20, 2004.

19. "Evan Bayh to deliver inaugural talk in lecture series named for his father," Indiana University, http://newscenter.iupui.edu/index.php?id=5815 (accessed September 3, 2018).

20. Melissa Isaacson, "Birch Bayh: A Senator Who Changed Lives," ESPN Chicago, May 3, 2012.

21. Birch Bayh quotations are taken the author's interviews with Birch Bayh in twelve sessions from 2012 to 2015.

22. "Report Shows Academia–Industry Technology Transfer Contributed up to $1.18 Trillion to U.S. Economy," *Business Wire*, March 18, 2015, https://www.bio.org/media/press-release/report-shows-academia-industry-technology-transfer-contributed-118-trillion-us-e.

23. http://westminstercollege.com/titleix/?detail=16259&parent=16244.

24. A portion of the Title IX summary from Susan Estrich, "37 Words and Birch Bayh's Courage," *Star Democrat* (Easton, MD), June 22, 2012, page A8.

25. The impact of Washington DC's subway: "Metro Debuts," *Washingtonian* 50th Anniversary Issue, October 2015, page 62.

26. Lynne Mann quotations provided in writing to the author.

27. John Rector quotations provided in writing to the author.

28. Abby Saffold quotations provided in writing to the author.

29. Louis Mahern, interview with the author on May 26, 2015.

30. Jeff Smulyan, interview with the author on May 26, 2015.

31. David Rubenstein quotations from an interview with the author on July 24, 2015.

32. Allan Rachles, interview with the author on May 27, 2015.

33. Diane Meyer, interview with the author on August 6, 2015.

34. Tom Connaughton, interview with the author on August 27, 2015.

35. Ray Scheele as quoted in Kate Cruikshank, *The Art of Leadership: A Companion to an Exhibition from the Senatorial Papers of Birch Bayh*, Lilly Library, Indiana University, January 29–May 5, 2007, page 75.

36. Lee Hamilton, interview with the author on May 28, 2015.

37. Richard Lugar, interview with the author on June 3, 2015.

38. Patrick Leahy, interview with the author on December 8, 2015.

39. Henry Clay as quoted in Lincoln Caplan, "The White-Supremacist Lineage of a Yale College," *Atlantic*, October 5, 2015, page 8.

40. Quotation about George H. W. Bush from Jon Meacham, *Destiny and Power: The American Odyssey of George Herbert Walker Bush*, as quoted in Ruth Marcus, "What Every President Should Learn from Bush 41," *Washington Post*, November 29, 2015, page 25.

Bibliography

Baker, Bobby, and Larry L. King. *Wheeling and Dealing.* New York: W. W. Norton, 1978.
Baker, Ross K. *Friend and Foe in the U.S. Senate,* New York: Free Press, 1980.
Bartlett, John. *Bartlett's Familiar Quotations.* Boston: Little, Brown, 1980.
Bayh, Birch. *One Heartbeat Away.* Indianapolis, IN: Bobbs-Merrill, 1968.
Bayh, Marvella, and Mary Lynn Kotz. *Marvella.* New York: Harcourt Brace Jovanovich, 1979.
Beschloss, Michael. *Taking Charge: The Johnson White House Tapes 1963–1964* (audio CD). New York: Simon & Schuster, 1998.
———. *Reaching for Glory: Lyndon Johnson's Secret White House Tapes, 1964–1965.* New York: Simon & Schuster, 2002.
Broder, David S. and Woodward, Bob, *The Man Who Would Be President: Dan Quayle,* Simon & Schuster, 1992.
Boomhower, Ray E. *Robert F. Kennedy and the 1968 Indiana Primary.* Bloomington, IN: Indiana University Press, 2008.
Caro, Robert A. *The Passage of Power.* New York: Alfred A. Knopf, 2012.
Cruikshank, Kate. *The Art of Leadership: A Companion to an Exhibition from the Senatorial Papers of Birch Bayh.* Bloomington, IN: Lilly Library, Indiana University, 2007.
Dallek, Robert. *Camelot's Court.* New York: HarperCollins, 2013.
De Ferrari, John. *Historic Restaurants of Washington, DC.* Charleston, SC: American Palate, 2013.
Dean, John W. III. *Blind Ambition.* New York: Pocket Books, 1976.
Drew, Elizabeth. *Washington Journal: The Events of 1973–1974.* New York: Macmillan, 1974.
———. *American Journal: The Events of 1976.* New York: Vintage Books, 1976.
Frank, John P. *Clement Haynsworth, the Senate, and the Supreme Court.* Charlottesville: University of Virginia Press, 1991.
Gardner, Gerald, editor. *The Quotable Mr. Kennedy.* New York: Eagle Books, 1962.
Harris, Richard. *Decision.* New York: E. P. Dutton, 1970.
Holzer, Harold, and Norton Garfinkle. *A Just and Generous Nation.* New York: Basic Books, 2015.
Humes, James C. *The Wit & Wisdom of Winston Churchill.* New York: Harper Perennial, 1995.
Kennedy, Edward M. *True Compass: A Memoir,* New York: Hachette Book Group, 2009.
Lepore, Jill. *Book of Ages.* New York: Alfred A. Knopf, 2013.

Maraniss, David. *They Marched into Sunlight*. New York: Simon & Schuster Paperbacks, 2003.

McGinniss, Joe. *Heroes*. New York: Viking, 1976.

Meyersohn, Maxwell, editor. *The Wit and Wisdom of Franklin D. Roosevelt*. Boston: Beacon Press, 1950.

Miller, Merle. *Plain Speaking*. New York: Berkley, 1973.

Morris, Edmund. *Colonel Roosevelt*. New York: Random House, 2010.

National Collegiate Athletic Association (NCAA). *Sporting Chance: The Lasting Legacy of Title IX*. NCAA, 2012, http://www.ncaa.org/about/resources/inclusion/sporting-chance-lasting-legacy-title-ix

Nixon, Richard M. *RN: The Memoirs of Richard Nixon*. Norwalk, CT: Easton Press, 1988.

Paddock, Geoff. *Indiana Political Heroes*. Indianapolis: Indiana Historical Society Press, 2008.

Pickett, William B. *Homer E. Capehart: A Senator's Life, 1897–1979*. Indianapolis: Indiana Historical Society Press, 1990.

Ranney, Austin, editor. *The American Elections of 1980*. Washington, DC: American Enterprise Institute, 1981.

Remini, Robert V. *The Life of Andrew Jackson*. New York: Penguin, 1990.

Risen, Clay. *The Bill of the Century* New York: Bloomsbury Press, 2014.

Sautter, Chris. *Hey Look Him Over*. DVD. Sautter Films, 2005.

Scheele, Raymond H. *Larry Conrad of Indiana*. Bloomington: Indiana University Press, 1997.

Shapiro, Ira. *The Last Great Senate*. Public Affairs, 2012.

Shenk, Joshua Wolf. *Lincoln's Melancholy*. New York: Houghton Mifflin, 2005.

Shrum, Bob. *No Excuses: Confessions of a Serial Campaigner*. New York: Simon & Schuster, 2007.

Sidey, Hugh. *John F. Kennedy, President*. New York: Atheneum, 1963.

Sorenson, Ted, *Kennedy*. New York: Harper & Row, 1965.

———. *Counselor*. New York: HarperCollins, 2008.

Stoner, Andrew E. *Legacy of a Governor: The Life of Indiana's Frank O'Bannon*. Bloomington, IN: Rooftop Publishing, 2006.

Vernetti, Michael. *Senator Howard Cannon of Nevada: A Biography*. Reno: University of Nevada Press, 2015.

White, Ronald C. Jr. *The Eloquent President* New York: Random House, 2005.

White, Theodore H. *The Making of the President 1960*. New York: Atheneum, 1961.

———. *The Making of the President 1964*. New York: Atheneum, 1965.

———. *The Making of the President 1972*. New York: Atheneum, 1973.

Wise, Bill M. *The Wisdom of Sam Ervin*, New York: Ballantine Books, 1973.

Witcover, Jules, *Marathon: The Pursuit of the Presidency 1972–1976*, New York: Viking, 1977.

———. *Joe Biden: A Life of Trial and Redemption*. New York: William Morrow, 2010.

Index

Note: Italicized page numbers represent photographs.

ROBERT BLAEMIRE began working for Senator Birch Bayh while a freshman in college and remained on his staff for the next thirteen years. After the Bayh election defeat in 1980, Blaemire formed a political action committee, the Committee for American Principles, to help combat the negative influence of the New Right in American politics. In 1982, he began a long career providing political computer services for Democratic candidates and progressive organizations. An early participant in building a business that has grown in prominence in politics—the rise of big data—he owned and managed his own company, Blaemire Communications, for seventeen years. Born in Indiana, he lives in Bethesda, Maryland, and has two sons and a daughter-in-law.